Apoptosis and Autoimmunity

Edited by
M. Herrmann and J. R. Kalden

Further titles of interest

Stuhler, G. / Walden, P. (Eds.)

Cancer Immune Therapy
Current and Future Strategies

2000, ISBN 3-527-30441-X

Stewart, C. C. / Nicholson, J. K. A. (Eds.)

Immunophenotyping
Cytometric Cellular Analysis

2000, ISBN 0-471-23957-7

Freshney, R. I.

Culture of Animal Cells
A Manual of Basic Technique

2000, ISBN 0-471-34889-9

Krauss, G. / Cooper, B. L. / Schönbrunner, N.

Biochemistry of Signal Transduction and Regulation
Building Blocks and Fine Chemicals

2001, ISBN 3-527-30378-2

Apoptosis and Autoimmunity

Edited by
M. Herrmann and J. R. Kalden

Prof. Dr. Dr. Joachim R. Kalden
University of Erlangen-Nuremberg
Department of Internal Medicine III
and Institute for Clinical Immunology
Krankenhausstraße 12
91054 Erlangen, Germany

Dr. Dr. Martin Herrmann
University of Erlangen-Nuremberg
Institute for Clinical Immunology
Krankenhausstraße 12
91054 Erlangen, Germany

Library of Congress Card No.: applied for

British Library Cataloguing-in-Publication Data: A catalogue record for this book is available from the British Library.

Bibliographic information published by Die Deutsche Bibliothek
Die Deutsche Bibliothek lists this publication in the Deutsche Nationalbibliografie; detailed bibliographic data is available in the Internet at <http://dnb.ddb.de>.

Printed in the Federal Republic of Germany
Printed on acid-free paper

Composition K+V Fotosatz GmbH, Beerfelden
Printing betz-druck gmbh, Darmstadt
Bookbinding Großbuchbinderei J. Schäffer GmbH & Co. KG, Grünstadt

ISBN 3-527-30442-8

Preface

More than 100 years ago Paul Ehrlich coined the expression 'horror autotoxicus' implicating the existence of autoimmune diseases. After the first description of an autoimmune disease of the thyroid, it soon became obvious that autoimmunity in principle is a self-limiting process, which in certain situations might proceed in an autoaggressive disease situation, when stringent control mechanisms have failed or are dysregulated. Despite of extensive research activities over the past decades, the etiology of autoimmune diseases is still enigmatic. Different hypotheses have been postulated, although these only partially explain the phenomenon of 'autoimmunity'.

More recently, the relationship between autoimmunity and apoptosis has been the focus of much research activity. Apoptosis as a genetically predetermined process is not only a vital mechanism sustaining homeostasis in the regulation of immune reactivity, but, in addition to being an important factor in general cell physiology, produces pronounced morphological changes of cells and the breakdown of cellular constituents by nucleolytic and proteolytic cleavage, resulting in the persisting presence of potential autoantigens. This book presents an up-to-date discussion on apoptosis and its role in autoimmunity.

We would like to express our appreciation and gratitude to the authors for their outstanding contributions and cooperation. We also gratefully acknowledge the continuous support of Andreas Sendtko and his colleagues at Wiley-VCH in the realization of this book.

Erlangen, July 2002

Martin Herrmann
Joachim R. Kalden

Preface

Contents

List of Contributors

Amendola, Alessandra
Universita degli Studi di Roma
'Tor Vergata'
Dipartimento di Biologia
Via della Ricerca Scientifica
00133 Roma
Italy

Baumann, Irith
Universität Erlangen-Nürnberg
Pathologisches Institut
Krankenhausstraße12
91054 Erlangen
Germany

Berden, Jo H.
University of Nijmegen
Departments of Medicine
Division of Nephrology
Intensive Care Medicine
PO Box 9101
6500 HB Nijmegen
The Netherlands

Beyer, Thomas D.
Universität Erlangen-Nürnberg
Institut für Klinische Immunologie
Medizinische Klinik III
Krankenhausstraße 12
91054 Erlangen
Germany

Casciola-Rosen, Livia
The Johns Hopkins University
School of Medicine
Departments of Medicine
and Dermatology
720 Rutland Avenue
Baltimore, MD 21205-2196
USA

Casiano, Carlos A.
Loma Linda University
Department of Biochemistry
and Microbiology
Division of Microbiology
and Molecular Genetics
Center for Molecular Biology
and Gene Therapy
and Department of Medicine
Section of Rheumatology
11085 Campus Street
Mortensen Hall 132
Loma Linda, CA 92350
USA

Chimini, Giovanna
Centre d'Immunologie
INSERM/CNRS de Marseille-Luminy
Case 906
Parc Scientifique de Luminy
13288 Marseille Cedex 09
France

Dieker, Jürgen W.C.
University of Nijmegen
Departments of Medicine
Division of Nephrology
Intensive Care Medicine
PO Box 9101
6500 HB Nijmegen
The Netherlands

Elkon, Keith B.
Hospital for Special Surgery
Weill Medical College
of Cornell University
1300 York Avenue
Box 144
New York, NY 10021
USA

Devitt, Andrew
University of Nottingham
Medical School
School of Biomedical Sciences
Queen's Medical Centre
D Floor
Nottingham NG7 2UH
UK

Firestein, Gary S.
University of California
MEDCTR/Rheumatology
9500 Gilman Drive BSB 5064
La Jolla, CA 92093-0656
USA

Fischer, Alain
INSERM Unit 429
Hopital Necker-Enfants Malades
149 rue de Sevres
75743 Paris Cedex 15
France

Flavell, Richard A.
Yale University
School of Medicine
Howard Hughes Medical Institute and
Section of Immunobiology
330 Cedar Street
PO Box 208011
New Haven, CT 06520
USA

Gaipl, Udo S.
Universität Erlangen-Nürnberg
Institut für Klinische Immunologie
Medizinische Klinik III
Krankenhausstraße 12
91054 Erlangen
Germany

Girkontaite, Irute
Universität Erlangen-Nürnberg
Institut für Klinische Immunologie
Medizinische Klinik III
Krankenhausstraße 12
91054 Erlangen
Germany

Gregory, Christopher D.
University of Nottingham
Medical School
School of Biomedical Sciences
Queen's Medical Centre
D Floor
Nottingham NG7 2UH
UK

Herrmann, Martin
Universität Erlangen-Nürnberg
Institut für Klinische Immunologie
Medizinische Klinik III
Krankenhausstraße 12
91054 Erlangen
Germany

Holder, Sheldon
Loma Linda University
Department of Biochemistry
and Microbiology
Division of Microbiology
and Molecular Genetics
Center for Molecular Biology
and Gene Therapy
11085 Campus Street
Mortensen Hall 132
Loma Linda, CA 92350
USA

Jenne, Lars
Universität Erlangen-Nürnberg
Dermatologische Klinik mit Poliklinik
Hartmannstraße 14
91052 Erlangen
Germany

Kalden, Joachim R.
Universität Erlangen-Nürnberg
Institut für Klinische Immunologie
Medizinische Klinik III
Krankenhausstraße 12
91054 Erlangen
Germany

Kolowos, Wasilis
Universität Erlangen-Nürnberg
Institut für Klinische Immunologie
Medizinische Klinik III
Krankenhausstraße 12
91054 Erlangen
Germany

Kuenkele, Susanne
Universität Erlangen-Nürnberg
Institut für Klinische Immunologie
Medizinische Klinik III
Krankenhausstraße 12
91054 Erlangen
Germany

Lakhani, Saquib
Yale University
Section of Immunobiology
330 Cedar Street
PO Box 208011
New Haven, CT 06520
USA

Le Deist, Françoise
INSERM Unit 429
Hopital Necker-Enfants Malades
149 rue de Sevres
75743 Paris Cedex 15
France

Lorenz, Hanns-Martin
Universität Erlangen-Nürnberg
Institut für Klinische Immunologie
Medizinische Klinik III
Krankenhausstraße 12
91054 Erlangen
Germany

Lu, Binfeng
Yale University
School of Medicine
Howard Hughes Medical Institute and
Section of Immunobiology
330 Cedar Street
PO Box 208011
New Haven, CT 06520
USA

Manfredi, Angelo A.
Instituto Scientifico Ospedale
San Raffaele
Department of Internal Medicine
via Olgettina 60
20132 Milano
Italy

Mevorach, Dror
The Hebrew University
Hadassah Medical Center
Ein Karem
Jerusalem 91120
Israel

Molinaro, Christine
Loma Linda University
Department of Biochemistry
and Microbiology
Division of Microbiology
and Molecular Genetics
Center for Molecular Biology
and Gene Therapy
11085 Campus Street
Mortensen Hall 132
Loma Linda, CA 92350
USA

Piacentini, Mauro
Universita degli Studi di Roma
'Tor Vergata'
Dipartimento di Biologia
Via della Ricerca Scientifica
00133 Roma
Italy

Rieux-Laucat, Frédéric
INSERM Unit 429
Hopital Necker-Enfants Malades
149 rue de Sevres
75743 Paris Cedex 15
France

Rigot, Véronique
Centre d'Immunologie
INSERM/CNRS de Marseille-Luminy
Case 906
Parc Scientifique de Luminy
13288 Marseille Cedex 09
France

Rosen, Antony
The Johns Hopkins University
School of Medicine
Departments of Medicine
and Pathology
Cell Biology and Anatomy
720 Rutland Avenue
Baltimore, MD 21205-2196
USA

Rovere-Querini, Patrizia
Instituto Scientifico Ospedale
San Raffaele
Department of Internal Medicine
via Olgettina 60
20132 Milano
Italy

Sauter, Birte
Universität Erlangen-Nürnberg
Dermatologische Klinik mit Poliklinik
Hartmannstraße 14
91052 Erlangen
Germany

Schuler, Gerold
Universität Erlangen-Nürnberg
Dermatologische Klinik mit Poliklinik
Hartmannstraße 14
91052 Erlangen
Germany

Utz, Paul J.
Stanford University
School of Medicine
Department of Medicine
Division of Immunology and Rheumatology
CCSR Building
Room 2215A
300 Pasteur Drive
Stanford, CA 94305-5166
USA

van der Vlag, Johan
University of Nijmegen
Departments of Medicine
Division of Nephrology
Intensive Care Medicine
PO Box 9101
6500 HB Nijmegen
The Netherlands

Voll, Reinhard E.
Universität Erlangen-Nürnberg
Institut für Klinische Immunologie
Medizinische Klinik III
Krankenhausstraße 12
91054 Erlangen
Germany

Wu, Xiwei
Loma Linda University
Department of Biochemistry
and Microbiology
Division of Microbiology
and Molecular Genetics
Center for Molecular Biology
and Gene Therapy
11085 Campus Street
Mortensen Hall 132
Loma Linda, CA 92350
USA

Yamanishi, Yuji
University of California
MEDCTR/Rheumatology
9500 Gilman Drive BSB 5064
La Jolla, CA 92093-0656
USA

Part 1
General Features of Apoptosis

1
Apoptosis and Autoimmunity

Keith Elkon

1.1
Introduction

Whereas the appearance of cells dying by apoptosis has been recognized for decades [1], if not more than a century [2], the understanding that cell survival and death are under stringent control is relatively new. The detailed elucidation of the biochemistry of apoptosis that has emerged over the last decade is nothing short of astounding and has altered the way in which we think about disease. Traditionally, diseases are classified according to the organ system effected – cardiovascular, endocrine, neurological, etc. However, it is much more useful for biomedical investigators to think about the mechanisms responsible for the diseases. In this respect, the simple reclassification according to whether diseases are associated with too little or too much cell death is highly informative [3, 4]. Cancers are clearly caused by enhanced cell growth and survival, whereas a large number of neurodegenerative diseases are caused by premature cell death of specific neurons. Amyotrophic lateral sclerosis (ALS), and Alzheimer's and Parkinson's diseases as well as diseases associated with polyglutamate repeats demonstrate intracellular protein aggregates that, most likely, trigger apoptosis through the mitochondrial pathway [5]. As will be discussed below, abnormal cell death is intimately involved in the pathogenesis of many autoimmune disorders, in part because apoptosis is a key facet of immunologic homeostasis and immune regulation (reviewed in [6]).

Descriptions of the biochemistry of apoptosis are detailed elsewhere in this book. Fig. 1.1 shows a simple overview of the process. There are four key steps in the death program: (1) initiation of death either through a professional death receptor or through the mitochondria, (2) activation of effector caspases, (3) execution of death including activation of nucleases and cell membrane changes, and (4) phagocytosis and removal of the corpse. Each of these steps is highly regulated, and, in most cases, both activators and inhibitors have been identified.

The causes and the pathogenesis of autoimmune diseases are complex. In some diseases such as systemic lupus erythematosus (SLE) and insulin-dependent diabetes mellitus (IDDM), a fairly strong genetic component is modulated by environmental factors [7, 8]. In fact, the increasing prevalence of IDDM in Western countries may well be associated with increased hygiene and reduced exposure to

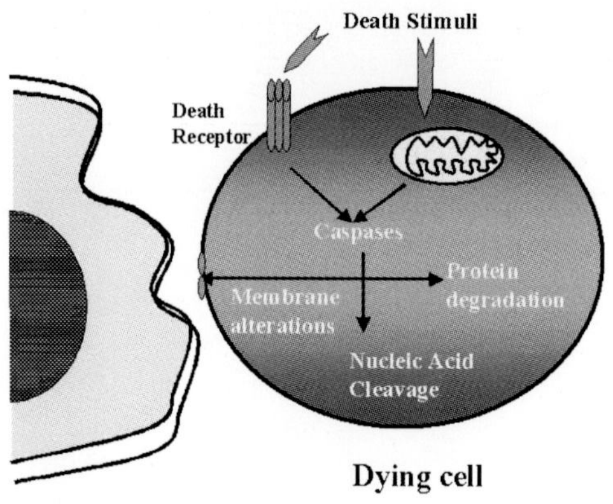

Fig. 1.1 Schematic of the major steps in apoptosis. Death is induced through professional death receptors (e.g. Fas/APO-1/CD95) on the cell surface, or by Bak- or Bax-mediated damage to the mitochondria (e.g. by drugs, ultraviolet radiation, genotoxic injury). Caspases are activated and cleave more than 100 substrates in the cell. These proteolytic changes lead to a variety of morphologic and functional alterations within the cell. Amongst the most important are cell surface membrane exposure of ligands for phagocytosis, collapse of the nuclear membrane, activation of nucleases, and inactivation transcriptional and protein synthesis machinery. Finally, the dying cell is recognized by specific receptors on phagocytes and is degraded. Further details are provided in subsequent chapters in this volume.

Tab. 1.1 Autoimmune diseases associated with targeted cell destruction

Disease	*Cell killed*
Diabetes (IDDM)	pancreatic β cell
Multiple sclerosis	oligodendrocyte
Hashimoto's thyroiditis	thyrocyte
Sjögren's syndrome	acinus and ductal cells
Polymyositis	myocyte
? ulcerative gastrointestinal diseases	intestinal cells
? primary biliary cirrhosis	bile duct cells

Tab. 1.2 Autoimmune diseases associated with enhanced cell survival/proliferation

Disease	*Cells/tissue*
Rheumatoid arthritis (RA)	pannus
Scleroderma	fibroblasts
Autoimmune lymphoproliferative syndrome (ALPS)/Canale–Smith syndrome (CSS)	cells of the immune system
Thyrotoxicosis (Graves' disease)	thyrocyte

Tab. 1.3 Autoimmune diseases possibly associated with abnormal processing of dying cells

Disease	*Cells/tissue*
SLE	cells of the immune system

common microorganisms. Despite the varied pathogenesis, most of the autoimmune diseases can be compartmentalized into diseases associated with targeted cell death (Tab. 1.1), enhanced cell growth (Tab. 1.2) or abnormal processing (Tab. 1.3).

1.2 Autoimmune Diseases Associated with Targeted Cell Destruction

The clinical manifestations of the diseases listed in Tab. 1.1 are caused by the death and subsequent loss of function of specific cells within a tissue or organ. The key questions relevant to this text are as follows.

1.2.1 What is the Mode of Cell Death?

In their pure forms, death by apoptosis occurs through defined biochemical pathways (programmed cell death), whereas death by necrosis is not programmed. This implies that, through intentional manipulation of the pathways, apoptotic death can either be induced or arrested, with obvious therapeutic implications. It is therefore vital to define the mode of cell death in organ specific autoimmune diseases.

The criteria for distinguishing between apoptosis and necrosis are now well described [9] and it would seem relatively straightforward to determine which process occurs in organ-specific autoimmune diseases. *In situ* staining for DNA strand breaks (TUNEL) is almost invariably positive in the effected organ in specific autoimmune diseases, but this test cannot be considered specific for apoptosis. A key distinction between apoptosis and necrosis is the lack of inflammation following apoptotic cell death. By this criterion alone, most autoimmune diseases would be associated with necrosis.

Another confounding factor relating to establishing the mode of cell death is that apoptosis/necrosis cannot simply be inferred by the identification of the cell type within the lesion. For example, the cell types most closely associated with a death effector function, CD8 T cells and natural killer (NK), could cause cell death by the perforin/granzyme pathway (apoptosis and necrosis), Fas ligand (FasL; apoptosis) or by the release of tumor necrosis factor (TNF)-α (apoptosis or necrosis). CD4 T cells can kill by FasL, yet expression of FasL may serve as a chemotactic signal for neutrophils [10] which, once activated, would almost certainly cause necrotic cell death. Even when a disease is clearly caused by defective apoptosis (the *lpr* mouse or humans with ALPS/CSS), is the massive infiltration of lymphocytes in multiple organs a failure of normal homeostasis ('lymphoaccumulation'), a manifestation of inflammation or both?

While necrotic cell death does occur in established disease, very little information is available on the mode of cell death in the preclinical stage of human diseases. It is the earliest changes that are likely to be most informative with regard to disease pathogenesis. Models of the human disease, IDDM in rodents, have been particularly informative in this regard. At around 2 weeks of age, there is a massive wave of spontaneous apoptosis of β cells in the pancreas [11]. Whereas this seems to occur at the same rate in non-obese diabetic (NOD) and control mice, it has been reported that the handling of the apoptotic cells may be different in the pre-diabetic mice [11]. Could this abnormality set in motion the unrelenting attack of the immune system involving both apoptosis and necrosis? In another disease, congenital heart block in the infants of mothers with SLE, could the exposure of Ro or La (SSA/SSB) on apoptotic cardiomyocytes in the developing heart allow autoantibodies to initiate disease [12]?

1.2.2 What Cells and What Effector Pathways are Responsible for Cell Death?

It is likely that the ultimate destruction of cells shown in Tab. 1.1 and, in some cases, the whole organ results from a concerted attack by all of the components of the immune system – macrophages, lymphocytes, neutrophils, dendritic cells, NK cells. Efforts to implicate a single cell type have generally been unsuccessful, although immunohistologic studies in humans and adoptive transfer studies in mice indicate that CD8 T cells are important effectors. As in infectious diseases, it is likely that components of the innate immune system (macrophages, dendritic cells, NK cells and, possibly, neutrophils) initiate the recruitment of lymphocytes and that CD4 T cells function in antigen recognition and/or cytokine priming of CD8 T cells.

As discussed above, each cell type can kill by multiple effector pathways and this topic has been reviewed in the context of organ specific autoimmune diseases previously [13]. Unfortunately, no clear consensus is available in individual diseases. In mouse models of IDDM, evidence supporting the involvement of perforin/granzyme, FasL, TNF-α and lymphotoxin have all been reported. More detailed discussion of these effector pathways are discussed elsewhere in this volume. The key question, at present, is whether attenuation of any one of these pathways will be sufficient to treat disease in the dramatic way in which TNF-α blockade has been of therapeutic value in RA and Crohn's disease.

1.3
Autoimmune Diseases Associated with Enhanced Cell Growth and Survival

Understanding the basic mechanisms responsible for *prevention* of cell death is just as important as understanding how cells die in different autoimmune disorders. Some pertinent examples will be discussed.

Graves' disease is a form of thyrotoxicosis associated with an autoantibody (LATS) that acts as an agonist for the thyroid stimulating hormone (TSH) receptor on thyrocytes. Reports that TSH exerts an anti-apoptotic effect by downmodulating Fas expression and that this effect is mimicked by IgG from patients with Graves' disease [14, 15] is appealing. However, other investigators have failed to confirm changes in Fas expression following exposure of thyrocytes to TSH *in vitro*. More recently, Stassi *et al.* [16] reported that T cell in Graves' disease express T_h2 cytokines (IL-4 and IL-10) in contrast to T cells in Hashimoto's disease that express T_h1 cytokines [interferon (IFN)-γ]. These authors propose that the two diametrically opposed diseases are a consequence of the effects of the cytokines – IFN-γ causes enhanced susceptibility to Fas-mediated apoptosis (Hashimoto's thyroiditis), whereas T_h2 cytokines promote Fas resistance through expression of c-FLIP and Bcl-X_L (Graves' disease).

Although the initiating events are poorly understood, two rheumatic diseases associated with activation and growth of fibroblasts are scleroderma and RA. Transforming growth factor (TGF)-β is the cytokine most closely associated with the skin thickening and fibrosis observed in scleroderma (reviewed in [17]). It is of considerable importance to determine how a cytokine such as TGF-β can be associated with growth of cells such as fibroblasts, but with cell cycle arrest and/or apoptosis of lymphocytes [18]. RA is characterized by growth of tissue called the pannus that invades cartilage and bone. Growth of the pannus is driven by a number of cytokines and growth factors, but the striking therapeutic effect of TNF-α blockade, suggests that TNF-α is a necessary component of this pathological process. In most cells, TNF-α activates NF-κB, which in turn activates a cascade of inflammatory mediators as well as anti-apoptotic pathways. Recent results suggest that NF-κB attenuates apoptosis through blockade of the JNK pathway and may interfere with GADD45 [19, 20]. Since TNF-α can also induce cell death (hence its name), elucidating how the signal transduction pathways diverge will have major implications for therapies of inflammatory disorders.

1.4
Autoimmune Diseases Associated with Abnormal Processing of Dying Cells

As shown in Fig. 1.1, the fourth and final component of the apoptotic pathway is the ingestion and 'safe' disposal of the dying cell. The basic mechanisms involved in the clearance of apoptotic cells and the possible links to autoimmunity are discussed elsewhere in this book. The evidence linking SLE to defective clearance of apoptotic cells (Tab. 1.3) is indirect, and derives mostly from knowledge of autoantibody specificities, the mechanisms responsible for the clearance of dying cells and information regarding macrophage signal transduction and cytokine release.

Autoantibodies may have greater specificity for cellular components that are either chemically modified or become more accessible to the immune system following death of the cell. The evidence to support this idea include: (1) phosphatidylserine, the negatively charged phospholipid that translocates to the outside of the cell is antigenic for some anticardiolipin autoantibodies – possibly following oxidation [21], (2) nucleosomes (a product of activation of DNases during the apoptotic process) are antigenic for B and T cells in SLE [22, 23], nucleosomes are deposited in the glomeruli (see Chapter 18), (3) some autoantigens are translocated to apoptotic blebs, some autoantigens are cleaved by caspases and/or granzyme B (see Chapter 15), and (4) hyperimmunization of mice with apoptotic cells results in an increase in antiphospholipid autoantibodies [24].

In vitro and *in vivo* studies support the idea that defective phagocytosis of dying cells promotes autoimmunity. Recent studies have shown that early complement components [25] as well as acute-phase proteins such as CRP [26] bind to, and promote the phagocytosis of [26], apoptotic cells. Mice that have disruption of genes encoding C1q [27] or the acute-phase protein serum amyloid P [28] develop a lupus-like disease. Of considerable interest, C1q-deficient mice demonstrate increased numbers of apoptotic cells in the glomeruli [27]. Together, these observations make a compelling case that the increased susceptibility of patients with early complement component deficiencies for the development of lupus is associated with delayed clearance of dying cells. It remains to be determined why the kidney is an important target organ and whether similar defects account for lupus that is not associated with inherited complement deficiencies.

Observations relating the quality of macrophage cytokine response to the nature of the dying cell [29–31] provide the critical link between cell death and the potential for autoimmunity. Specifically, *in vitro* studies revealed that the uptake of apoptotic cells induced the expression of immunosuppressive cytokines such as TGF-β1, prostaglandin E_2 and, possibly, IL-10 by macrophages (see Chapter 12). These cytokines are known to dampen the immune response to self or foreign antigens. Additional evidence supporting an anti-inflammatory role of apoptotic cells is that the administration of apoptotic cells promotes the resolution of inflammation *in vivo* [32]. In contrast, necrotic cells provoke a pro-inflammatory response associated with the release of TNF-α. TNF-α not only provokes an immune response to the antigens phagocytosed (self-antigens in this case), but also promotes the maturation of macrophages to dendritic cells, the cell type most efficient in antigen presentation [33].

The receptors and ligands involved in the recognition of apoptotic cells by phagocytes are discussed elsewhere in this book. Whereas blockade of many of these receptor/ligand systems reduce phagocytosis *in vitro*, deletion of the genes encoding these proteins rarely cause an obvious increase in apoptotic cells or autoimmune diseases in mice. This suggests a redundancy in function of many of the receptor/ligands identified to date. In contrast, knockout of the gene encoding the MER kinase, a member of the Tyro 3 receptor tyrosine kinase family, caused both an increase in the numbers of apoptotic cells in the thymus and spleen as well as a lupus-like autoimmune disease [34]. This mouse model therefore provides perhaps the most persuasive evidence linking defective clearance of apoptotic cells to lupus-like disease. It will be important to determine how MER is linked to apoptotic cell recognition and to identify its downstream signal transduction pathways.

As discussed above, macrophages provide signals that dictate the response of lymphocytes. Regardless of how the immune response to self-antigens is initiated, the receptors that become engaged on macrophages continue to orchestrate the immune response. For example, Manfredi *et al.* [35] have shown that when anti-cardiolipin antibodies engage Fcγ receptors, they promote a TNF-α dominated pro-inflammatory response. Thus, once initiated the antigen antibody complexes most likely fuel the inflammatory process and continue to promote immune responses to self.

1.5 Conclusions

Elucidation of the cell biology and biochemistry of cell death has led to some of the most important breakthroughs in the understanding autoimmunity in decades. Insight into how cells kill and how cells die provides knowledge that can be translated into therapeutic action. The literature contains many examples of successful blockade of death effectors or receptors with antibodies or soluble receptors as well as attenuation of biochemical pathways of apoptosis (e.g. with cell-permeable tetrapeptide inhibitors of caspases [36]) in animal models. As exciting as these observations are in experimental animals, the fact that either 'too little or too much apoptosis' is associated with diseases, indicates that therapy must either be short lived or cell specific when considering application to spontaneous autoimmune diseases in humans. Furthermore, as emphasized above, it is likely that multiple cells and death pathways are operational in established autoimmune diseases suggesting that therapy administered early is much more likely to effective.

TNF-α inhibitors provide the most striking example of single molecule administration with therapeutic efficacy in patients with rheumatoid arthritis and Crohn's disease. Do TNF-α inhibitors work by blocking growth, inflammation or by inducing apoptosis of monocytes [37, 38]? This very important question should be resolved prior to the administration of small molecule inhibitors of TNF *in vivo.* Can the TNF success story be replicated in other diseases?

'Apoptosis and autoimmunity' has been a bidirectional learning process – each has informed the other. For example, the discovery of Fas mutations in *lpr* mice led to the elucidation of the concept of 'activation-induced cell death' and the understanding that activated cells were eliminated at sites of inflammation by suicide and fratricide [39]. This process is vital to the maintenance of peripheral tolerance [40]. Similarly, the almost invariable association between C1q deficiency and lupus coupled with elucidation of the responses of phagocytes to their meals described above have provided fundamental new insight into problem of 'self/

non-self' discrimination. At this pace, it is likely that additional basic discoveries in apoptosis will be uncovered, leading to further insight and improved therapy of many human autoimmune disorders.

1.6
References

1 Kerr JFR, Wyllie AH, Currie AR. *Br J Cancer* **1972**, 26, 239–57.
2 Lockshin RA. *Cell Death Different* **1997**, 4, 347–51.
3 Carson DA, Ribeiro JR. *Lancet* **1993**, 341, 1251–54.
4 Thompson CB. *Science* **1995**, 267, 1456–62.
5 Guegan C, Vila M, Rosoklija G, Hays AP, Przedborski S. *J Neurosci* **2001**, 21, 6569–76.
6 Elkon K. Immunologic tolerance and apoptosis. In: Rich RR FT, Kotzin B, Shearer W, Schroeder HW, ed. *Clinical Immunology*, 2nd edn. London: Harcourt International **2001**: 11.1–8.
7 Wakeland EK, Liu K, Graham RR, Behrens TW. *Immunity* **2001**, 15, 397–408.
8 Todd JA. *BioEssays* **1999**, 21, 164–74.
9 Duvall E, Wyllie AH. *Immunol Today* **1986**, 7, 115.
10 Chen JJ, Sun Y, Nabel GJ. *Science* **1998**, 282, 1714–7.
11 Trudeau JD, Dutz JP, Arany E, Hill DJ, Fieldus WE, Finegood DT. *Diabetes* **2000**, 49, 1–7.
12 Tran HB, Ohlsson M, Beroukas D, Hiscock J, Bradley J, Buyon JP, Gordon TP. *Arthritis Rheum* **2002**, 46, 202–8.
13 Ohsako S, Elkon KB. *Cell Death Different* **1999**, 6, 13–21.
14 Kawakami A, Eguchi K, Matsuoka N, Tsuboi M, Kawabe Y, Ishikawa N, Ito K, Nagataki S. *Endocrinology* **1996**, 137, 3163–9.
15 Kawakami A, Eguchi K, Matsuoka N, Tsuboi M, Urayama S, Kawabe Y, Tahara K, Ishikawa N, Ito K, Nagataki S. *Clin Exp Immunol* **1997**, 110, 434–9.
16 Stassi G, Di Liberto D, Todaro M, Zeuner A, Ricci-Vitiani L, Stoppacciaro A, Ruco L, Farina F, Zummo G, De Maria R. *Nat Immunol.* **2000**, 1, 483–8.
17 Denton CP, Abraham DJ. *Curr Opin Rheumatol* **2001**, 13, 505–11.
18 Chaouchi N, Arvanitakis L, Auffredou MT, Blanchard DA, Vazquez A, Sharma S. *Oncogene* **1995**, 11, 1615–22.
19 Tang G, Minemoto Y, Dibling B, Purcell NH, Li Z, Karin M, Lin A. *Nature* **2001**, 414, 313–7.
20 De Smaele E, Zazzeroni F, Papa S, Nguyen DU, Jin R, Jones J, Cong R, Franzoso G. *Nature* **2001**, 414, 308–13.
21 Horkko S, Miller E, Dudl E, Reaven P, Curtiss LK, Zvaifler NJ, Terkeltaub R, Pierangeli SS, Branch DW, Palinski W, Witztum JL. *J Clin Invest* **1996**, 98, 815–25.
22 Burlingame RW, Rubin RL, Balderas RS, Theofilopoulos AN. *J Clin Invest* **1993**, 91, 1687–96.
23 Mohan C, Adams S, Stanik V, Datta SK. *J Exp Med* **1993**, 177, 1367–81.
24 Mevorach D, Zhou J-L, Song X, Elkon KB. *J Exp Med* **1998**, 188, 387–92.
25 Korb LC, Ahearn JM. *J Immunol* **1997**, 158, 4525–8.
26 Gershov D, Kim S, Brot N, Elkon KB. *J Exp Med* **2000**, 192, 1353–64.
27 Botto M, Dell'Agnola C, Bygrave AE, Thompson EM, Cook T, Petry F, Loos M, Pandolfi PP, Walport MJ. *Nat Genet* **1998**, 19, 56–9.
28 Bickerstaff MCM, Botto M, Hutchinson WL, Herbert J, Tennent GA, Bybee A, Mitchell DA, Cook HT, Butler PJG, Walport MJ, Bepys MB. *Nat Med* **1999**, 5, 694–7.
29 Stern M, Savill J, Haslett C. *Am J Pathol* **1996**, 149, 911–21.
30 Fadok VA, Bratton DL, Konowal A, Freed PW, Westcott JY, Henson PM. *J Clin Invest* **1998**, 101, 890–8.
31 Voll RE, Herrmann M, Roth EA, Stach C, Kalden JR, Girkontaite I. *Nature* **1997**, 390, 350–1.

32 Huynh ML, Fadok VA, Henson PM. *J Clin Invest* **2002**, 109, 41–50.
33 Banchereau J, Steinman RM. *Nature* **1998**, 392, 245–52.
34 Scott RS, McMahon EJ, Pop SM, Reap EA, Caricchio R, Cohen PL, Earp HS, Matsushima GK. *Nature* **2001**, 411, 207–11.
35 Manfredi AA, Rovere P, Galati G, Heltai S, Bozzolo E, Soldini L, Davoust J, Balestrieri G, Tincani A, Sabbadini MG. *Arthritis Rheum* **1998**, 41, 205–14.
36 Garcia-Calvo M, Peterson EP, Leiting B, Ruel R, Nicholson DW, Thornberry NA. *J Biol Chem* **1998**, 273, 32608–13.
37 Lugering A, Schmidt M, Lugering N, Pauels HG, Domschke W, Kucharzik T. *Gastroenterology* **2001**, 121, 1145–57.
38 ten Hove T, van Montfrans C, Peppelenbosch MP, van Deventer SJ. *Gut* **2002**, 50, 206–11.
39 Green DR, Scott DW. *Curr Opin Immunol* **1994**, 6, 476–87.
40 Van Parijs L, Abbas AK. *Science* **1998**, 280, 243–8.

2
Caspase Knockouts: Matters of Life and Death

Saquib Lakhani, Binfeng Lu and Richard A. Flavell

2.1
Death, Development and Immune Function

Apoptosis has been defined as programmed cell death, a term which, in contrast to necrosis, implies the active participation of a dying cell in its own death. This process can be initiated in response to a range of intrinsic and extrinsic signals. In developing vertebrates and invertebrates, apoptosis serves to carefully regulate cell number and tissue formation. This physiologic program continues in mature organisms, working to maintain a constant cellular homeostasis. There is growing evidence, however, that cell death by apoptosis also plays a role in a variety of disease states. Tissue and organ injury in response to environmental stressors, such as hypoxia, trauma and infection, are thought to be mediated in part through apoptotic cell death. Insufficient apoptosis has been linked to conditions such as cancer, while excess apoptosis is felt to contribute to the pathogenesis of some chronic degenerative diseases.

In the immune system, the importance of apoptosis is borne out by observations that dysregulated apoptosis can lead to a variety of disorders including autoimmune disease, lymphoid tumors or immune suppression from excessive lymphocyte death. Experimental evidence has revealed that programmed cell death plays an important role in several steps throughout both the maturation and subsequent functional life of T and B cells. During early development, apoptosis contributes to the generation of a functional repertoire of mature cells by eliminating lymphocytes that fail to express an antigen receptor (death by neglect). Expression of both cytokine receptors and the pre-T cell receptor (TCR) is required for expansion and differentiation, and in the absence of survival signals delivered via these receptors, the $CD4^{-}CD8^{-}$ cells undergo apoptosis (reviewed in [1]). Apoptosis is also the method of elimination of T cell during the next phase of development, positive and negative selection. During this stage, CD4/CD8 double-positive thymocytes that express TCR of intermediate affinity for peptide ligands in the context of MHC molecules differentiate into either CD4 or CD8 single-positive cells. T cells that express TCR with low affinity for peptide, however, fail to receive sufficient survival stimuli and die by neglect. On the other end of the spectrum from positive selection, an overly high-affinity TCR interaction with self-MHC causes

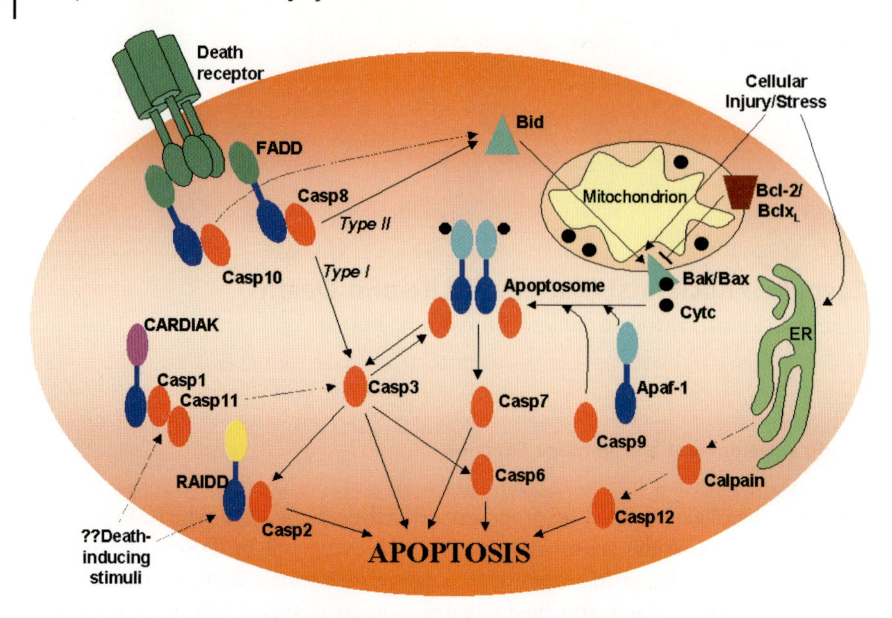

Fig. 2.1 Apoptotic pathways involving caspases. Established pathways appear as solid lines. Dashed lines indicate possible pathways.

negative selection, the clonal deletion of these potentially autoreactive cells (reviewed in [2]). In a similar fashion, B cells that recognize self-antigens are eliminated in the bone marrow by apoptosis. Apoptosis, therefore, serves as an important tool in the development of lymphocytes, ensuring the central elimination of both non-functional and potentially autoreactive cells.

Programmed cell death is also a critical component of the mature immune system. The clonal expansion of antigen-specific lymphocytes is a central feature of the adaptive immune response. However, once an infection has been successfully cleared, it is important to efficiently remove these activated, proliferating lymphocytes to avoid risking adverse effects such as an autoimmune response or potential malignancy. In this situation, apoptosis functions to maintain homeostasis by culling activated lymphocytes [3]. For $CD4^+$ T cells, this aim is achieved either by activation-induced cell death (AICD), a process of inducing apoptosis through repetitive TCR stimulation, or through death by cytokine withdrawal, in which lymphocytes die from a lack of trophic factors [4]. Likewise, activated B cells can be induced to undergo apoptosis through B cell receptor (BCR) stimulation [5].

Our understanding of the mechanisms of apoptosis in these and other paradigms has rapidly advanced over the last few years (Fig. 2.1), and although we have developed a more complete understanding of its molecular events, numerous aspects remain to be elucidated. Much of our appreciation of the intricacies of apoptosis has evolved from the study of animals lacking specific caspases (*c*ystinyl *asp*artate prote*ases*) and other related proteins.

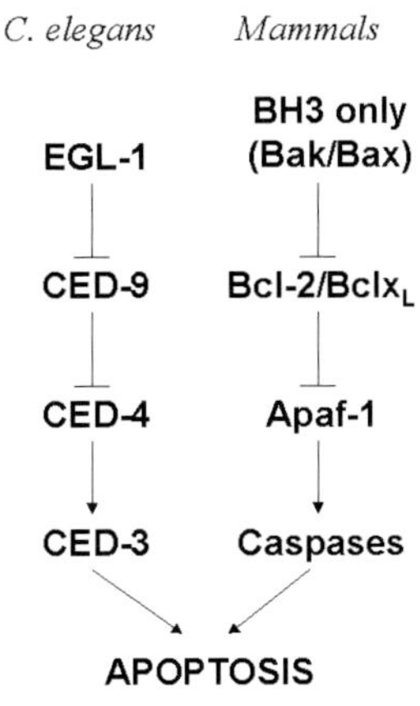

Fig. 2.2 Apoptosis-related molecules in *C. elegans* and mammals.

2.2 Apoptotic Pathways: from Nematode to Mammals

The crucial role of caspases in cell death was first recognized through studies of their homologues in *Caenorhabditis elegans*. During development, 131 of this nematode's 1090 somatic cells die by apoptosis. Screens of *C. elegans* with defective apoptosis initially revealed loss-of-function mutations of the proteins CED-3 (for *c*ell *d*eath abnormal) and CED-4, demonstrating their necessity for apoptosis [6]. Subsequently, the protein CED-9 has been identified as an inhibitor of apoptosis functioning upstream of CED-3 and CED-4. Loss-of-function CED-9 mutations cause embryonic lethality due to excess apoptosis, a phenotype that is reversible by loss-of-function CED-3 or CED-4 mutations [7]. Finally, loss-of-function mutations of the protein EGL-1, which binds to CED-9, also suppress apoptosis [8]. The scheme that has emerged (Fig. 2.2) places CED-9 as a negative regulator of apoptosis that works by binding to and suppressing CED-4 [9]. When EGL-1 is induced, it displaces this complex, allowing CED-4 to bind to CED-3 and resulting in apoptosis.

Analysis of these proteins identified CED-3 as a caspase, with homology to the mammalian interleukin (IL)-1β-converting enzyme (ICE, also known as caspase-1) [10]. The recognition that CED-3 is crucial for apoptosis prompted the search for other mammalian caspases that could play a similar role. While the general scheme for apoptosis that has emerged for higher organisms is conserved from *C. elegans*, there have been 14 mammalian caspases identified to date and the resulting pathways are accordingly more complex. In addition to the caspases, mammalian counterparts have also been found for the other members of the *C. elegans* apoptotic pathway (Fig. 2.2) [8, 11, 12].

2.3
Triggering a Killer: General Aspects of Caspase Activation

Although there are no fixed criteria that define apoptosis, a number of changes in cellular morphology can help distinguish it from necrotic cell death. In the apoptotic cell, chromosomes condense, the nucleus fragments, cytoplasmic volume decreases, organelles compact, the cell membrane fuses with the endoplasmic reticulum and the cell finally fragments into numerous 'apoptotic bodies', which are engulfed by surrounding cells [13]. These morphologic changes are accompanied by other subcellular indicators of apoptosis, including the exposure of phosphatidylserine on the external surface of the cell membrane and a decrease in mitochondrial transmembrane potential [14]. An important characteristic that results from this distinctive cellular packaging is that apoptosis lacks the inflammation and potential for injury to surrounding cells that is seen with necrosis. The subcellular changes that transpire in apoptosis result, at least in part, from the cleavage of specific subcellular proteins by caspases. Caspases have among the most stringent substrate specificities of all proteases, always cleaving on the carboxyl side of aspartate residues [15]. Their name also hints to the presence of a conserved cysteine residue found in the peptide motif QACXG, which, along with a conserved glycine and histidine, contribute to the catalytic site [16]. Caspases exist in cells as inactive monomeric zymogens (Fig. 2.3), consisting of an N-terminal prodomain, a large subunit and a small subunit. Processing occurs at sites between the domains that, consistent with the ability of pro-caspases to autoactivate or to be activated by other caspases, contain aspartate residues. Cleavage of two zymogen precursors at these sites releases the subunits, which then form the functional heterotetrameric enzyme, a complex of two large and two small subunits with two separate active sites [17, 18].

The nature of the prodomain (Tab. 2.1), classified as either short (caspase-3, -6, -7 and -14) or long (the remainder) is thought to influence the route of caspase ac-

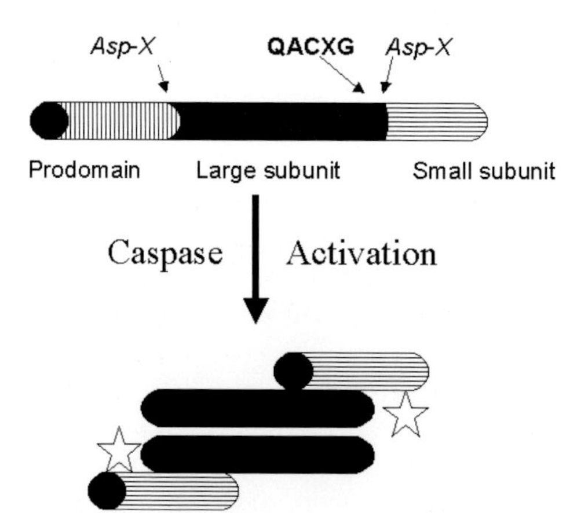

Fig. 2.3 Caspase zymogen cleavage and assembly of active heterotetramer. Asp-X indicates aspartate residues at processing sites; QACXG indicates catalytic cysteine residue; stars indicate location of enzyme active site.

Tab. 2.1 Characterization of murine caspases with long prodomains

Caspase	*Interaction domain*	*Adaptor molecule*
1	CARD	CARDIAK
2	CARD	RAIDD/CRADD
4	CARD	?
5	CARD	?
8	DED	FADD
9	CARD	Apaf-1
10	DED	FADD
11	CARD	?
12	CARD	?
13	CARD	?

tivation. The long prodomains have been found to contain either death effector domains (DED) or caspase recruitment domains (CARD) that are capable of interacting with adapter molecules and result in clustering of pro-caspases. This close proximity can enhance a low, intrinsic autocatalytic activity of the zymogen, allowing it to cleave and activate itself. For example, oligomerization of the Fas receptor (Apo1, CD95) recruits the adapter molecule FADD (Fas-associated death domain protein, also called Mort-1) through the interaction of their death domains (DD) [19, 20]. The other end of FADD contains a DED, which can then interact with a DED on pro-caspase-8, recruiting it to the death-inducing signaling complex (DISC) [21, 22]. Once recruited, pro-caspase-8 can activate itself through this 'induced proximity' [23]. Caspase-10, which also has a DED, is similarly thought to interact with FADD [24]. Known interactions of CARD include those mediating the association of pro-caspase-9 with Apaf-1 [25] and pro-caspase-2 with RAIDD/CRADD [26, 27]. Caspases with long prodomains have frequently been termed 'initiators' and, once activated, are thought to cleave caspases with short domains, termed 'effectors', which then carry out the apoptotic program. It has become clear, however, that such a distinction is an oversimplification. For example, there is evidence that caspase-3 can cleave upstream caspase-8 and -9 [28], and also that it may be capable of autoactivation [29].

2.4
Caspase-1 and -11: More than Mediators of Inflammatory Cytokines?

Caspase-1, or ICE, was the first member of this family to be cloned [30]. It is a cytoplasmic protease that is capable of converting the 34-kDa inactive precursor of IL-1 to its mature 17-kDa form, and can also process the cytokine precursor of IL-18 [interferon (IFN)-inducing factor] [31]. The discovery of the serine/threonine kinase RIP2/CARDIAK/RICK and demonstration that it can bind and activate caspase-1 processing [32] provide a possible mechanism for caspase-1 regulation.

Upon receipt of a pro-inflammatory stimulus, RIP2 engages caspase-1 through interaction of their respective CARD, causing oligomerization of pro-caspase-1. These zymogens then undergo autoprocessing to generate large (p20) and small (p10) subunits which combine to form the active protease. In addition to its ability to process the pro-inflammatory cytokines IL-1β and IL-18, caspase-1, like all caspases, is able to induce apoptosis when overexpressed in cultured cells [33]. Unlike mice with targeted deletions in some of the other caspase genes, however, caspase-1$^{-/-}$ mice do not have any developmental defects [34, 35]; instead, they are deficient in IL-1β and IL-18 production.

Mouse caspase-11 is most homologous to human caspase-4 [36]. Overexpression of caspase-11 in Rat-1 and HeLa cells induces apoptosis, which can be inhibited by CrmA and Bcl-2 [36]. The expression of caspase-11 is highly inducible by LPS, suggesting that it may have a regulatory role in both apoptosis and inflammatory responses [36]. Caspase-11 does not process pro-IL-1 directly, but overexpression of caspase-11 stimulates processing of pro-IL-1 by caspase-1 [36].

Analysis of knockout mice established the critical role of caspase-1 in acute inflammatory responses. As compared to wild-type animals, caspase-1$^{-/-}$ mice are resistant to the effects of lipopolysaccharide (LPS)-induced shock, showing improved survival and, in addition to the absence of mature IL-1β, decreased production of IL-1α, IL-6 and tumor necrosis factor (TNF)-α [34, 35]. Interestingly, caspase-11$^{-/-}$ mice show a similar survival advantage in LPS-induced shock [37]. The observations that caspase-11 is absolutely required for caspase-1 activity and that they exist in the same protein complex inside cells suggest a mechanism whereby interaction of the two molecules is required for activity and cytokine processing.

Although caspase-1$^{-/-}$ and -11$^{-/-}$ thymocytes are normally susceptible to apoptosis from dexamathosone and γ-irradiation exposure, both are partially resistant to Fas-induced apoptosis [35, 37]. This raises the issue of their involvement in negative selection, possibly via CD30, a TNF family member suspected to play a role in negative selection [38, 39]. It has been shown using H-Y transgenic mice, in which the TCR recognizes a Y chromosome antigen resulting in clonal deletion by negative selection in the male thymus, that CD30 knockout causes a partial deficiency in eliminating self-reactive transgenic CD8^{+} T cells in male mice [39]. Additionally, overexpressing CD30 in the mouse thymus results in enhance program cell death by TCR and CD30 crosslinking reagents. In this situation, caspase-1 is activated by CD30 and may thereby mediate negative selection. A critical role of CD30 in negative selection is arguable, however, due to recent studies showing that CD30-deficient mice are able to carry out negative selection of both anti-self CD4^{+} and CD8^{+} T cells [40]. Still, further examination of this aspect of negative selection is warranted, perhaps by employing the caspase-1$^{-/-}$ or -11$^{-/-}$ backgrounds to generate TCR transgenic mice that recognize endogenous antigen, either a peptide presented by MHC or superantigen, and monitor the deletion of T cells in such mice. Likewise, future study should also address the role of caspase-1 and -11 in peripheral deletion of activated cells, although there is no report of systemic autoimmune conditions in these mice, making any role they may have in peripheral tolerance likely to be less significant.

In addition to its involvement in inflammation, caspase-11 has been shown to play a role in pathological death in neurons. In a model of brain injury induced by middle cerebral artery occlusion, caspase-11 knockout mice were found to have a reduction in caspase-3 activation and brain apoptosis [41]. Cultured oligodendrocytes from caspase-11 knockout mice also showed reduced caspase-3 activation and cell death in response to hypoxia, IFN-γ and anti-Fas stimulation [42, 43]. Furthermore, caspase-11 knockout mice were found to have significant resistance to development of MOG peptide-induced experimental autoimmune encephalitis [42]. In addition, caspase-11 is able to activate caspase-1 and -3 by direct cleavage *in vitro* [41]. Although these data may indicate that caspase-11 is important in apoptosis via modulation of caspase-3 activation, it has remained difficult to definitively separate this effect from alterations in cytokine levels seen in caspase-11 knockouts. The possibility that caspase-11 deficiency protects against apoptosis due to altered cytokine regulation will need to be more conclusively addressed in the future.

2.5 Caspase-8 and the FAS Signaling Pathway

A subset of TNF receptor (TNF-R) family members that includes Fas, TNF-RI, DR3 (TRAMP, wsl-1, APO-3, LARD), DR4 (TRAIL-R1, APO-2), DR5 (TRAIL-R2, TRICK2, KILLER) and DR6 (reviewed in [44]) is involved in transducing signals that result in cell death and are therefore referred to as 'death receptors'. Fas-induced apoptosis is triggered by binding of its natural ligand, Fas ligand (FasL), resulting in the rapid recruitment of FADD to the cytoplasmic membrane. FADD then brings in pro-caspase-8, also called FLICE (FADD-like IL-1-converting enzyme) or MACH, via their homologous DED, forming the DISC (reviewed in [5]). Next, proteolytic cleavage of pro-caspase-8, presumably an autocatalyzed reaction promoted by the close proximity of multiple caspase-8 zymogens recruited to the DISC, results in its activation and subsequent release from the DISC into the cytoplasm [23]. Depending on the cell context, the downstream signal of caspase-8 is propagated in one of two ways (Fig. 2.1). In so-called type I cells [45], induction of apoptosis is accompanied by activation of large amounts of caspase-8 by the DISC. Caspase-8 then rapidly cleaves and activates caspase-3, leading to the effector stage of apoptosis. In type II cells, on the other hand, DISC formation is strongly reduced and activation of caspase-8 and -3 occurs following the loss of mitochondrial transmembrane potential (see Chapter 1). In this variant of the caspase cascade, caspase-8 cuts and activates the pro-apoptotic Bcl-2 family member Bid [46]. Truncated Bid induces mitochondria pore formation via Bak or Bax [47], resulting in the unleashing of pro-apoptotic molecules such as cytochrome *c* [48] and Smac/DIABLO [49, 50]. Cytochrome *c* can subsequently form a complex with Apaf-1 and pro-caspase-9 in the cytoplasm, to form the apoptosome (reviewed in [5]), which can then activate effector caspases such as caspase-3 and -7. Of note, although mitochondrial activation can occur in both type I and II cells, it is not necessary for apoptosis in type I cells.

Despite diverging at a later stage, both of these mechanisms involve caspase-8 as an initiator caspase in death receptor signaling. This linear model has been tested by developing caspase-$8^{-/-}$ mice. In contrast to Fas-, TNF-αRI- or TNF-αRII-deficient mice, which all develop normally to adulthood, caspase-8 deficiency causes prenatal lethality [51] with two particularly striking features: impaired heart muscle development and congested accumulation of erythrocytes (hyperemia). This indicates that caspase-8 mediates key developmental steps, either through death receptors other than Fas and TNF-α, or through some combination of these receptors. Wild-type and caspase-8 embryonic fibroblasts were used to show that the activation of JNK and NF-κB by death receptors such as Fas and TNF-α receptors is not affected by caspase-8 deficiency. However, the apoptosis induced by these receptors is totally blocked in caspase-8 knockout fibroblasts, showing that caspase-8 is a critical initiator caspase for transducing apoptosis signals from these death receptors. Consistent with the model that FADD is a key adaptor molecule that recruits caspase-8 to the death receptors, the phenotype of $FADD^{-/-}$ mice is very similar to that of caspase-$8^{-/-}$ mice [51]. Interestingly, Bid-deficient mice develop normally but are resistant to Fas-mediated hepatocyte apoptosis, indicating that the developmental effects of caspase-8 deficiency likely occur in a type I fashion through direct activation of caspase-3 by caspase-8, without involvement of Bid [40].

Members of death receptors have been implicated in negative selection in the thymus [52]. However, deletion of autoreactive thymocytes occurs normally in Lpr (Fas mutation) [53], $Fas^{-/-}$ [54], TNF-$RI^{-/-}$ [55] and TNF-$RII^{-/-}$ mice [56], and is only partially impaired in $CD30^{-/-}$ mice [39], indicating that negative selection cannot be attributed to ligation of any one of these receptors alone. Therefore, conditional deletion of caspase-8 or its partner FADD, a strategy that presumably blocks the signaling of many death receptors, may be informative in addressing the role of death receptors in negative selection. The role of caspase-8 in lymphoid development has also not been well established using the knockout model due to the prenatal lethality of these mice. Interestingly, however, T cell-specific FADD-deficient mice show a proliferative defect of thymocytes between the $CD4^-CD8^-$ and $CD4^+CD8^+$ stage [57]. Furthermore, analysis of transgenic mice expressing a dominant-negative mutant FADD in the thymus provided the surprising result of enhanced thymocyte negative selection [58]. The effect of the caspase-8/FADD module on the homeostasis of peripheral lymphocytes appears to be a balance of hypo-responsiveness and defective apoptosis. FADD deficiency causes decreased activation in peripheral T cells which is associated with a defective co-stimulatory response [58, 59]. On the other hand, it is also observed that activated peripheral T cells are present in higher proportion in $RAG1^{-/-}$ animals reconstituted with $FADD^{-/-}$ ES cells [59], likely due to a failure of AICD or cytokine withdrawal-mediated death. Since activated T cells have prolonged survival [60], they may potentially accumulate in aging animals, consistent with the autoimmune syndromes seen in TNF-RI/Fas double knockouts [61] and, in a strain-dependent fashion, Fas and FasL mutants [62]. It is likely, therefore, that signaling pathways operating through FADD do not lead exclusively to apoptosis, but under certain circumstances can promote cell survival and proliferation. Whether such pro-survival pathways would involve caspase-8 is unclear. A selective knockout strategy that induces deletion of caspase-8 within cells undergoing negative selection, for example, would help clarify this issue.

2.6 Caspase-3: The Chief Executioner?

Caspase-3 has assumed a position as a central executioner caspase in mammals, though not as indispensable as its counterpart CED-3 which is absolutely required for programmed cell death in *C. elegans*. This assumed role of caspase-3 is supported by the convergence of both death-receptor and mitochondrial-mediated death pathways at caspase-3 activation as well as by the wide range of potential caspase-3 substrates. Caspase-3$^{-/-}$ mice can survive to birth, but they exhibit perinatal mortality as a result of defects in brain development that correlate with a decrease in levels of apoptosis [63, 64].

In contrast to the essential role caspase-3 plays in neuronal development, it seems to be dispensable for the developing lymphoid system, as caspase-3$^{-/-}$ mice display normal T and B cell development [63, 64]. Apoptosis of thymocytes induced by anti-CD3 crosslinking or anti-Fas is unaltered in the absence of caspase-3 [63], suggesting that at least some aspect of negative selection is not affected by caspase-2. An involvement of caspases in thymocyte apoptosis is supported, however, by previous studies that employed fetal thymic organ culture and the pharmacological pan-caspase inhibitor zVAD-fmk. In these experiments, the authors observed inhibition of deletion of thymocytes induced either by anti-CD3, dexamethasone, or antigenic peptide *in vitro*. Moreover, caspase-3 activation was detected specifically during apoptosis induced by TCR stimulation, but not during spontaneous cell death [65, 66]. It has been established recently [67] that compensatory caspase activation is a mechanism for mammalian cells to use to induce apoptosis in the absence of a given key caspase. The lack of defects in negative selection in caspase-3 mice can therefore be due to activation of other alternative caspases, such as caspase-6 and -7.

Both p53 and Bcl-2 have been shown to be important in mediating death by neglect during T cell development in the thymus (see Section 2.7) [68, 69]. However, T cell development in caspase-3-deficient mice appears normal. Additionally, studies utilizing a strategy that block all caspase activities in the thymus failed to reveal a deficiency in this type of cell death [70]. These data argue that a pathway that is not dependent on caspases, but instead on p53 and Bcl-2 may be required for death by neglect in T cell development.

After infection, professional antigen-presenting cells, dendritic cells in particular, present antigenic peptides of the infectious agents and drive T cells into clonal expansion. After a clonal expansion phase and resolution of infection, the number of antigen-reactive lymphocytes must decline until the pool of lymphoid cells reaches a baseline level. This is achieved by AICD, a balanced fine-tuning between growth/expansion and death by apoptosis which classically occurs via death receptors typified by Fas (reviewed in [71]). Activated T cells from caspase-$3^{-/-}$ mice show a dramatic deficiency in AICD [64]. After challenge with superantigen staphylococcal enterotoxin A (SEA), caspase-$3^{-/-}$ peripheral T lymphocytes showed reduced cell death compared with wild-type controls. In addition, activated caspase-$3^{-/-}$ lymphocytes showed an increased viability when treated with anti-CD3 or anti-Fas antibody [64]. These results indicate a requirement for caspase-3 for AICD in peripheral T cells. The deficiency of the Fas signaling pathway has been linked direct to systemic autoimmune conditions such as lupus (reviewed in [5]). Therefore, further studies should focus on the effect of caspase deficiency in this kind of autoimmune condition.

In addition to AICD, there are many other ways that cause the death of peripheral T and B cells. The death of peripheral T cells may be due to a high intensity of TCR signaling [72], the withdrawal of growth factors such as IL-2 or IL-15 from T cells [73, 74], or the absence of a tonic TCR signal on naïve T cells [75]. Future studies should also address whether these types of cell death are affected by caspase-3 mutation.

2.7 Caspase-9: Mitochondrial Activation and the Apoptosome

The caspase-9 activation pathway was discovered after the observation that the addition of ATP, or preferably dATP, to cell extracts prepared from normally growing cells initiates an apoptotic program, as measured by caspase-3 activation and DNA fragmentation [48]. Subsequent biochemical studies identified a protein complex consisting of Apaf-1 and cytochrome *c* that, upon hydrolysis of ATP or dATP, is able to recruit and activate pro-caspase-9 (reviewed in [76]). Although caspases are usually activated by cleavage, proteolytic processing of pro-caspase-9 does not significantly increase its catalytic activity [77, 78]. Rather, the key requirement for caspase-9 activation is its association with its protein cofactors, Apaf-1 and cytochrome *c*. Together they form the active holoenzyme, often referred to as the apoptosome (reviewed in [79]).

The *in vivo* significance of the apoptosome has been demonstrated by studying mice deficient in either caspase-9 or Apaf-1 [80–82]. Both of these knockouts caused a perinatal lethality starting at post-conception day 16.5. Brain malformations with protruding neural tissues were observed in both caspase-$9^{-/-}$ and Apaf-$1^{-/-}$ mice, and TUNEL assays revealed a lack of apoptosis in Apaf-$1^{-/-}$ and caspase-$9^{-/-}$ knockouts. Additionally, the supernumerary cells were post-mitotic, suggesting that these cells were not proliferating tumors. These phenotypes are reminiscent of those of caspase-$3^{-/-}$ mice, and indeed, caspase-3 activation is abolished in the developing brain in caspase-$9^{-/-}$ embryos [81]. Further establishing a central role of the apoptosome in programmed cell death is the cytochrome *c* knockout [83]. These animals die prenatally at embryonic day 8.5, presumably due to the defect in aerobic metabolism, but analysis of knockout embryo-derived cell lines reveal a resistance to apoptosis induced by ultraviolet (UV) irradiation, serum withdrawal and staurosporine. Interestingly, however, cytochrome $c^{-/-}$ cells appear more sensitive to TNF-α-induced death.

In contrast to the dramatic effect of caspase-9 and Apaf-1 deficiency on neuronal development, T cell development appears to be normal in these mice. However, when Apaf-$1^{-/-}$ and caspase-$9^{-/-}$ thymocytes are treated with stimuli such as dexamethasone, etoposide and γ-irradiation, both show resistance to apoptosis [80, 81]. In contrast, Apaf-$1^{-/-}$, caspase-$9^{-/-}$ and wild-type thymocytes are equally susceptible to Fas-mediated apoptosis, suggesting that Fas-mediated cell death in thymocytes is independent of the Apaf1/caspase-9 pathway. The lack of gross developmental defects in T cell development also suggests that Apaf-1 and caspase-9 are not involved in death by neglect in thymus.

Due to the timing of the lethality caused by caspase-9 and Apaf-1 mutation, it is not possible to address the significance of this pathway in peripheral cell death using these models. Since the lymphoid systems are largely normal, further experiments using $Rag1^{-/-}$ mice reconstituted with fetal liver cells derived from either caspase-$9^{-/-}$ or Apaf-$1^{-/-}$ embryos will aid in answering questions about the role of this pathway in programmed cell death of the peripheral lymphoid system. This is of particular interest because recent studies have shown an involvement of a positive feedback loop mediated by mitochondria during tumor-induced death of activated T cells [84, 85]. These studies demonstrate that loss of mitochondrial transmembrane potential, an event apparently independent of death receptor signaling, leads to a series of events including cleavage of Bid, that in turn further increase mitochondrial permeability and lead to apoptosome activation. Therefore, it is important to examine specifically whether the Apaf-1/caspase-9 pathway is involved in the death of activated T cells, despite the demonstration that Fas-mediated thymocyte death is not affected by these mutations.

In embryonic fibroblasts expressing c-Myc and Ras, Apaf-1 and caspase-9 mutations can also block p53-induced cell death, a process which is felt to play a role in the negative selection of developing thymocytes [86], placing Apaf-1 and caspase-9 downstream in a p53-induced apoptosis pathway [87]. Therefore, p53 may mediate negative selection through activation of Apaf1 and caspase-9. Interestingly, p53 may also mediate death by neglect at the pre-TCR stage because its deficiency partially restores development to $CD4^{+}CD8^{+}$ pre-T cells in SCID, $Rag1^{-/-}$ and $Rag2^{-/-}$ mice [68, 88–90]. Furthermore, loss of function of p53 or pro-apoptotic Bcl-2 family members results in thymus hyperplasia and thymoma [91–93].

The molecular mechanisms underlying apoptosis in the immune system have also been examined among the Bcl-2 family, whose members modulate the mitochondrial-directed activation of Apaf1 and caspase-9. Deficiency of the anti-apoptotic Bcl-2 protein results in early postnatal lethality, but initially normal lymphocyte development [94]. At variable times after birth, however, these mice undergo fulminant lymphocyte apoptosis with involution of lymphoid organs, consistent with susceptibility to uncontrolled apoptosis in the face of activation. In contrast, Bcl-x-deficient mice, which lack both the pro-apoptotic Bcl-x_S and the anti-apoptotic Bcl-x_L, are embryonic lethal with severe excessive neuronal apoptosis and reduced survival of immature lymphocytes [95]. Furthermore, chimeras of $Rag2^{-/-}$ mice with

$Bcl\text{-}x^{-/-}$ embryonic stem cells revealed reduced survival of $CD4^+CD8^+$ thymocytes, but no effect on survival of mature single-positive cells. Deficiency in the pro-apoptotic Bcl-2 family member Bim, as expected, resulted in increased numbers of $CD4^-CD8^-$ and mature single-positive lymphocytes, although, surprisingly, there was a reduction in $CD4^+CD8^+$ cells [96]. Bim-deficient mice also demonstrated increased leukocyte numbers, peripheral lymphoid organ hyperplasia and an age-dependent autoimmune syndrome. Although examination of the immunologic consequences of caspase-9 deficiency has not been rigorously examined due to the high degree of embryonic lethality, the lack of a clearer effect is perplexing in light of the phenotypes seen from deficiency of these Bcl-2 family members. This implies the possible existence of a mitochondrial-mediated death pathway independent of caspase-9, perhaps functioning by compensatory activation of other caspases, such as caspase-2 (see Section 2.8), to counter caspase-9 deficiency [67].

2.8
Caspase-2: A Duality of Function

Caspase-2 (NEDD2) was first identified as an mRNA species highly expressed in the developing mouse brain and then down-regulated in the adult [97]. Our understanding of the position of caspase-2 in apoptotic pathways is complicated by the fact that it possesses a long prodomain, suggesting a role as an upstream initiator. Its prodomain contains a CARD, which allows it to efficiently interact with the adapter molecule CRADD/RAIDD, though the physiologic significance of this interaction remains unclear [26, 27]. CRADD/RAIDD possesses a DD, which allows it to interact with RIP, an enzyme that in turn associates with the TNF-RI via the adapter molecule TRADD [26, 98]. This raises the question of whether the CARD of caspase-2 may function to recruit it to a DISC-like complex, similar to caspase-8. Caspase-2-deficient embryonic fibroblasts, however, are sensitive to TNF-α-mediated apoptosis, implying that, at least in this cell type, RAIDD/caspase-2 interactions are not required [99]. A second possibility, raised by the fact that caspase-2 activation is decreased in $Apaf\text{-}1^{-/-}$ cells, is that caspase-2 participates in the formation of an apoptosome-like complex [80]. Conversely, evidence from cell-free extracts stimulated with cytochrome *c* indicates that caspase-2 can be activated by caspase-3 in a caspase-9-dependent fashion, placing it as a potential downstream effector rather than an upstream initiator [100].

An interesting property of caspase-2 is its potential ability to either induce or antagonize apoptosis through alternatively spliced forms [101]. Caspase-2_L, the prevailing form, is pro-apoptotic. On the other hand, caspase-2_S, a truncated form generated by insertion of an early stop codon, is anti-apoptotic. Although anti-apoptotic effects have been identified for alternatively spliced forms of other caspases, deletion of caspase-2 in mice is the only example of a caspase knockout with evidence of both pro- and anti-apoptotic effects [99]. Ovarian germ cells from mice almost exclusively express the caspase-2_L form. In the caspase-2 knockout mice, these germ cells were resistant to apoptosis from both normal ovarian developments, in which one-half to two-thirds of germ cells normally die, and from exposure to doxorubicin. In facial motor neurons, however, where both caspase-2_L and -2_S are thought to be expressed, the knockout mice showed reduced cell numbers during late embryogenesis, implying an increase in apoptosis. By postnatal day 7, however, the knockout has the same cell numbers as wild-type, implying a decrease in apoptosis. Finally, caspase-$2^{-/-}$ sympathetic neurons were found to undergo apoptosis more effectively than wild-type neurons in response to growth factor withdrawal. These results suggest that the effects of caspase-2 are variable, depending not only on the particular tissue but also on the stage of development.

Caspase-2 (like caspase-12, see below) has also been shown to play a critical role in neuronal apoptosis induced by β-amyloid, a protein that is found to accumulate in the brains of Alzheimer's disease patients and is felt to contribute to the disease's pathophysiology. Antisense oligonucleotides to caspase-2 inhibit β-amyloid-induced apoptosis in PC-12 cells, and sympathetic neurons from caspase-2 knockout mice are resistant to β-amyloid-induced apoptosis, even after extended exposure [102].

Evaluation of lymphocyte apoptosis in caspase-2 knockout mice centered on a paradigm for cytotoxic T lymphocytes (CTL), which can induce apoptosis in target cells through the release of the serine protease granzyme B and perforin. Consistent with the ability of granzyme B to cleave caspase-2 *in vitro*, B lymphoblasts from caspase-2 knockout mice were resistant to apoptosis induced by granzyme B/perforin, but not to anti-Fas, doxorubicin, etoposide, γ-irradiation or staurosporine [99]. Although CTL-mediated apoptosis is felt to be an important mechanism for removal of activated B cells following an immune response, to date these animals have not been reported to have signs of autoimmune disease.

2.9
Caspase-12: Responding to Stress

As an organelle responsible for aspects of protein synthesis, folding and trafficking, the endoplasmic reticulum (ER) is sensitive to alterations in protein homeostasis. Perturbation of this system, such as by accumulation of malfolded proteins or from a variety of other insults that affect the ER, including oxidative stress, glycosylation inhibitors or calcium ionophores, results in the activation of the ER stress response, which can ultimately lead to apoptosis [103]. Caspase-12 was found to localize to the ER, raising the possibility of involvement in the ER stress response [104]. Caspase-$12^{-/-}$ mice are viable with no apparent developmental abnormalities. Knockout thymocytes were normally susceptible to apoptosis induced by anti-Fas and dexamethasone and, similarly, knockout embryonic fibroblasts are normally susceptible to apoptosis from anti-Fas, TNF-α and staurosporine. When exposed to ER stress-inducing stimuli, such as brefeldin A, tunicamycin and thapsigargin, however, caspase-$12^{-/-}$ embryonic fibroblasts showed resistance to apoptosis. Furthermore, using an *in vivo* model of tunicamycin-mediated renal epithelial cell apoptosis, knockout mice were shown to have reduced apoptosis and damage to renal tubular cells as well as a survival advantage. Interestingly, primary cortical neurons from caspase-12$^{-/-}$ mice (like caspase-2, see above) are resistant to β-amyloid-induced apoptosis, implying that β-amyloid exerts its effects, at least in part, by causing ER stress. The route leading from ER stress to caspase-12 activation, however, remains unclear. There is evidence, however, suggesting that release of intracellular calcium stores associated with ER stress activates the protease calpain, which in turn can activate caspase-12 and lead to apoptosis [105].

2.10 Compensatory Caspase Activation: A Caveat to Knockout Analysis

Although the generation of knockout animals has vastly increased our knowledge of the cellular apoptotic machinery, one difficulty in assessing the contributions of individual proteins has been the ability to preserve function by the compensatory activation of related proteins. This has been shown in a number of paradigms, such as a model of Fas-mediated cell death in hepatocytes, which are type II cells and therefore rely on the mitochondrial pathway to execute apoptosis. Interestingly, hepatocytes from caspase-9 or caspase-3 knockout mice could still undergo cell death from anti-Fas antibody by the activation of alternative caspases, but the patterns of caspase activation were different between wild-type and knockouts [67]. Caspase-3$^{-/-}$ hepatocytes showed activation of caspase-6 and -7, not seen in the wild-type, while caspase-9$^{-/-}$ hepatocytes showed activation of caspase-2 and -6, also not seen in the wild-type. Perhaps the most striking *in vivo* example of this quandary is the phenotype of the double knockout of the two pro-apoptotic Bcl-2 family members Bax and Bak. Whereas a single knockout of either Bax or Bak produces viable mice with only mild apoptotic defects, the double Bax/Bak knockout results in a severe phenotype consisting of perinatal lethality, persistence of interdigital webbing, central nervous system hyperplasia, splenomegaly and significant resistance to Fas-induced hepatocyte apoptosis [106, 107]. In addition, double Bax/Bak knockout embryonic fibroblasts are resistant to staurosporine, etoposide, UV radiation and serum withdrawal as well as the ER stress-inducers tunicamycin, thapsigargin and brefeldin A, but not to TNF-α [106].

A related illustration of this caveat is the potential for variation between mice strains. The caspase-3 knockout, for example, is embryonic lethal in 100% of offspring on one pure background strain [123]. On a different background, however, survival of knockout mice reaches approximately two-thirds of expected Mendelian inheritance levels. Thus, there are likely to be strain-specific caspase activity-modifying factors which may affect interpretation of results. These data reinforce the fact that it is difficult to exclude a particular caspase as significant for a cell death paradigm if its function can be assumed by another caspase. In the future, we will need to continue to study combinations of knockouts to gain a clearer understanding of apoptotic pathways. Furthermore, it is important to realize that future pharmacotherapeutic approaches to diseases that target caspase pathways may need to inhibit more than one caspase to achieve the desired effect.

2.11 Caspases: More than Simple Killers

Although the focus of caspase research has been on their roles in apoptosis, there is growing evidence that caspases, besides caspase-1 and -11, can play other roles in the immune system. There are recent reports that caspase-3 cleavage happens during T cell stimulation in the absence of apoptosis and that the poly-caspase inhibitor zVAD-fmk can block proliferation, MHC class II expression and blasting during stimulation of peripheral blood lymphocytes. Moreover, T cell activation triggers the selective processing and activation of downstream caspase-3, -6 and -7, but not caspase-1, -2 or -4, as demonstrated even in intact cells using a cell-permeable fluorescent substrate [108–110]. TCR stimulation activates caspase-8, but not caspase-9, and, most importantly, caspase activity results in a selective substrate cleavage, since poly(ADP-ribose) polymerase (PARP), lamin B and Wee1 kinase, but not DNA fragmentation factor (DFF45) or replication factor C (RFC140), are processed. In addition, inhibitors of caspase activity block anti-CD3-induced proliferation and IL-2 production by human T cells. Thus, caspase activation may be an early physiological response in viable, stimulated lymphocytes, and appears to be involved in early steps of lymphocyte activation [108–110].

One model suggests that caspase activation during TCR triggering occurs by stimulation of death receptors such as Fas, TNF-related apoptosis-inducing ligand (TRAIL) and TNF-R [111]. This notion is supported by the fact that Fas activation can induce cell proliferation in some experimental system [112, 113]. It was recently shown that stimulation of Jurkat T cells with FasL or the closely related TRAIL leads to up-regulation of the proto-oncogene c-*fos*, a target gene of mitogenic stimuli [111]. In addition, FADD, a molecule upstream of initiator caspases, was shown to be involved in T cell activation. T lymphocytes deficient in FADD or expressing a dominant-negative mutant allele of FADD were not only resistant to FasL-mediated apoptosis, but also defective in their proliferative capacity in response to TCR stimulation [58, 114–116].

Selection of substrates in a cell context-dependent manner seems to be one of the mechanisms by which caspases take on a role other than apoptosis. One example is the series of well-orchestrated actions of caspases during erythroid differentiation that do not involve apoptosis [110, 117]. The production of red blood cells follows the sequential formation, from erythroid progenitors, of proerythroblasts and basophilic (immature erythroblasts), polychromatophilic and orthochromatic erythroblasts (mature erythroblasts), and enucleated red cells [117]. Immature erythroid cells express several death receptors whose ligands are produced by mature erythroblasts. Exposure of erythroid progenitors to mature erythroblasts or death-receptor ligands resulted in caspase-mediated degradation of the transcription factor GATA-1, which is associated with impaired erythroblast development [110]. The effect of differentiation arrest is independent of cell death and is stimulated by death receptor activation [110]. In this manner, caspase-mediated cleavage of GATA-1 may represent an important negative control mechanism to control differentiation from immature erythroblasts to mature erythroblasts. Caspase activation is also involved in terminal differentiation from mature erythroblasts to enucleated red cells. The morphology of erythroblasts changes dramatically during terminal erythroid differentiation, including condensation of chromatin, loss of organelles and enucleation. Some of these morphological changes share similarities with features occurring during apoptosis as a consequence of caspase activation. Caspase inhibitors arrest the maturation of human erythroid progenitors at early stages of differentiation, before nucleus and chromatin condensation. Effector caspases such as caspase-3 are transiently activated through the mitochondrial pathway during erythroblast differentiation, and cleave proteins involved in nucleus integrity (lamin B) and chromatin condensation (acinus) without inducing cell death and cleavage of GATA-1 [110]. Therefore activation of caspases and selection of specific substrates are critical for terminal differentiation of erythroblasts.

Caspases, therefore, appear to function during differentiation using at least two related modes of action: they can cleave transcription factors that program the cell fate or they can deplete specific proteins to reorganize the cell's internal structure. These same mechanisms are likely used in other cell types, especially during lymphocyte differentiation. During memory T cell generation, caspases may be involved in both the process of terminal differentiation to memory cells and the massive apoptosis of blast T cells, depending on their substrates. Another example is SATB1, a transcription factor that has been shown to be critical for T and B cell differentiation [118–120], but is also processed by caspase-6 in activated T cells [121, 122]. It would be interesting to study whether this phenomenon holds up *in vivo* in caspase-6-deficient mice. This aspect of caspase biology will certainly attract more active research in the future.

2.12 Concluding Remarks

The use of targeted gene ablation of caspases and other apoptosis-related proteins in mice has vastly increased our understanding of the molecular mechanisms of programmed cell death. Still, many crucial details remain to be elucidated, and knockout animals will likely continue to serve as a central tool for future studies. These include the need to use developmental stage- and tissue-specific conditional knockouts to progress past the barriers of embryonic lethality. Examination of compound knockouts should aim to address questions of compensatory mechanisms among related proteins. Assessment of strain-specific variations in knockout phenotypes may serve to reveal novel modulators of the apoptotic pathways. The study of caspase-deficient mice should also help to reveal other, non-apoptotic functions of these enzymes. These critical animal models allow us to understand the function of specific proteins, helping to identify potential targets for manipulation to address a variety of disease states.

2.13 References

1 Haks, M.C., Oosterwegel, M.A., Blom, B., Spits, H.M. and Kruisbeek, A.M. Cell-fate decisions in early T cell development: regulation by cytokine receptors and the pre-TCR. *Semin Immunol* **1999**, 11, 23–37.

2 Jameson, S.C. and Bevan, M.J. T cell selection. *Curr Opin Immunol* **1998**, 10, 214–9.

3 Strasser, A., Harris, A.W., Huang, D.C., Krammer, P.H. and Cory, S. Bcl-2 and Fas/APO-1 regulate distinct pathways to lymphocyte apoptosis. *EMBO J* **1995**, 14, 6136–47.

4 Jenkins, M.K., Khoruts, A., Ingulli, E., Mueller, D.L., McSorley, S.J., Reinhardt, R.L., Itano, A. and Pape, K.A. *In vivo* activation of antigen-specific CD4 T cells. *Annu Rev Immunol* **2001**, 19, 23–45.

5 Krammer, P.H. CD95's deadly mission in the immune system. *Nature* **2000**, 407, 789–95.

6 Ellis, H.M. and Horvitz, H.R. Genetic control of programmed cell death in the nematode *C. elegans*. *Cell* **1986**, 44, 817–29.

7 Hengartner, M.O., Ellis, R.E. and Horvitz, H.R. *Caenorhabditis elegans* gene *ced-9* protects cells from programmed cell death. *Nature* **1992**, 356, 494–9.

8 Conradt, B. and Horvitz, H.R. The *C. elegans* protein EGL-1 is required for programmed cell death and interacts with the Bcl-2-like protein CED-9. *Cell* **1998**, 93, 519–29.

9 Chinnaiyan, A.M., O'Rourke, K., Lane, B.R. and Dixit, V.M. Interaction of CED-4 with CED-3 and CED-9, a molecular framework for cell death. *Science* **1997**, 275, 1122–6.

10 Xue, D., Shaham, S. and Horvitz, H.R. The *Caenorhabditis elegans* cell-death protein CED-3 is a cysteine protease with substrate specificities similar to those of the human CPP32 protease. *Genes Dev* **1996**, 10, 1073–83.

11 Zou, H., Henzel, W.J., Liu, X., Lutschg, A. and Wang, X. Apaf-1, a human protein homologous to *C. elegans* CED-4, participates in cytochrome *c*-dependent activation of caspase-3. *Cell* **1997**, 90, 405–13.

12 Hengartner, M.O. and Horvitz, H.R. *C. elegans* cell survival gene *ced-9* encodes a functional homolog of the mammalian proto-oncogene *bcl-2*. *Cell* **1994**, 76, 665–76.

13 Kerr, J.F., Wyllie, A.H. and Currie, A.R. Apoptosis: a basic biological phenomenon with wide-ranging implications in tissue kinetics. *Br J Cancer* **1972**, 26, 239–57.

14 Fadok, V.A., Voelker, D.R., Campbell, P.A., Cohen, J.J., Bratton, D.L. and Henson, P.M. Exposure of phosphatidylserine on the surface of apoptotic lymphocytes triggers specific recognition and removal by macrophages. *J Immunol* **1992**, 148, 2207–16.

15 Thornberry, N.A., Bull, H.G., Calaycay, J.R., Chapman, K.T., Howard, A.D., Kostura, M.J., Miller, D.K., Molineaux, S.M., Weidner, J.R., Aunins, J., *et al.* A novel heterodimeric cysteine protease is required for interleukin-1 beta processing in monocytes. *Nature* **1992**, 356, 768–74.

16 Stennicke, H.R. and Salvesen, G.S. Catalytic properties of the caspases. *Cell Death Different* **1999**, 6, 1054–9.

17 Ramage, P., Cheneval, D., Chvei, M., Graff, P., Hemmig, R., Heng, R., Kocher, H.P., Mackenzie, A., Memmert, K., Revesz, L., *et al.* Expression, refolding, and autocatalytic proteolytic processing of the interleukin-1 beta-converting enzyme precursor. *J Biol Chem* **1995**, 270, 9378–83.

18 Wilson, K.P., Black, J.A., Thomson, J.A., Kim, E.E., Griffith, J.P., Navia, M.A., Murcko, M.A., Chambers, S.P., Aldape, R.A., Raybuck, S.A., *et al.* Structure and mechanism of interleukin-1 beta converting enzyme. *Nature* **1994**, 370, 270–5.

19 Chinnaiyan, A.M., O'Rourke, K., Tewari, M. and Dixit, V.M. FADD, a novel death domain-containing protein, inter-

acts with the death domain of Fas and initiates apoptosis. *Cell* **1995**, 81, 505–12.

20 Boldin, M.P., Varfolomeev, E.E., Pancer, Z., Mett, I.L., Camonis, J.H. and Wallach, D.A novel protein that interacts with the death domain of Fas/APO1 contains a sequence motif related to the death domain. *J Biol Chem* **1995**, 270, 7795–8.

21 Medema, J.P., Scaffidi, C., Kischkel, F.C., Shevchenko, A., Mann, M., Krammer, P.H. and Peter, M.E. FLICE is activated by association with the CD95 death-inducing signaling complex (DISC). *EMBO J* **1997**, 16, 2794–804.

22 Kischkel, F.C., Hellbardt, S., Behrmann, I., Germer, M., Pawlita, M., Krammer, P.H. and Peter, M.E. Cytotoxicity-dependent APO-1 (Fas/CD95)-associated proteins form a death-inducing signaling complex (DISC) with the receptor. *EMBO J* **1995**, 14, 5579–88.

23 Muzio, M., Stockwell, B.R., Stennicke, H.R., Salvesen, G.S. and Dixit, V.M. An induced proximity model for caspase-8 activation. *J Biol Chem* **1998**, 273, 2926–30.

24 Kischkel, F. C., Lawrence, D. A., Tinel, A., LeBlanc, H., Virmani, A., Schow, P., Gazdar, A., Blenis, J., Arnott, D. and Ashkenazi, A. Death receptor recruitment of endogenous caspase-10 and apoptosis initiation in the absence of caspase-8. *J Biol Chem* **2001**, 2, 2.

25 Li, P., Nijhawan, D., Budihardjo, I., Srinivasula, S.M., Ahmad, M., Alnemri, E.S. and Wang, X. Cytochrome *c* and dATP-dependent formation of Apaf-1/caspase-9 complex initiates an apoptotic protease cascade. *Cell* **1997**, 91, 479–89.

26 Ahmad, M., Srinivasula, S.M., Wang, L., Talanian, R.V., Litwack, G., Fernandes-Alnemri, T. and Alnemri, E.S. CRADD, a novel human apoptotic adaptor molecule for caspase-2, and FasL/tumor necrosis factor receptor-interacting protein RIP. *Cancer Res* **1997**, 57, 615–9.

27 Duan, H. and Dixit, V. M. RAIDD is a new 'death' adaptor molecule. *Nature* **1997**, 385, 86–9.

28 Woo, M., Hakem, A., Elia, A.J., Hakem, R., Duncan, G.S., Patterson, B.J. and Mak, T.W. *In vivo* evidence that caspase-3 is required for Fas-mediated apoptosis of hepatocytes. *J Immunol* **1999**, 163, 4909–16.

29 Roy, S., Bayly, C.I., Gareau, Y., Houtzager, V.M., Kargman, S., Keen, S.L., Rowland, K., Seiden, I.M., Thornberry, N.A. and Nicholson, D.W. Maintenance of caspase-3 proenzyme dormancy by an intrinsic 'safety catch' regulatory tripeptide. *Proc Natl Acad Sci USA* **2001**, 98, 6132–7.

30 Kostura, M.J., Tocci, M.J., Limjuco, G., Chin, J., Cameron, P., Hillman, A.G., Chartrain, N.A. and Schmidt, J.A. Identification of a monocyte specific preinterleukin 1 beta convertase activity. *Proc Natl Acad Sci USA* **1989**, 86, 5227–31.

31 Gu, Y., Kuida, K., Tsutsui, H., Ku, G., Hsiao, K., Fleming, M.A., Hayashi, N., Higashino, K., Okamura, H., Nakanishi, K., Kurimoto, M., Tanimoto, T., Flavell, R.A., Sato, V., Harding, M.W., Livingston, D.J. and Su, M.S. Activation of interferon-gamma inducing factor mediated by interleukin-1beta converting enzyme. *Science* **1997**, 275, 206–9.

32 Thome, M., Hofmann, K., Burns, K., Martinon, F., Bodmer, J. L., Mattmann, C. and Tschopp, J. Identification of CARDIAK, a RIP-like kinase that associates with caspase-1. *Curr Biol* **1998**, 8, 885–8.

33 Lemaigre-Dubreuil, Y., Brugg, B., Chianale, C., Delhaye-Bouchaud, N. and Mariani, J. Over-expression of interleukin-1 beta-converting enzyme mRNA in staggerer cerebellum. *Neuroreport* **1996**, 7, 1777–80.

34 Li, P., Allen, H., Banerjee, S., Franklin, S., Herzog, L., Johnston, C., McDowell, J., Paskind, M., Rodman, L., Salfeld, J., *et al.* Mice deficient in IL-1 beta-converting enzyme are defective in production of mature IL-1 beta and resistant to endotoxic shock. *Cell* **1995**, 80, 401–11.

35 Kuida, K., Lippke, J.A., Ku, G., Harding, M.W., Livingston, D.J., Su, M.S. and Flavell, R.A. Altered cytokine export and apoptosis in mice deficient in interleukin-1 beta converting enzyme. *Science* **1995**, 267, 2000–3.

36 Wang, S., Miura, M., Jung, Y., Zhu, H., Gagliardini, V., Shi, L., Greenberg,

A. H. and Yuan, J. Identification and characterization of Ich-3, a member of the interleukin-1beta converting enzyme (ICE)/Ced-3 family and an upstream regulator of ICE. *J Biol Chem* **1996**, 271, 20580–7.

37 Wang, S., Miura, M., Jung, Y. K., Zhu, H., Li, E. and Yuan, J. Murine caspase-11, an ICE-interacting protease, is essential for the activation of ICE. *Cell* **1998**, 92, 501–9.

38 Chiarle, R., Podda, A., Prolla, G., Podack, E. R., Thorbecke, G. J. and Inghirami, G. CD30 overexpression enhances negative selection in the thymus and mediates programmed cell death via a Bcl-2-sensitive pathway. *J Immunol* **1999**, 163, 194–205.

39 Amakawa, R., Hakem, A., Kundig, T. M., Matsuyama, T., Simard, J. J., Timms, E., Wakeham, A., Mittruecker, H. W., Griesser, H., Takimoto, H., Schmits, R., Shahinian, A., Ohashi, P., Penninger, J. M. and Mak, T. W. Impaired negative selection of T cells in Hodgkin's disease antigen CD30-deficient mice. *Cell* **1996**, 84, 551–62.

40 DeYoung, A. L., Duramad, O. and Winoto, A. The TNF receptor family member CD30 is not essential for negative selection. *J Immunol* **2000**, 165, 6170–3.

41 Kang, S. J., Wang, S., Hara, H., Peterson, E. P., Namura, S., Amin-Hanjani, S., Huang, Z., Srinivasan, A., Tomaselli, K. J., Thornberry, N. A., Moskowitz, M. A. and Yuan, J. Dual role of caspase-11 in mediating activation of caspase-1 and caspase-3 under pathological conditions. *J Cell Biol* **2000**, 149, 613–22.

42 Hisahara, S., Yuan, J., Momoi, T., Okano, H. and Miura, M. Caspase-11 mediates oligodendrocyte cell death and pathogenesis of autoimmune-mediated demyelination. *J Exp Med* **2001**, 193, 111–22.

43 Shibata, M., Hisahara, S., Hara, H., Yamawaki, T., Fukuuchi, Y., Yuan, J., Okano, H. and Miura, M. Caspases determine the vulnerability of oligodendrocytes in the ischemic brain. *J Clin Invest* **2000**, 106, 643–53.

44 Locksley, R. M., Killeen, N. and Lenardo, M. J. The TNF and TNF receptor superfamilies: integrating mammalian biology. *Cell* **2001**, 104, 487–501.

45 Scaffidi, C., Fulda, S., Srinivasan, A., Friesen, C., Li, F., Tomaselli, K. J., Debatin, K. M., Krammer, P. H. and Peter, M. E. Two CD95 (APO-1/Fas) signaling pathways. *EMBO J* **1998**, 17, 1675–87.

46 Li, H., Zhu, H., Xu, C. J. and Yuan, J. Cleavage of BID by caspase-8 mediates the mitochondrial damage in the Fas pathway of apoptosis. *Cell* **1998**, 94, 491–501.

47 Korsmeyer, S. J., Wei, M. C., Saito, M., Weiler, S., Oh, K. J. and Schlesinger, P. H. Pro-apoptotic cascade activates BID, which oligomerizes BAK or BAX into pores that result in the release of cytochrome *c*. *Cell Death Different* **2000**, 7, 1166–73.

48 Liu, X., Kim, C. N., Yang, J., Jemmerson, R. and Wang, X. Induction of apoptotic program in cell-free extracts: requirement for dATP and cytochrome *c*. *Cell* **1996**, 86, 147–57.

49 Verhagen, A. M., Ekert, P. G., Pakusch, M., Silke, J., Connolly, L. M., Reid, G. E., Moritz, R. L., Simpson, R. J. and Vaux, D. L. Identification of DIABLO, a mammalian protein that promotes apoptosis by binding to and antagonizing IAP proteins. *Cell* **2000**, 102, 43–53.

50 Du, C., Fang, M., Li, Y., Li, L. and Wang, X. Smac, a mitochondrial protein that promotes cytochrome *c*-dependent caspase activation by eliminating IAP inhibition. *Cell* **2000**, 102, 33–42.

51 Varfolomeev, E. E., Schuchmann, M., Luria, V., Chiannilkulchai, N., Beckmann, J. S., Mett, I. L., Rebrikov, D., Brodianski, V. M., Kemper, O. C., Kollet, O., Lapidot, T., Soffer, D., Sobe, T., Avraham, K. B., Goncharov, T., Holtmann, H., Lonai, P. and Wallach, D. Targeted disruption of the mouse Caspase-8 gene ablates cell death induction by the TNF receptors, Fas/Apo1, and DR3 and is lethal prenatally. *Immunity* **1998**, 9, 267–76.

52 Kishimoto, H., Surh, C. D. and Sprent, J. **1998**, A role for Fas in negative selection of thymocytes *in vivo*. *J Exp Med* 187, 1427–38.

53 Sidman, C. L., Marshall, J. D. and Von Boehmer, H. Transgenic T cell receptor interactions in the lymphoproliferative and autoimmune syndromes of lpr and gld mutant mice. *Eur J Immunol* **1992**, 22, 499–504.

54 Adachi, M., Suematsu, S., Kondo, T., Ogasawara, J., Tanaka, T., Yoshida, N. and Nagata, S. Targeted mutation in the Fas gene causes hyperplasia in peripheral lymphoid organs and liver. *Nat Genet* **1995**, 11, 294–300.

55 Rothe, J., Lesslauer, W., Lotscher, H., Lang, Y., Koebel, P., Kontgen, F., Althage, A., Zinkernagel, R., Steinmetz, M. and Bluethmann, H. Mice lacking the tumour necrosis factor receptor 1 are resistant to TNF-mediated toxicity but highly susceptible to infection by *Listeria monocytogenes*. *Nature* **1993**, 364, 798–802.

56 Erickson, S. L., de Sauvage, F. J., Kikly, K., Carver-Moore, K., Pitts-Meek, S., Gillett, N., Sheehan, K. C., Schreiber, R. D., Goeddel, D. V. and Moore, M. W. Decreased sensitivity to tumour-necrosis factor but normal T cell development in TNF receptor-2-deficient mice. *Nature* **1994**, 372, 560–3.

57 Kabra, N. H., Kang, C., Hsing, L. C., Zhang, J. and Winoto, A. T cell-specific FADD-deficient mice: FADD is required for early T cell development. *Proc Natl Acad Sci USA* **2001**, 98, 6307–12.

58 Newton, K., Harris, A. W., Bath, M. L., Smith, K. G. and Strasser, A. A dominant interfering mutant of FADD/MORT1 enhances deletion of autoreactive thymocytes and inhibits proliferation of mature T lymphocytes. *EMBO J* **1998**, 17, 706–18.

59 Zhang, J., Kabra, N. H., Cado, D., Kang, C. and Winoto, A. FADD-deficient T cells exhibit a disaccord in regulation of the cell cycle machinery. *J Biol Chem* **2001**, 276, 29815–8.

60 Byrne, J. A., Butler, J. L. and Cooper, M. D. Differential activation requirements for virgin and memory T cells. *J Immunol* **1988**, 141, 3249–57.

61 Zhou, T., Edwards, C. K., 3rd, Yang, P., Wang, Z., Bluethmann, H. and Mountz, J. D. Greatly accelerated lymphadenopathy and autoimmune disease in *lpr* mice lacking tumor necrosis factor receptor I. *J Immunol* **1996**, 156, 2661–5.

62 Cohen, P. L. and Eisenberg, R. A. *Lpr* and *gld*: single gene models of systemic autoimmunity and lymphoproliferative disease. *Annu Rev Immunol* **1991**, 9, 243–69.

63 Kuida, K., Zheng, T. S., Na, S., Kuan, C., Yang, D., Karasuyama, H., Rakic, P. and Flavell, RA. Decreased apoptosis in the brain and premature lethality in CPP32-deficient mice. *Nature* **1996**, 384, 368–72.

64 Woo, M., Hakem, R., Soengas, M. S., Duncan, G. S., Shahinian, A., Kagi, D., Hakem, A., McCurrach, M., Khoo, W., Kaufman, S. A., Senaldi, G., Howard, T., Lowe, S. W. and Mak, T. W. Essential contribution of caspase-3/CPP32 to apoptosis and its associated nuclear changes. *Genes Dev* **1998**, 12, 806–19.

65 Alam, A., Braun, M. Y., Hartgers, F., Lesage, S., Cohen, L., Hugo, P., Denis, F. and Sekaly, R. P. Specific activation of the cysteine protease CPP32 during the negative selection of T cells in the thymus. *J Exp Med* **1997**, 186, 1503–12.

66 Clayton, L. K., Ghendler, Y., Mizoguchi, E., Patch, R. J., Ocain, T. D., Orth, K., Bhan, A. K., Dixit, V. M. and Reinherz, E. L. T cell receptor ligation by peptide/MHC induces activation of a caspase in immature thymocytes: the molecular basis of negative selection. *EMBO J* **1997**, 16, 2282–93.

67 Zheng, T. S., Hunot, S., Kuida, K., Momoi, T., Srinivasan, A., Nicholson, D. W., Lazebnik, Y. and Flavell, R. A. Deficiency in caspase-9 or caspase-3 induces compensatory caspase activation. *Nat Med* **2000**, 6, 1241–7.

68 Jiang, D., Lenardo, M. J. and Zuniga-Pflucker, C. p53 prevents maturation to the $CD4^+CD8^+$ stage of thymocyte differentiation in the absence of T cell receptor rearrangement. *J Exp Med* **1996**, 183, 1923–8.

69 Linette, G. P., Grusby, M. J., Hedrick, S. M., Hansen, T. H., Glimcher, L. H. and Korsmeyer, S. J. Bcl-2 is upregulated at the $CD4^+$ $CD8^+$ stage during positive selection and promotes thymocyte differentiation at several control points. *Immunity* **1994**, 1, 197–205.

70 Zhang, J., Mikecz, K., Finnegan, A. and Glant, T. T. Spontaneous thymocyte apoptosis is regulated by a mitochondrion-mediated signaling pathway. *J Immunol* **2000**, 165, 2970–4.

71 Budd, R.C. Activation-induced cell death. *Curr Opin Immunol* **2001**, 13, 356–62.

72 Mixter, P.F., Russell, J.Q., Morrissette, G.J., Charland, C., Aleman-Hoey, D. and Budd, R.C. A model for the origin of TCR-alphabeta$^+$ CD4$^-$CD8$^-$B220$^+$ cells based on high affinity TCR signals. *J Immunol* **1999**, 162, 5747–56.

73 Lodolce, J.P., Boone, D.L., Chai, S., Swain, R.E., Dassopoulos, T., Trettin, S. and Ma, A. IL-15 receptor maintains lymphoid homeostasis by supporting lymphocyte homing and proliferation. *Immunity* **1998**, 9, 669–76.

74 Cantrell, D.A. and Smith, K.A. The interleukin-2 T cell system: a new cell growth model. *Science* **1984**, 224, 1312–6.

75 Pestano, G.A., Zhou, Y., Trimble, L.A., Daley, J., Weber, G.F. and Cantor, H. Inactivation of misselected CD8 T cells by CD8 gene methylation and cell death. *Science* **1999**, 284, 1187–91.

76 Budihardjo, I., Oliver, H., Lutter, M., Luo, X. and Wang, X. Biochemical pathways of caspase activation during apoptosis. *Annu Rev Cell Dev Biol* **1999**, 15, 269–90.

77 Rodriguez, J. and Lazebnik, Y. Caspase-9 and APAF-1 form an active holoenzyme. *Genes Dev* **1999**, 13, 3179–84.

78 Stennicke, H.R., Deveraux, Q.L., Humke, E.W., Reed, J.C., Dixit, VM. and Salvesen, G.S. Caspase-9 can be activated without proteolytic processing. *J Biol Chem* **1999**, 274, 8359–62.

79 Hengartner, M.O. The biochemistry of apoptosis. *Nature* **2000**, 407, 770–6.

80 Yoshida, H., Kong, Y.Y., Yoshida, R., Elia, A. J., Hakem, A., Hakem, R., Penninger, J.M. and Mak, T.W. Apaf1 is required for mitochondrial pathways of apoptosis and brain development. *Cell* **1998**, 94, 739–50.

81 Kuida, K., Haydar, T.F., Kuan, C.Y., Gu, Y., Taya, C., Karasuyama, H., Su, M. S., Rakic, P. and Flavell, R.A. Reduced apoptosis and cytochrome *c*-mediated caspase activation in mice lacking caspase-9. *Cell* **1998**, 94, 325–37.

82 Hakem, R., Hakem, A., Duncan, G.S., Henderson, J.T., Woo, M., Soengas, M.S., Elia, A., de la Pompa, J.L., Kagi, D., Khoo, W., Potter, J., Yoshida, R., Kaufman, S.A., Lowe, S.W., Penninger, J.M. and Mak, T.W. Differential requirement for caspase-9 in apoptotic pathways *in vivo*. *Cell* **1998**, 94, 339–52.

83 Li, K., Li, Y., Shelton, J.M., Richardson, J.A., Spencer, E., Chen, Z.J., Wang, X. and Williams, R.S. Cytochrome *c* deficiency causes embryonic lethality and attenuates stress-induced apoptosis. *Cell* **2000**, 101, 389–99.

84 Gastman, B.R., Yin, X.M., Johnson, D.E., Wieckowski, E., Wang, G. Q., Watkins, S.C. and Rabinowich, H. Tumor-induced apoptosis of T cells: amplification by a mitochondrial cascade. *Cancer Res* **2000**, 60, 6811–7.

85 Hildeman, D.A., Mitchell, T., Teague, T.K., Henson, P., Day, B.J., Kappler, J. and Marrack, P.C. Reactive oxygen species regulate activation-induced T cell apoptosis. *Immunity* **1999**, 10, 735–44.

86 Zhu, J.W., DeRyckere, D., Li, F.X., Wan, Y.Y. and DeGregori, J. A role for E2F1 in the induction of ARF, p53, and apoptosis during thymic negative selection. *Cell Growth Different* **1999**, 10, 829–38.

87 Soengas, M.S., Alarcon, R.M., Yoshida, H., Giaccia, A.J., Hakem, R., Mak, T.W. and Lowe, S.W. Apaf-1 and caspase-9 in p53-dependent apoptosis and tumor inhibition. *Science* **1999**, 284, 156–9.

88 Nacht, M., Strasser, A., Chan, Y.R., Harris, A.W., Schlissel, M., Bronson, R.T. and Jacks, T. Mutations in the p53 and SCID genes cooperate in tumorigenesis. *Genes Dev* **1996**, 10, 2055–66.

89 Guidos, C.J., Williams, C.J., Grandal, I., Knowles, G., Huang, M.T. and Danska, J.S. V(D)J recombination activates a p53-dependent DNA damage checkpoint in scid lymphocyte precursors. *Genes Dev* **1996**, 10, 2038–54.

90 Bogue, M.A., Zhu, C., Aguilar-Cordova, E., Donehower, L.A. and Roth, D.B. p53 is required for both radiation-induced differentiation and rescue of V(D)J rearrangement in scid mouse thymocytes. *Genes Dev* **1996**, 10, 553–65.

91 Knudson, C.M., Johnson, G.M., Lin, Y. and Korsmeyer, S.J. Bax accelerates tu-

morigenesis in p53-deficient mice. *Cancer Res* **2001**, 61, 659–65.
92 Donehower, L.A., Harvey, M., Slagle, B.L., McArthur, M.J., Montgomery, C.A., Jr, Butel, J.S. and Bradley, A. Mice deficient for p53 are developmentally normal but susceptible to spontaneous tumours. *Nature* **1992**, 356, 215–21.
93 Ranger, A.M., Malynn, B.A. and Korsmeyer, S.J. Mouse models of cell death. *Nat Genet* **2001**, 28, 113–8.
94 Veis, D.J., Sorenson, C.M., Shutter, J.R. and Korsmeyer, S.J. Bcl-2-deficient mice demonstrate fulminant lymphoid apoptosis, polycystic kidneys, and hypopigmented hair. *Cell* **1993**, 75, 229–40.
95 Motoyama, N., Wang, F., Roth, K.A., Sawa, H., Nakayama, K., Negishi, I., Senju, S., Zhang, Q., Fujii, S., *et al.* Massive cell death of immature hematopoietic cells and neurons in Bcl-x-deficient mice. *Science* **1995**, 267, 1506–10.
96 Bouillet, P., Metcalf, D., Huang, D.C., Tarlinton, D.M., Kay, T.W., Kontgen, F., Adams, J.M. and Strasser, A. Proapoptotic Bcl-2 relative Bim required for certain apoptotic responses, leukocyte homeostasis, and to preclude autoimmunity. *Science* **1999**, 286, 1735–8.
97 Kumar, S., Kinoshita, M., Noda, M., Copeland, N.G. and Jenkins, N.A. Induction of apoptosis by the mouse Nedd2 gene, which encodes a protein similar to the product of the *Caenorhabditis elegans* cell death gene *ced-3* and the mammalian IL-1 beta-converting enzyme. *Genes Dev* **1994**, 8, 1613–26.
98 Hsu, H., Huang, J., Shu, H.B., Baichwal, V. and Goeddel, D.V. TNF-dependent recruitment of the protein kinase RIP to the TNF receptor-1 signaling complex. *Immunity* **1996**, 4, 387–96.
99 Bergeron, L., Perez, G.I., Macdonald, G., Shi, L., Sun, Y., Jurisicova, A., Varmuza, S., Latham, K.E., Flaws, J.A., Salter, J.C., Hara, H., Moskowitz, M.A., Li, E., Greenberg, A., Tilly, J.L. and Yuan, J. Defects in regulation of apoptosis in caspase-2-deficient mice. *Genes Dev* **1998**, 12, 1304–14.
100 Slee, E.A., Harte, M.T., Kluck, R.M., Wolf, B.B., Casiano, C.A., Newmeyer, D.D., Wang, H.G., Reed, J.C., Nicholson, D.W., Alnemri, E.S., Green, D.R. and Martin, S.J. Ordering the cytochrome c-initiated caspase cascade: hierarchical activation of caspases-2, -3, -6, -7, -8 and -10 in a caspase-9-dependent manner. *J Cell Biol* **1999**, 144, 281–92.
101 Wang, L., Miura, M., Bergeron, L., Zhu, H. and Yuan, J. *Ich-1*, an *Ice/ced-3*-related gene, encodes both positive and negative regulators of programmed cell death. *Cell* **1994**, 78, 739–50.
102 Troy, C.M., Rabacchi, S.A., Friedman, W.J., Frappier, T.F., Brown, K. and Shelanski, M.L. Caspase-2 mediates neuronal cell death induced by beta-amyloid. *J Neurosci* **2000**, 20, 1386–92.
103 Welihinda, A.A., Tirasophon, W. and Kaufman, R.J. The cellular response to protein misfolding in the endoplasmic reticulum. *Gene Expr* **1999**, 7, 293–300.
104 Nakagawa, T., Zhu, H., Morishima, N., Li, E., Xu, J., Yankner, B.A. and Yuan, J. Caspase-12 mediates endoplasmic-reticulum-specific apoptosis and cytotoxicity by amyloid-beta. *Nature* **2000**, 403, 98–103.
105 Nakagawa, T. and Yuan, J. Cross-talk between two cysteine protease families. Activation of caspase-12 by calpain in apoptosis. *J Cell Biol* **2000**, 150, 887–94.
106 Wei, M.C., Zong, W.X., Cheng, E.H., Lindsten, T., Panoutsakopoulou, V., Ross, A.J., Roth, KA., MacGregor, G.R., Thompson, C.B. and Korsmeyer, S.J. Proapoptotic BAX and BAK: a requisite gateway to mitochondrial dysfunction and death. *Science* **2001**, 292, 727–30.
107 Lindsten, T., Ross, A.J., King, A., Zong, W.X., Rathmell, J.C., Shiels, H. A., Ulrich, E., Waymire, K.G., Mahar, P., Frauwirth, K., Chen, Y., Wei, M., Eng, V.M., Adelman, D.M., Simon, M.C., Ma, A., Golden, J.A., Evan, G., Korsmeyer, S.J., MacGregor, G.R. and Thompson, C.B. The combined functions of proapoptotic Bcl-2 family members bak and bax are essential for normal development of multiple tissues. *Mol Cell* **2000**, 6, 1389–99.
108 Kennedy, N.J., Kataoka, T., Tschopp, J. and Budd, R.C. Caspase activation is required for T cell proliferation. *J Exp Med* **1999**, 190, 1891–6.

109 Alam, A., Cohen, L.Y., Aouad, S. and Sekaly, R.P. Early activation of caspases during T lymphocyte stimulation results in selective substrate cleavage in nonapoptotic cells. *J Exp Med* **1999**, 190, 1879–90.

110 De Maria, R., Zeuner, A., Eramo, A., Domenichelli, C., Bonci, D., Grignani, F., Srinivasula, S.M., Alnemri, E. S., Testa, U. and Peschle, C. Negative regulation of erythropoiesis by caspase-mediated cleavage of GATA-1. *Nature* **1999**, 401, 489–93.

111 Siegmund, D., Mauri, D., Peters, N., Juo, P., Thome, M., Reichwein, M., Blenis, J., Scheurich, P., Tschopp, J. and Wajant, H. Fas-associated death domain protein (FADD) and caspase-8 mediate up-regulation of c-Fos by Fas ligand and tumor necrosis factor-related apoptosis-inducing ligand (TRAIL) via a FLICE inhibitory protein (FLIP)-regulated pathway. *J Biol Chem* **2001**, 276, 32585–90.

112 Jelaska, A. and Korn, J.H. Anti-Fas induces apoptosis and proliferation in human dermal fibroblasts: differences between foreskin and adult fibroblasts. *J Cell Physiol* **1998**, 175, 19–29.

113 Freiberg, R.A., Spencer, D.M., Choate, K.A., Duh, H.J., Schreiber, S.L., Crabtree, G.R. and Khavari, P.A. Fas signal transduction triggers either proliferation or apoptosis in human fibroblasts. *J Invest Dermatol* **1997**, 108, 215–9.

114 Yeh, W.C., Pompa, J.L., McCurrach, M.E., Shu, H.B., Elia, A.J., Shahinian, A., Ng, M., Wakeham, A., Khoo, W., Mitchell, K., El-Deiry, W.S., Lowe, S.W., Goeddel, D.V. and Mak, T.W. FADD: essential for embryo development and signaling from some, but not all, inducers of apoptosis. *Science* **1998**, 279, 1954–8.

115 Zhang, J., Cado, D., Chen, A., Kabra, N.H. and Winoto, A. Fas-mediated apoptosis and activation-induced T cell proliferation are defective in mice lacking FADD/Mort1. *Nature* **1998**, 392, 296–300.

116 Walsh, C.M., Wen, B.G., Chinnaiyan, A.M., O'Rourke, K., Dixit, V.M. and Hedrick, S.M. A role for FADD in T cell activation and development. *Immunity* **1998**, 8, 439–49.

117 Zermati, Y., Garrido, C., Amsellem, S., Fishelson, S., Bouscary, D., Valensi, F., Varet, B., Solary, E. and Hermine, O. Caspase activation is required for terminal erythroid differentiation. *J Exp Med* **2001**, 193, 247–54.

118 Lechner, O., Lauber, J., Franzke, A., Sarukhan, A., von Boehmer, H. and Buer, J. Fingerprints of anergic T cells. *Curr Biol* **2001**, 11, 587–95.

119 Glynne, R., Ghandour, G., Rayner, J., Mack, D.H. and Goodnow, C.C. B-lymphocyte quiescence, tolerance and activation as viewed by global gene expression profiling on microarrays. *Immunol Rev* **2000**, 176, 216–46.

120 Alvarez, J.D., Yasui, D.H., Niida, H., Joh, T., Loh, D.Y. and Kohwi-Shigematsu, T. The MAR-binding protein SATB1 orchestrates temporal and spatial expression of multiple genes during T cell development. *Genes Dev* **2000**, 14, 521–35.

121 Galande, S., Dickinson, L.A., Mian, I.S., Sikorska, M. and Kohwi-Shigematsu, T. SATB1 cleavage by caspase-6 disrupts PDZ domain-mediated dimerization, causing detachment from chromatin early in T cell apoptosis. *Mol Cell Biol* **2001**, 21, 5591–604.

122 Gotzmann, J., Meissner, M. and Gerner, C. The fate of the nuclear matrix-associated-region-binding protein SATB1 during apoptosis. *Cell Death Different* **2000**, 7, 425–38.

123 Leonard, J.R., Klocke, B.I., D'sa, C., Flavell, R.A. and Roth, K.A. Strain-dependent neurodevelopmental abnormalities in caspase-3-deficient mice. *J Neuropath Exp Neurol* **2002**, 61, 6731.

Part 2
Clearance of Apoptotic Cells

3
Anti-inflammatory and Immunoregulatory Effects of Apoptotic Cells

Reinhard E. Voll, Martin Herrmann, Irute Girkontaite, Christian Stach, Wasilis Kolowos and Joachim R. Kalden

3.1
Introduction

When single cells started to organize themselves to form multicellular organisms in which individual cells became more and more specialized and differentiated, these new organisms must have faced a novel problem: what to do with cells that are not needed any more, misplaced, injured, infected or mutated and therefore dangerous for the organism. In order to maintain the function of the organism, the removal of damaged or supernumerary cells should happen quietly, rapidly and without inflammation and organ damage. The biologic answers to this complex tasks were suicide of unnecessary cells, and cannibalism by their neighbors and specialized phagocytes [1, 2].

Cellular suicide, termed apoptosis, can be initiated by an intrinsic program at a strictly determined stage of the development (e.g. during development of *Caenorhabditis elegans*) or can be triggered by extrinsic stimuli. Multiple stimuli and signaling pathways lead to the activation of an evolutionary conserved cell death program, which results in modifications of the cell membrane, caspase-mediated cleavage of many proteins and DNase-mediated cleavage of the nuclear DNA.

A characteristic feature of apoptotic in contrast to necrotic cell death is the absence of an inflammatory response and tissue damage, at least in most circumstances. The absence of inflammation at sites of increased apoptosis was commonly explained by the fact that cells undergoing apoptosis maintain their membrane integrity and express phagocyte recognition signals, which target apoptotic cells for phagocytosis prior to cell lysis [2]. Thereby, noxious contents such as proteolytic enzymes, which could provoke inflammation and tissue damage, are not released from apoptosing cells [2]. In addition, it has been observed that pro-inflammatory mediators such as thromboxane B_2, which are normally induced in macrophages upon phagocytosis, are not secreted after engulfment of apoptotic cells [3].

These mechanisms could explain the absence of inflammation and tissue damage at sites of apoptosis; however, it remained unclear, why many inducers of apoptotic cell death such as ultraviolet (UV) light and X-rays efficiently ameliorate numerous inflammatory diseases [4, 5]. Therefore, we began to investigate whether apoptotic cells simply fail to provide pro-inflammatory signals or whether they also actively suppress an inflammatory response. Recent results from our and other laboratories demonstrate that apoptotic cells in fact inhibit the production of pro-inflammatory cytokines and induce the production of anti-inflammatory and immunosuppressive cytokines in monocytes/macrophages. This chapter provides an overview about the effects of apoptotic cells on the function of monocytes/macrophages and – briefly – dendritic cells (DCs). Cell surface receptors and intracellular signaling pathways, which may mediate the anti-inflammatory effects of apoptotic cells, are summarized. Finally, the implications of the anti-inflammatory and immunomodulatory properties of apoptotic cells on normal physiology and the pathogenesis of diseases are discussed.

3.2
Anti-inflammatory Effects of Apoptotic Cells on Monocytes/Macrophages

Monocytes are generated in the bone marrow from the myeloid stem cells. After a relatively short circulation in the peripheral blood most monocytes evade into the tissues and differentiate into macrophages. Phenotypically and functionally different subsets of monocytes and macrophages have been described; however, the physiological functions of these different subsets are not yet fully understood. Monocytes/macrophages display as, in addition to polymorphonuclear leukocytes, the main phagocytic cells in the body. They play an important role as antigen-presenting cells (APCs), since they phagocytose pathogens, digest them and are able to present pathogen-derived peptides within major histocompatibility complex class II (MHC II) molecules to T helper cells. Activated macrophages express both MHC II and co-stimulatory molecules such as B7-1/B7-2 and thereby play an important role in the activation process of memory T cells. However, in contrast to DCs, macrophages cannot efficiently activate naïve T lymphocytes.

Macrophages may be the most efficient eaters of apoptotic cells, although many cells including fibroblasts and endothelial cells participate in the uptake of dying neighboring cells. The engulfment of apoptotic cells – presumably via a tether and tickle mechanism which involves a variety of receptors – occurs through macropinocytosis [6, 7]. Characteristic features include membrane ruffling, formation of fluid filled phagosomes, and the involvement of the Rho family GTPases Cdc42 and Rac, which link surface receptors to actin cytoskeletal organization [7, 8].

To investigate whether apoptotic cells modify the immune response, especially the function of monocytes/macrophages, we analyzed the cytokine secretion patterns of peripheral blood mononuclear cells (PBMCs) and purified monocytes in the absence and in the presence of apoptotic cells. Apoptosis was induced in autologous peripheral blood lymphocytes (PBLs) by UV irradiation and in murine CTLL-2 cells by IL-2 deprivation. Without stimulation, PBMCs and monocytes did not produce measurable amounts of cytokines. Therefore, PBMCs and monocytes were stimulated either with bacterial lipopolysaccharides (LPS), which are potent stimulators of cytokine secretion from monocytes and macrophages, or with heat-aggregated human immunoglobulins mimicking immune complexes. In the presence of apoptotic cells, but not in the presence of freshly prepared paraformaldehyde-fixed PBLs, PBMCs as well as purified monocytes produced substantially more of the anti-inflammatory cytokine interleukin (IL)-10, but markedly less of the pro-inflammatory cytokines tumor necrosis factor (TNF)-α, IL-1β and IL-12 (Figs. 3.1 and 3.2) [9]. Apoptotic cells themselves did not release measurable amounts of cytokines (not shown). This anti-inflammatory effect of apoptotic cells was independent of the apoptosis-inducing stimulus and of the origin of apoptotic cells: UV- or γ-irradiated human PBL or human or murine cell lines or IL-2-deprived murine CTLL-2 cells or human T cell clones (Figs. 3.1 and 3.3, and data not shown). The anti-inflammatory effects of apoptotic cells appeared to be conserved between mammalian species, since they could be mediated by both murine and human apoptotic cells. The incubation time of monocytes with apoptotic cells prior to LPS stimulation influenced the changes in the cytokine secretion pattern – whereas the induction of IL-10 was highest without preincubation, the relative inhibition of pro-inflammatory cytokine secretion increased with the preincubation time (Fig. 3.2) [9].

Monocytes or PBMCs activated with heat-aggregated human immunoglobulins instead of LPS also secreted more IL-10 and less TNF-α when exposed to apoptotic cells, while IL-1β and IL-12 were undetectable under those conditions (data not shown). Just the presence of apoptotic cells, however, may not be sufficient to induce IL-10 expression in monocytes, since unstimulated monocytes cultured with endotoxin-free autologous serum produced virtually no detectable amounts of cytokines, regardless of the presence of apoptotic cells (data not shown). Possibly, even non-activated monocytes may produce low amounts of IL-10, which are below the detection threshold of our ELISA, but still may exert significant biological effects. In contrast, just the presence of apoptotic cells can induce well measurable concentrations of IL-10 in human monocyte-derived macrophages (own unpublished observations and [10]).

Using human monocyte-derived macrophages, we also observed a marked decrease in LPS-induced IL-1β and TNF-α secretion in the presence of apoptotic cells. The IL-10 secretion was usually significantly increased by the addition of apoptotic cells; however, this was somewhat dependent on the experimental conditions. Fadok *et al.* investigated the influence of apoptotic cells on the cytokine secretion pattern of macrophages in more detail. They observed that *in vitro* differentiated monocyte-derived macrophages after phagocytosis of apoptotic neutrophils secreted less IL-1β, IL-8, IL-10, granulocyte macrophage colony stimulating factor (GM-CSF), TNF-α, leukotriene C_4 and thromboxane B_2 upon LPS-activation. In contrast, the production of transforming growth factor (TGF)-β1, prostaglandin E_2 and platelet-activating factor (PAF) was increased [11].

In different experimental settings, however, apoptotic cells seem to induce transcription of pro-inflammatory cytokines. Kurosaka *et al.* observed that upon interaction with apoptotic cells phagocytes produce not only anti-inflammatory cytokines, but also pro-inflammatory cytokines such as IL-1β, IL-8 and MIP-2 [12–14]. Pro-inflammatory cytokine production was detected in phorbol myristate acetate-

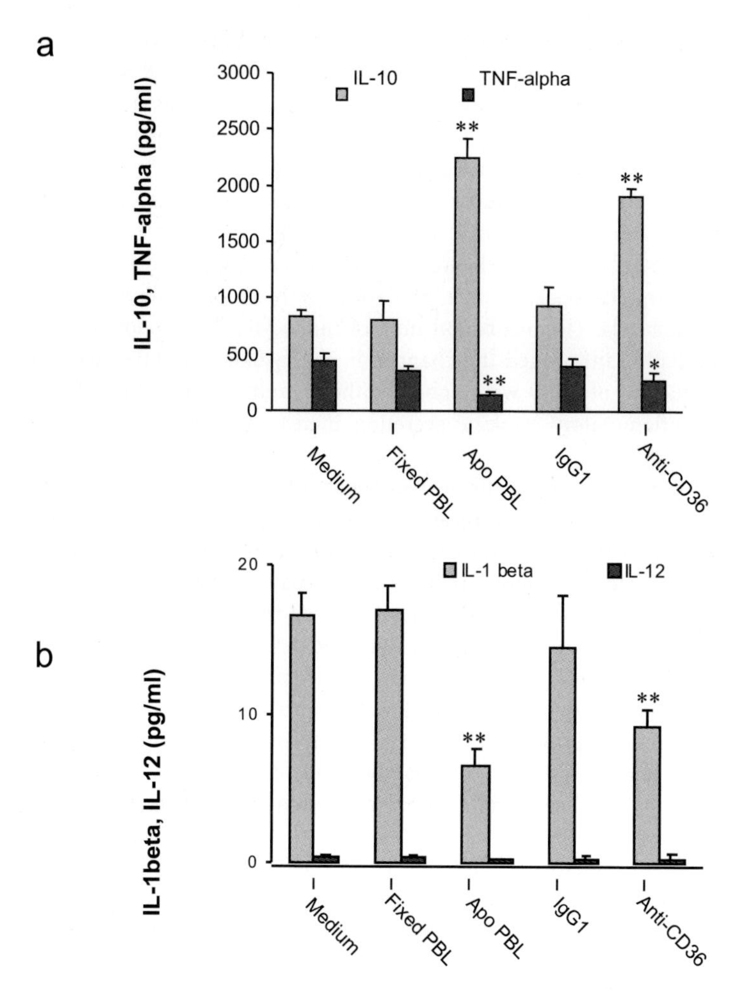

Fig. 3.1 Modulation of cytokine secretion in LPS-activated PBMCs by apoptotic cells and anti-CD36 antibodies. Concentrations of IL-10, TNF-α (a), IL-1β and IL-12 (b) in supernatants of PBMCs activated for 16 h with LPS in the absence (Medium), or in the presence of either freshly prepared paraformaldehyde-fixed autologous PBLs (Fixed PBL), autologous UV-irradiated apoptotic PBL (Apo PBL), isotype-matched control monoclonal antibodies (IgG1) or CD36-specific antibodies (Anti-CD36). Results shown represent the means of quadruplicate cultures; standard deviations are indicated. Asterisks indicate $^{*}P<0.05$ or $^{**}P<0.01$ in comparison to medium control. Representative results of at least four independent experiments are shown.

Monocytes

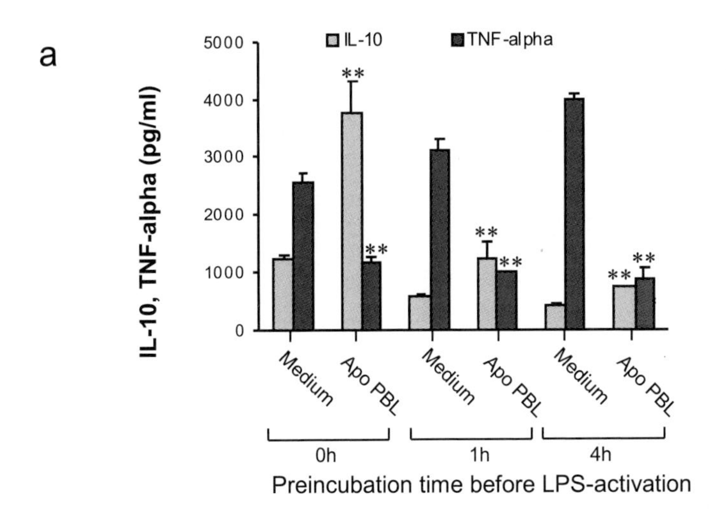

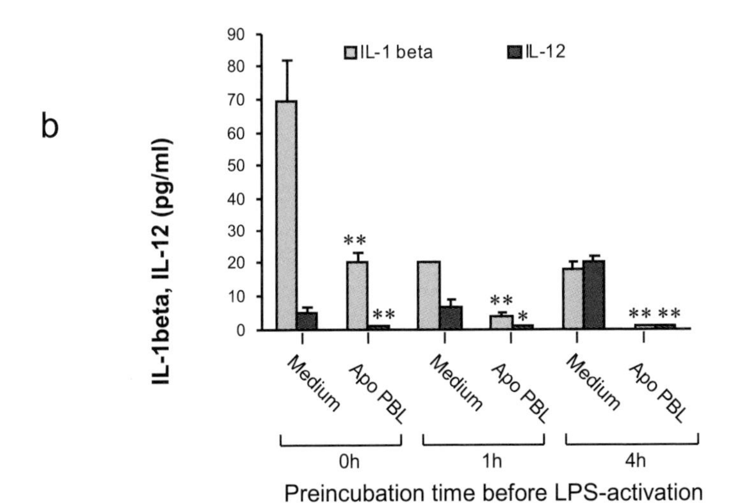

Fig. 3.2 Influence of the incubation time of monocytes with apoptotic cells prior to LPS stimulation on the cytokine secretion pattern. Results shown are the means of quadruplicate cultures; standard deviations are indicated. Asterisks indicate $^{*}P<0.05$ or $^{**}P<0.01$ in comparison to medium control. Representative results of at least three independent experiments are shown.

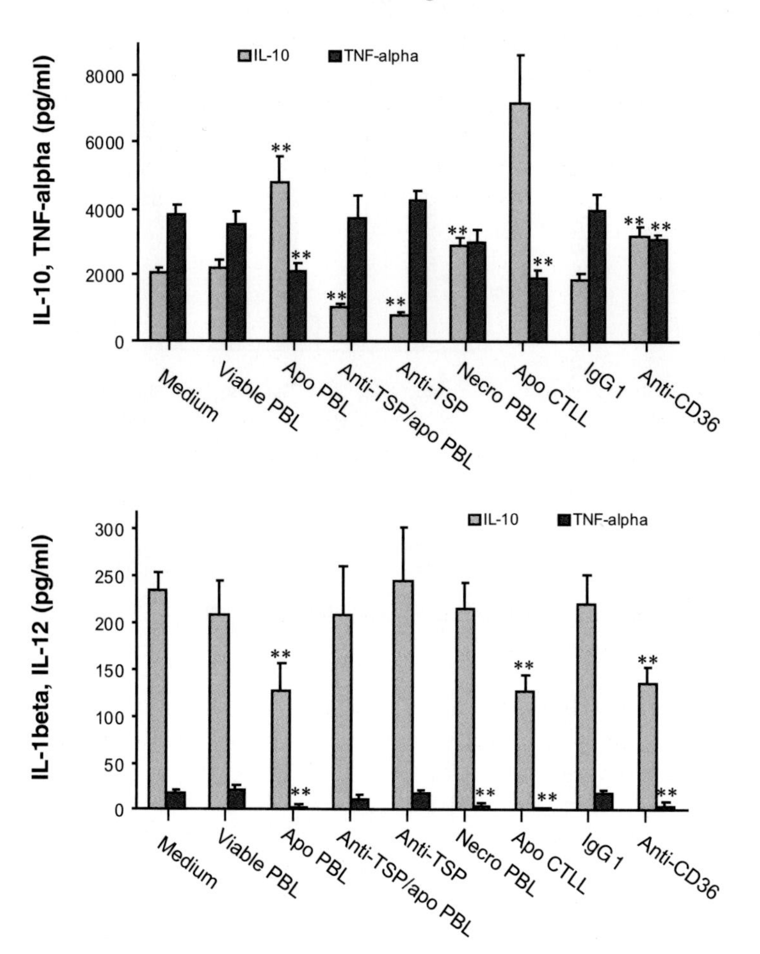

Fig. 3.3 Concentrations of cytokines in supernatants of elutriated monocytes activated for 16 h with LPS in the absence (Medium), or in the presence of autologous living PBLs, autologous UV-irradiated apoptotic PBLs (ApoPBL), apoptotic PBLs together with thrombospondin-1-specific antibody ahTSP-1 (Anti-TSP/ApoPBL), ahTSP-1 (Anti-TSP), IL-2-deprived apoptotic CTLL-2 cells (Apo CTLL), isotype-matched control antibodies (IgG1) or CD36-specific antibody FA6-152 (Anti-CD36). Paraformaldehyde-fixed, apoptotic and necrotic PBLs and apoptotic CTLL-2 cells themselves did not release detectable amounts of cytokines upon LPS stimulation (not shown). Results shown represent the means of quadruplicate cultures; standard deviations are indicated. Asterisks indicate $^{*}P<0.05$ or $^{**}P<0.01$ in comparison to medium control. Representative results of at least three independent experiments are shown.

treated THP-1 cells, human monocyte-derived macrophages, murine thioglycollate-induced peritoneal exudate cells, Kupffer cells and alveolar macrophages [12–14]. Therefore, it might be that apoptotic cells can also cause a minor pro-inflammatory stimulus, which is normally overcome by anti-inflammatory signals. Most likely, different experimental settings may account for these apparently inconsistent results. Another possible explanation for these seemingly contradictory results might be the contamination of apoptotic cells with primary or secondary necrotic cells, which can display pro-inflammatory effects [10]. Generally, apoptotic cell preparations contain a least some necrotic cells, with increasing numbers of secondary necrotic cells during prolonged incubation. In addition, Kurosaka *et al.* recently described that apoptotic cells caused a potentiation of anti-inflammatory and suppression of pro-inflammatory cytokine production by macrophages in the presence of human serum or human IgG [15], as had been observed by us and other investigators. Moreover, they demonstrated that human IgG increases the anti-inflammatory effect of apoptotic cells by engagement of Fcγ receptor I [15]. Thus, IgG and Fcγ receptor I appear to be critically involved in the generation of an efficient anti-inflammatory signal by apoptotic cells.

3.3 The Role of Anti-inflammatory Cytokines for the Inhibition of Pro-inflammatory Cytokine Production

To investigate whether the observed changes in the cytokine secretion pattern are secondary to the enhanced production and/or activation of IL-10 or TGF-β, which is proteolytically cleaved to its active form by thrombospondin-1, we performed the cytokine secretion assays in the presence of neutralizing antibodies against IL-10 or TGF-β, respectively. Neutralization of IL-10 largely, but not completely, restored pro-inflammatory cytokine secretion in the presence of apoptotic cells, whereas the neutralization of TGF-β displayed only a minor influence on the cytokine secretion pattern (unpublished data). These results indicate that the observed changes in cytokine secretion of monocytes are predominantly caused by increased IL-10 secretion and only marginally by increased TGF-β activity. However, these data do not exclude that the secretion and/or activation of TGF-β is increased in the presence of apoptotic cells. Moreover, there is evidence that increased TGF-β production by macrophages crucially contributes to the inhibition and immunomodulatory effects of apoptotic cells [6, 11].

In macrophages, the contribution of increased IL-10 secretion to the anti-inflammatory and immunomodulatory effects of apoptotic cells is somewhat contradictory, indicating that induction of IL-10 secretion from macrophages by apoptotic cells might depend on the experimental conditions. Fadok et al. did not detect increased IL-10 concentrations after exposure of macrophages to apoptotic cells using human monocyte-derived macrophages stimulated with LPS [11]. In contrast, we and others observed increased IL-10 concentrations after exposure to apoptotic cells in human or in murine macrophages either with or without stimulation (unpublished results) [10]. Most importantly, IL-10 appears to be critical to mediate immunosuppressive effects of apoptotic cells *in vivo*, since the immunogenicity of apoptotic tumor cells was strongly increased in a mouse tumor vaccination model in case of IL-10-deficient mice [16].

In monocyte-derived macrophages, TGF-β1, prostaglandin E_2 and PAF appeared to be involved in the inhibition of pro-inflammatory cytokine secretion, since anti-TGF-β1 antibodies, indometacin or PAF receptor antagonists restored cytokine production of LPS-activated macrophages that had been exposed to apoptotic cells [11].

In addition to the paracrine inhibition of pro-inflammatory cytokine production, there is evidence for additional mechanisms directly inhibiting the production of pro-inflammatory cytokines in monocytes/macrophages exposed to apoptotic cells. First, we observed that individual or even combined addition of neutralizing antibodies to IL-10 and TGF-β could not completely restore cytokine secretion compared to macrophages incubated with the same neutralizing antibodies, but in the absence of apoptotic cells. Second, Cocco and Ucker described an immediate inhibition of TNF-α and IL-6 secretion in macrophages exposed to apoptotic cells, which was detectable already 2 h after LPS activation. These results suggest that apoptotic cells exert an immediate and direct anti-inflammatory effect on the engulfing macrophage [17].

In summary, there is clear evidence that apoptotic cells not only passively avoid, but actively suppress, pro-inflammatory responses by monocytes/macrophages. However, massive apoptosis, overwhelming the clearance system, or an impaired engulfment of apoptotic cells by macrophages, may result in accumulation of secondary necrotic cells and thereby may exert pro-inflammatory effects and promote autoimmunity [18].

3.4 Monocyte/Macrophage Receptors receiving the Anti-inflammatory Signal from Apoptotic Cells

The observation of an anti-inflammatory effect of apoptotic cells raises the question, by which mechanisms might this anti-inflammatory effect be mediated. Changes in the architecture of the plasma membrane, such as phosphatidylserine (PS) exposure on the outer leaflet, occur soon after the initiation of apoptosis and serve as phagocyte recognition signals [2, 19, 20]. The trimeric glycoprotein thrombospondin, which is secreted by various cells including platelets, macrophages and monocytes, may form a 'molecular bridge' between apoptotic cells and phagocytes [2, 21, 22]: on apoptotic cells, thrombospondin binds to anionic sites that are not yet defined; on monocytes/macrophages, thrombospondin interacts with the thrombospondin receptor (CD36) and with the vitronectin receptor ($\alpha_v\beta_3$), which are both crucially involved in recognition and engulfment of apoptotic cells [2, 21–23]. To investigate whether these receptors also transduce 'anti-inflammatory signals' to monocytes, we performed blocking and activation studies using monoclonal antibodies against CD36, $a_v\beta_3$ and thrombospondin-1, and RGD peptides which block thrombospondin binding to $a_v\beta_3$. Incubation of PBMCs and monocytes with the CD36-specific antibody FA6-152 (10) prior to LPS stimulation mimicked the anti-inflammatory effect of apoptotic cells (Figs 3.1 and 3.3). FA6-152 recognizes an epitope located within the primary thrombospondin binding site of CD36. This domain of CD36 has been implicated in the phagocytosis of apoptotic neutrophils [24]. Furthermore, thrombospondin-specific antibodies (ahTSP-1) were able to restore TNF-a and IL-1β secretion, and to decrease IL-10 secretion from monocytes in the presence of apoptotic cells (Fig. 3.3). These findings suggest that the anti-inflammatory state in monocytes is induced by apoptotic cells binding CD36 via thrombospondin. Thrombospondin-specific antibodies increased TNF-a and decreased IL-10 production even in the absence of apoptotic cells (Fig. 3.3), indicating the presence of either some spontaneously apoptosing cells or other thrombospondin/CD36 ligands such as activated platelets. Interestingly, activated platelets that release and bind thrombospondin on their surfaces also suppressed pro-inflammatory cytokine secretion from monocytes (unpublished data).

Engagement of the vitronectin receptor $a_v\beta_3$ by antibodies (LM609) or RGD peptides reduced the anti-inflammatory effect of apoptotic cells only marginally, if at all, and displayed no agonistic activity (data not shown).

Although CD36 ligation clearly mimicked the anti-inflammatory effects of apoptotic cells, the CD36-mediated changes in the cytokine secretion patterns of monocytes were always somewhat weaker than those induced by apoptotic cells themselves. In addition, humans that lack the CD36 antigen due to genetic mutations do not display an obvious phenotype in respect to impaired clearance of apoptotic cells, increased inflammatory disorders or autoimmunity. Therefore, we suspected that there might be more than one receptor transducing the anti-inflammatory signal into monocytes/macrophages, as there is more than one receptor system mediating the engulfment of apoptotic cells.

Recently, Fadok *et al.* identified a PS receptor, which is expressed on the surfaces of macrophages, fibroblasts and epithelial cells. Although other surface molecules with potential PS receptor function have been previously described, the uptake of apoptotic cells via this novel PS receptor was stereospecifically inhibited by PS and its structural analogs, but not by other anionic phospholipids. Transfection of the PS receptor gene into T and B cells conferred the capacity to recognize and engulf apoptotic cells in a PS-specific manner. Most importantly, a monoclonal antibody raised against the PS receptor induced an anti-inflammatory state in macrophages [25]. Therefore, at least two receptors, which are engaged by apoptotic cells, can transduce an anti-inflammatory signal. However, we would not be surprised if even more receptors, still to be identified, ensure that apoptosis does not cause inflammation, tissue destruction and autoimmunity.

3.5
Intracellular Signaling Events Causing the Anti-inflammatory State in Macrophages

Since induction of IL-10 and suppression of TNF-α, IL-1β and IL-12 are caused by agents that increase intracellular cAMP concentrations (e.g. prostaglandin E_2, pentoxifylline, isobutyl-methylxanthine and dibutyryl-cAMP) [26–28], we investigated whether ligation of CD36 influenced cAMP levels in monocytes. Since CD36 ligation caused a sustained elevation of intracellular cAMP levels in purified peripheral blood monocytes, we propose that the CD36 signal transduction pathway is either coupled to the adenylate cyclase or elevation of intracellular cAMP is caused indirectly in a autocrine/paracrine manner, e.g. by prostaglandin E_2 production. However, the fact that a marked increase in intracellular cAMP levels is observed already less than 5 min after CD36 ligation argues for a direct rather than an indirect, autocrine/paracrine mechanism. Therefore, elevation of intracellular cAMP may provide the molecular basis for the modification of the cytokine secretion pattern induced by CD36 ligation.

Investigating the anti-inflammatory response in a murine macrophage cell line, McDonald *et al.* found that neither uptake of apoptotic cells nor exposure to TGF-β modulated NF-κB or AP-1 DNA binding activity. Although the exact mechanism of pro-inflammatory cytokine transcription could not be defined, TNF-α production appeared to be translationally downregulated, whereas other pro-inflammatory cytokines and chemokines appeared to be inhibited at the level of transcription [46].

3.6
Apoptotic Cells Impair MHC Class II Surface Expression on Monocytes

We also studied the influence of apoptotic cells on MHC class II surface expression on monocytes and cytokine secretion from T cells. After phagocytosis of bacteria and viruses, monocytes/macrophages degrade proteins to peptides that can then bind to MHC class II molecules and be presented to T_h cells. Normally, blood monocytes express only low levels of MHC class II molecules on their surface, but MHC class II expression increases during differentiation, culture *in vitro* or upon stimulation, most potently with interferon (IFN)-γ. In the presence of either CD36-specific antibodies or apoptotic cells, MHC class II surface expression on LPS-activated monocytes was significantly reduced (not shown). Increased IL-10 secretion, induced by apoptotic cells, may play a role in the decreased expression of MHC class II, since IL-10-neutralizing antibodies could partially restore the MHC class II expression on monocytes (unpublished results). Therefore, apoptotic cells should impair the ability of monocytes/macrophages to present antigens efficiently to T_h cells.

3.7
Influence of Apoptotic Cells on DC Function in Allogeneic MLR

DCs are one of the key players of the immune response. Only DCs can efficiently activate naïve T cells and play a critical role for the initiation of immune responses. Since sufficient amounts of primary DCs are difficult to isolate, many investigators use *in vitro* differentiation systems for the generation of DCs.

When monocytes are cultured for 6 days in tissue culture medium in the presence of human IL-4 and GM-CSF they develop an immature DC-like phenotype, and are referred to as monocyte-derived immature DCs (iDCs). These cells are potent stimulators of allogeneic T cells in 'mixed lymphocyte reactions' (MLR). The stimulatory capacity of iDCs for allogeneic T cells is significantly reduced if the iDCs are pre-incubated with apoptotic cells 2 days before the addition of allogeneic T cells (Fig. 3.4, column 2). In contrast, pre-incubation of iDCs with necrotic cells did not reduce the allogeneic MLR. In some cases we observed an even higher stimulation index if iDCs had been pre-incubated with necrotic cells (Fig. 3.4, column 3).

As shown in Fig. 3.5, monocytes cultured for 3 days with 500 U of IL-4 (referred to as APCs) secrete IL-10 if they are incubated with apoptotic cells. In contrast, necrotic cells did not induce IL-10 secretion of APCs. The addition of high concentrations of GM-CSF to the culture abrogated the IL-10 inducing capacity of apoptotic cells in a dose dependent manner. Culture of monocytes in the presence of IL-4 and GM-CSF induces the maturation of monocytes into monocyte-derived iDCs. Hence, the apoptotic cell-dependent IL-10 response may represent an im-

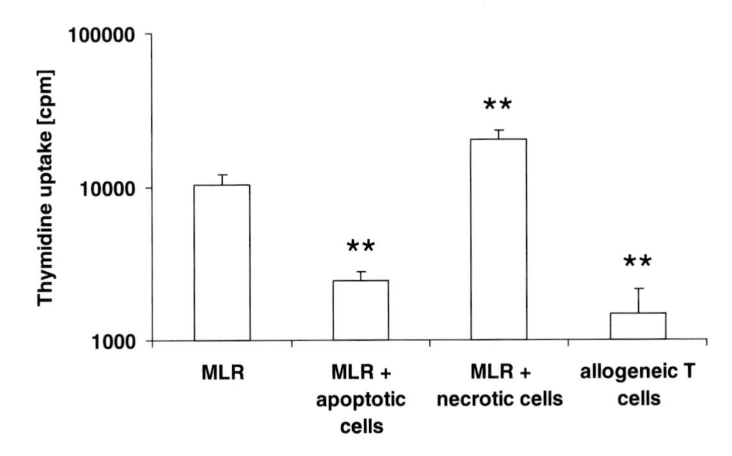

Fig. 3.4 $CD14^+$ cells were cultured for 6 days in tissue culture medium containing 10% fetal calf serum and 500 U of both recombinant human IL-4 and GM-CSF. Then 10^4 DCs were incubated with 10^4 UV irradiated (apoptotic) or heat-treated [30 min 56°C (necrotic)] Hacat cells for 2 days. Then MLR was performed with 10^5 allogeneic $CD14^-$ cells for 24 h. The experiment was performed in decaplicates. All *P* values are <0.0001.

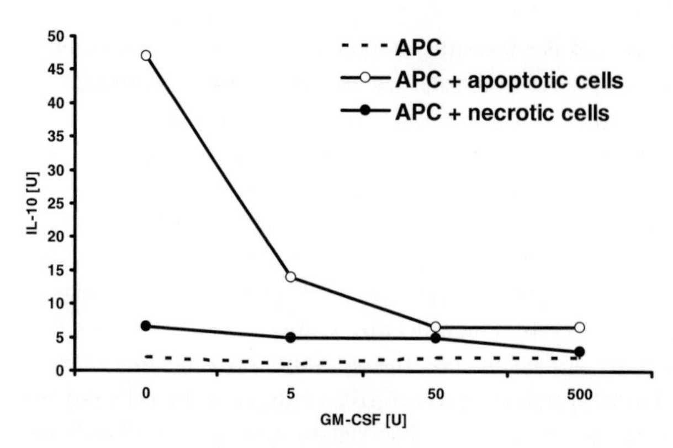

Fig. 3.5 Difference between monocytes/macrophages and immature DCs differentiated *in vitro* using IL-4 and GM-CSF on ability to secrete IL-10 upon encounter with apoptotic cells. Monocytes cultured for 3 days with 500 U of IL-4 (referred to as APCs) secrete IL-10 if they are incubated with apoptotic cells. In contrast, necrotic cells display no influence on the IL-10 secretion of APCs. However, if GM-CSF is added to the culture, the IL-10-inducing capacity of apoptotic cells is abrogated in a dose-dependent fashion. Culture of monocytes in the presence of IL-4 and GM-CSF induces the maturation of monocytes into monocyte-derived iDCs. The apoptotic cell-dependent IL-10 response may, therefore, represent an important functional difference between monocytes and early iDCs.

portant functional difference between monocytes and early iDCs. The influences of apoptotic versus necrotic cells on DC function will be discussed in detail in another chapter by Jenne and Sauter (Chapter 12).

3.8 The Presence of Apoptotic Cells can Shift the T_h Cell Response towards T_h2

Based on their cytokine secretion profile, T_h cells can be classified in functionally different subsets. T_h1 cells secrete IFN-γ, IL-2 and TNF-α, and are crucially involved in cell-mediated immunity and inflammatory reactions [29, 30]. T_h2 cells secrete mainly IL-4, IL-5, IL-6, IL-10 (also produced by human T_h1 cells) and IL-13, and have important immunoregulatory functions and support the humoral immune response [29, 30]. IL-12 is important for IFN-γ expression and the differentiation of naïve T cells to T_h1 cells, whereas IL-10 favors the differentiation into T_h2 cells, mostly by inhibiting IL-12 production [29]. Therefore, we tested whether the inverse regulation of IL-10 versus IL-12 caused by CD36 ligation may alter the T_h1/T_h2 balance. If PBMCs have been exposed to CD36-specific antibodies or autologous apoptotic cells prior to stimulation with antigen, production of the T_h1 cytokine IFN-γ was drastically decreased. The concentration of the T_h2 cytokine

IL-4 was usually reduced in comparison to control samples when measured 48 h after stimulation, unchanged when measured after 80 h and usually increased at later time points. This results in an increased IL-4:IFN-γ ratio, consistent with a shift in the cytokine secretion pattern from T_h1 towards T_h2. IFN-γ production in response to LPS containing bacteria (*Escherichia coli*) was markedly inhibited in all experimental settings, whereas significant inhibition of IFN-γ production in response to tetanus toxoid was largely dependent on additional monocyte activation, e.g. by LPS. This result suggests that monocyte-derived factors, namely IL-10, mediate the preferential suppression of IFN-γ. Indeed, neutralizing antibodies to IL-10 partially restored IFN-γ production in PBMCs incubated with anti-CD36 antibodies or apoptotic cells [9] (Voll *et al.*, manuscript in preparation). Therefore, IL-10 may be an important mediator of the immunoregulatory effects induced by apoptotic cells. However, inhibition of TNF-α and IL-12 early after activation seemed to be IL-10 independent, and might be directly mediated by elevated levels of cAMP [26–28].

3.9 Apoptotic Cells Suppress Delayed-type Hypersensitivity (DTH) *In Vivo*

Cell-mediated immunity is crucially dependent on IL-12 and IFN-γ production and can be measured by the DTH reaction. Based on our *in vitro* results, apoptotic cells were expected to increase IL-10 and inhibit IFN-γ secretion also *in vivo*, and thereby cell-mediated immunity should be inhibited. To test this hypothesis, mice were subcutaneously immunized with sheep red blood cells (SRBCs), either alone or in the presence of living or apoptotic syngeneic spleen cells. After 6 days a DTH reaction was induced by injection of SRBCs into foot pads. Mice immunized in the presence of apoptotic cells showed a significantly reduced DTH reaction, indicating an impaired induction of cell-mediated immunity (Fig. 3.6a). Further experiments addressed the question whether apoptotic cells also modify the inflammatory effector phase of established cell-mediated immunity. Therefore, a DTH reaction was induced in SRBC-immunized mice by injection of SRBCs into the left ears and SRBCs together with either apoptotic or necrotic syngeneic spleen cells into the right ears. The presence of apoptotic, but not necrotic syngeneic cells significantly reduced the inflammatory ear swelling caused by the DTH reaction (Fig. 3.6b). Similar results were obtained using allogeneic splenocytes and thymocytes (not shown).

3.10 Necrosis and Inflammation

In vivo, necrotic cell death is usually associated with inflammation, tissue injury and immune responses, even autoimmunity (e.g. Dressler's syndrome after heart surgery or myocardial infarctions). However, as apoptotic cells, necrotic cells ex-

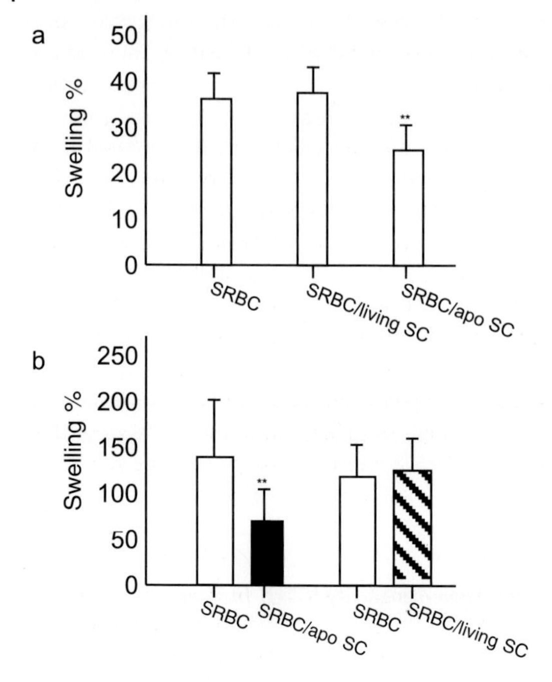

Fig. 3.6 Impaired DTH response in mice co-injected with apoptotic cells during immunization. (a) Mice were subcutaneously immunized with SRBCs alone (SRBC), or SRBCs together with either syngeneic living (SRBC/living SC) or apoptotic spleen cells (SRBC/apo SC). A DTH reaction was induced by injection of SRBCs into foot pads. Percent swelling of SRBC-injected foot pads compared to PBS-injected foot pads after 24 h is shown. (b) Local inhibition of the DTH reaction by apoptotic cells. Mice immunized with SRBCs were injected with SRBCs alone into the left ears and SRBCs together with either apoptotic or necrotic syngeneic spleen cells into the right ears. Results shown are the means + standard deviation with $n=6$ in representative experiments. $^{**}P<0.01$ in comparison to untreated controls.

pose the 'eat me signal' PS on there surface and, therefore, should engage the PS receptor, which mediates an anti-inflammatory signal. In addition, necrotic cells may also bind to the thrombospondin receptor CD36 (unpublished data), which also transduces an anti-inflammatory signal. However, in the case of necrotic cell death, pro-inflammatory intracellular constituents such as heat shock proteins, possibly also unmethylated CpG-rich mitochondrial DNA and other pro-inflammatory mediators are released, and may dominate over the anti-inflammatory signals. Fadok *et al.* suggest an additional mechanism, by which necrotic cells reduce anti-inflammatory signaling via the PS receptor. Proteases or Annexin V released from necrotic cells, especially granulocytes, may efficiently cleave the PS receptor off the membrane or mask the PS on the surfaces of necrotic cells, respectively, thereby preventing anti-inflammatory signaling [31–33].

3.11 Implications of the Anti-inflammatory and Immunomodulatory Effects of Apoptotic Cells for Health and Disease

The data presented here suggest that apoptotic cells induce an 'anti-inflammatory state' in monocytes by selectively decreasing pro-inflammatory and increasing anti-inflammatory cytokine expression. This anti-inflammatory effect is likely to be mediated by ligation of CD36 and the PS receptor. However, we cannot rule out the involvement of additional receptors inducing the anti-inflammatory state. Apoptotic cells may also provide an anti-inflammatory signal to other $CD36^+$ cells such as DCs [33], microvascular endothelial cells and keratinocytes in diseased skin [34].

Conditions which result in increased apoptotic cell death *in vivo* are associated with suppression of inflammation and cell-mediated immunity. Physiologically, the anti-inflammatory and immunoregulatory effects of apoptotic cells may contribute to the non-inflammatory clearance of apoptotic cells and help to avoid organ damage. In addition, activation of autoreactive T cells by peptides from ingested apoptotic cells might be prevented. Therefore, the anti-inflammatory and immunosuppressive clearance of apoptotic cells by macrophages appears to be an important mechanism to prevent or control autoreactivity and autoimmunity.

3.11.1 Apoptosis and Pregnancy

During embryo implantation, uterine epithelial cells surrounding the blastocyst undergo apoptosis [35], and thus may form an anti-inflammatory milieu and prevent immunological rejection of the embryo. Moreover, the increased IL-10 expression detectable at the interface between maternal and fetal tissues may contribute to the systemic reduction of cell-mediated immunity during pregnancy [29, 36].

3.11.2 Apoptosis and Irradiation

Moderate irradiation with γ- or X-rays or UV light causes apoptosis, and exerts strong anti-inflammatory and immunosuppressive effects [4, 5, 37]. In particular, UV exposure of the skin potently induces IL-10 expression by immunosuppressive tolerance-inducing $CD36^+$ $CD11^+$ macrophages [38, 39]. The secretion of anti-inflammatory mediators may contribute to the UV-mediated amelioration of inflammatory skin diseases such as psoriasis.

3.11.3 Apoptosis and Cancer

Most tumors display not only an increased proliferation rate, but also an increased rate of apoptosis. Therefore, the presence of apoptotic cell material [40] could impair an effective cell-mediated immune response against cancer cells and represent an important immune escape mechanism. We have demonstrated that addition of annexin V, which binds with high affinity to PS on apoptotic cell surfaces, markedly increases the humoral immunogenicity of apoptotic cells [41]. Currently, we are investigating the potential use of annexin V as an adjuvant for cancer vaccines.

3.11.4 Apoptosis and Infections

Many viral infections, including HIV, cause extensive apoptosis [22, 42] which may contribute to the impaired immune response during and after such infections. Moreover, *Plasmodium falciparum*-infected erythrocytes bind to CD36. This may provide an important mechanism by which the parasite impairs the host defense. During *Trypanosoma cruzi* infection, the parasites usually survive within macrophages, which appear to not be adequately activated or activatable, respectively [43]. Since *T. cruzi* primarily infects neutrophils, in which they induce apoptotic cell death, they may enter macrophages using apoptotic neutrophils as a Trojan horse. The apoptotic neutrophils may force the engulfing macrophage into an anti-inflammatory state, which prevents efficient killing mechanisms of the intracellular parasites. A similar mechanism may also work in other parasite infections which might employ apoptotic 'Trojan horses' as anti-inflammatory door-openers in a similar way.

3.11.5 Apoptosis and Blood Transfusions

Immunosuppression after transfusion of allogeneic and autologous blood might be, at least partially, caused by apoptotic leukocytes [44]. In addition, aged erythrocytes themselves expose PS and thereby may engage the PS receptor, resulting in the secretion of immunosuppressive cytokines [45]. Therapeutically, it should be possible to interfere with the immunosuppressive effects of apoptotic cells by blocking the interaction with CD36 or by interfering with the CD36-linked signal transduction pathway. On the other hand, agonistic CD36 ligands such as specific antibodies, apoptotic cells or induction of apoptosis *in vivo* may be helpful in controlling graft rejection, inflammation and autoimmunity. Similarily, blocking the interaction between PS and the PS-receptor by annexin V appears to partialially neutralize the anti-inflammatory effects of apoptotic cells. In contrast, agonistic ligands of the PS-receptor should display immunosuppressive and anti-inflammatory effects.

3.12
References

1 Duvall, E., A.H. Wyllie and R.G. Morris, Macrophage recognition of cells undergoing programmed cell death (apoptosis). *Immunology* **1985**, 56, 351–8.
2 Savill, J., *et al.*, Phagocyte recognition of cells undergoing apoptosis. *Immunol Today* **1993**, 14, 131–6.
3 Meagher, L.C., *et al.*, Phagocytosis of apoptotic neutrophils does not induce macrophage release of thromboxane B2. *J Leukoc Biol* **1992**, 52, 269–73.
4 Abel, E.A., Phototherapy. *Dermatol Clin* **1995**, 13, 841–9.
5 Trott, K.R., Therapeutic effects of low radiation doses. *Strahlenther Onkol* **1994**, 170, 1–12.
6 Hoffmann, P.R., *et al.*, Phosphatidylserine (PS) induces PS receptor-mediated macropinocytosis and promotes clearance of apoptotic cells. *J Cell Biol* **2001**, 155, 649–59.
7 Somersan, S. and N. Bhardwaj, Tethering and tickling: a new role for the phosphatidylserine receptor. *J Cell Biol* **2001**, 155, 501–4.
8 Leverrier, Y., *et al.*, Cutting edge: the Wiskott-Aldrich syndrome protein is required for efficient phagocytosis of apoptotic cells. *J Immunol* **2001**, 166, 4831–4.
9 Voll, R.E., *et al.*, Immunosuppressive effects of apoptotic cells. *Nature* **1997**, 390, 350–351.
10 Gough, M.J., *et al.*, Macrophages orchestrate the immune response to tumor cell death. *Cancer Res* **2001**, 61, 7240–7.
11 Fadok, V.A., *et al.*, Macrophages that have ingested apoptotic cells *in vitro* inhibit pro-inflammatory cytokine production through autocrine/paracrine mechanisms involving TGF-beta, PGE2, and PAF. *J Clin Invest* **1998**, 101, 890–8.
12 Kurosaka, K., N. Watanabe and Y. Kobayashi, Production of pro-inflammatory cytokines by phorbol myristate acetate-treated THP-1 cells and monocyte-derived macrophages after phagocytosis of apoptotic CTLL-2 cells. *J Immunol* **1998**, 161, 6245–9.
13 Kawagishi, C., *et al.*, Cytokine production by macrophages in association with phagocytosis of etoposide-treated P388 cells *in vitro* and *in vivo*. *Biochim Biophys Acta* **2001**, **1541**, 221–30.
14 Kurosaka, K., N. Watanabe and Y. Kobayashi, Production of pro-inflammatory cytokines by resident tissue macrophages after phagocytosis of apoptotic cells. *Cell Immunol* **2001**, 211, 1–7.
15 Kurosaka, K., N. Watanabe and Y. Kobayashi, Potentiation by human serum of anti-inflammatory cytokine production by human macrophages in response to apoptotic cells. *J Leukoc Biol* **2002**, 71, 950–6.
16 Ronchetti, A., *et al.*, Immunogenicity of apoptotic cells *in vivo*: role of antigen load, antigen-presenting cells, and cytokines [In process citation]. *J Immunol* **1999**, 163, 130–6.
17 Cocco, R.E. and D.S. Ucker, Distinct modes of macrophage recognition for apoptotic and necrotic cells are not specified exclusively by phosphatidylserine exposure. *Mol Biol Cell* **2001**, 12, 919–30.
18 Herrmann, M., R.E. Voll and J.R. Kalden, Etiopathogenesis of systemic lupus erythematosus. *Immunol Today* **2000**, 21, 424–6.
19 Verhoven, B., *et al.*, Regulation of phosphatidylserine exposure and phagocytosis of apoptotic T lymphocytes [In process citation]. *Cell Death Different* **1999**, 6, 262–70.
20 Verhoven, B., R.A. Schlegel and P. Williamson, Mechanisms of phosphatidylserine exposure, a phagocyte recognition signal, on apoptotic T lymphocytes. *J Exp Med* **1995**, 182, 1597–601.
21 Savill, J., *et al.*, Thrombospondin cooperates with CD36 and the vitronectin receptor in macrophage recognition of neutrophils undergoing apoptosis. *J Clin Invest* **1992**, 90, 1513–22.
22 Akbar, A.N., *et al.*, The specific recognition by macrophages of $CD8^+$, $CD45RO^+$ T cells undergoing apoptosis: a mechanism for T cell clearance during resolution of viral infections. *J Exp Med* **1994**, 180, 1943–7.

23 Savill, J., *et al.*, Vitronectin receptor-mediated phagocytosis of cells undergoing apoptosis. *Nature* **1990**, 343, 170–3.

24 Navazo, M. D., *et al.*, Identification of a domain (155–183) on CD36 implicated in the phagocytosis of apoptotic neutrophils. *J Biol Chem* **1996**, 271, 15381–5.

25 Fadok, V. A., *et al.*, A receptor for phosphatidylserine-specific clearance of apoptotic cells. *Nature* **2000**, 405, 85–90.

26 Han, J., P. Thompson and B. Beutler, Dexamethasone and pentoxifylline inhibit endotoxin-induced cachectin/tumor necrosis factor synthesis at separate points in the signaling pathway. *J Exp Med* **1990**, 172, 391–4.

27 van der Pouw Kraan, T. C., *et al.*, Prostaglandin-E_2 is a potent inhibitor of human interleukin 12 production. *J Exp Med* **1995**, 181, 775–9.

28 Haraguchi, S., R. A. Good and N. K. Day, Immunosuppressive retroviral peptides: cAMP and cytokine patterns. *Immunol Today* **1995**, 16, 595–603.

29 Mosmann, T. R., Properties and functions of interleukin-10. *Adv Immunol* **1994**, 56, 1–26.

30 Paul, W. E. and R. A. Seder, Lymphocyte responses and cytokines. *Cell* **1994**, 76, 241–51.

31 Fadok, V. A., *et al.*, Differential effects of apoptotic versus lysed cells on macrophage production of cytokines: role of proteases. *J Immunol* **2001**, 166, 6847–54.

32 Henson, P. M., D. L. Bratton and V. A. Fadok, The phosphatidylserine receptor: a crucial molecular switch? *Nat Rev Mol Cell Biol* **2001**, 2, 627–33.

33 Lenz, A., *et al.*, Human and murine dermis contain dendritic cells. Isolation by means of a novel method and phenotypical and functional characterization. *J Clin Invest* **1993**, 92, 2587–96.

34 Begany, A., *et al.*, Expression of thrombospondin-1 (TSP1) and its receptor (CD36) in healthy and diseased human skin. *Acta Derm Venereol* **1994**, 74, 269–72.

35 Parr, E. L., H. N. Tung and M. B. Parr, Apoptosis as the mode of uterine epithelial cell death during embryo implantation in mice and rats. *Biol Reprod* **1987**, 36, 211–25.

36 Wegmann, T. G., *et al.*, Bidirectional cytokine interactions in the maternal-fetal relationship: is successful pregnancy a T_h2 phenomenon? *Immunol Today* **1993**, 14, 353–6.

37 Kripke, M. L., Immunological unresponsiveness induced by ultraviolet radiation. *Immunol Rev* **1984**, 80, 87–102.

38 Kang, K., *et al.*, $CD11b^+$ macrophages that infiltrate human epidermis after *in vivo* ultraviolet exposure potently produce IL-10 and represent the major secretory source of epidermal IL-10 protein. *J Immunol* **1994**, 153, 5256–64.

39 Meunier, L., Z. Bata-Csorgo and K. D. Cooper, In human dermis, ultraviolet radiation induces expansion of a $CD36^+$ $CD11b^+$ $CD1^-$ macrophage subset by infiltration and proliferation; $CD1^+$ Langerhans-like dendritic antigen-presenting cells are concomitantly depleted. *J Invest Dermatol* **1995**, 105, 782–8.

40 Staunton, M. J. and E. F. Gaffney, Tumor type is a determinant of susceptibility to apoptosis. *Am J Clin Pathol* **1995**, 103, 300–7.

41 Stach, C. M., *et al.*, Treatment with annexin V increases immunogenicity of apoptotic human T-cells in BALB/c mice. *Cell Death Different* **2000**, 7, 911–5.

42 Oyaizu, N. and S. Pahwa, Role of apoptosis in HIV disease pathogenesis. *J Clin Immunol* **1995**, 15, 217–31.

43 Freire-de-Lima, C. G., *et al.*, Uptake of apoptotic cells drives the growth of a pathogenic trypanosome in macrophages. *Nature* **2000**, 403, 199–203.

44 Greenburg, A. G., Benefits and risks of blood transfusion in surgical patients. *World J Surg* **1996**, 20, 1189–93.

45 Fadok, V. A., *et al.*, A receptor for phosphatidylserine-specific clearance of apoptotic cells [see Comments]. *Nature* **2000**, 405, 85–90.

46 McDonald P. P., *et al.*, Transcriptional and translational regulation of inflammatory mediator production by endogenous TGF-β in macrophages that have ingested apoptotic cells. *J Immunol* **1999**, 163, 6164–72.

4
Complement and Apoptosis

Dror Mevorach

4.1
Introduction

Both programmed cell death and complement are tightly regulated systems, mainly orchestrated by activation of proteases. The complement system has usually been described as a defense mechanism against invading organisms and, although its role in clearing debris was suggested a long time ago, only recently has its role in the clearance of apoptotic material become appreciated. Membranes of cells undergoing apoptosis were shown to bind fragments of complement that facilitated their clearance via specific receptors, establishing its role in homeostasis in addition to defense mechanisms. On the other hand, complement-mediated cell lysis that traditionally has been presented as a classical example of necrotic cell death was recently suggested to have pro- or anti-apoptotic effects, at low dosages of the membrane attack complex (MAC).

This chapter describes and discusses the reports covering the various aspects of the interface between complement and apoptosis, and its possible relevance to autoimmune diseases, inflammatory conditions and homeostasis.

4.2
Programmed Cell Death (PCD)

PCD plays an important role in development and tissue homeostasis [1, 2]. Normal and altered PCD were found to be fundamental processes in the pathogenesis of numerous diseases, including cancer, AIDS, neurodegenerative disorders, ischemia and autoimmune syndromes. The term apoptosis that was coined in 1972 by Kerr *et al.* [3] means 'falling leaves' in Greek, and describes a de-adhesiveness of cells that undergo morphological changes like cell shrinkage, plasma membrane blebbing and chromatin condensation with intact membrane. This is subsequently followed by cellular fragmentation into apoptotic bodies. Cell death by necrosis, named by some authors 'accidental cell death' or non-PCD, has morphologically distinct features from those observed in cells undergoing apoptotic cell death, with cytoplasmic and mitochondrial swelling [4]. In addition, the membrane of apoptotic cells remains almost intact until the late stages, in contrast to the damaged membrane of a necrotic cell. Inflammation has been traditionally considered to be a byproduct of necrotic cell death, but not of apoptosis [5]. However, there is growing support for the claim that exceptions do exist and apoptosis of some cell types by certain inducers can be associated with inflammatory signals [6–8], whereas necrosis may be non-pro-inflammatory under certain conditions [9]. According to the circumstances, apoptotic cells are cleared by neighboring cells, professional phagocytes or both. Several receptors, including integrins, receptor(s) for lectin(s), phosphatidylserine (PS) receptor, scavenger receptors and CD14, were described as having a role in the clearance of apoptotic cells (reviewed in [10]).

Apoptosis is a genetically controlled process responding to a variety of signals – receptor- or non-receptor mediated, extracellular or intracellular. The signal for apoptosis may originate by ligands of the described death receptors, e.g. tumor necrosis factor (TNF) receptor I, Fas (CD95), DR3/WSL and TRAIL/APO-2L, or upon DNA damage, growth factor removal or exposure to toxic chemicals. In both cases, mitochondria are believed to play a central role in the propagation of PCD. Accordingly, some of the major inhibitors of apoptosis, i.e. the anti-apoptotic members of the Bcl-2 family, act at the level of the mitochondrial membrane [11–13]. Another set of inhibitors of apoptosis proteins (IAPs) binds directly to and inhibits specific caspases [14].

The apoptotic executional process is based on a cascade of cytoplasmic cysteine proteinases, known as caspases. Their name is derived from the fact that they all cleave their substrates after aspartic acid residues [15–17]. The family of human caspases contains over 14 evolutionary conserved homologous proteases [15, 16]. The caspases are synthesized as zymogens (pro-caspases) and are activated by cleavage at specific aspartic acid residues. Following removal of an N-terminal fragment, two caspase subunits assemble into a heterotetramer to form the active enzyme. Casapases can be further divided to initiators that are more pattern-specific and effectors that designate a common pathway. For example caspase-8 is an initiator caspase in the CD95-dependent pathway and capase-9 is an initiator in the mitochondrial-dependent pathway. Both pathways utilize caspase-3 and -7 as effector caspases. Caspases are known to cleave, in addition to other caspases, numerous specific distinct cellular proteins, including structural and regulatory proteins whose degradation leads to the morphological changes that characterize apoptotic cell death (reviewed in [16]). Thus, digestion of nuclear lamin permits nuclear budding [18, 19], cleavage of fodrin and gelsolin leads to membrane blebbing [20], and cleavage of ICAD leads to internucleosomal DNA fragmentation [21].

In certain cell types the caspase cascade is necessary and sufficient for the execution of apoptosis [22]. However, apoptotic cell death of other cells requires the release of apoptogenic proteins from damaged mitochondria [23]. Two of the pro-apoptotic mitochondrial proteins that have been well described are cytochrome *c* and apoptosis-inducing factor (AIF) [23]. Cytochrome *c*, once released into the cytoplasm, interacts with the adaptor protein Apaf-1 and with pro-caspase-9, leading to caspase-9 activation and induction of a death signal [24, 25]. AIF initiates apoptosis by activation of caspase-3 [26]. Non-apoptotic forms of PCD have been described [27, 28]. In terms of its morphology (cytoplasmic vacuolization), at least one of them [27] resembled the morphology of necrotic cell death. For additional and detailed information on apoptosis activation and regulation, the reader should refer to recent reviews [1, 2, 29, 30].

4.3 Complement

The initial descriptions of the complement system were of a heat-labile factor 'complementing' bacteriolytic activity of antibodies [31]. The complement system is usually described as a defense mechanism against invading organisms and although its role in clearing debris was appreciated a long time ago, only recently [32, 34] has its role in the clearance of apoptotic material become appreciated. The complement system consists of a group of 13 soluble plasma proteins (Tab. 4.1), 15 regulatory proteins (Tab. 4.2) and 10–11 known receptors (Tab. 4.3). The proteins of the complement system interact with one another in three different enzymatic cascades, termed the classical, alternative and lectin pathways (Fig. 4.1 and Tab. 4.1). Activation of the system leads to opsonization by proteins and generation of C3 convertase that cleaves C3. The C3 convertase structure depends on the mode of activation (Fig. 4.1).

The classical activation pathway, suggested by Erlich and others, was described initially as a consequence of opsonization by antibodies to a pathogen. However, we now know that it can be activated in the absence of an antibody by nucleic acid, phospholipids, lipopolysaccharide (LPS) and apoptotic cells.

Either receptor-dependent or -independent, complement–cell interactions lead to variable responses like cytokine secretion, phagocytosis, mitosis or cell arrest and proliferation or cell death. Opposite effects then depend on cell types, modes of activation and the activity of different regulatory mechanisms. Activated complement components may induce *de novo* synthesis or shut-off of proteins, phosphorylation or de-phosphorylation of proteins and synthesis or degradation of DNA. Complement activation on the surface and/or in the vicinity of cells generates a mixture of active compounds and the outcome of these complement–cell interactions will be governed by their local concentration, and by many other factors such as growth factors, hormones, cytokines and toxic substances.

Complement activation on surfaces results in the generation of convertases that further cleave C3 (C3 convertase) or C5 (C5 convertase) to initiate terminal complex assembly (Fig. 4.1). In addition, activated complement components and fragments are generated, some of which are recognized by specific receptors present on the surface of many cell types. Receptors for C3a, C5a, C3b and iC3b have been well studied and characterized (Tab. 4.3) [35, 36]. Review articles on various aspects of complement activation can be found in Volanakis and Frank [31]. In contrast to these activated complement proteins, the terminal complement complexes C5b–7,

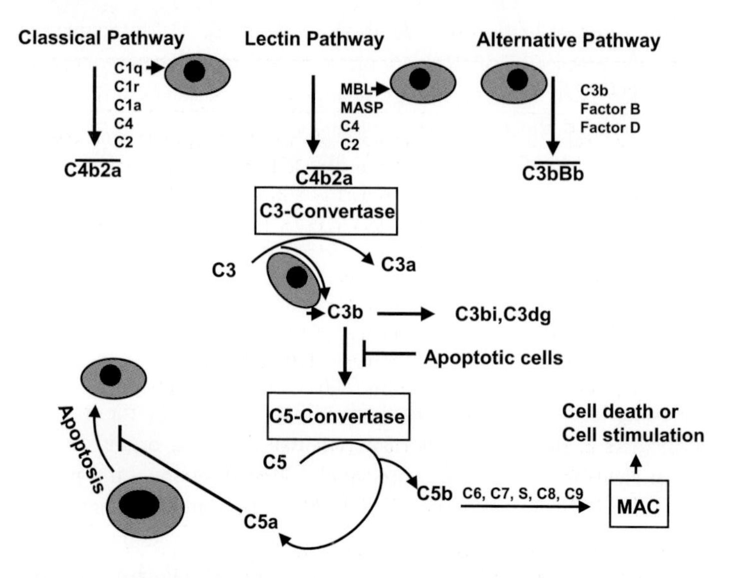

Fig. 4.1 Complement activation. Complement activation in humans is initiated either directly or following opsonization by a variety of molecules. The three pathways that are recognized so far, the classical, lectin and the alternative pathways, all lead to the formation of C3 convertase. C4b2a is the convertase of the classical and lectin pathways, and C3bBb of the alternative pathway. C3b, the cleavage product of C3, can form C5 convertase that cleaves C5A or undergo degradation. If C5 convertase is formed, MAC may be generated in the presence of C6–9 and factor S. Apoptotic cells were shown to activate both pathways, but without formation of C5 convertase, under normal conditions. Binding of factors like C1q and MBL was suggested to mediate clearance via specific receptors. MAC may induce either death or stimulatory effects, depending on dosage, cells and circumstances.

C5b–8 and MAC are known to bind to cell membranes directly via hydrophobic regions in the absence of a known receptor [37, 38]. The assembled complement MAC contains one molecule of C5b, C6, C7 and C8, and several C9 molecules. At low copy numbers per cell (non-lytic or sublytic doses of MAC), the MAC is known to have stimulatory activities on many cell types [31, 39, 40]. However, at high copy numbers per cell (a lytic dose of MAC) the MAC induces a loss in membrane integrity and rapid necrotic type cell death [41, 42]. The morphology of complement-mediated cell death, as studied with Ehrlich ascites tumor cells at the ultrastructural level, is characterized by swelling of mitochondria, dilation of the rough endoplasmic reticulum, disruption of the Golgi complex and of the plasma and nuclear membranes, and heterochromatin disappearance [42]. Cells vary markedly in their threshold for lysis by MAC. The factors that contribute to cell resistance to the damage inflicted by the MAC have been recently reviewed [43]. It is noteworthy that cell resistance depends on both extracellular protection by membrane complement regulatory proteins and proteases, and intracellular damage repair processes involving protein kinases and heat-shock proteins [44–46].

Tab. 4.1 Proteins of the complement system.

Component	*Pathway*	*Structure*
C1	classical	a complex of three subunits, C1q (460 kDa) and two C1r (80 kDa), C1s (80 kDa)
MBL–MASP	lectin	a complex of MBL and MASP
C4	classical	3 chains; α (97 kDa), β (75kDa), γ (33kDa)
C2	classical	1 chain (102 kDa)
FB	alternative	1 chain (93 kDa)
FD	alternative	1 chain (24 kDa)
Properdin	alternative	oligomers of 1 chain (53 kDa)
C3	common	2 chains; α (110 kDa), β (75 kDa)
C5	terminal	2 chains; α (115 kDa), β (75 kDa)
C6	terminal	1 chain (120 kDa)
C7	terminal	1 chain (110 kDa)
C8	terminal	3 chains; α (65 kDa), β (65kDa), γ (22kDa)
C9	terminal	1 chain (69 kDa)

Proteins of the complement system can be divided according to the initial (classical, lectin and alternative) and common (terminal) pathways of activation. Their basic structure is indicated. C1q [106], MBL [96], C4, C2 [32, 34] and C3 [32, 103] were shown to bind to apoptotic cells. The alternative pathway [32, 98], classical pathway [32, 34, 103] or both [32] were shown to be possible pathways of activation. MBL was shown to bind to apoptotic cells but activation via the lectin pathway was not clearly shown [96]. FB = Factor B, FD = Factor D, MBL–MASP = mannan binding lectin (MBL) in complex with the zymogen form of serine protease MASP.

Tab. 4.2 Regulatory proteins of the complement system.

Component	*Regulatory function*
C1in (P, 200 µg/ml)	serine protease inhibitor; inhibition of activated C1
C4bp (P, 200–300)	CP: C4b → C4d/C4c and decay acceleration of convertase
FH (P, 200–600)	AP: C3b → iC3b and decay acceleration of convertase
FI (P, 30–40)	serine protease. Inactivation of C3 convertase AP: +Fh, CP: +C4bp:
S protein (P, 500)	prevention of MAC
CPN (P, 30)	C3a and C5a inhibitor
MCP (CD46) (M)	cleavage of C4b (CP) cleavage of C3b (AP) DAF (M) decay acceleration of convertase, CP and AP
CR1 (CD 35) (M)	cleavage of C4b (CP), Cleavage of C3b (AP) decay acceleration of convertase, CP and AP
CD59 (M)	Blocks binding of C9 and formation of MAC

Plasma (P) and membrane bound proteins (M) that regulate complement activation. AP = alternative pathway, C1in = C1 inhibitor, C4bp = C4 binding protein, CP = classical pathway, CPN = caboxypeptidase N, DAF = decay-accelerating factor, fH = Factor H, fI = Factor I, MCP = membrane cofactor protein.

Tab. 4.3 Membrane receptors for complement fragments.

Receptor	*Size and distribution*	*Ligand*	*Clearance of apoptotic material*
C1qR-gl-h	33, 80 kDa	C1q-gl-h	
C1qR-coll-r (calreticulin)	70 kDa, LY, MO, MA, EN	C1q-coll-r lectins (MBL)	+
C1qR-coll-r	100 kDa, MO, NE	C1q-coll-r	
CR1 (CD35)	180–200 kDa	C3b/C4b	?
CR2 (CD21)	145 kDa	C3d	
CR3 (CD11b/CD18)	αM (165 kDa/β2 (95 kDa)	iC3b	+
CR4 (CD11c/CD18)	αX (150 kDa/β2 (95 kDa)	iC3b	+
C3aR (CD88)	60 kDa	C3a	
C4 aR	NA kDa	C4a	
fHR	NA kDa	fH	
C5b–7R	NA kDa	C5b67	

C1qR-gl-h (calreticulin), and the integrin receptors, CR3 (CD11b/CD18) and CR4 (CD18/CD11c) were shown to have a role in removal of apoptotic material [32, 96–98]. Coll-r=collagenous region, gl-h= globular heads, MA=macrophages LK=leukocytes, LY=lymphocytes, MA=macrophages, MO=monocytes, NA=not available, NE=neutrophils, PL=platelets.

4.4 Complement and Apoptosis

PCD may be divided into two distinct sequential processes: execution and removal of dying cells. The two are linked together and *ex vivo* examination of tissues undergoing a high rate of apoptosis depicts apoptotic cells within phagocytes [47]. The prevailing notion is that uptake of apoptotic cells via specific receptors into phagocytes results in disposal of cellular debris without an induction of inflammation [10, 48, 49]. In Section 4.4.1 we focus on the possible role of complement in the execution phase of apoptosis and necrosis, and in Section 4.4.2 we discuss the role complement plays in the removal of dead corpses.

4.4.1 Role of Complement in the Execution Phase

Recent works suggest a role for complement activation in the execution phase of PCD. Cell death induced by complement MAC appears morphologically distinct from apoptosis, especially when comparing the progression of death at the level of cytoplasmic organelles [42]. However, certain inducers of cell death may induce apoptotic-type cell death at low concentration and necrotic-type cell death at higher concentrations [50–52].

Lytic MAC doses are known to produce DNA fragmentation [53]. The question of internucleosomal DNA fragmentation under lytic complement conditions was addressed recently *in vitro* [54]. DNA fragmentation, as determined by TUNEL, was evident as early as 30 min after addition of serum and DNA ladders were observed later, between 2 and 11 h after addition of serum. Importantly, these changes required a high (30%) serum concentration and occurred in cells with impaired plasma membrane integrity, i.e. dead cells. Apoptosis was identified in these studies by the use of criteria that may not be regarded as definite indicators of apoptotic cell death [55–57]. Cragg *et al.* proposed that the DNA fragmentation and laddering observed in complement-lysed cells (e.g. Raji, Daudi and EHRB cells) is mediated by serum DNase I which enters the cells through complement-induced pores [54]. The data presented in the ultrastructural studies [42, 54] best supports the conclusion that complement-induced death is necrotic in essence with the possible involvement of apoptotic features produced by external or secondary (e.g. oxygen radicals) factors.

Activated complement proteins, other than MAC, may also contribute to apoptotic cell death. Rosenkranz *et al.* demonstrated that neutrophils incubated with a Cubrophan membrane and serum undergo apoptosis which is prevented by removal of C3, but not of C5, from the serum [58]. Heat inactivation of the serum also reduced the serum capacity to induce neutrophil apoptosis. Uwai *et al.* reported that Factor B and its fragment Bb are apoptogenic to HL-60 cells [59]. The apoptogenic effect of Factor B may be enhanced by the addition of C3 or Factor D and is inhibited by antibodies directed to Factor B, CD35 or CD11c. Interestingly, Factor B was identified in this study to be a growth-inhibiting factor produced by HL-60 cells, activating caspase-3 and DNA fragmentation in these cells. The involvement of Fas and TNF pathways in the Bb-induced apoptotic pathway was excluded. Finally, Farkas *et al.* showed that an oligomeric form of a C5a-derived peptide may cause apoptosis (TUNEL staining) of human neuroblastoma cells [60]. This activity is C5a receptor dependent.

Complement MAC contributes to tissue injury following myocardial ischemia and reperfusion (MI/R) [61–63]. Using a rat model of MI/R, Vakeva *et al.* analyzed the cell death type, and concluded that cell loss in MI/R results from necrotic and apoptotic-type cell death [64]. By using anti-C5 antibody therapy, it was found that both necrotic and apoptotic (DNA fragmentation and laddering) events require C5, suggesting that MAC is the cause of the apoptotic-type cell death in tissue following MI/R.

The involvement of complement in kidney tissue damage accompanying autoimmune and inflammatory diseases is widely accepted ([31] and reviewed in [65]). To investigate the factors involved in kidney diseases, several animal models have been developed. Perfusion of rat kidneys with anti-Thy-1 antibodies leads to the development of a complement-dependent nephritis accompanied by a rapid loss of mesangial cells [66, 67]. Sato *et al.* used this rat model to study the type of mesangial cell death. Cells with condensed chromatin and cells stained by TUNEL were observed in glomeruli of affected rats [68]. Such cells were not seen in glomeruli of C6-deficient rats infused with Thy-1 antibodies. Therefore, it was concluded that MAC is involved in induction of apoptosis in mesangial cells. Shimizu *et al.* used the same nephritis model and also described complement-dependent mesangial cell death characterized by chromatin condensation, TUNEL staining and DNA ladder formation [69]. Based on ultrastructural analysis of the dead cells by transmission electron microscopy and clear indications of necrotic-type death in the mesangial cell cytoplasm, they suggested that complement induced both necrotic and apoptotic cell death in the same cell. In contrast, based on an extended ultrastructural study of mesangial cells in rats developing acute nephritis upon administration of anti-Thy-1 antibody, Mosley *et al.* concluded that these cells undergo a typical necrotic cell death. Chromatin condensation was observed, but there was also loss of nuclear membrane, cell swelling, degeneration of cytoplasmic organelles, and release of chromatin and organelles into the interstitium [70]. Hughes *et al.* used two other rat models of induced experimental glomerulonephritis and followed death of glomerular endothelial cells [71]. Future studies will have to further address whether complement MAC may induce, under pathological conditions, cell death by apoptosis.

Sublytic and non-lytic doses of MAC have been shown, over the years, to induce numerous cell stimulatory effects ([31] and reviewed in [37, 40]). In 1992, Reiter *et al.* demonstrated that sublytic MAC doses can also induce enhanced cell protection from lytic MAC doses [72]. This finding was later extended to include several other protein pore-formers like perforin, mellitin, streptolysin O and staph-α-toxin [73]. The process of complement-induced protection depends on calcium ion influx and activation of protein kinase C [43], and the extracellularly regulated protein kinase ERK [46].

The possibility that sublytic MAC can induce cell protection from apoptosis was first examined with oligodendrocytes (OLG). OLG treated with sublytic MAC doses are stimulated to enter the cell cycle by induction of c-Jun through activation of the c-Jun N-terminal kinase pathway [74]. Upon *in vitro* culture, OLG spontaneously die by apoptosis. However, treatment of OLG with sublytic MAC reduces the percentage of dead cells [75]. Interestingly, sublytic MAC was shown to increase the amount of Bcl-2 in OLG and to reduce activation of apoptosis-associated caspase-3 activation [75]. In addition, sublytic MAC can protect OLG from TNF-α-mediated apoptosis by prevention of caspase-3 activation [75]. A similar study was performed recently with Schwann cells in culture [76]. Again, sublytic MAC was shown to activate in Schwann cells, DNA synthesis and a shift into S or G_2/M phases of cell cycle, and to induce cell proliferation. Like OLG, Schwann cells undergo during culture spontaneous apoptosis that may be inhibited by treatment with sublytic MAC. Dashiell *et al.* concluded that the MAC is a potent Schwann cell trophic factor capable of stimulating mitogenesis and apoptosis rescue. Whether or not the complement system is protecting glial cells *in vivo* from apoptotic death is still not known. However, clearly, such a protective activity may have a major impact on axonal remyelinization in the central and peripheral nervous systems in neurological diseases or following axonal damage.

Another complement component that appears to have an anti-apoptotic effect is the anaphylatoxin C5a. Complement activation in the brain has been mainly implicated in neurodegeneration in diseases like Alzheimer's disease (AD) [77]. However, surprisingly, C5-deficient mice are more susceptible to excitotoxic brain injury [78]. Pasinetti *et al.* presented *in vivo* and *in vitro* results that suggest a role for C5a in neuroprotection [79]. C5a reduces neuronal apoptosis in the hippocampal pyramidal layer of mice injected with kainic acid in the intra-brain ventricles [80]. C5a also protects *in vitro* primary cultured neurons from cell death induced by glutamate [81]. As neuronal cell death was measured in the latter reports by Trypan blue exclusion [80] or release of lactic dehydrogenase [81], it is possible that C5a prevented progression of secondary necrosis or inhibited a non-apoptotic effect of glutamate. The fact that C5a could reduce glutamate-induced activation of caspase-3 and staurosporin-induced activation of caspase-3 and -8 [79] supports an anti-apoptotic effect for C5a. Ultrastructural studies will provide a more definite identification of the earliest apoptotic events affected by C5a. The protective effect of C5a depends on a signaling cascade, initiated following engagement of the C5a receptor and activation of ERK2 [81].

An earlier study suggested that C5a may also protect neutrophils from apoptotic cell death [82]. The biological relevance and extent of these findings are yet unclear, but clearly multi-factorial and complex mechanisms appear to be involved in the intersection between complement and apoptosis.

4.4.2
Complement Activation by Apoptotic Cells

The mechanisms whereby apoptotic cells are identified, taken up and degraded by phagocytes are not well understood. Interestingly, at least seven of the 12 genes involved in PCD in the nematode *Caenorhabditis elegans* participate in the removal of apoptotic bodies [83, 84]. The function of these genes (*ced-1, -2, -5, -6, -7, -10* and *-12*) has only been partially characterized. One group consists of *ced-2, -5, -10* and *-12* [83, 85]. Analysis of this group of genes in *C. elegans* indicates their role in organizing and controlling cytoskeletal rearrangement during cell migration and engulfment of apoptotic cells [86–88]. The *ced-2* mammalian homolog was identified as *crkII* [86], *ced-5* encodes a homolog of human DOCK 180, a protein involved in receptor signaling and surface extension [89], while *ced-10* and *-12* analogs are *rac* [86] and *elmo* [87], respectively. The second group includes *ced-1, -6/gulp* and *-7/abc1*, and may be related to recognition functions. Recently, CED-1 was cloned and identified to be a transmembrane protein sharing sequence homology with the human scavenger receptor SREC [90]. However, SREC fail to show any homology to the cytoplasmic region of CED-1. *ced-6* encodes an adaptor molecule with a phosphotyrosine-binding domain acting in signal transduction that mediates engulfment of apoptotic cells [91]. Expression of human *ced-6* as a transgene rescues the engulfment defect [92], showing perhaps conservation of function between worms and humans. CED-7 is similar to the ABC1 cassette transporter, and may be involved in the homotypic interaction between the cell surface of dying and engulfing cells [93]. A recent work showed interaction between CED-6/gulp and CED-1 and CD91/LRP, and suggested interaction of these proteins during engulfment [94]. Interestingly, all six genes, as well as *ced-12*, are also required for the clearance of necrotic cells [95].

Publications over the past decade have attributed to a role integrins, scavenger receptors, PS receptor, CD14, ABC1 cassette transporter and C1q/CD91 receptors in the uptake of apoptotic cells by macrophages (reviewed in [10, 96]). As suggested by the multitude of receptors involved, the mechanisms underlying the recognition, engulfment and phagocytosis of apoptotic cells by macrophages are complex. Complement opsonins like C1q, mannose-binding lectin (MBL) and iC3b, appear to have a role in the uptake of apoptotic cells, possibly by interaction with C1qR/CD91, CR3 (CD11b/CD18) and/or CR4 (CD11c/CD18) on phagocytes [9, 32, 34, 96–98]. Complement factors were also suggested to bind to late apoptotic or necrotic cells [99]. Serum factors were shown to increase the uptake of apoptotic cells by human macrophages by 3- to 10-fold [32]. Contrasting data has been recently reported by Ren *et al.*, who claimed that CR3 and CR4 are not essential for uptake of aged neurophils by macrophages [100]. No difference in ingestion of apoptotic neutrophils was found between macrophages derived from CD11b- or CD18-deficient and wild-type mice. Unfortunately, the latter report did not provide any clear indication of the level of complement-mediated opsonization of the aged neutrophils employed in the interaction assay [100]. CR3 and CR4 are expected to have an impact on phagocytosis only after efficient opsonization of the apoptotic cells by iC3b [101]. Therefore, further experimentation is required to clarify the role of complement receptors in uptake of apoptotic cells by macrophages.

Complement activation by apoptotic cells has been shown in a variety of cells: human T lymphocytes, human lymphoma cell lines [32, 102], human umbilical vein endothelial cells (HUVEC) [103, 104], human lung adenocarcinoma cells [105], as well as in murine thymocytes [32]. Complement activation may be initiated via the classical, alternative and/or lectin pathway, and the first reports [102, 104] concluded that apoptotic cells activate the alternative pathway of complement. Later, Korb and Ahearn described binding of C1q, the first component of the classical pathway, to apoptotic keratinocytes [106], and Mevorach *et al.* reported activation of the classical pathway by human apoptotic leukocytes and murine thymocytes [32]. Interestingly, when using a heterologous system of murine apoptotic thymocytes and human complement, both alternative and classical pathways appear to be activated. This was supported by the use of sera deficient in C1q, C2 or C4 (classical pathway components), factor B (an alternative pathway component) or C3 (for both pathways) [32]. Apoptotic HUVEC cells subjected *in vitro* to hypoxia and re-oxygenation or starvation become activators of the classical pathway of complement [103]. This is prevented by treatment of the cells with the pan-caspase inhibitor z-VAD. Remarkably, apoptosis of HUVEC cells [104] or human neutrophils is associated with downregulation of the complement membrane regulatory molecules decay-accelerating factor (DAF or CD55) and CD59 [107], whereas apoptosis of human lung adenocarcinoma cells is associated with down regulation of DAF, membrane cofactor protein (MCP or CD46) and complement receptor type 1 (CR1 or CD35) [105]. Reduced levels of CD59 were found also *in vivo* in $CD8^+$ T lymphocytes collected from Epstein–Barr virus-induced acute infectious mononucleosis patients [108], and, more recently, from systemic lupus erythematosus (SLE) and Sjögren's syndrome patients [109]. Interestingly, the $CD4^+$ T lymphocytes from the same patients contained normal levels of CD59. Whether or not CD59 and the other complement membrane regulatory proteins play any role in the apoptotic process is still an open question. Another, yet unresolved, issue is whether or not apoptotic cells lacking these membrane proteins are by nature more susceptible to complement-mediated attack.

To date, little is known about complement-activating molecules on apoptotic cells or acceptors onto which activated complement components are deposited. One of the early molecular events in apoptosis is an externalization of PS from the inner to the outer face of the plasma membrane [9, 110]. PS exposure on apoptotic cells is associated with C3b deposition on the surface of apoptotic cells. Only cells exposing PS have C3 fragments (mainly iC3b) on their surface [32]. This claim has been recently reinforced by Mold and Morris who also found a good correlation between PS^+ and $C3b^+$ cells in a population of apoptotic HUVEC cells treated with complement [103]. However, the role PS exposure on apoptotic cells plays in the apoptotic process and in complement activation is still an open question. In support of a possible direct role for PS in complement activation, PS micelles have been shown to activate complement both *in vitro* [111, 112] and *in vivo* [113], and loss of membrane asymmetry of red blood cells was shown to be associated with complement activation [114, 115]. In most reports of complement activation by phospholipids, the activation occurred via the alternative pathway [116]. Cardiolipin, a phospholipid that shares some similarities in structure and charge with PS, was shown to activate complement via the classical pathway in an antibody-independent mechanism [117–119]. Thus, liposomes that share similarities with the membrane of apoptotic cells have the capacity to activate complement through either the classical or alternative complement pathways [120]. Finally, complement activation may also be regulated by binding to apoptotic cells of proteins such as natural or autoimmune antibodies [9, 33], β_2-glycosylphosphatidylinositol (β_2-GPI) [9, 121–123], C-reactive protein (CRP) [124], serum amyloid P and other proteins (reviewed in [9]).

Complement activation by apoptotic cells imposes the risk of generation of the MAC and cell lysis, and consequently development of an inflammatory process. However, Mevorach *et al.* showed that the complement system undergoes only a limited activation by apoptotic lymphocytes, Jurkat cells and murine thymocytes, and that C5- or C9-deficient serum are as efficient as C-sufficient sera in opsonization of apoptotic cells and in their subsequent clearance by phagocytes [32]. Given that CD59 expression was reported to be reduced on some apoptotic cells, what protects them from the MAC? One possible explanation is that C3b deposited on apoptotic cells is rapidly converted into iC3b [32, 104, 108]. In the absence of MCP and perhaps also CR1, the inactivation of C3b is largely dependent on Factor H [125]. It is possible that the surface of apoptotic cells permits a better regulation of C3b by Factors H and I. Protection of apoptotic cells from MAC by clusterin has also been proposed [126].

CRP is an acute phase reactant synthesized following tissue injury and inflammation (reviewed in [31]). It binds to nucleated cells and activates complement activation that is restricted to formation of the C3 convertase [127]. C5 convertase formation and MAC generation are probably blocked by Factor H that binds directly to CRP [128]. A role has been proposed recently for CRP in activation of the classical complement pathway on apoptotic Jurkat cells and in limiting, together with Factor H, the stage of complement activation to the C3b deposition and iC3b generation stages [124]. Hence, CRP may support non-inflammatory clearance of apoptotic cells and, under normal conditions, complement activation on the surface of apoptotic cells will facilitate phagocytosis but not cell lysis. However, the rules of the game may change when conditions change. The intensity of complement activation may increase at sites of inflammation and in the presence of autoantibodies. Will the mechanism(s) that protect the apoptotic cells from lysis still be as effective? It is envisaged that in the presence of exogenous complement activators such as autoantibodies, immune complexes or bacterial polysaccharides, apoptotic cells will undergo MAC-mediated necrosis to some extent.

Tagging the surface of apoptotic cells with C3b and iC3b may not only promote efficient clearance of apoptotic cells, but may also exert anti-inflammatory responses. Binding and phagocytosis via macrophage CD11b/CD18 does not trigger leukotriene release [129] or a respiratory burst [130, 131]. Furthermore, ligation of CD11b/CD18 and other complement receptors may actually be immunosuppressive by down-regulating interleukin (IL)-12 and interferon (IFN)-γ production by human monocytes [132]. It seems likely, therefore, that the pro- and anti-inflammatory consequences of complement activation will depend upon the specific ligands that are involved and the co-receptors that are engaged. Phagocytosis of apoptotic cells was shown to suppress the release of granulocyte macrophage colony stimulating factor, IL-1β, IL-8, TNF-α, and thromboxane B_2, but not transforming growth factor-β and prostaglandin E_2 [48]. IL-10 has anti-inflammatory properties manifested by suppression of the release of pro-inflammatory cytokines such as IL-1, IL-6, IL-8 and TNF-α. However, contrasting results were obtained when the production of IL-10following phagocytosis of apoptotic cells was evaluated. Both increased [133] and suppressed [48] release of IL-10 following ingestion of apoptotic cells by human macrophages was found. The latter studies were performed in the absence of serum; therefore, the effect of complement on IL-10 release could not have been evaluated. Under certain conditions, phagocytosis of apoptotic cells may also lead to production of pro-inflammatory cytokines, such as IL-8, by phorbol myristate acetate-activated macrophages [7].

4.5 Apoptosis, Complement and Autoimmunity

The significance of the complement system to uptake of apoptotic cells *in vivo* may be illustrated in the pathogenesis of SLE (reviewed in [9, 134–136]). It has long been appreciated that DNA and histones are major autoantigens in SLE. However, more recently it has become clear that DNA–histone complexes, i.e. nucleosomes, are preferred targets for autoantibodies in SLE [137]. How do nucleosomes and several other intracellular antigens become immunogenic in SLE patients? Part of the answer to this question may lie in the process of apoptosis. During apoptosis, the plasma membrane of apoptotic cells undergoes blebbing accompanied by other membrane transformations, culminating in shedding of apoptotic bodies. Thus, exposure of keratinocytes to ultraviolet B-mediated apoptosis induces cell surface expression of Ro and La, nucleosomes, and ribosomes, possibly due to translocation of certain intracellular particles to the apoptotic surface blebs [138]. PS, restricted to the inner membrane leaflet in a viable cell, translocates during apoptosis to the apoptotic cell surface [139]. β_2-GPI bound to PS on apoptotic cells [123] is a major autoantigen for anti-phospholipid antibodies in SLE and *in vivo* experiments in mice showed transient elevations of anti-phospholipid antibodies following high-dose injection of apoptotic cells [33]. Therefore, it has been envisaged that SLE patients are triggered by surface exposed intracellular macromolecules translocated to the cell surface during apoptosis [33, 138]. The next question that arises then is why do SLE patients mount an immune response to the apoptotic material? One possible explanation is an impaired capacity of SLE patients to clear apoptotic cells that serve as efficient immunogens. This is based on the findings that macrophages of SLE patients show *in vitro* reduced uptake of apoptotic cells [134, 136], and high level of circulating nucleosomes in peripheral blood of SLE patients [140]. Furthermore, reduction of DNase I and increased nucleosomes level was suggested to have a critical role in the initiation of human SLE [141]. Interestingly, increased spontaneous apoptosis of lymphocytes [142, 143] and monocytes [136] was documented in patients with SLE, and immunization of mice with a high dose of syngeneic apoptotic cells was shown to induce autoantibodies in normal strains of mice [33]. How is complement involved in this process? Homozygous deficiency of C1q in man and mice shows the strongest single gene association with development of an SLE-like disease (reviewed in [144, 145]). Furthermore, multiple TUNEL$^+$ cells are found in the kidneys of C1q-deficient mice developing an SLE-like disease [144]. Clearance of apoptotic cells *in vitro* [32] and *ex vivo* was suggested to depend both on C1q and C4 [34]. Yet, C1q was not essential for *in vivo* clearance of sunburn apoptotic cells [146]. It will be interesting to see what role has MBL and C1qR/CD91 [96] in *in vivo* removal of apoptotic cells. C2- and Factor B-deficient mice do not develop autoantibodies or an SLE-like disease, suggesting that C3 activation is not essential for protection from this disease development [147]. However, this does not disprove the claim [32] that C3 activation is involved in clearance of apoptotic cells. In fact, using *ex vivo* models of apoptotic cell clearance, Taylor *et al.* concluded that, depending on the cell type studied, complement mediated apoptotic cell clearance is either effected directly by C1q (via candidate C1q receptors) or by C1q as an activator of the classical pathway [34]. The acceptor molecule (s) for C1q on apoptotic cells has not been identified yet, but, apparently, C1q binds through its globular head to the surface blebs developing on apoptotic keratinocytes [106], endothelial cells and peripheral blood mononuclear cells [148]. These results act as the basis for the hypothesis implicating the role of early classical pathway components, together with phagocytic cells, in the disposal of potentially hazardous immunogens from the body. Early complement deficiencies, acting together with genetic and environmental susceptibility factors, thus increase susceptibility to an SLE-like disease due to diminished clearance of dying cells. In addition, diminished clearance of immune complexes associated with complement-deficient states could cause an inflammatory damage and activation of antigen-presenting cells [149]. An alternative hypothesis, although not mutually exclusive, proposes a role for the early classical pathway components in maintenance of self-tolerance by removing or silencing self-reactive B lymphocytes [150]. Interestingly, in mice, C4 inhibits autoimmunity through a mechanism independent of complement receptors CR1 and CR2 [151]. What happens to apoptotic material when clearance is impaired? Will B cell tolerance be impaired [152]? Is it necrosis that evokes a pro-inflammatory immune response by dendritic cells [153]? Is it lysis of apoptotic material (secondary necrosis) and not primary necrosis that induces autoimmunity [9]? What is the role of autoantibodies, antinuclear, anti-DNA and anti-C1q, on one hand, or natural autoantibodies, on the other hand, in induction or prevention of autoimmunity?

Better understanding of homeostasis, degenerative (atherosclerotic) processes and inflammatory conditions will hopefully shed more light on the role of innate immunity in conditions we today call 'autoimmunity'.

4.6 References

1. Hengartner M.O. The biochemistry of apoptosis. *Nature* **2000**, 407, 770–6.
2. Vaux D.L. and Korsmeyer S.J. Cell death in development. *Cell* **1999**, 96, 245–54.
3. Kerr J.F., Wyllie A.H. and Currie A.R. Apoptosis: a basic biological phenomenon with wide-ranging implications in tissue kinetics. *Br J Cancer* **1972**, 26, 239–57.
4. Walker N.I., Harmon B.V., Gobe G.C. and Kerr J.F.R. Patterns of cell death. *Methods Achiev Exp Pathol.* **1988**, 13, 18–54.
5. Wyllie A.H., Kerr J.F. and Currie A.R. Cell death: the significance of apoptosis. *Int Rev Cytol* **1980**, 68, 251–306.
6. Kurosaka K., Watanabe N. and Kobayashi Y. Production of proinflammatory cytokines by resident tissue macrophages after phagocytosis of apoptotic cells. *Cell Immunol* **2001**, 10, 1–7.
7. Kurosaka K., Watanabe N. and Kobayashi Y. Production of proinflammatory cytokines by phorbol myristate acetate-treated THP-1 cells and monocyte-derived macrophages after phagocytosis of apoptotic CTLL-2 cells. *J Immunol* **1998**, 161, 6245–9.
8. Kawagishi C., Kurosaka K., Watanabe N. and Kobayashi Y. Cytokine production by macrophages in association with phagocytosis of etoposide-treated P388 cells *in vitro* and *in vivo*. *Biochim Biophys Acta* **2001**, 19, 221–30.
9. Mevorach D. Opsonization of apoptotic cells. Implications for uptake and autoimmunity. *Ann NY Acad Sci* **2000**, 926, 226–35.
10. Savill J. and Fadok V. Corpse clearance defines the meaning of cell death. *Nature* **2000**, 407, 784–8.
11. Harris M.H. and Thompson C.B. The role of the Bcl-2 family in the regulation of outer mitochondrial membrane permeability. *Cell Death Different* **2000**, 7, 1182–91.
12. Korsmeyer S.J. BCL-2 gene family and the regulation of programmed cell death. *Cancer Res* **1999**, 59, 1693–700s.
13. Van der Heiden M.G. and Thompson C.B. Bcl-2 proteins: regulators of apoptosis or of mitochondrial homeostasis? *Nat Cell Biol* **1999**, 1, E209–16.
14. Ekert P.G., Silke J. and Vaux D.L. Caspase inhibitors. *Cell Death Different* **1999**, 6, 1081–6.

15 Nicholson D.W. Caspase structure, proteolytic substrates, and function during apoptotic cell death. *Cell Death Different* **1999**, 6, 1028–42.

16 Slee E.A., Adrain C. and Martin S.J. Serial killers: ordering caspase activation events in apoptosis. *Cell Death Different* **1999**, 6, 1067–74.

17 Thornberry N.A. Caspases: a decade of death research. *Cell Death Different* **1999**, 6, 1023–7.

18 Buendia B., Santa-Maria A. and Courvalin J.C. Caspase-dependent proteolysis of integral and peripheral proteins of nuclear membranes and nuclear pore complex proteins during apoptosis. *J Cell Sci* **1999**, 112, 1743–53.

19 Takahashi A., Alnemri E.S., Lazebnik Y.A., Fernandes-Alnemri T., Litwack G., Moir R.D., Goldman R.D., Poirier G.G., Kaufmann S.H. and Earnshaw W.C. Cleavage of lamin A by Mch2 alpha but not CPP32: multiple interleukin 1 beta-converting enzyme-related proteases with distinct substrate recognition properties are active in apoptosis. *Proc Natl Acad Sci USA* **1996**, 93, 8395–400.

20 Kothakota S., Azuma T., Reinhard C., Klippel A., Tang J., Chu K., McGarry T.J., Kirschner M.W., Koths K., Kwiatkowski D.J. and Williams L.T. Caspase-3-generated fragment of gelsolin: effector of morphological change in apoptosis. *Science* **1997**, 278, 294–8.

21 Nagata S. Apoptotic DNA fragmentation. *Exp Cell Res* **2000**, 256, 12–8.

22 Scaffidi C., Fulda S., Srinivasan A., Friesen C., Li F., Tomaselli K.J., Debatin K.M., Krammer P.H. and Peter M.E. Two CD95 (APO-1/Fas) signaling pathways. *EMBO J* **1998**, 17, 1675–87.

23 Kroemer G. and Reed J.C. Mitochondrial control of cell death. *Nat Med* **2000**, 6, 513–9.

24 Cain K., Bratton S.B., Langlais C., Walker G., Brown D.G., Sun X.M. and Cohen G.M. Apaf-1 oligomerizes into biologically active approximately 700-kDa and inactive approximately 1.4-MDa apoptosome complexes. *J Biol Chem* **2000**, 275, 6067–70.

25 Zou H., Li Y., Liu X. and Wang X. An APAF-1.cytochrome *c* multimeric complex is a functional apoptosome that activates procaspase-9. *J Biol Chem* **1999**, 274, 11549–56

26 Susin S.A., Lorenzo H.K., Zamzami N., Marzo I., Snow B.E., Brothers G.M., Mangion J., Jacotot E., Costantini P., Loeffler M., Larochette N., Goodlett D.R., Aebersold R., Siderovski D.P., Penninger J.M. and Kroemer G. Molecular characterization of mitochondrial apoptosis-inducing factor. *Nature* **1999**, 397, 441–6.

27 Amarante-Mendes G.P., Finucane D.M., Martin S.J., Cotter T. G., Salvesen G.S. and Green D.R. Anti-apoptotic oncogenes prevent caspase-dependent and independent commitment for cell death. *Cell Death Different* **1998**, 5, 298–306.

28 Sperandio S., de Belle I. and Bredesen D.E. An alternative, nonapoptotic form of programmed cell death. *Proc Natl Acad Sci USA* **2000**, 97, 14376–81.

29 Green D.R. Apoptotic pathways: paper wraps stone blunts scissors. *Cell* **2000**, 102, 1–4.

30 Kumar S., ed. *Apoptosis: Mechanisms and Role in Disease.* Berlin: Springer-Verlag **1998**.

31 Volanakis J.E. and Frank M.M., eds. *The Human Complement System in Health and Disease.* Marcel Dekker: New York 1998.

32 Mevorach D., Mascarenhas J.O., Gershov D. and Elkon K.B. Complement-dependent clearance of apoptotic cells by human macrophages. *J Exp Med* **1998**, 188, 2313–20.

33 Mevorach D., Zhou J.L., Song X. and Elkon K.B. Systemic exposure to irradiated apoptotic cells induces autoantibody production. *J Exp Med* **1998**, 188, 387–92.

34 Taylor P.R., Carugati A., Fadok V.A., Cook H.T., Andrews M., Carroll M. C., Savill J. S., Henson P.M., Botto M. and Walport M.J. A hierarchical role for classical pathway complement proteins in the clearance of apoptotic cells *in vivo.* *J Exp Med* **2000**, 192, 359–66.

35 Ehlers M.R. CR3: a general purpose adhesion-recognition receptor essential for innate immunity. *Microbes Infect* **2000**, 2, 289–94.

36 Ember J.A. and Hugli T.E. Complement factors and their receptors. *Immunopharmacology* **1997**, 38, 3–15.
37 Morgan B.P. Regulation of the complement membrane attack pathway. *Crit Rev Immunol* **1999**, 19, 173–98.
38 Muller-Eberhard H.J. The membrane attack complex of complement. *Annu Rev Immunol* **1986**, 4, 503–28.
39 Morgan B.P. Complement membrane attack on nucleated cells: resistance, recovery and non-lethal effects. *Biochem J* **1989**, 264, 1–14.
40 Shin M.L. and Carney D.F. Cytotoxic action and other metabolic consequences of terminal complement proteins. *Prog Allergy* **1988**, 40, 44–81.
41 Koski C.L., Ramm L.E., Hammer C.H., Mayer M.M. and Shin M.L. Cytolysis of nucleated cells by complement: cell death displays multi-hit characteristics. *Proc Natl Acad Sci USA* **1983**, 80, 3816–20.
42 Papadimitriou J.C., Drachenberg C.B., Shin M.L. and Trump B.F. Ultrastructural studies of complement mediated cell death: a biological reaction model to plasma membrane injury. *Virchows Arch* **1994**, 424, 677–85.
43 Jurianz K., Ziegler S., Garcia-Schuler H., Kraus S., Bohana-Kashtan O., Fishelson Z. and Kirschfink M. Complement resistance of tumor cells: basal and induced mechanisms. *Mol Immunol* **1999**, 36, 929–39.
44 Fishelson Z., Hochman I., Greene L.E. and Eisenberg E. Contribution of heat shock proteins to cell protection from complement-mediated lysis. *Int Immunol* **2001**, 13, 983–91.
45 Kraus S. and Fishelson Z. Cell desensitization by sublytic C5b–9 complexes and calcium ionophores depends on activation of protein kinase C. *Eur J Immunol* **2000**, 30, 1272–80.
46 Kraus S., Seger R. and Fishelson Z. Involvement of the ERK mitogen-activated protein kinase in cell resistance to complement-mediated lysis. *Clin Exp Immunol* **2001**, 123, 366–74.
47 Surh C.D. and Sprent J. T-cell apoptosis detected *in situ* during positive and negative selection in the thymus. *Nature* **1994**, 372, 100–3.
48 Fadok V.A., Bratton D.L., Konowal A., Freed P.W., Westcott J.Y. and Henson P.M. Macrophages that have ingested apoptotic cells *in vitro* inhibit proinflammatory cytokine production through autocrine/paracrine mechanisms involving TGF-beta, PGE_2, and PAF. *J Clin Invest* **1998**, 101, 890–8.
49 Mevorach D. The immune response to apoptotic cells. *Ann NY Acad Sci* **1999**, 887, 191–8.
50 Bonfoco E., Krainc D., Ankarcrona M., Nicotera P. and Lipton S.A. Apoptosis and necrosis: two distinct events induced, respectively, by mild and intense insults with *N*-methyl-D-aspartate or nitric oxide/superoxide in cortical cell cultures. *Proc Natl Acad Sci USA* **1995**, 92, 7162–6.
51 Lennon S.V., Martin S.J. and Cotter T.G. Dose-dependent induction of apoptosis in human tumour cell lines by widely diverging stimuli. *Cell Prolif* **1991**, 24, 203–14.
52 Tidball J.G., Albrecht D.E., Lokensgard B.E. and Spencer M.J. Apoptosis precedes necrosis of dystrophin-deficient muscle. *J Cell Sci* **1995**, 108, 2197–204.
53 Shipley W.U., Baker A.R. and Colten H.R. DNA degradation in mammalian cells following complement-mediated cytolysis. *J Immunol* **1971**, 106, 576–9.
54 Cragg M.S., Howatt W.J., Bloodworth L., Anderson V.A., Morgan B.P. and Glennie M.J. Complement mediated cell death is associated with DNA fragmentation. *Cell Death Different* **2000**, 7, 48–58.
55 Collins R.J., Harmon B.V., Gobe G.C. and Kerr J.F. Internucleosomal DNA cleavage should not be the sole criterion for identifying apoptosis. *Int J Radiat Biol* **1992**, 61, 451–3.
56 Nishiyama K., Kwak S., Takekoshi S., Watanabe K. and Kanazawa I. *In situ* nick end-labeling detects necrosis of hippocampal pyramidal cells induced by kainic acid. *Neurosci Lett* **1996**, 212, 139–42.
57 Thomas L.B., Gates D.J., Richfield E.K., O'Brien T.F., Schweitzer J.B. and Steindler D.A. DNA end labeling (TUNEL) in Huntington's disease and other neuropathological conditions. *Exp Neurol* **1995**, 133, 265–72.

58 Rosenkranz A. R., Peherstorfer E., Kormoczi G. F., Zlabinger G. J., Mayer G., Horl W. H. and Oberbauer R. Complement-dependent acceleration of apoptosis in neutrophils by dialyzer membranes. *Kidney Int* **2001**, 59 (suppl 78), S216–20.

59 Uwai M., Terui Y., Mishima Y., Tomizuka H., Ikeda M., Itoh T., Mori M., Ueda M., Inoue R., Yamada M., Hayasawa H., Horiuchi T., Niho Y., Matsumoto M., Ishizaka Y., Ikeda K., Ozawa K. and Hatake K. A new apoptotic pathway for the complement factor B-derived fragment Bb. *J Cell Physiol* **2000**, 185, 280–92.

60 Farkas I., Baranyi L., Liposits Z. S., Yamamoto T. and Okada H. Complement C5a anaphylatoxin fragment causes apoptosis in TGW neuroblastoma cells. *Neuroscience* **1998**, 86, 903–11.

61 Hill J. H. and Ward P. A. The phlogistic role of C3 leukotactic fragments in myocardial infarcts of rats. *J Exp Med* **1971**, 133, 885–900.

62 Vakeva A., Laurila P. and Meri S. Regulation of complement membrane attack complex formation in myocardial infarction. *Am J Pathol* **1993**, 143, 65–75.

63 Weisman H. F., Bartow T., Leppo M. K., Marsh H. C., Jr, Carson G. R., Concino M. F., Boyle M. P., Roux K. H., Weisfeldt M. L. and Fearon D. T. Soluble human complement receptor type 1: *in vivo* inhibitor of complement suppressing post-ischemic myocardial inflammation and necrosis. *Science* **1990**, 249, 146–51.

64 Vakeva A. P., Agah A., Rollins S. A., Matis L. A., Li L. and Stahl G. L. Myocardial infarction and apoptosis after myocardial ischemia and reperfusion: role of the terminal complement components and inhibition by anti-C5 therapy. *Circulation* **1998**, 97, 2259–67.

65 Couser W. G. Pathogenesis of glomerular damage in glomerulonephritis. *Nephrol Dial Transplant* **1998**, 13, 10–5.

66 Brandt J., Pippin J., Schulze M., Hansch G. M., Alpers C. E., Johnson R. J., Gordon K. and Couser W. G. Role of the complement membrane attack complex (C5b–9) in mediating experimental mesangioproliferative glomerulonephritis. *Kidney Int* **1996**, 49, 335–43.

67 Yamamoto T. and Wilson C. B. Complement dependence of antibody-induced mesangial cell injury in the rat. *J Immunol* **1987**, 138, 3758–65.

68 Sato T., van Dixhoorn M. G., Prins F. A., Mooney A., Verhagen N., Muizert Y., Savill J., van Es L. A. and Daha M. R. The terminal sequence of complement plays an essential role in antibody-mediated renal cell apoptosis. *J Am Soc Nephrol* **1999**, 10, 1242–52.

69 Shimizu A., Masuda Y., Kitamura H., Ishizaki M., Ohashi R., Sugisaki Y. and Yamanaka N. Complement-mediated killing of mesangial cells in experimental glomerulonephritis: cell death by a combination of apoptosis and necrosis. *Nephron* **2000**, 86, 152–60.

70 Mosley K., Collar J. and Cattell V. Mesangial cell necrosis in Thy 1 glomerulonephritis – an ultrastructural study. *Virchows Arch* **2000**, 436, 567–73.

71 Hughes J., Nangaku M., Alpers C. E., Shankland S. J., Couser W. G. and Johnson R. J. C5b–9 membrane attack complex mediates endothelial cell apoptosis in experimental glomerulonephritis. *Am J Physiol Renal Physiol* **2000**, 278, F747–57.

72 Reiter Y., Ciobotariu A. and Fishelson Z. Sublytic complement attack protects tumor cells from lytic doses of antibody and complement. *Eur J Immunol* **1992**, 22, 1207–13.

73 Reiter Y., Ciobotariu A., Jones J., Morgan B. P. and Fishelson Z. Complement membrane attack complex, perforin, and bacterial exotoxins induce in K562 cells calcium-dependent cross-protection from lysis. *J Immunol* **1995**, 155, 2203–10.

74 Rus H. G., Niculescu F. and Shin M. L. Sublytic complement attack induces cell cycle in oligodendrocytes. *J Immunol* **1996**, 156, 4892–900.

75 Soane L., Rus H. G., Niculescu F. and Shin M. L. Inhibition of oligodendrocyte apoptosis by sublytic C5b–9 is associated with enhanced synthesis of bcl-2 and mediated by inhibition of caspase-3 activation. *J Immunol* **1999**, 163, 6132–8.

76 Dashiell S. M., Rus H. and Koski C.L. Terminal complement complexes concomitantly stimulate proliferation and rescue of Schwann cells from apoptosis. *Glia* **2000**, 30, 187–98.

77 Akiyama H., Barger S., Barnum S., Bradt B., Bauer J., Cole G. M., Cooper N.R., Eikelenboom P., Emmerling M., Fiebich B.L., Finch C.E., Frautschy S., Griffin W.S., Hampel H., Hull M., Landreth G., Lue L., Mrak R., Mackenzie I. R., McGeer P.L., O'Banion M.K., Pachter J., Pasinetti G., Plata-Salaman C., Rogers J., Rydel R., Shen Y., Streit W., Strohmeyer R., Tooyoma I., Van Muiswinkel F.L., Veerhuis R., Walker D., Webster S., Wegrzyniak B., Wenk G. and Wyss-Coray T. Inflammation and Alzheimer's disease. *Neurobiol Aging* **2000**, 21, 383–421.

78 Pasinetti G.M., Tocco G., Sakhi S., Musleh W.D., DeSimoni M.G., Mascarucci P., Schreiber S., Baudry M. and Finch C.E. Hereditary deficiencies in complement C5 are associated with intensified neurodegenerative responses that implicate new roles for the C-system in neuronal and astrocytic functions. *Neurobiol Dis* **1996**, 3, 197–204.

79 Mukherjee P. and Pasinetti G.M. The role of complement anaphylatoxin C5a in neurodegeneration: implications in Alzheimer's disease. *J Neuroimmunol* **2000**, 105, 124–30.

80 Osaka H., Mukherjee P., Aisen P.S. and Pasinetti G.M. Complement-derived anaphylatoxin C5a protects against glutamate-mediated neurotoxicity. *J Cell Biochem* **1999**, 73, 303–11.

81 Mukherjee P. and Pasinetti G.M. Complement anaphylatoxin C5a neuroprotects through mitogen-activated protein kinase-dependent inhibition of caspase 3. *J Neurochem* **2001**, 77, 43–9.

82 Lee A., Whyte M.K. and Haslett C. Inhibition of apoptosis and prolongation of neutrophil functional longevity by inflammatory mediators. *J Leukoc Biol* **1993**, 54, 283–8.

83 Ellis R.E. and Horvitz H.R. Two *C. elegans* genes control the programmed deaths of specific cells in the pharynx. *Development* **1991**, 112, 591–603.

84 Hedgecock E.M., Sulston J.E. and Thomson J.N. Mutations affecting programmed cell deaths in the nematode *Caenorhabditis elegans. Science* **1983**, 220, 1277–9.

85 Reddien P.W., Cameron S. and Horvitz H.R. Phagocytosis promotes programmed cell death in *C. elegans. Nature* **2001**, 412, 198–202

86 Reddien P.W. and Horvitz H.R. CED-2/CrkII and CED-10/Rac control phagocytosis and cell migration in *Caenorhabditis elegans. Nat Cell Biol* **2000**, 2:131

87 Gumienny T.L., Brugnera E., Tosello-Trampont A.C., Kinchen J.M., Haney L.B., Nishiwaki K., Walk S.F., Nemergut M.E., Macara I.G., Francis R., Schedl T., Qin Y., Van Aelst L., Hengartner M.O. and Ravichandran K.S. CED-12/ELMO, a novel member of the CrkII/Dock180/Rac pathway, is required for phagocytosis and cell migration. *Cell* **2001**, 107, 27–41.

88 Levrrier Y and Rideley A.J. Requirement for Rho GTPases and PI 3-kinases during apoptotic cell phagocytosis by macrophages. *Curr Biol* **2001**, 11, 195–9.

89 Wu Y.C. and Horvitz H.R. *C. elegans* phagocytosis and cell-migration protein CED-5 is similar to human DOCK180. *Nature* **1998**, 392, 501–4.

90 Zhou Z., Hartwieg E. and Horvitz H.R. CED-1 is a transmembrane receptor that mediates cell corpse engulfment in *C. elegans.* Cell **2001**, 104, 43–56.

91 Liu Q.A. and Hengartner M.O. Candidate adaptor protein CED-6 promotes the engulfment of apoptotic cells in *C. elegans. Cell* **1998**, 93, 961–72.

92 Liu Q.A. and Hengartner, M.O. Human CED-6 encodes a functional homologue of the Caenorhabditis elegans engulfment protein CED-6. *Curr Biol* **1999**, 9, 1347–50.

93 Wu Y.C. and Horvitz H.R. The *C. elegans* cell corpse engulfment gene *ced-7* encodes a protein similar to ABC transporters. *Cell* **1998**, 93, 951–60.

94 Su, H.P., Nakada-Tsukui, K., Tosello-Trampont, A.C., Li, Y., Bu, G., Henson P.M. and Ravichandran, K.S. Interaction of CED-6/GULP, an adapter protein involved in engulfment of apoptotic cells

with CED-1 and CD91/low density lipoprotein receptor- related protein (LRP). *J Biol Chem* **2002**, 277, 11772–9.

95 Chung S., Gumienny T. L., Hengartner M. O. and Driscoll M. A common set of engulfment genes mediates removal of both apoptotic and necrotic cell corpses in *C. elegans. Nat Cell Biol* **2000**, 2, 931–7.

96 Ogden, C. A., deCathehneau, A. et al. C1q and mannose binding lectin engagement of cell surface calreticulin and CD91 initiates macropinocytosis and uptake of apoptotic cells. *J Exp Med* **2001**, 194, 781–95.

97 Godson C., Mitchell S., Harvey K., Petasis N. A., Hogg N. and Brady H. R. Cutting edge: lipoxins rapidly stimulate nonphlogistic phagocytosis of apoptotic neutrophils by monocyte-derived macrophages. *J Immunol* **2000**, 164, 1663–7.

98 Takizawa F., Tsuji S. and Nagasawa S. Enhancement of macrophage phagocytosis upon iC3b deposition on apoptotic cells. *FEBS Lett* **1996**, 397, 269–72.

99 Gaipl U. S., Kuenkele S., Voll R. E., Beyer T. D., Kolowos W., Heyder P., Kalden J. R. and Herrmann M. Complement binding is an early feature of necrotic and a rather late event during apoptotic cell death. *Cell Death Different* **2001**, 8, 327–34.

100 Ren Y., Stuart L., Lindberg F. P., Rosenkranz A. R., Chen Y., Mayadas T. N. and Savill J. Nonphlogistic clearance of late apoptotic neutrophils by macrophages: efficient phagocytosis independent of beta (2) integrins. *J Immunol* **2001**, 166, 4743–50.

101 Ross G. D. Regulation of the adhesion versus cytotoxic functions of the Mac- 1/CR3/alphaMbeta2-integrin glycoprotein. *Crit Rev Immunol* **2000**, 20, 197–222.

102 Matsui H., Tsuji S., Nishimura H. and Nagasawa S. Activation of the alternative pathway of complement by apoptotic Jurkat cells. *FEBS Lett* **1994**, 351, 419–22.

103 Mold C. and Morris C. A. Complement activation by apoptotic endothelial cells following hypoxia/reoxygenation. *Immunology* **2001**, 102, 359–64.

104 Tsuji S., Kaji K. and Nagasawa S. Activation of the alternative pathway of human complement by apoptotic human umbilical vein endothelial cells. *J Biochem (Tokyo)* **1994**, 116, 794–800.

105 Hara T., Matsumoto M., Tsuji S., Nagasawa S., Hiraoka A., Masaoka T., Kodama K., Horai T., Sakuma T. and Seya T. Homologous complement activation on drug-induced apoptotic cells from a human lung adenocarcinoma cell line. *Immunobiology* **1996**, 196, 491–503.

106 Korb L. C. and Ahearn J. M. C1q binds directly and specifically to surface blebs of apoptotic human keratinocytes: complement deficiency and systemic lupus erythematosus revisited. *J Immunol* **1997**, 158, 4525–8.

107 Jones J. and Morgan B. P. Apoptosis is associated with reduced expression of complement regulatory molecules, adhesion molecules and other receptors on polymorphonuclear leucocytes: functional relevance and role in inflammation. *Immunology* **1995**, 86, 651–60.

108 Kawano M., Tsunoda S., Koni I., Mabuchi H., Muramoto H., Yachie A. and Seki H. Decreased expression of 20-kD homologous restriction factor (HRF20, CD59) on T lymphocytes in Epstein–Barr virus (EBV)-induced infectious mononucleosis. *Clin Exp Immunol* **1997**, 108, 260–5.

109 Tsunoda S., Kawano M., Koni I., Kasahara Y., Yachie A., Miyawaki T. and Seki H. Diminished expression of CD59 on activated $CD8^+$ T cells undergoing apoptosis in systemic lupus erythematosus and Sjögren's syndrome. *Scand J Immunol* **2000**, 51, 293–9.

110 Fadok V. A., Bratton D. L., Frasch S. C., Warner M. L. and Henson P. M. The role of phosphatidylserine in recognition of apoptotic cells by phagocytes. *Cell Death Different* **1998**, 5, 551–62.

111 Comis A. and Easterbrook-Smith S. B. Inhibition of serum complement haemolytic activity by lipid vesicles containing phosphatidylserine. *FEBS Lett* **1986**, 197, 321–7.

112 Tomasko M. A. and Chudwin D. S. Complement activation in sickle cell disease: a liposome model. *J Lab Clin Med* **1988**, 112, 248–53.

113 Liu D., Liu F. and Song Y.K. Recognition and clearance of liposomes containing phosphatidylserine are mediated by serum opsonin. *Biochim Biophys Acta* **1995**, 1235, 140–6.

114 Test S.T. and Mitsuyoshi J. Activation of the alternative pathway of complement by calcium-loaded erythrocytes resulting from loss of membrane phospholipid asymmetry. *J Lab Clin Med* **1997**, 130, 169–82.

115 Wang R.H., Phillips G., Jr, Medof M.E. and Mold C. Activation of the alternative complement pathway by exposure of phosphatidylethanolamine and phosphatidylserine on erythrocytes from sickle cell disease patients. *J Clin Invest* **1993**, 92, 1326–35.

116 Mold C. Effect of membrane phospholipids on activation of the alternative complement pathway. *J Immunol* **1989**, 143, 1663–8.

117 Kovacsovics T., Tschopp J., Kress A. and Isliker H. Antibody-independent activation of C1, the first component of complement, by cardiolipin. *J Immunol* **1985**, 135, 2695–700.

118 Kovacsovics T. J., Peitsch M. C., Kress A. and Isliker H. Antibody-independent activation of C1. I. Differences in the mechanism of C1 activation by nonimmune activators and by immune complexes: C1r-independent activation of C1s by cardiolipin vesicles. *J Immunol* **1987**, 138, 1864–70.

119 Peitsch M.C., Kovacsovics T.J., Tschopp J. and Isliker H. Antibody-independent activation of C1. II. Evidence for two classes of nonimmune activators of the classical pathway of complement. *J Immunol* **1987**, 138, 1871–6.

120 Chonn A., Cullis P.R. and Devine D.V. The role of surface charge in the activation of the classical and alternative pathways of complement by liposomes. *J Immunol* **1991**, 146, 4234–41.

121 Balasubramanian K. and Schroit A.J. Characterization of phosphatidylserine-dependent beta2-glycoprotein I macrophage interactions. Implications for apoptotic cell clearance by phagocytes. *J Biol Chem* **1998**, 273, 29272–7.

122 Chonn A., Semple S.C. and Cullis P.R. Beta 2 glycoprotein I is a major protein associated with very rapidly cleared liposomes *in vivo*, suggesting a significant role in the immune clearance of 'non-self' particles. *J Biol Chem* **1995**, 270, 25845–9.

123 Price B.E., Rauch J., Shia M.A., Walsh M.T., Lieberthal W., Gilligan H.M., O'Laughlin T., Koh J.S. and Levine J.S. Anti-phospholipid autoantibodies bind to apoptotic, but not viable, thymocytes in a beta 2-glycoprotein I-dependent manner. *J Immunol* **1996**, 157, 2201–8.

124 Gershov D., Kim S., Brot N. and Elkon K.B. C-Reactive protein binds to apoptotic cells, protects the cells from assembly of the terminal complement components, and sustains an antiinflammatory innate immune response: implications for systemic autoimmunity. *J Exp Med* **2000**, 192, 1353–64.

125 Pangburn M.K. Host recognition and target differentiation by factor H, a regulator of the alternative pathway of complement. *Immunopharmacology* **2000**, 49, 149–57.

126 French L.E., Wohlwend A., Sappino A.P., Tschopp J. and Schifferli J.A. Human clusterin gene expression is confined to surviving cells during *in vitro* programmed cell death. *J Clin Invest* **1994**, 93, 877–84.

127 Berman S., Gewurz H. and Mold C. Binding of C-reactive protein to nucleated cells leads to complement activation without cytolysis. *J Immunol* **1986**, 136, 1354–9.

128 Mold C., Gewurz H. and Du Clos T.W. Regulation of complement activation by C-reactive protein. *Immunopharmacology* **1999**, 42, 23–30.

129 Aderem A.A., Wright S.D., Silverstein S.C. and Cohn Z.A. Ligated complement receptors do not activate the arachidonic acid cascade in resident peritoneal macrophages. *J Exp Med* **1985**, 161, 617–22.

130 Wright S.D. and Silverstein S.C. Receptors for C3b and C3bi promote phagocytosis but not the release of toxic oxygen

from human phagocytes. *J Exp Med* **1983**, 158, 2016–23.

131 Yamamoto K. and Johnston R.B., Jr. Dissociation of phagocytosis from stimulation of the oxidative metabolic burst in macrophages. *J Exp Med* **1984**, 159, 405–16.

132 Marth T. and Kelsall B.L. Regulation of interleukin-12 by complement receptor 3 signaling. *J Exp Med* **1997**, 185, 1987–95.

133 Voll R.E., Herrmann M., Roth E.A., Stach C., Kalden J.R. and Girkontaite I. Immunosuppressive effects of apoptotic cells. *Nature* **1997**, 390, 350–1.

134 Herrmann M., Voll R.E., Zoller O.M., Hagenhofer M., Ponner B.B. and Kalden J.R. Impaired phagocytosis of apoptotic cell material by monocyte-derived macrophages from patients with systemic lupus erythematosus. *Arthritis Rheum* **1998**, 41, 1241–50.

135 Herrmann M., Voll R.E. and Kalden J.R. Etiopathogenesis of systemic lupus erythematosus. *Immunol Today* **2000**, 21, 424–6.

136 Shoshan Y., Shapira I., Toubi E., Frolkis I., Yaron M. and Mevorach D. Accelerated Fas-mediated apoptosis of monocytes and maturing macrophages from patients with systemic lupus erythematosus: relevance to *in vitro* impairment of interaction with iC3b-opsonized apoptotic cells. *J Immunol* **2001**, 167, 5963–9.

137 Mohan C., Adams S., Stanik V. and Datta S.K. Nucleosome: a major immunogen for pathogenic autoantibody-inducing T cells of lupus. *J Exp Med* **1993**, 177, 1367–81.

138 Casciola-Rosen L.A., Anhalt G. and Rosen A. Autoantigens targeted in systemic lupus erythematosus are clustered in two populations of surface structures on apoptotic keratinocytes. *J Exp Med* **1994**, 179, 1317–30.

139 Verhoven B., Schlegel R.A. and Williamson P. Mechanisms of phosphatidylserine exposure, a phagocyte recognition signal, on apoptotic T lymphocytes. *J Exp Med* **1995**, 182, 1597–601.

140 Rumore P.M. and Steinman C.R. Endogenous circulating DNA in systemic lupus erythematosus. Occurrence as multimeric complexes bound to histone. *J Clin Invest* **1990**, 86, 69–74.

141 Napirei M., Karsunky H., Zevnik B., Stephan H., Mannherz H.G. and Moroy T. Features of systemic lupus erythematosus in DNase1-deficient mice. *Nat Genet* **2000**, 25, 177–81.

142 Emlen W., Niebur J. and Kadera R. Accelerated *in vitro* apoptosis of lymphocytes from patients with systemic lupus erythematosus. *J Immunol* **1994**, 152, 3685–92.

143 Perniok A., Wedekind F., Herrmann M., Specker C. and Schneider M. High levels of circulating early apoptic peripheral blood mononuclear cells in systemic lupus erythematosus. *Lupus* **1998**, 7, 113–8.

144 Botto M., Dell'Agnola C., Bygrave A.E., Thompson E.M., Cook H.T., Petry F., Loos M., Pandolfi P.P. and Walport M.J. Homozygous C1q deficiency causes glomerulonephritis associated with multiple apoptotic bodies. *Nat Genet* **1998**, 19, 56–9.

145 Morgan B.P. and Walport M.J. Complement deficiency and disease. *Immunol Today* **1991**, 12, 301–6.

146 Pickering M.C., Fischer S., Lewis M.R., Walport M.J., Botto M. and Cook H.T. Ultraviolet-radiation-induced keratinocyte apoptosis in C1q-deficient mice. *J Invest Dermatol* **2001**, 117, 52–60.

147 Mitchell D.A., Taylor P.R., Cook H.T., Moss J., Bygrave A.E., Walport M.J. and Botto M. Cutting edge: C1q protects against the development of glomerulonephritis independently of C3 activation. *J Immunol* **1999**, 162, 5676–9.

148 Navratil J.S., Watkins S.C., Wisnieski J.J. and Ahearn J.M. The globular heads of C1q specifically recognize surface blebs of apoptotic vascular endothelial cells. *J Immunol* **2001**, 166, 3231–9.

149 Walport M.J., Davies K.A. and Botto M. C1q and systemic lupus erythematosus. *Immunobiology* **1998**, 199, 265–85.

150 Gommerman J.L. and Carroll M.C. Negative selection of B lymphocytes: a novel role for innate immunity. *Immunol Rev* **2000**, 173, 120–30.

151 Chen Z., Koralov S. B. and Kelsoe G. Complement C4 inhibits systemic autoimmunity through a mechanism independent of complement receptors CR1 and CR2. *J Exp Med* **2000**, 192, 1339–52.

152 Baumann I., Kolowos W., Voll R. E., Manger B., Gaipl U., Neuhuber W. L., Kirchner T., Kalden J. R., Herrmann M. Impaired uptake of apoptotic cells into tingible body macrophages in germinal centers of patients with systemic lupus erythematosus. *Arthritis Rheum* **2002**, 46, 191–201.

153 Sauter B., Albert M. L., Francisco L., Larsson M., Somersan S. and Bhardwaj N. Consequences of cell death: exposure to necrotic tumor cells, but not primary tissue cells or apoptotic cells, induces the maturation of immunostimulatory dendritic cells. *J Exp Med* **2000**, 191, 423–34.

154 Botto M., Taylor P. R., Lachmann P. J. and Walport M. J. Systemic lupus erythematosus, complement deficiency, and apoptosis. *Adv Immunol* **2000**, 76, 227–324.

155 Thompson C. B. Apoptosis in the pathogenesis and treatment of disease. *Science* **1995**, 267, 1456–62.

156 Tosello-Trampont, A. C., Brugnera, E., Ravichandran, K. S. Evidence for a conserved role for CRKII and Rac in engulfment of apoptotic cells. *J Biol Chem* **2001**, 176, 13797–802.

157 Franc N. C., Heitzler P., Ezekowitz R. A. and White K. Requirement for croquemort in phagocytosis of apoptotic cells in *Drosophila*. *Science* **1999**, 284, 1991–4.

5
Soluble Factors that Bind to Dying Cells Control the Outcome of Corpse Disposal: The Role of Pentraxins, Collectins and Autoantibodies

Patrizia Rovere-Querini

5.1
Introduction

Over the last few years, receptors for apoptotic cells have been identified on the surface of the phagocyte. They include integrins, scavenger receptors, lectins, the phosphatidylserine (PS) receptor and the lipopolysaccharide receptor CD14. Dedicated studies allowed the dissection of signals on dying cells that activate the recognition machinery required for engulfment (the so-called 'eat me' signals). Even the simultaneous blockade of multiple interactions does not abrogate phagocytosis [1]. This suggests that phagocytes use more than one receptor-based recognition mechanism. The cooperation of independent, low-affinity receptors guarantees rapid uptake and reduces escape possibilities [2–4]. Recent evidence also implicates soluble molecules as crucial non-redundant players in the clearance of dying cells. The more cogent demonstrations come from *in vivo* studies in genetically modified animals.

5.2
Soluble Factors Involved in Apoptotic Cell Recognition and Internalization

5.2.1
Corpse Clearance at Rest: Collectins

Mice bearing a genetic deletion of the first component of the classical pathway of complement activation, C1q, highlight the link between the clearance of dying cells and the immune homeostasis. C1q binds to membranes of keratinocytes undergoing apoptosis *in vitro* [5] and *in vivo* [6]. Apoptotic glomerular cells accumulate in the kidneys of C1q deficient animals [7]. A consistent number of animals (25%) also develop glomerulonephritis. Inflammatory peritoneal macrophages from C1q and C4 deficient mice fail to clear *in vivo* syngeneic apoptotic thymocytes [8], and the addition of the purified complement fraction corrects the defect. Resident peritoneal macrophages of C1q-deficient, but not of C4-deficient, mice also dispose apoptotic thymocytes with a reduced efficiency [8].

Endothelial cells undergoing apoptosis as a consequence of hypoxia/reoxygenation bind to C1q *in vitro* and activate the complement cascade [9]. This does not necessarily mean assembly of the C5–C9 terminal complement components [membrane attack complex (MAC)] involved in cell lysis and direct tissue damage. Gershov *et al.* recently proposed that C-reactive protein (CRP) limits MAC assembly on the surface of dying cells, protecting tissues in which extensive cell death is taking place [10]. Interestingly caspase inhibitors, which prevent apoptotic but not necrotic cell death of hypoxic/reoxygenated endothelial cells, *in vivo* abolish complement activation [9]. Therefore the sequence of events ensuing the delivery of an apoptotic stimulus into living tissues comprises: (1) caspase activation, (2) 'eat me' signals exposure, including C1q binding, and (3) phagocytic clearance of dying cells.

The dying cell membrane, after caspase activation, acquires discrete domains in which C1q and other related molecules, like the mannose-binding lectin (MBL) co-cluster. Both C1q and MBL use the globular heads to bind to apoptotic cell membranes, while interact via the collagenous tail with phagocyte receptors. Ogden *et al.* suggest that this interaction *per se* is endowed with a relatively low affinity. Clustering would be required to ensure downstream signaling via the phagocyte receptors, necessary to ensure internalization. Furthermore, phagocytosing macrophages secrete MBL and C1q *in vitro*, while these factors possibly derive from plasma *in vivo*. Convincing evidence implicates calreticulin as the moiety bridging C1q on the apoptotic cell membrane to CD91 receptors on the phagocyte membrane [11]. On the other hand, both anti-CD91 and anticalreticulin antibodies quantitatively inhibited the internalization of particulates.

Not all tissues depend on C1q for proper removal of dying cells. C1q binds to keratinocytes undergoing apoptosis after ultraviolet exposure of wild-type animals. However, C1q-deficient mice dispose of apoptotic keratinocytes efficiently and do not develop autoantibodies [6]. Furthermore, phagocytes from peripheral tissues differ, e.g. alveolar macrophages phagocytose with similar efficiency in the presence or the absence of the complement fraction [12]. However, macrophages phagocytosing apoptotic cells can produce and secrete C1q *in vitro* [11], making the relevance of the latter information difficult to evaluate.

C1q is not a lectin. However, it is highly homologous to members of a family of proteins named collectins – pattern recognition proteins that bind to microbial surface carbohydrates, interfering with infectivity and targeting pathogens for rapid clearance by immune cells. In addition to MBL, this protein family includes the surfactant proteins A (SP-A) and D (SP-D), 'defense collagens' produced in the lung, a district continuously exposed to pathogens. C1q, MBL and SP-A all interact with calreticulin via the S domain (residues 160–283) [13]. Recently, Schagat *et al.* reported that SP-A and SP-D, which bind to both bacteria and lipid vesicles enhancing their phagocytosis by alveolar macrophages, also recognize cells undergoing apoptosis. However, they rely for recognition on a domain, which is unrelated to their lectin functions. Upon recognition, SP-A and SP-D potentiate the *in vitro* disposal of apoptotic neutrophils by alveolar macrophages [12]. Taken, together these data indicate that collectins, besides their well-characterized role in innate immunity [14], are a key player in the disposal of dying cells.

5.2.2
Corpse Clearance at Rest: Cationic Factors and Other PS-binding Moieties

The loss of PS asymmetry and its exposure on the outer cell membrane characterize cells undergoing early apoptosis. It is instrumental for the rapid internalization and clearance of apoptotic cells. Specific receptors [15], class B scavenger receptors [16] and CD14 [17] can directly bind to PS (although they are not specific for the phospholipid). Charge modification, and the overall subversion of the plasma membrane architecture recruit bridging molecules like thrombospondin (TSP) and β_2-glycoprotein I (β_2-GPI) to the site of recognition and internalization of apoptotic cells (discussed in [18]). TSP is a glycoprotein bearing multiple RGD sequences, present in a variety of tissues. TSP binds to a still unidentified site on apoptotic cell membranes and acts as a molecular bridge with the vitronectin receptor ($\alpha_V\beta_3$) and CD36 [19]. A functional involvement in the clearance of early apoptotic cells has been unequivocally demonstrated [19].

Neutrophils that undergo later phases of apoptosis bind to TSP with a substantially higher efficiency [20]. This is not surprising, since there is growing consensus on the observation that apoptotic cells membrane changes as the apoptotic program proceeds (discussed in [21, 22]). This possibly reflects well-defined molecular events, including the oxidation of exposed phospholipid residues [23]. From this point of view, the swift phagocytosis by surrounding cells eliminates apoptotic cells before they reach further complexity.

Most PS-binding molecules efficiently associate to dying cells. This is the case of Annexin V (AxV) [24], gas6 [25] and β_2-GPI [26–30]. β_2-GPI binds with similar efficiency to early or late apoptotic cells. β_2-GPI (or apolipoprotein H) is a single-chain, highly glycosylated plasma protein (average concentration in healthy subjects blood is around 200 µg/ml) of relatively unknown function (discussed in [31]). β_2-GPI binds via the cationic portion of its fifth short consensus repeat domain to PS [32]. *In vivo*, it preferentially binds to PS mainly exposed by activated or senescent platelets and damaged erythrocytes (for recent review, see [33]). Balasubramanian *et al.* showed that macrophages uptake more efficiently lipid vesicles and PS-expressing cells (apoptotic thymocytes and red blood cell ghosts) in the presence of β_2-GPI. While basal uptake of radioactive apoptotic mouse thymocytes by peritoneal thioglycollate-elicited C3H mice macrophages was higher than 40%, addition of the cofactor increased uptake to 50% [27, 28]. We failed, in a different assay, to reveal such an enhancement when human monocyte-derived macrophages and apoptotic leukemia cells were co-incubated in the presence of β_2-GPI [29]. However, we consistently observed a 5–8% increase in the phagocytosis of human activated platelets by macrophages when β_2-GPI was added (Bondanza and Rovere-Querini, unpublished). The mechanism involved in the facilitatory effect of β_2-GPI is still under debate: it may be related to the alteration of the overall charge properties of the anionic surfaces or to a still unidentified phagocyte receptor that is, however, distinct from CD36, CD68 and CD14 [28].

The availability of high concentration of a weakly cationic cofactor, like the β_2-GPI, in the circulating blood warrants that exposed anionic charges are buffered, thus preventing the activation of the coagulation cascade and the initiation of thrombosis. Other cationic factors that bind with high affinity to PS on apoptotic cell membranes have been well characterized, including AxV [24]. The shape of AxV is a concave disk: the convex side contains both the phospholipid and the Ca^{2+} binding sites [34]. AxV forms two-dimensional crystalline arrays on membranes exposing anionic phospholipids, thus yielding a 'lattice' that prevents the activation of the coagulation cascade (discussed by [31]). AxV therefore provides a soluble anticoagulant shield, ready to be assembled when anionic phospholipids are exposed. This function is prominent in the placental circulation, particularly on surfaces lining the intervillous spaces in contact with the maternal blood. Its failure correlates with placental infarctions and abortions in experimental models, and AxV expression is severely affected in pre-eclamptic placentas [35].

Gas6 is a ligand of the Axl/Mer/Tyro3 receptor tyrosine kinase family and a member of the vitamin K-dependent protein family. It binds to PS exposing surfaces via its N-terminal domain, which is enriched in γ-carboxyglutamic acid residues. It bridges PS exposing surfaces to the membranes of cells expressing its specific receptors via its exposed C-terminal globular G domains. $Gas6^{-/-}$ animals are protected against thromboembolic events and their activated platelets fail to form stable macroaggregates. These observations suggest a role of Gas6 in linking adjacent thrombocytes [36]. It also bridges PS exposing apoptotic cells to phagocytes, an event that associates to increased *in vitro* clearance [25]. The relevance of the receptors activated by Gas6 recognition for the physiological clearance of dying cells has been substantiated in elegant *in vivo* studies, relying on engineered animals bearing mutations of Gas6 receptor tyrosine kinases. These mice fail to properly clear cell corpses, which accumulate in peripheral tissues [37, 38] (discussed in [39]). Corpse accumulation is further aggravated when apoptotic cells or their constituents are administered acutely to experimental animals [39].

5.2.3
Corpse Clearance during Acute Inflammation: Pentraxins

Uningested corpses have been suggested to represent a threat to tissue homeostasis. For example, they release their content into the tissue environment, causing direct tissue damage [40] and possibly favoring the onset of autoimmunity [39, 41]. Different mechanisms contribute to limit the noxious effect of uninternalized apoptotic cells. Tissue transglutaminase, which exerts a finely tuned role in the regulation of the apoptotic machinery, is responsible for the protein crosslinks that prevent the leakage of intracellular constituents from late apoptotic cells [42, 43]. Phagocytes involved in the clearing of dying cells release immunosuppressive cytokines, including interleukin (IL)-10 and transforming growth factor (TGF)-β_1 [44–46]. Therefore, under normal conditions, apoptotic cells that escaped phagocytosis are 'sealed' and the environment is enriched in immunosuppressive cytokines.

Acute inflammation causes the synchronous death of parenchyma and infiltrating cells. Therefore, the scavenger system of inflamed tissues must cope with a sudden increase of the phagocytic load, in an environment in which the anti-inflammatory effect of phagocytic clearance is by definition overwhelmed. 'Uningested' dying cells often persist and release their intracellular constituents that in turn influence the environment: elastase and serine proteases released by apoptotic polymorphonucleophils (PMNs) that escaped phagocytosis induce the release of inflammatory factors by macrophages [47]. Only partially characterized released moieties, possibly including heat shock proteins, elicit the functional maturation of the most potent antigen presenting cells, the dendritic cells (DCs) [48]. Accordingly, the cytoplasm of apoptotic cells has an adjuvant activity that favors the generation of cytotoxic T lymphocytes in response to particulate antigens (see [49, 50] and our unpublished results). The issue of the molecular identity and functional consequences of the release of intracellular innate adjuvants is described in better detail in Chapter 14. However, it is important to consider that dying cells that have escaped phagocytosis in mammals release factors that: (1) maintain inflammation and (2) favor the initiation of immune responses in the absence of pathogens.

Persistent corpses evolve *in vivo* towards late apoptosis [51, 52]. Recent studies demonstrate that several reactants that are generated during inflammation share the ability to bind to uningested corpses. These factors possibly represent a protective feedback mechanism, aimed at limiting their noxious effect.

Pentraxins are acute-phase proteins usually characterized by cyclic pentameric structure that are conserved during the phylogenesis [53, 54]. Short pentraxins, like CRP or the serum amyloid P component (SAP), are produced in the liver in response to several cytokine, in particular IL-6. PTX3 is the prototypic long pentraxin, structurally related to, but distinct from, CRP and SAP. PTX3 is produced at extra-hepatic sites. Primary pro-inflammatory factors, like IL-1β and tumor necrosis factor (TNF)-α, induce the release of PTX3 in peripheral tissues by endothelial cells and monocytes [55–59].

Therefore, pathogens and/or cytokines present in inflamed tissues cause the immediate local production of pentraxins like PTX3. This early event is followed by the production in the liver of impressive amounts of short pentraxins. Short pentraxins possibly fulfill systemically functions similar to those exerted by tissue pentraxins [58].

The function of pentraxins includes amplification of innate resistance against microbial infections and regulation of the scavenging of DNA released from dying cells [60]. However, binding of SAP enables microbes to evade phagocytosis by neutrophils and treatments with drugs that inhibit SAP binding prolong the survival of mice injected with Gram-negative bacteria [61] . The data indicate that SAP-coated organisms evade recognition *in vivo* and are more pathogenic, suggesting an anti-opsonic role for SAP. SAP binds to Gram-negative bacteria or their constituents, inhibits the deposition of C1q and strongly diminishes the phagocytosis of bacteria *in vitro* [62], while SAP$^{-/-}$ mice display an unusual resistance to the *in vivo* injection of high doses of bacterial endotoxin [63].

Both short pentraxins (SAP and CRP) and PTX3 bind to cells undergoing apoptosis [10, 64–66]. The ligand for pentraxins on the apoptotic cell has not yet been

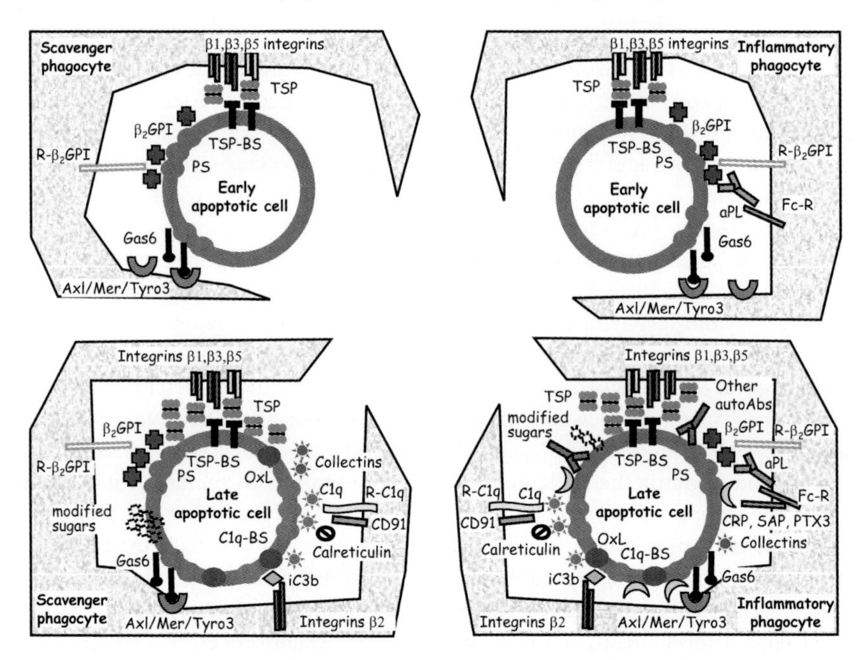

Fig. 5.1 The dying cell interacts with soluble factors, acquiring further complexity during late apoptosis and ongoing inflammation. Top panels represent soluble factors interacting with early apoptotic cells, which have been involved in the clearance by scavenger phagocytes (left) or by inflammatory phagocytes (right). A whole array of soluble factors associates with cells undergoing later phases of the apoptotic program (bottom), recruiting novel receptor–ligand interactions when the clearance is performed by scavenger phagocytes (left) or during inflammatory and/or autoimmune conditions (right). TSP, thrombospondin; BS, binding site; β_2GPI, β_2-glycoprotein I; R, receptor; aPL, anti-phospholipid antibodies; OxL, oxidized lipids; CRP, C-reactive protein; SAP, serum amyloid P component.

identified. It localizes to the membrane, and well-characterized ligands like chromatin, small nuclear ribonucleoproteins or C1q have been excluded. In different *in vitro* systems (Jurkat leukemia cells for CRP, SAP and PTX3, and activated human lymphocytes and PMNs for PTX3), pentraxins preferentially recognize late apoptotic cells [10, 65, 66] (Fig. 5.1). Since PS exposure is a hallmark of early apoptosis, direct recognition of PS is not a likely candidate. To the best of my knowledge, among the factors that bind to apoptotic cells, pentraxins and C1q only bind to late apoptotic cells [67]. Soluble factors recognizing late apoptotic cells possibly offer a second chance for safe clearance *in vivo*. For example, in patients with systemic lupus erythematosus (SLE) the defective clearance of apoptotic cells unveils autoimmune features [68, 69]. Breathnach *et al.* showed that globular deposits of nuclear material that persist in the dermis of SLE patients contain SAP [70].

Notably, the accumulation of uncleared cells is not an exclusive feature of SLE patients. Extensive tissue apoptosis, evolving *in vivo* towards post-apoptotic necrosis, occurs after ischemia of the central nervous system and contributes to tissue damage (e.g. [71]). Nucleosomes, which are generated exclusively during apoptosis, become detectable only if early apoptotic cells are not efficiently removed in living parenchimas. They consistently accumulate in patients' blood after antineoplastic chemotherapy [72]. A role for persisting nucleosomes and nucleosome-associated moieties in the maintenance of autoimmunity has been convincingly proposed [73–75].

Recent studies have offered clues on the molecular events underlying the protective effect pentraxins exert towards autoimmunity induced by dying cells or their constituents [60]. PTX3 binds to apoptotic cells also in the absence of plasma cofactors, a situation that mimics apoptotic cell clearance in the context of solid parenchymas. PTX3 bound to apoptotic cells plays an anti-opsonic role [65] similar to that described above for SAP and pathogens [61]. In particular, it prevents internalization of PTX3-coated apoptotic cells by antigen-presenting DC, while leaving apparently unaffected the binding. These data well fit with the recent demonstration of a functional hierarchy among receptors involved in the recognition and in the actual internalization of apoptotic cells [22, 76].

PTX3 is produced at the site of acute cell death *in vivo* [77] and is the first characterized molecule that behaves as a 'do not eat me' signal for apoptotic cells [22]. This is possibly relevant for the pathogenesis of chronic inflammatory diseases characterized by the selective involvement of small vessels, like anti-neutrophil cytoplasmic autoantibodies (ANCA)-associated vasculitis. These diseases are characterized by leukocytoclasia, i.e. by the persistence of uncleared cell debris in the perivascular parenchymas of involved sites [78]. Despite the failed clearance of cell corpses that leukocytoclasia reflects, vasculitis patients seldom develop clear-cut features of systemic autoimmunity. In particular they do not develop antibodies targeting antigens modified, redistributed or cleaved during apoptotic cell death. We recently found that PTX3, which fails to increase during flares of SLE, is an independent indicator of the activity of small vessel vasculitis [59]. Further studies are necessary to verify whether PTX3 produced by activated endothelial cells or infiltrating mononuclear cells in vasculitis lesions [59] plays a role in the protection against the spreading of the autoimmune response.

Hindering the access to DC is not the unique mechanism by which pentraxins bound to apoptotic cells limit the initiation of autoimmune diseases. Indeed, CRP bound to apoptotic cells in the presence of serum has been reported to increase the phagocytosis *in vitro* of apoptotic Jurkat cells by monocyte-derived macrophages and to promote the synthesis of the anti-inflammatory cytokine TGF-β1 [10]. Therefore, pentraxins play diverse overlapping functions, including increase clearance by scavenger macrophages and interference with internalization by professional antigen-presenting cells. The role of Fc receptors, which have been recently proposed to mediate the leukocyte recognition of short pentraxins, remains to be established [79, 80]. The intracellular signaling events that follow pentraxins recognition are still poorly characterized. Romero *et al.* demonstrated that immunoglobulin complexes and pentraxin complexes, although interacting with similar receptors on PMNs, mediate different functional outcomes, since pentraxin complexes fail to activate PMNs [81].

5.3 Corpse Clearance in Autoimmune Patients: Autoantibodies

The administration of large numbers of apoptotic cells causes the transient development of antinuclear and antiphospholipid antibodies (aPL) [82, 83]. Two lines of evidences derive from this and following studies. (1) The phenomenon is strictly dose dependent, since it requires that dying cells overwhelm the scavenging abilities of injected animals. (2) The response is skewed towards nuclear antigens and membrane phospholipids, i.e. autoantigens frequently targeted with low affinity during acute infectious diseases. These features possibly represent the hallmark of a more extensive brake of tolerance towards the antigens dying cells contain. We recently observed that the immunization of autoimmunity resistant or susceptible mice results in the generation of a similar pattern of autoantibodies. However, the elicited autoimmune response is maintained in susceptible animals only and results in the spreading of the autoimmune response to a progressively more ample array of autoantigens. Finally, the modalities of the targeting of dying cells to antigen presenting phagocytes *in vivo* and *in vitro* appear to be a limiting factors, further shaping the epitope hierarchy and the kinetics of the autoimmune response in injected animals (Rovere-Querini *et al.*, unpublished).

At least a fraction of the antibodies elicited by apoptotic cells bind to cell corpses. This event probably fulfils a protective function, increasing the ability of Fc receptor-positive phagocytes to recognize and clear apoptotic cells [84]. For example, aPL selectively bind to apoptotic cells [26, 29, 30, 85–88]. These antibodies preferentially recognize epitopes induced by the binding of cationic soluble factors, like the β_2-GPI to the apoptotic cell membranes. Well-characterized anti-DNA autoantibodies also efficiently recognize epitopes on β_2-GPI revealed by the binding to apoptotic cells [88]. This finding bears immediate relevance for the processes that sustain autoimmunity once initiated [88]. Phospholipid epitopes *per se* are also involved. For example, cardiolipin, usually incorporated in the inner mitochondrial membrane, translocates in association with specific mitochondrial glycoproteins into membranes of apoptotic blebs [87]. This complex is specifically and selectively recognized by aPL purified from autoimmune patients. Restricted groups of epitopes are implicated, which have also been shown to cluster on apoptotic cell membranes in the absence of β_2-GPI [86].

Macrophages do not secrete pro-inflammatory factors when they phagocytose apoptotic cells. The engagement of PS receptor(s) and other scavenger receptors, like CD36, results in the preferential release of immunosuppressive cytokines (see above). Macrophages dispose of activated PS exposing platelets in a similar 'silent' manner, while immature DCs simply ignore them [89, 90].

The presence of circulating autoantibodies that bind to dying cell membranes, however, does not only enhance the efficiency of corpse clearance. It also modifies it qualitatively. Elegant experiments from the laboratory of Alan Schroit clearly demonstrate that animals with circulating aPL dispose PS-expressing particulate substrates in a biased, pro-inflammatory and pro-thrombotic, manner [91]. These results well fit with the *in vitro* demonstration that aPL behave as potent *bona fide* opsonins *in vitro* for DC and macrophages: they increase the efficiency of phagocytosis, possibly via the recruitment of Fc receptors in recognition and clearance, and trigger the release of TNF-α and IL-1β.

Opsonization of dying cells by aPL may be involved in the association between these antibodies and increased risk of autoimmunity, described since the 1950s. Moore *et al.* showed in large cohorts of subjects followed for decades that false positive tests for syphilis are associated with the increased risk to develop SLE, rheumatoid arthritis and possibly sarcoidosis [92]. This belief has been supported by the results of more recent studies (discussed in [30]).

Other autoantibodies may as well skew the clearance of dying cells. Fetal cardiac myocytes undergo apoptosis during heart remodeling. Their selective recognition by maternal anti-Ro/SSA and anti-La/SSB antibodies contributes to the pathogenesis of the congenital heart block [93, 94]. Ro and La become accessible to antibodies in myocytes undergoing late apoptosis, a result that agrees with those reported in apoptotic keratinocytes [95]. Maternal antibodies also transiently determine photosensitivity in children with neonatal lupus, a feature that associates with the failed clearance of apoptotic keratinocytes. These observations suggest that autoantibodies bind to apoptotic cells and skew their clearance towards a proinflammatory outcome. Accordingly, apoptotic fetal cardiocytes elicit the release of TNF-α by phagocytosing macrophages only in the presence of opsonizing anti-Ro and anti-La antibodies purified by affinity chromatography from autoimmune patient sera [94]. The inflammatory damage of the developing heart parenchyma would ultimately cause permanent scarring of the conducting system, leading to congenital heart block. The outcome is different in tissues with higher regenerative capacity, like the skin. Accordingly, photosensitivity disappears with the disappearance of maternal autoantibodies [96].

While immunization with apoptotic cells almost always causes the *in vivo* development of autoantibodies, diverse autoantibodies are likely to be generated depending on the pathway of immunization and on the characteristics of the immunizing cells. Immunization with apoptotic lymphoma cells elicits the production of tumor-specific antibodies [97]. Immunization with non-transformed syngeneic cells elicits '*bona fide*' autoantibodies, which recognize endogenous molecules selectively expressed by the dying cell. This is the case of vaccination with PMNs. Patry *et al.* showed that rats immunized with apoptotic PMNs develop ANCA [98]. ANCA, which recognize proteins contained in primary azurophil granules of PMNs and lysosomes of monocytes, are closely associated and possibly involved in the pathogenesis of systemic small vessels vasculitis, like Wegener's granulomatosis, Churg Strauss syndrome and microscopic polyangiitis [99–103]. Of interest, immunized rats apparently fail to develop antinuclear antibodies (ANA). This observation well fits with clinical data. For example, only a minority of ANCA$^+$ vasculitis patients have detectable ANA, while patients with SLE and related disorders, which develop both ANA and aPL quite frequently, only occasionally have detectable ANCA [104]. The molecular constraints explaining the discrepancy are poorly understood. We recently observed that the adjuvant activities contained in necrotic cells [105, 106] are apparently missing in the nuclei of short living PMNs

(Rovere-Querini *et al.*, submitted). The lack of proper adjuvants at the site of PMN cell death may impair antigen presentation by DC [48], biasing the response towards PMN antigens expressed at the cell surface upon activation and eventually released, like proteinase 3 (PR3) and myeloperoxidase (MPO).

The precise role anti-PR3 and anti-MPO antibodies play in the clearance of dying PMNs is still under debate. Like activated PMNs, apoptotic PMNs also express MPO and PR3 at the cell surface [107–110]. Moosig *et al.* showed that human monocyte-derived macrophages phagocytose with higher efficiency and release pro-inflammatory factors including TNF-α when challenged with apoptotic PMNs in the presence of anti-PR3 antibodies [111]. TNF-α levels are elevated in vasculitis patients and the cytokine has been involved in the 'priming' of PMNs (an event which is associated with the translocation of PR3 and MPO to the plasma membrane). Therefore, Moosig *et al.* suggested that opsonization by ANCA may play a perpetuating role in the pathogenesis of ANCA-associated vasculitis. In support, Harper *et al.* found that ANCA may exert independent actions: they accelerate apoptosis of primed PMN, via generation of reactive oxygen species, and prompt post-apoptotic necrosis [109]. This accelerated apoptosis actually reduces the 'safe window' for clearance. However, upon binding to already apoptotic PMN, ANCA favor *in vitro* clearance and pro-inflammatory cytokine generation. Opsonization also associates with the generation of the PMN chemotactic factor IL-8 [110].

5.4 Conclusions

Studies *in vitro*, in which well-characterized phagocytes are challenged with populations of dying cells at a defined phase of the apoptotic program, were instrumental to our understanding of apoptotic cell phagocytosis as an active process. Phagocytosis is dependent on receptor–ligand pairs and is followed by defined signaling events. However, this reductionistic approach may have underscored the role of soluble factors in the clearance of dying cells in living tissues [112]. Evidence in the last few years demonstrates that molecules constitutively present or generated during acute and chronic inflammation play both opsonic and anti-opsonic roles. They tightly control the functional outcome of dying cell clearance. Many other involved factors are probably still unknown. Studies in the future years will eventually unravel their complex role in maintaining tolerance towards antigens contained in dying cells and indicate whether they may be valuable for molecular therapies aimed at modulating the outcome of apoptotic cell clearance in humans.

5.5 Acknowledgements

The Author is supported by the Ministero della Sanitá and by the AIRC.

5.6 References

1 Giles, K.M., S.P. Hart, C. Haslett, A.G. Rossi and I. Dransfield. An appetite for apoptotic cells? Controversies and challenges. *Br J Haematol* **2000**, 109, 1–12.

2 Platt, N., H. Suzuki, T. Kodama and S. Gordon. Apoptotic thymocyte clearance in scavenger receptor class A-deficient mice is apparently normal. *J Immunol* **2000**, 164, 4861–7.

3 Chimini, G. Engulfing by lipids: a matter of taste? *Cell Death Different* **2001**, 8, 548.

4 Hengartner, M.O. Apoptosis: corralling the corpses. *Cell* **2001**, 104, 325–8.

5 Korb, L.C. and J. M. Ahearn. C1q binds directly and specifically to surface blebs of apoptotic human keratinocytes: complement deficiency and systemic lupus erythematosus revisited. *J Immunol* **1997**, 158, 4525–8.

6 Pickering, M.C., S. Fischer, M.R. Lewis, M.J. Walport, M. Botto and H.T. Cook. Ultraviolet-radiation-induced keratinocyte apoptosis in C1q-deficient mice. *J Invest Dermatol* **2001**, 117, 52–8.

7 Botto, M., C. Dell'Agnola, A.E. Bygrave, E.M. Thompson, H.T. Cook, F. Petry, M. Loos, P.P. Pandolfi and M.J. Walport. Homozygous C1q deficiency causes glomerulonephritis associated with multiple apoptotic bodies. *Nat Genet* **1998**, 19, 56–9.

8 Taylor, P.R., A. Carugati, V. A. Fadok, H.T. Cook, M. Andrews, M.C. Carroll, J.S. Savill, P.M. Henson, M. Botto and M.J. Walport. A hierarchical role for classical pathway complement proteins in the clearance of apoptotic cells *in vivo*. *J Exp Med* **2000**, 192, 359–66.

9 Mold, C. and C.A. Morris. Complement activation by apoptotic endothelial cells following hypoxia/reoxygenation. *Immunology* **2001**, 102, 359–64.

10 Gershov, D., S. Kim, N. Brot and K.B. Elkon. C-Reactive protein binds to apoptotic cells, protects the cells from assembly of the terminal complement components and sustains an antiinflammatory innate immune response: implications for systemic autoimmunity. *J Exp Med* **2000**, 192, 1353–64.

11 Ogden. C.A., A. deCathelineau, P.R. Hoffmann, D. Bratton, B. Ghebrehiwet, V.A. Fadok and P.M. Henson. C1q and mannose binding lectin engagement of cell surface calreticulin and CD91 initiates macropinocytosis and uptake of apoptotic cells. *J Exp Med* **2001**, 194, 781–95.

12 Schagat, T.L., J.A. Wofford and J.R. Wright. Surfactant protein A enhances alveolar macrophage phagocytosis of apoptotic neutrophils. *J Immunol* **2001**, 166, 2727–33.

13 Stuart, G.R., N. J. Lynch, A.J. Day, W.J. Schwaeble and R.B. Sim. The C1q and collectin binding site within C1q receptor (cell surface calreticulin). *Immunopharmacology* **1997**, 38, 73–80.

14 Holmskov, U.L. Collectins and collectin receptors in innate immunity. *APMIS Suppl* **2000**, 100, 1–59.

15 Fadok, V.A., D.L. Bratton, D.M. Rose, A. Pearson, R.A. Ezekewitz and P.M. Henson. A receptor for phosphatidylserine-specific clearance of apoptotic cells. *Nature* **2000**, 405, 85–90.

16 Savill, J., N. Hogg, Y. Ren and C. Haslett. Thrombospondin cooperates with CD36 and the vitronectin receptor in macrophage recognition of neutrophils undergoing apoptosis. *J Clin Invest* **1992**, 90, 1513–22.

17 Devitt, A., O.D. Moffatt, C. Raykundalia, J.D. Capra, D.L. Simmons and C.D. Gregory. Human CD14 mediates recognition and phagocytosis of apoptotic cells. *Nature* **1998**, 392, 505–9.

18 Savill, J. and V. Fadok. Corpse clearance defines the meaning of cell death. *Nature* **2000**, 407, 784–8.

19 Savill, J., I. Dransfield, N. Hogg and C. Haslett. Vitronectin receptor-mediated phagocytosis of cells undergoing apoptosis. *Nature* **1990**, 343, 170–3.

20 Hart, S.P., J.A. Ross, K. Ross, C. Haslett and I. Dransfield. Molecular characterization of the surface of apoptotic neutrophils: implications for functional downregulation and recognition by phagocytes. *Cell Death Different* **2000**, 7, 493–503.

21 Fadok, V.A., D.L. Bratton and P.M. Henson. Phagocyte receptors for apoptotic cells: recognition, uptake and consequences. *J Clin Invest* **2001**, 108, 957–62.

22 Henson, P.M., D.L. Bratton and V.A. Fadok. Apoptotic cell removal. *Curr Biol* **2001**, 11, R795–805.

23 Chang, M.K., C. Bergmark, A. Laurila, S. Horkko, K. H. Han, P. Friedman, E.A. Dennis and J.L. Witztum. Monoclonal antibodies against oxidized low-density lipoprotein bind to apoptotic cells and inhibit their phagocytosis by elicited macrophages: evidence that oxidation-specific epitopes mediate macrophage recognition. *Proc Natl Acad Sci USA* **1999**, 96, 6353–8.

24 Martin, S.J., C.P. Reutelingsperger, A.J. McGahon, J.A. Rader, R.C. van Schie, D.M. LaFace and D.R. Green. Early redistribution of plasma membrane phosphatidylserine is a general feature of apoptosis regardless of the initiating stimulus: inhibition by overexpression of Bcl-2 and Abl. *J Exp Med* **1995**, 182, 1545–56.

25 Ishimoto, Y., K. Ohashi, K. Mizuno and T. Nakano. Promotion of the uptake of PS liposomes and apoptotic cells by a product of growth arrest-specific gene, gas6. *J Biochem (Tokyo)* **2000**, 127, 411–7.

26 Price, B.E., J. Rauch, M.A. Shia, M.T. Walsh, W. Lieberthal, H.M. Gilligan, T. O'Laughlin, J.S. Koh and J.S. Levine. Anti-phospholipid autoantibodies bind to apoptotic, but not viable, thymocytes in a beta 2-glycoprotein I-dependent manner. *J Immunol* **1996**, 157, 2201–8.

27 Balasubramanian, K., J. Chandra and A.J. Schroit. Immune clearance of phosphatidylserine-expressing cells by phagocytes. The role of beta2-glycoprotein I in macrophage recognition. *J Biol Chem* **1997**, 272, 31113–7.

28 Balasubramanian, K. and A. J. Schroit. Characterization of phosphatidylserine-dependent beta2-glycoprotein I macrophage interactions. Implications for apoptotic cell clearance by phagocytes. *J Biol Chem* **1998**, 273, 29272–7.

29 Manfredi, A.A., P. Rovere, S. Heltai, G. Galati, G. Nebbia, A. Tincani, G. Balestrieri and M.G. Sabbadini. Apoptotic cell clearance in Systemic Lupus Erythematosus: II. Role for the beta2-glycoprotein I. *Arthritis Rheum* **1998**, 41, 215–23.

30 Manfredi, A.A., P. Rovere, G. Galati, S. Heltai, E. Bozzolo, L. Soldini, J. Davoust, G. Balestrieri, A. Tincani and M.G. Sabbadini. Apoptotic cell clearance in systemic lupus erythematosus. I. Opsonization by antiphospholipid antibodies. *Arthritis Rheum* **1998**, 41, 205–14.

31 Rand, J.H. Molecular pathogenesis of the antiphospholipid syndrome. *Circ Res* **2002**, 90, 29–37.

32 Bouma, B., P. G. de Groot, J.M. van den Elsen, R.B. Ravelli, A. Schouten, M.J. Simmelink, R.H. Derksen, J. Kroon and P. Gros. Adhesion mechanism of human beta(2)-glycoprotein I to phospholipids based on its crystal structure. *EMBO J* **1999**, 18, 5166–74.

33 Bondanza, A. and P. Rovere-Querini. Haematological autoimmunity in Systemic Lupus Erythematosus: anti-phospholipid antibodies at the cross-road. In: Pandalai, S. G., ed. *Recent Research Developments in Immunology.* Trivandrum: Transworld Research Network **2001**, 3, 187–201.

34 Voges, D., R. Berendes, A. Burger, P. Demange, W. Baumeister and R. Huber. Three-dimensional structure of membrane-bound annexin V. A correlative electron microscopy-X-ray crystallography study. *J Mol Biol* **1994**, 238, 199–213.

35 Shu, F., M. Sugimura, N. Kanayama, H. Kobayashi, T. Kobayashi and T. Terao. Immunohistochemical study of annexin V expression in placentae of preeclampsia. *Gynecol Obstet Invest* **2000**, 49, 17–23.

36 Angelillo-Scherrer, A., P. de Frutos, C. Aparicio, E. Melis, P. Savi, F. Lupu, J. Arnout, M. Dewerchin, M. Hoylaerts, J. Herbert, D. Collen, B. Dahlback and P. Carmeliet. Deficiency or inhibition of Gas6 causes platelet dysfunction and protects mice against thrombosis. *Nat Med* **2001**, 7, 215–21.

37 Lu, Q. and G. Lemke. Homeostatic regulation of the immune system by receptor tyrosine kinases of the Tyro 3 family. *Science* **2001**, 293, 306–11.

38 Scott, R.S., E.J. McMahon, S.M. Pop, E.A. Reap, R. Caricchio, P.L. Cohen, H.S. Earp and G.K. Matsushima. Phagocytosis and clearance of apoptotic cells is mediated by MER. *Nature* **2001**, 411, 207–11.

39 Rosen, A. and L. Casciola-Rosen. Clearing the way to mechanisms of autoimmunity. *Nat Med* **2001**, 7, 664–5.

40 Kurts, C., J.F. Miller, R.M. Subramaniam, F.R. Carbone and W.R. Heath. Major histocompatibility complex class I-restricted cross-presentation is biased towards high dose antigens and those released during cellular destruction. *J Exp Med* **1998**, 188, 409–14.

41 Ren Y and Savill J. Apoptosis: the importance of being eaten. *Cell Death Different* **1998**, 5, 563–8.

42 Piredda L, A. Amendola, V. Colizzi, P.J. Davies, M.G.Farrace, M. Fraziano, V. Gentile, I. Uray, M. Piacentini and L. Fesus. Lack of tissue transglutaminase protein cross linking leads to leakage of macromolecules from dying cells: relationship to the development of autoimmunity in MRL*lpr*/*lpr* mice. *Cell Death Different* **1997**, 4, 463–72.

43 Piacentini, M. and V. Colizzi. Tissue transglutaminase: apoptosis versus autoimmunity. *Immunol Today* **1999**, 20, 130–4.

44 Voll, R.E., M. Herrmann, E.A. Roth, C. Stach, J.R. Kalden and I. Girkontaite. Immunosuppressive effects of apoptotic cells. *Nature* **1997**, 390, 350–1.

45 Fadok, V.A., D. Bratton, S.C. Frasch, H.L. Warner and P.M. Henson. The role of phosphatidylserine in the recognition of apoptotic cells by phagocytes. *Cell Death Different* **1998**, 5, 551–62.

46 Henson, P.M., D.L. Bratton and V.A. Fadok. The phosphatidylserine receptor: a crucial molecular switch? *Nat Rev Mol Cell Biol* **2001**, 2, 627–33.

47 Fadok, V.A., D.L. Bratton, L. Guthrie and P.M. Henson. Differential effects of apoptotic versus lysed cells on macrophage production of cytokines: role of proteases. *J Immunol* **2001**, 166, 6847–54.

48 Larsson, M., J.F. Fonteneaou and N. Bhardwaj. Dendritic cells resurrect antigens from dead cells. *Trends Immunol* **2001**, 22, 142–8.

49 Shi, Y., W. Zheng and K.L. Rock. Cell injury releases endogenous adjuvants that stimulate cytotoxic T cell responses. *Proc Natl Acad Sci USA* **2000**, 97, 14590–5.

50 Shi, Y. and K.L. Rock. Cell death releases endogenous adjuvants that selectively enhance immune surveillance of particulate antigens. *Eur J Immunol* **2002**, 32, 155–62.

51 Fadok, V.A., P.P. McDonald, D.L. Bratton and P.M. Henson. Regulation of macrophage cytokine production by phagocytosis of apoptotic and post-apoptotic cells. *Biochem Soc Trans* **1998**, 26, 653–6.

52 Hamon, Y., C. Broccardo, O. Chambenoit, M.F. Luciani, F. Toti, S. Chaslin, J. M. Freyssinet, P. F. Devaux, J. McNeish, D. Marguet and G. Chimini. ABC1 promotes engulfment of apoptotic cells and transbilayer redistribution of phosphatidylserine. *Nat Cell Biol* **2000**, 2, 399–406.

53 Steel, D.M. and A.S. Whitehead. The major acute phase reactants: C-reactive protein, serum amyloid P component and serum amyloid A protein. *Immunol Today* **1994**, 15, 81–8.

54 Gewurz, H., X.H. Zhang and T.F. Lint. Structure and function of the pentraxins. *Curr Opin Immunol* **1995**, 7, 54–64.

55 Lee, G.W., T.H. Lee and J. Vilcek. TSG-14, a tumor necrosis factor- and IL-1-inducible protein, is a novel member of the pentaxin family of acute phase proteins. *J Immunol* **1993**, 150, 1804–12.

56 Alles, V.V., B. Bottazzi, G. Peri, J. Golay, M. Introna and A. Mantovani. Inducible expression of PTX3, a new member of the pentraxin family, in human mononuclear phagocytes. *Blood* **1994**, 84, 3483–93.

57 Introna, M., V. V. Alles, M. Castellano, G. Picardi, L. De Gioia, B. Bottazzai, G. Peri, F. Breviario, M. Salmona, L. De Gregorio, T. A. Dragani, N. Srinivasan, T. L. Blundell, T. A. Hamil-

TON and A. MANTOVANI. Cloning of mouse ptx3, a new member of the pentraxin gene family expressed at extrahepatic sites. *Blood* **1996**, 87, 1862–72.

58 BOTTAZZI, B., V. VOURET-CRAVIARI, A. BASTONE, L. DE GIOIA, C. MATTEUCCI, G. PERI, F. SPREAFICO, M. PAUSA, C. D'ETTORRE, E. GIANAZZA, A. TAGLIABUE, M. SALMONA, F. TEDESCO, M. INTRONA and A. MANTOVANI. Multimer formation and ligand recognition by the long pentraxin PTX3. Similarities and differences with the short pentraxins C-reactive protein and serum amyloid P component. *J Biol Chem* **1997**, 272, 32817–23.

59 FAZZINI, F., G. PERI, A. DONI, G. DELL'ANTONIO, E. DAL CIN, E. BOZZOLO, F. D'AURIA, L. PRADERIO, G. CIBODDO, M.G. SABBADINI, A.A. MANFREDI, A. MANTOVANI and P. ROVERE-QUERINI. PTX3 in small-vessel vasculitides: an independent indicator of disease activity produced at sites of inflammation. *Arthritis Rheum* **2001**, 44, 2841–50.

60 BICKERSTAFF, M.C., M. BOTTO, W.L. HUTCHINSON, J. HERBERT, G.A. TENNENT, A. BYBEE, D.A. MITCHELL, H.T. COOK, P.J. BUTLER, M.J. WALPORT and M.B. PEPYS. Serum amyloid P component controls chromatin degradation and prevents antinuclear autoimmunity. *Nat Med* **1999**, 5, 694–7.

61 NOURSADEGHI, M., M.C. BICKERSTAFF, J.R. GALLIMORE, J. HERBERT, J. COHEN and M.B. PEPYS. Role of serum amyloid P component in bacterial infection: protection of the host or protection of the pathogen. *Proc Natl Acad Sci USA* **2000**, 97, 14584–9.

62 DE HAAS, C.J., E.M. VAN LEEUWEN, T.VAN BOMMEL, J. VERHOEF, K.P. VAN KESSEL and J.A. VAN STRIJP. Serum amyloid P component bound to gram-negative bacteria prevents lipopolysaccharide-mediated classical pathway complement activation. *Infect Immun* **2000**, 68, 1753–9.

63 SOMA, M., T. TAMAOKI, H. KAWANO, S. ITO, M. SAKAMOTO, Y. OKADA, Y. OZAKI, S. KANBA, Y. HAMADA, T. ISHIHARA and S. MAEDA. Mice lacking serum amyloid P component do not necessarily develop severe autoimmune disease. *Biochem Biophys Res Commun* **2001**, 286, 200–5.

64 HINTNER, H., J. BOOKER, J. ASHWORTH, J AUBOCK, M.B. PEPYS and S.M. BREATHNACH. Amyloid P component binds to keratin bodies in human skin and to isolated keratin filament aggregates *in vitro*. *J Invest Dermatol* **1988**, 91, 22–8.

65 ROVERE, P., G. PERI, F. FAZZINI, B. BOTTAZZI, A. DONI, A. BONDANZA, V.S. ZIMMERMANN, C. GARLANDA, U. FASCIO, M.G. SABBADINI, C. RUGARLI, A. MANTOVANI and A.A. MANFREDI. The long pentraxin PTX3 binds to apoptotic cells and regulates their clearance by antigen-presenting dendritic cells. *Blood* **2000**, 96, 4300–6.

66 FAMILIAN, A., B. ZWART, H.G. HUISMAN, I. RENSINK, D. ROEM, P.L. HORDIJK, L.A. AARDEN and C.E. HACK. Chromatin-independent binding of serum amyloid P component to apoptotic cells. *J Immunol* **2001**, 167, 647–54.

67 GAIPL, U.S., S. KUENKELE, R.E. VOLL, T.D. BEYER, W. KOLOWOS, P. HEYDER, J.R. KALDEN and M. HERRMANN. Complement binding is an early feature of necrotic and a rather late event during apoptotic cell death. *Cell Death Different* **2001**, 8, 327–34.

68 HERRMANN, M., R.E. VOLL, O.M. ZOLLER, M. HAGENHOFER, B.B. PONNER and J.R. KALDEN. Impaired phagocytosis of apoptotic cell material by monocyte-derived macrophages from patients with systemic lupus erythematosus. *Arthritis Rheum* **1998**, 41, 1241–50.

69 BAUMANN, I., W. KOLOWOS, R.E. VOLL, B. MANGER, U. GAIPL, W.L. NEUHUBER, T. KIRCHNER, J.R. KALDEN and M. HERRMANN. Impaired uptake of apoptotic cells into tingible body macrophages in germinal centers of patients with systemic lupus erythematosus. *Arthritis Rheum* **2002**, 46, 191–201.

70 BREATHNACH, S.M., H. KOFLER, N. SEPP, J. ASHWORTH, D. WOODROW, M.B. PEPYS and H. HINTNER. Serum amyloid P component binds to cell nuclei *in vitro* and to *in vivo* deposits of extracellular chromatin in systemic lupus erythematosus. *J Exp Med* **1989**, 170, 1433–8.

71 RABUFFETTI, M., C. SCIORATI, G. TAROZZO, E. CLEMENTI, A.A. MANFREDI and M. BELTRAMO. Inhibition of caspase-1-

like activity by Ac-Tyr-Val-Ala-Asp- chloromethyl ketone induces long-lasting neuroprotection in cerebral ischemia through apoptosis reduction and decrease of proinflammatory cytokines. *J Neurosci* **2000**, 20, 4398–404.

72 Holdenrieder, S., P. Stieber, H. Bodenmuller, M. Busch, G. Fertig, H. Furst, A. Schalhorn, N. Schmeller, M. Untch and D. Seidel. Nucleosomes in serum of patients with benign and malignant diseases. *Int J Cancer* **2001**, 95, 114–20.

73 Berden, J.H., R. Licht, M.C. van Bruggen and W.J. Tax. Role of nucleosomes for induction and glomerular binding of autoantibodies in lupus nephritis. *Curr Opin Nephrol Hypertens* **1999**, 8, 299–306.

74 Amoura, Z., S. Koutouzov and J.C. Piette. The role of nucleosomes in lupus. *Curr Opin Rheumatol* **2000**, 12, 369–73.

75 Licht, R., M.C. van Bruggen, B. Oppers-Walgreen, T.P. Rijke and J.H. Berden. Plasma levels of nucleosomes and nucleosome-autoantibody complexes in murine lupus: effects of disease progression and lipopolysaccharide administration. *Arthritis Rheum* **2001**, 44, 1320–30.

76 Hoffmann, P.R., A.M. deCathelineau, C.A. Ogden, Y. Leverrier, D.L. Bratton, D.L. Daleke, A.J. Ridley, V.A. Fadok and P.M. Henson. Phosphatidylserine (PS) induces PS receptor-mediated macropinocytosis and promotes clearance of apoptotic cells. *J Cell Biol* **2001**, 155, 649–59.

77 Ravizza, T., D. Moneta, B. Bottazzi, G. Peri, C. Garlanda, E. Hirsch, G. J. Richards, A. Mantovani and A. Vezzani. Dynamic induction of the long pentraxin PTX3 in the CNS after limbic seizures: evidence for a protective role in seizure-induced neurodegeneration. *Neuroscience* **2001**, 105, 43–53.

78 Weidner, N. Introduction. Vasculitis. *Semin Diagn Pathol* **2001**, 18, 1–2.

79 Stein, M.P., C. Mold and T. W. Du Clos. C-reactive protein binding to murine leukocytes requires Fc gamma receptors. *J Immunol* **2000**, 164, 1514–20.

80 Bharadwaj, D., C. Mold, E. Markham and T.W. Du Clos. Serum amyloid P component binds to Fc gamma receptors and opsonizes particles for phagocytosis. *J Immunol* **2001**, 166, 6735–41.

81 Romero, I.R., C. Morris, M. Rodriguez, T.W. Du Clos and C. Mold. Inflammatory potential of C-reactive protein complexes compared to immune complexes. *Clin Immunol Immunopathol* **1998**, 87, 155–62.

82 Mevorach, D., J.L. Zhou, X. Song and K.B. Elkon. Systemic exposure to irradiated apoptotic cells induces autoantibody production. *J Exp Med* **1998**, 188, 387–92.

83 Levine, J.S., R. Subang, J.S. Koh and J. Rauch. Induction of anti-phospholipid autoantibodies by beta2-glycoprotein I bound to apoptotic thymocytes. *J Autoimmun* **1998**, 11, 413–24.

84 Rovere, P., M.G. Sabbadini, A. Bondanza, V.S. Zimmermann, F. Fazzini, C. Rugarli and A.A. Manfredi. Remnants of suicidal cells fostering systemic autoaggression: apoptosis in the origin and maintenance of autoimmunity. *Arthritis Rheum* **2000**, 43, 1663–72.

85 Casciola-Rosen, L., A. Rosen, M. Petri and M. Schlissel. Surface blebs on apoptotic cells are sites of enhanced procoagulant activity: implications for coagulation events and antigenic spread in systemic lupus erythematosus. *Proc Natl Acad Sci USA* **1996**, 93, 1624–9.

86 Pittoni, V., C.T. Ravirajan, S. Donohoe, S.J. MacHin, P.M. Lydyard and D.A. Isenberg. Human monoclonal anti-phospholipid antibodies selectively bind to membrane phospholipid and beta2-glycoprotein I (beta2-GPI) on apoptotic cells. *Clin Exp Immunol* **2000**, 119, 533–43.

87 Sorice, M., A. Circella, R. Misasi, V. Pittoni, T. Garofalo, A. Cirelli, A. Pavan, G.M. Pontieri and G. Valesini. Cardiolipin on the surface of apoptotic cells as a possible trigger for antiphospholipids antibodies. *Clin Exp Immunol* **2000**, 122, 277–84.

88 Cocca, B.A., S.N. Seal, P. D'Agnillo, Y.M. Mueller, P.D. Katsikis, J. Rauch, M. Weigert and M.Z. Radic. Structural basis for autoantibody recognition of phosphatidylserine – beta 2 glycoprotein

I and apoptotic cells. *Proc Natl Acad Sci USA* **2001**, 98, 13826–31.

89 BONDANZA, A., A.A. MANFREDI, V.S. ZIMMERMANN, M. IANNACONE, A. TINCANI, G. BALESTRIERI, M.G. SABBADINI and P.R. QUERINI. Anti-beta2 glycoprotein I antibodies cause inflammation and recruit dendritic cells in platelet clearance. *Thromb Haemost* **2001**, 86, 1257–63.

90 BONDANZA, A., M.G. SABBADINI, F. PELLEGATTA, V.S. ZIMMERMANN, A. TINCANI, G. BALESTRIERI, A.A. MANFREDI and P. ROVERE. Anti-beta2 glycoprotein I antibodies prevent the De-activation of platelets and sustain their phagocytic clearance. *J Autoimmun* **2000**, 15, 469–77.

91 DOMBROSKI, D., K. BALASUBRAMANIAN and A.J. SCHROIT. Phosphatidylserine expression on cell surfaces promotes antibody-dependent aggregation and thrombosis in beta2-glycoprotein I-immune mice. *J Autoimmun* **2000**, 14, 221–9.

92 MOORE, J.E. and C.F. MOHR. Biologically false positive serologic tests for syphlis. *J Am Med Ass* **1952**, 150, 467–73.

93 MIRANDA, M.E., C.E. TSENG, W. RASHBAUM, R.L. OCHS, C.A. CASIANO, F. DI DONATO, E.K. CHAN and J.P. BUYON. Accessibility of SSA/Ro and SSB/La antigens to maternal autoantibodies in apoptotic human fetal cardiac myocytes. *J Immunol* **1998**, 161, 5061–9.

94 MIRANDA-CARUS, M.E., A.D. ASKANASE, R.M. CLANCY, F. DI DONATO, T.M. CHOU, M.R. LIBERA, E.K. CHAN and J.P. BUYON. Anti-SSA/Ro and anti-SSB/La autoantibodies bind the surface of apoptotic fetal cardiocytes and promote secretion of TNF-alpha by macrophages. *J Immunol* **2000**, 165, 5345–51.

95 CASCIOLA-ROSEN, L.A., G. ANHALT and A. ROSEN. Autoantigens targeted in systemic lupus erythematosus are clustered in two populations of surface structures on apoptotic keratinocytes. *J Exp Med* **1994**, 179, 1317–30.

96 WALTUCK, J. and J.P. BUYON. Autoantibody-associated congenital heart block: outcome in mothers and children. *Ann Intern Med* **1994**, 120, 544–51.

97 RONCHETTI, A., P. ROVERE, G. IEZZI, G. GALATI, S. HELTAI, M.P. PROTTI, M.P. GARANCINI, A.A. MANFREDI, C. RUGARLI and M. BELLONE. Immunogenicity of apoptotic cells *in vivo*: role of antigen load, antigen-presenting cells and cytokines. *J Immunol* **1999**, 163, 130–6.

98 PATRY, Y.C., D.C. TREWICK, M. GREGOIRE, M.A. AUDRAIN, A.M. MOREAU, J.Y. MULLER, K. MEFLAH and V.L. ESNAULT. Rats injected with syngenic rat apoptotic neutrophils develop antineutrophil cytoplasmic antibodies. *J Am Soc Nephrol* **2001**, 12, 1764–8.

99 GROSS, W.L., A. TRABANDT and E. REINHOLD-KELLER. Diagnosis and evaluation of vasculitis. *Rheumatology (Oxford)* **2000**, 39, 245–52.

100 HARPER, L. and C.O. SAVAGE. Pathogenesis of ANCA-associated systemic vasculitis. *J Pathol* **2000**, 190, 349–59.

101 FALK, R.J., P.H. NACHMAN, S.L. HOGAN and J.C. JENNETTE. ANCA glomerulonephritis and vasculitis: a Chapel Hill perspective. *Semin Nephrol* **2000**, 20, 233–43.

102 PRESTON, G.A. and R.J. FALK. ANCA signaling: not just a matter of respiratory burst. *Kidney Int* **2001**, 59, 1981–2.

103 VAN DER GELD, Y.M., P.C. LIMBURG and C.G. KALLENBERG. Proteinase 3, Wegener's autoantigen: from gene to antigen. *J Leukoc Biol* **2001**, 69, 177–90.

104 CABRAL, A.R. and D. ALARCON-SEGOVIA. Autoantibodies in systemic lupus erythematosus. *Curr Opin Rheumatol* **1998**, 10, 409–16.

105 SAUTER, B., M.L. ALBERT, L. FRANCISCO, M. LARSSON, S. SOMERSAN and N. BHARDWAJ. Consequences of cell death: exposure to necrotic tumor cells, but not primary tissue cells or apoptotic cells, induces the maturation of immunostimulatory dendritic cells. *J Exp Med* **2000**, 191, 423–34.

106 GALLUCCI, S. and P. MATZINGER. Danger signals: SOS to the immune system. *Curr Opin Immunol* **2001**, 13, 114–9.

107 GILLIGAN, H.M., B. BREDY, H.R. BRADY, M.J. HEBERT, H.S. SLAYTER, Y. XU, J. RAUCH, M.A. SHIA, J.S. KOH and J.S. LEVINE. Antineutrophil cytoplasmic autoantibodies interact with primary granule constituents on the surface of apoptotic neutrophils in the absence of neutrophil priming. *J Exp Med* **1996**, 184, 2231–41.

108 YANG, J.J., R.H. TUTTLE, S.L. HOGAN, J.G. TAYLOR, B.D. PHILLIPS, R.J. FALK and J.C. JENNETTE. Target antigens for anti-neutrophil cytoplasmic autoantibodies (ANCA) are on the surface of primed and apoptotic but not unstimulated neutrophils. *Clin Exp Immunol* **2000**, 121, 165–72.

109 HARPER, L., Y. REN, J. SAVILL, D. ADU and C.O. SAVAGE. Antineutrophil cytoplasmic antibodies induce reactive oxygen-dependent dysregulation of primed neutrophil apoptosis and clearance by macrophages. *Am J Pathol* **2000**, 157, 211–20.

110 HARPER, L., P. COCKWELL, D. ADU and C. O. SAVAGE. Neutrophil priming and apoptosis in anti-neutrophil cytoplasmic autoantibody-associated vasculitis. *Kidney Int* **2001**, 59, 1729–38.

111 MOOSIG, F., E. CSERNOK, G. KUMANOVICS and W.L. GROSS. Opsonization of apoptotic neutrophils by anti-neutrophil cytoplasmic antibodies (ANCA) leads to enhanced uptake by macrophages and increased release of tumour necrosis factor-alpha (TNF-alpha). *Clin Exp Immunol* **2000**, 122, 499–503.

112 MANFREDI, A.A., M. IANNACONE, F. D'AURIA and P. ROVERE-QUERINI. The disposal of dying cells in living tissues. *Apoptosis* **2002**, 7, 153–61.

6
The Role of ATP-binding Cassette Transporters in the Clearance of Apoptotic Cells: A Tale of Two Systems

Véronique Rigot and Giovanna Chimini

6.1
Introduction

During development, programmed cell death or apoptosis acts in the shaping of tissues, the refinement of neuronal connections and the removal of unnecessary or damaged cells. These cells are recognized and eliminated by phagocytes before the release of their potentially harmful contents. This process plays a crucial role for the development of immunological tolerance, the suppression of inflammatory response and in tissue remodeling. A panoply of molecules has been identified both in mammals and *Caenorhabditis elegans* as required for the clearance of dying cells. Their cooperative action defines parallel and partially redundant molecular cascades not completely clarified as yet. In both model systems a molecule belonging to the structural family of ATP-binding cassette (ABC) transporters has been identified.

We will briefly introduce here the main features of this large and evolutionary conserved family of ABC proteins before discussing in detail the case of the mammalian ABCA1 and its relative in the nematode, CED-7.

6.2
The Family of ABC Transporters

ABC transporters are one of the largest families of membrane proteins [1, 2]. They drive the transport of a wide variety of substrates across cell membranes in an ATP-dependent fashion. In terms of structure, they contain a pair of nucleotide-binding domains (NBD) and two sets of membrane-anchoring domains (TM), typically composed by six transmembrane α helices. A diagnostic combination of consensus signatures has been defined as the hallmark of the family. This is located in the NBD and associates the Walker A and B motifs [3], shared by several ATP-binding proteins, with a specific C motif located just upstream of the B site [4]. ABC genes may encode a full, intrinsically symmetrical, product or a half transporter containing a single set of TM and NBD. In the latter case, the symmetry, which is essential for function, is achieved by the posttranslational association

Tab. 6.1 ABC transporter subclasses across the evolution (only sequences encoding TM domains have been considered).

	Mammals	*D. melanogaster*	*C. elegans*	*A. Thaliana*	*S. cerevisiae*
	51 ORFs 48 proteins	**46** sequences	**63** sequences	**129** ORFs 129 proteins	**31** ORFs 29 proteins
ABCA	**12**	**12**	**6**	**17**	
■⬭■⬭	ABCA1	CG6162	CED-7	AtAOH1 1	
■⬭	ABCA4			AtATH 16	
ABCB	**11**	**2**	**25**	**24**	**4**
■⬭■⬭	MDR1	CG9330	CO5A9.1		STE6
■⬭	TAP1	CG9281		AtMDR1 22 AtTAP1 2	
ABCC	**10**	**13**	**8**	**15**	**6**
		CG5944	C18C4.2	AtMRP1	YOR1
■⬭⬡■⬭	CFTR				
■■⬭■⬭	SUR				
ABCD	**4**	**2**	**5**	**2**	**5**
■⬭		CG12703	C44B7.8		
■⬭	ALD			AtPMP1	PxA1
ABCG	**6**	**14**	**9**	**11**	**1**
⬭■			CO5D10.3		
⬭■	ABCG5	White		AtWBC1	ADP1
PDR				**13**	**9**
⬭■⬭■				AtPDR1	PDR5

■, TM; ⬭, NBD; ⬡, Rdomain, ■ additional TM.

in homo- or heterodimers. The ABC protein superfamily is extremely conserved across evolution and represented in bacteria, yeast, nematodes, plants, insects and mammals [2]. To date, 48 ABC genes have been fully characterized in the human genome (www.humanabc.org). Seven subclasses have been defined, and named from A to G, on the basis of sequence homologies and structural peculiarities (Tab. 6.1).

In the yeast *Saccharomyces cerevisiae,* the 31 ABC genes can be grouped into six clusters or subfamilies with distinguished topologies and broadly corresponding to the mammalian subclasses B–G [5, 6]. In yeast, transporters bearing the structural features of the mammalian A class are remarkably absent.

This is not the case for *Arabidopsis thaliana,* where a significant expansion of the ABC family can be observed [7]. In fact, 129 ABC proteins organized in 13 subfamilies are present in plants as opposed to the 50 or so members in most of eukaryote model genomes [2]. All the mammalian subclasses are present; however, distinctive expansion or contraction in their relative size have been reported. Seventeen proteins belong to the A class, but only one (AtAOH1) is organized as a full transporter encoded by a single gene and is considered the homolog of mammalian ABCA1. The others A members are hemitransporters [7].

An analysis of *Drosophila melanogaster* genome revealed the presence of 46 ABC genes, split in seven subclasses [2] (http://flybase.bio.indiana.edu). In the fly, an expansion of the G class can be observed. In *C. elegans*, six out of the 63 sequences annotated as ABC proteins encode transporters of the A subclass and include CED-7 [8–10] (http://www.proteome.com/databases/index.html).

Fourteen of the human ABC genes are associated with genetic disorders featuring highly varied clinical syndromes [1, 2], including cystic fibrosis, adrenoleukodystrophy, pseudoxantoma elasticum, retinal degeneration, defects in lipid metabolism, insulin resistance, anemia and drug resistance.

This remarkably broad spectrum of metabolic pathways affected by ABC proteins stems directly from the diversified nature of the substrates that they actually can transport via the ATPase energized mechanism. As an example, cystic fibrosis transmembrane regulator (CFTR) is a chloride channel, P-glycoprotein confers resistance to chemotherapeutic drugs on tumor cells by an active outward pumping mechanism which lowers their intracellular concentration and ABCB4 transports phosphatidylcholine (PC) in the liver. However, it has to be noted that the implication of a given transporter in the pathogenesis of a disease and therefore in a precise metabolic pathway is suggestive but frequently insufficient to provide any useful information on the nature of its substrate [4, 11]. This illustrates how, to date, the identification of a specific substrate remains a tremendous challenge.

6.3 The Case of ABCA1 in Mammals

Together with other 11 proteins, ABCA1 defines the A subclass of mammalian transporters [2]. These can be split into two subgroups on the basis of sequence homology, phylogenetic analysis and gene organization. The first group consists of seven genes (ABCA1, 2, 3, 4, 7, 12 and 13) closely related to the prototype ABCA1. Only the first five have been fully characterized. They map to different chromosomes in the human genome and to the synthenic region in the mouse genome. The second group includes five genes organized into a head-to-tail cluster on human chromosome 17. In contrast to the first set of genes which originated before speciation, the latter probably arose by recent species-specific events of gene duplications. All the mammalian ABCA genes encode complete transporters with four domains of the type TM1/NBD1/TM2/NBD2.

Individual ABCA proteins are extremely highly conserved in mouse and man with interspecies identity frequently exceeding 95%. As mentioned before ABCA proteins are present in plants, *D. melanogaster* and *C. elegans* but absent from yeast [2] (Tab. 6.1).

The ABCA1 gene encompasses a genomic region larger than 100 kb on human chromosome 9q and on mouse chromosome 4 [13, 14]. The transcriptional regulation of ABCA1 turns out to be rather complex and incompletely understood at present. It is mediated by three clusters of transcriptional start sites positioned in the ABCA1 promoter [13–15]. Each of them seems to be preferentially, but not exclusively, used in a given set of tissues and responds to different sets of regulatory proteins. Upregulation of ABCA1 transcription has been observed after treatment with agonists of peroxisome proliferator-activated receptor γ/δ, which act on a precisely positioned LRR/RXR binding site [16–21], cAMP [22], oxysterols and retinoic acid [20]. The net effect on ABCA1 transcription of any of these inducers, however, is expected to vary in different sets of tissues accordingly to the starting equilibrium of each transcript.

Interestingly, a steroP dependent transcriptional regulation has been reported for two other ABCA members, i.e. ABCA2 and 7 [23–25]. Since this profile of inducers acts, as a rule, on the promoters of genes involved in lipid metabolism, it is legitimate to surmise a similar involvement for ABCA2 and 7. No experimental information addressing this point has, however, been provided so far.

The ABCA1 transcript and protein are ubiquitously expressed at low levels in most mouse and human tissues. ABCA1 transcripts are particularly abundant in the pregnant uterus and placenta, whereas an intermediate level of expression is detectable in the liver, lungs and adrenals [12, 26, 27].

The protein encoded by the ABCA1 gene is 2261 amino acids long both in mouse and human [12, 14, 28], and contains a minimum of 12 TM spanners as deduced from computer-based analysis of hydrophobicity plots.

As it is frequently the case for ABC transporters and more generally for polytrophic membrane proteins, a precise topological assessment of individual TM helices is extremely difficult. For this reason several laboratories, including our own, have set out to address experimentally the question. Consistent with the general structural model, both NBD are located intracellularly. The membrane-anchoring scheme is still unclear and a number of potential models have been proposed.

The data we obtained by an approach based on systematic epitope insertion suggest the model illustrated in Fig. 6.1. This proposes a symmetrical arrangement of six TM in the two halves of the proteins, consistent with the model suggested for ABCA4 [29]. In fact, the first hydrophobic segment which follows the starting methionine, now positioned at bp 84 in the sequence GB X75926, behaves, in our hands, as a signal anchor (Rigot et al., J. Lipid Res. in press 2002)) [30, 31]. Based on this model, the original feature of the A class of transporters, as opposed to all other mammalian classes, is the presence of a symmetrical and quite large extracellular loop lying between TM1 and 2 and TM7 and 8. This architecture is predicted to be conserved in all members and throughout evolution (Fig. 6.1). It is worth noting that the primary sequence in the loops is hypervariable; these may therefore specify functional diversifications among members of the A subclass.

The ABCA1 transporter resides on the cell plasma membrane, from which it has been reported to shuttle continuously and rapidly to intracellular vesicles [32, 33]. The functional significance of this trafficking along the intracellular endocytic pathway remains to be elucidated; however, it is possible that this reflects the involvement of ABCA1 in intracellular lipid metabolism [32, 34].

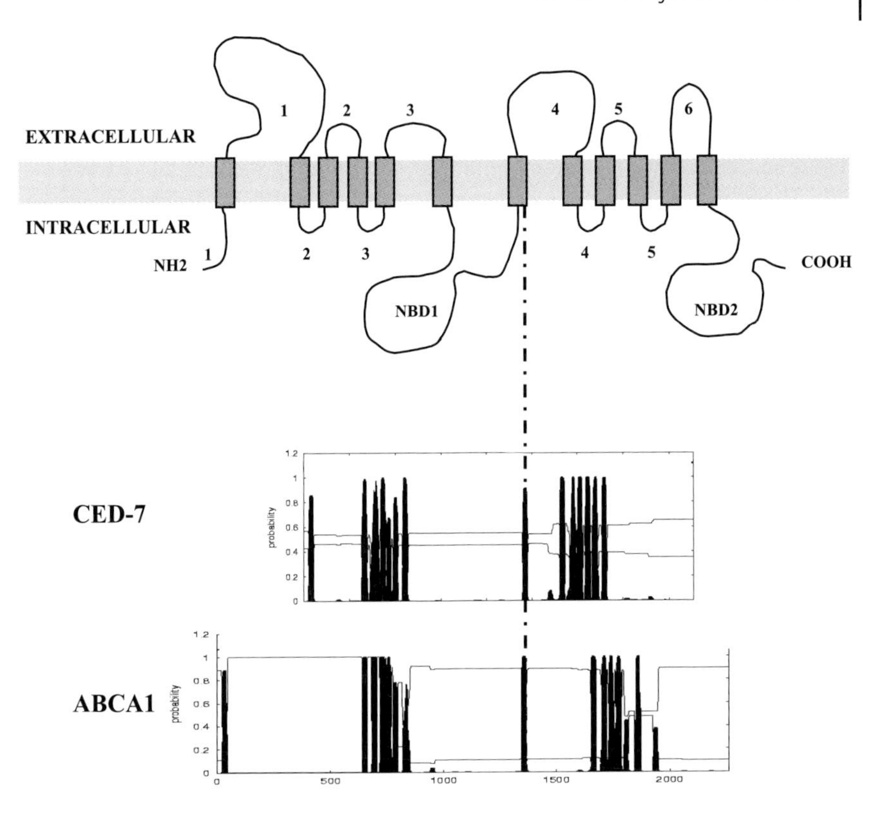

	I1	E1	I2	E2	I3	E3	NBD1	E4	I4	E5	I5	E6	NBD2
ABCA1	24	232	42	38	24	76	513	161	42	37	33	52	388
CED-7	24	210	20	16	2	54	503	139	20	15	11	52	373

Fig. 6.1 Topological model for ABC transporters of the A class. A schematic diagram illustrating the succession and position of TM α helices in ABCA1 is shown in the top panel. A similar topology is proposed for CED-7 on the basis of hydrophobicity plots (middle panel). A comparison of the length in amino acids of each loop is shown in the lower panel. E, extracellular; I, intracellular. Loops are numbered progressively from the N-terminus.

ABCA1 function affects at least two apparently unrelated systemic processes. On the one side stands its ability to promote the clearance of dying cells by phagocytes, our interest here, which involves the transporter in the control of body cell turnover. On the opposite side, ABCA1 plays a crucial role in lipid metabolism acting as a rate-limiting molecule in the first step of reverse cholesterol transport.

6.3.1
ABCA1 and Reverse Cholesterol Transport

This term indicates the metabolic pathway shuttling cholesterol from peripheral tissue to the liver via its loading onto high-density lipoprotein (HDL) particles [35, 36]. These arise from the successive maturation of lipid poor apolipoproteins with phospholipids and cholesterol originated from cell membranes. Experimental evidence clearly indicates now that only the initial step of HDL maturation, i.e. the phospholipidation of apoproteins, is under the control of the ABCA1 transporter [37, 38]. This was first suggested by the observation that spontaneous ABCA1 deficiency leads to Tangier disease [39–42], one of the best *in vivo* models for the study of the reverse cholesterol transport pathway. The disease is characterized by an absence, essentially complete, of α-migrating HDL from the plasma and storage of cholesteryl ester within focal accumulations of macrophages, notably the tonsils [43].

This systemic phenotype arises from a defect of peripheral tissues to release cellular lipids, as demonstrated by an impaired *in vitro* efflux, from Tangier cells, of newly synthesized free cholesterol onto lipid free apoA-I, the principal apolipoprotein on HDL, widely used as lipid acceptor in *in vitro* studies [44].

These initial observations were further confirmed and refined by several laboratories [45–47], and, more recently, the analysis of ABCA1 knockout animals [33, 48–50] provided direct evidence of the implication of ABCA1 in the active release of cellular lipids. These results prompted in addition an investigation into the role of the transporter as a candidate apoA-I receptor [51, 52]. In fact the mechanism of the interaction of apoA-I acceptor with cell membrane has long been elusive and two conflicting hypothesis are evoked. The first suggests that apoA-I insertion into the lipid bilayer is a mere consequence of its intrinsic lipid composition, while the second postulates the existence of a specific membrane receptor mediating surface binding [53, 54].

The recent availability of ABCA1 transfected cells has allowed us to test these models and confirm that indeed the presence of ABCA1 at the cell surface is sufficient to promote the docking of apoproteins [51, 52, 55]. However, this cannot be considered a direct receptor–ligand interaction since it relies on functional rather than structural criteria [55, 56]. On the basis of the observed ability of ABCA1 to act as a lipid floppase we propose that the transporter generates membrane domains of peculiar lipid architecture and chemicophysical properties, which *per se* will favor the insertion of acceptors into the membrane. This view is consistent with the physical interaction of ABCA1 with membrane lipids reported by Wang *et al.* [56] and also with the observed promiscuous ability of ABCA1 to promote effluxes to most classes of apoprotein acceptors [22, 57].

Whether or not a direct physical interaction between the ligand and ABCA1 also occurs is still matter of debate, but none of the data we could obtain so far supports this hypothesis.

6.3.2 ABCA1 and Engulfment

In mammals, the participation of ABCA1 transporter in engulfment was established in the mid-1990s by the description of a direct spatio-temporal relationship between the occurrence of developmental cell death and the *in situ* detection of ABCA1 transcripts [58]. In these areas, ABCA1 is expressed by macrophages actively engaged in the clearance of cell corpses. Consistently, in an *in vitro* situation, the inhibition of ABCA1 function by specific antibody greatly reduces the ability of peritoneal macrophages to phagocytose apoptotic thymocytes without affecting ingestion of yeast. These observations were further supported by (1) the finding of a persistence of apoptotic corpses during development in ABCA1 null animals, paralleled by a defective ability to ingest apoptotic prey of *ex vivo* macrophages, and (2) the acquisition of a phagocytic phenotype as a consequence of the forced expression of the transporter [33].

It is of note that the *in vivo* absence of ABCA1 does not affect the recruitment of embryonic macrophages in the areas of cell death or result in major developmental defects [33]. Indeed, the persistence of corpses is transient and matches perfectly the phenotype elicited by mutations in the engulfment genes in the nematode [59].

A similar result is also reported in PU-1 mice, in which the lack of macrophages leads to retarded clearance of corpses [60]. In this case the task is performed, albeit less efficiently, by the neighboring mesenchymal cells. Neither the transcript of ABCA1 nor that of any other ABCA transporter is detectable in these animals [60]. This formally proves the exclusive expression of ABCA1 in macrophages and indicates that only the expression of ABCA1, among ABCA transporters, is enhanced during engulfment in mammals.

As mentioned before, the presence or absence of ABCA1 in *in vitro* or *in vivo* systems, respectively, supports the involvement of ABCA1 in the two homeostatic pathways dealing with either cell or cholesterol turnover. These observations may reflect two parallel and independent functions of the transporters or arise secondarily from a single and common molecular mechanism.

We favor the latter hypothesis on the basis of the combined *in vitro* analysis of cells either overexpressing the transporter or derived from ABCA1 null animals [33]. This allowed us to determine that the absence or presence of ABCA1 at the plasma membrane perturbs the physiological arrangement of lipid species across the bilayer. In particular, we could demonstrate that the outward movement of phosphatidylserine is directly affected. However, since it is known that the movement of any phospholipid alters in cascade the distribution of the others [61–63], we postulate that membrane domains of peculiar composition and, hence, with peculiar biophysical features, are generated under the influence of the transporter. As present it is not known whether or how these domains are related to detergent-insoluble domains/rafts. These actively participate in the sorting of membrane proteins into molecular platforms, essential for the delivery of intracellular signals in mammalian cells [64–66].

6.4
The Case of CED-7 in *C. elegans*

During the development of a *C. elegans* hermaphrodite, 131 somatic cells and approximately 300 germ cells undergo programmed cell death [67]. Since no professional phagocyte exists in the worm, the dying cells are rapidly engulfed by their neighbors, whose identity varies in different individuals, in contrast to the fixed fate of the dying cell [68, 69]. Genetic screens for mutants containing unengulfed cell corpses led to the identification of seven genes that control engulfment *ced-1, -2, -5, -6, -7, -10* and *-12* [59, 70–72]. The visual quantification of phenotypes in single, double or triple combinations of mutants allowed the genes to be grouped into two parallel pathways [59]. The first is defined by the participation in a signaling cascade of CED-2, -5, -10 and -12 [70, 71]. Those which encode functional orthologs of the mammalian Crk-II, DOCK180, ELMO and Rac-GTPase, respectively [70, 73], are essential components of the intracellular signal reorganization occurring during both the engulfment of cell corpses and cell migration. Indeed, in mammals this cascade follows the surface stimulation of integrins [74, 75] and mutations in any of these genes in the worm leads, as an additional phenotype, to a defective migration of distal tip cells in the gonads [70, 73, 76].

ced-7 together with *ced-1* and -6 defines a second pathway exclusively involved in engulfment [59]. Although all these genes have been cloned, it is not clear yet whether their products physically interact. *ced-6* has, however, been positioned downstream on the basis of the phenotypic rescue of both *ced-7* and *-1* mutants. Consistently CED-6, which is conserved in mouse and man, bears the structural hallmark of an adaptor molecule containing a phosphotyrosine-binding domain [77–79].

Among the 63 sequences annotated as ABC proteins in the *C. elegans* genome, six bear the structural features of the ABCA subclass and include CED-7 [80] (Tab. 6.2) (http://www.proteome.com/databases/index.html). On the basis of sequence analysis, CED-7 bears an identity to all members of the A class related to ABCA1 (i.e. ABCA2, 3, 4 and 7) of 26–28% and an overall homology of 44–46% [81]. Not surprisingly, the best-conserved regions correspond to the NBD. It is of note, however, that similar scores are observed when comparing any of the mammalian ABCA transporters against any member of the A class in the worm and that, in contrast to mammals, the A members in the worm do not share extensive identity. With the exception of CED-7, none of the worm A transporters has been associated so far to a functional phenotype.

The *ced-7* gene encodes a transcript of 5.8 kb. This is translated into a protein of 1704 amino acids, which consists of two similar halves, with twice six TM domains and two NBD [80]. No topological study has been carried out to date; however, the CED-7 hydrophobicity plot is consistent with an arrangement of membrane α helices similar to that postulated for ABCA1 and 4 in mammals [29, 31]. On the basis of mosaic analysis, *ced-7* was shown to be required in both the prey that is the apoptotic cells and the phagocytes for an efficient engulfment [80]. This is at odds with the results obtained in mammals, where at least in the developing limb bud the ABCA1 transcript is detectable exclusively in macrophages

Tab. 6.2 Conservation of ABCA transporters between mammals and *C. elegans*.

	Percent identity/percent homology					
	CED-7	***ABCA1***	***ABCA2***	***ABCA3***	***ABCA4***	***ABCA7***
ABCA1	25/44	100/100	45/59	26/43	50/66	50/66
CED-7	100/100	25/44	28/46	27/46	26/45	26/44
C24F3.5	22/46	23/43	24/50	21/39	23/43	23/48
F12B6.1	26/44	30/46	29/44	30/46	29/44	29/44
F55G11.9	32/48	23/38	25/39	27/43	24/40	23/38
Y39D8C1	28/46	29/47	29/46	33/52	30/48	29/47
Y53C10A.9	33/51	24/42	25/42	27/46	26/45	26/43

ABCA1, ABCA2, ABCA4 and ABCA7 are mouse sequences, ABCA3 is human.

[58, 60]. It has to be emphasized, however, that the pre-existence of an active protein (ABCA1 or a related transporter) at work in the dying cell has not been addressed in the study.

All the *ced-7* mutant alleles turned out to encode prematurely truncated proteins and a thorough structure–function analysis has not been carried out yet. However, the relative impact of mutations impairing ATP binding and hydrolysis at the two NBD seems to be different [80]. Indeed, it was shown that the K586R mutation in NBD1 abolished the ability of CED-7 to engulf apoptotic corpses, whereas the same substitution in NBD2 has a limited effect. Such hierarchical orders of NBD are not new in the field of transporters and a similar result has been reported for CFTR. However, in mammals, but again in a slightly different and *in vitro* experimental setting, a K to M substitutions at either or both NBD led to the complete loss of the ABCA1-associated phenotype [33].

CED-7 is expressed at the plasma membrane of virtually all cells during embryonic development, but only on the surface of amphid sheath cells, pharyngeal intestinal valve and phasmid sheath cells in the larvae or adult worms [80]. Of interest is that none of the somatic cells expressing CED-7 appear to be involved in engulfment and so far it has not been possible to find a *ced-7*-induced phenotype associated with the specialized functions of these particular cells. No functional study has addressed the molecular function of CED-7 in the worm during engulfment; however, it is known that, whereas the absence of any other gene involved in engulfment does not affect the localization of CED-7, the absence of the ABC transporter alters dramatically that of the CED-1 receptor [80, 82].

The product of *ced-1*, which participates to the same molecular cascade, has been recently characterized [82]. It belongs to the family of scavenger receptors, multiligand receptors involved in mammals in the recognition/uptake of dying corpses [83–87]. CED-1 is normally expressed on the membrane of all cells, but during the engulfment process it redistributes on the membrane and clusters at the phagocytic extensions formed around the dying cell [82]. This lateral mobility of CED-1, which might highlight the requirement of a topological cluster of surface receptors to start efficiently engulfment, is abolished in *ced-7* mutant worms.

Zhou *et al.* [82] proposed that CED-7 may facilitate the physical contact between dying cells and engulfing cells, acting on dying cells to expose a 'eat me' signal at the cell surface and on the engulfing cells to facilitate the CED-1 recognition for its ligand.

6.5 The Model

The evidence for the involvement of ABC transporters during engulfment was provided independently and contemporarily in the two model systems. This is rather an exception to the rule. Usually, in fact, *C. elegans* has played a pioneering role in the engulfment and more generally in the apoptosis field: genes and their product being first characterized in this organism and only then the evidence provided for their functional conservation in mammals.

Membrane receptors, on the contrary, have been mainly if not exclusively identified in mammalian systems which experimentally rely on *in vitro* inhibitory studies [88]. This dichotomy stems from the intrinsic strengths and drawbacks of each system, and it is also perfectly illustrated by the complementary information derived from the studies of either ABCA1 or *ced-7*.

In fact, the possibility to analyze the phenotype at the cell level has allowed us to highlight the lipid-transporting activity of ABCA1 [33], whereas the powerful genetic approach combined with *in vivo* overexpression allowed us to assess the impact of *ced-7* on the topological distribution of membrane receptors [33]. This information actually fits together very well and if we take the liberty of transposing freely between the two systems, we can figure out a consistent working model for the function of ABC transporters in engulfment (Fig. 6.2).

The ability to modify even only locally and transiently the biophysical properties of the membrane by flipping lipids (ABCA1) between the two leaflets would be the initial step. This provides the perfect explanation as to how ABC transporters (CED-7) can influence the lateral mobility of surface receptors involved in the recognition of dying cells. The topological recruitment witnessed by the example of CED-1 in *C. elegans* can be then put in context with the notion that, in mammals, individual receptors display a limited affinity for the ligands on the surface of the prey. Hence, spatial recruitment may be the morphological counterpart of the requirement for a threshold of signals to trigger ingestion [89]. Whether this concerns all or only a subclass of the receptors involved in engulfment is as yet not known in either model system. In fact, in mammals, the behavior of receptors during engulfment or the influence of ABCA1 on their mobility has not been tested and in the nematode only *ced-1* encodes surface receptors. Similarly, we may ask whether topological reinforcement is an optimizing option featured only by professional phagocytes in mammals. These are, in fact, known to sense and react more rapidly to the presence of an apoptotic prey [90].

As an alternative approach we may also wish to test to which extent CED-7 and ABCA1 are interchangeable. Rescuing *ced-7* mutants by overexpression of ABCA1

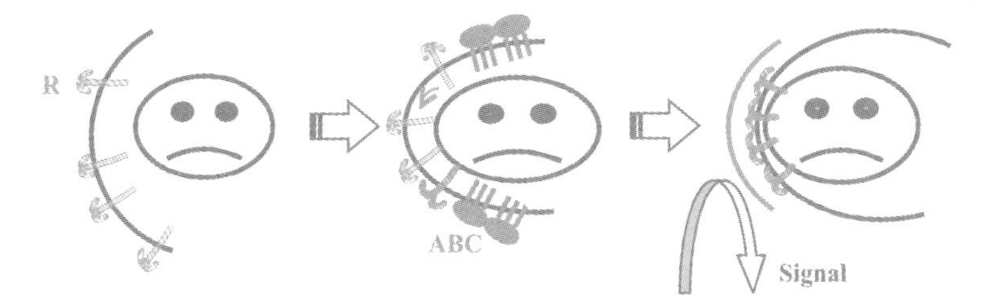

Fig. 6.2 ABC transporter function during engulfment. This diagram proposes a three-step model for engulfment based on the combination of evidence derived from mammals and *C. elegans*. In a basal situation, engulfment receptors (R) are homogeneously dispersed on the membrane of the phagocyte; upon contact with the apoptotic prey they redistribute along the contact region. This spatial recruitment is favored by the ABC transporter (ABC). Topological clustering allows the minimal threshold to trigger downstream signalization to be achieved.

may turn out to be deceptive because of the maternal effect in the nematode [59, 80]. We may also test whether and how CED-7 reproduces the phenotypes elicited by the expression of ABCA1. Preliminary results in this direction confirm that the two transporters share, as expected, the ability to promote engulfment in non phagocytic recipient cells (Hamon, personal communication). In addition, CED-7 possesses a lipid flipping activity similar to that of ABCA1, but fails to promote either docking of apoproteins or the consequent efflux of phospholipids from cell membranes. Whether this depends on structural peculiarities harbored in specific domains of each transporter or reflects evolutionary imprinted refinements of the function is currently under investigation. Unfortunately, the complete lack of information on sterols and lipid metabolism in the worm together with the inaccessibility of the nematode system to the study of membrane composition and dynamics hampers as for now a rapid progression in this direction.

6.6 References

1 Klein, I., *et al. Biochim Biophys Acta* **1999**, 1461, 237–62.
2 Dean, M., *et al. J Lipid Res* **2001**, 42, 1007–17.
3 Walker, J. E., *et al. EMBO J* **1982**, 8, 945–51.
4 Higgins, C. F. *Annu Rev Cell Biol* **1992**, 8, 67–113.
5 Taglicht, D. and S. Michaelis. *Methods Enzymol* **1998**, 292, 130–62.
6 Decottignies, A. and A. Goffeau. *Nat Genet* **1997**, 15, 137–45.
7 Sanchez-Fernandez, R., *et al. J Biol Chem* **2001**, 9, 9.
8 Broccardo, C., *et al. Biochim Biophys Acta* **1999**, 1461, 395–404.
9 Costanzo, M. C., *et al. Nucleic Acids Res* **2001**, 29, 75–9.
10 Costanzo, M. C., *et al. Nucleic Acids Res* **2000**, 28, 73–6.
11 Doige, C. A. and G. F.-L. Ames. *Annu Rev Microbiol* **1993**, 47, 291–319.
12 Luciani, M. F., *et al. Genomics* **1994**, 21, 150–9.
13 Qiu, Y., *et al. Genomics* **2001**, 73, 66–76.
14 Santamarina-Fojo, S., *et al. Proc Natl Acad Sci USA* **2000**, 97, 7987–92.

15 Pullinger, C.R., *et al. Biochem Biophys Res Commun* **2000**, 271, 451–5.
16 Oliver, W.R., Jr., *et al. Proc Natl Acad Sci USA* **2001**, 98, 5306–11.
17 Chinetti, G., *et al. Nat Med* **2001**, 7, 53–58.
18 Costet, P., *et al. J Biol Chem* **2000**, 275, 28240–5.
19 Chawla, A., *et al. Mol Cell* **2001**, 7, 161–71.
20 Tall, A.R., *et al. Nat Med* **2000**, 6, 1104–1105.
21 Venkateswaran, A., *et al. J Biol Chem* **2000**, 275, 14700–7.
22 Bortnick, A.E., *et al. J Biol Chem* **2000**, 275, 28634–40.
23 Kaminski, W.E., *et al. Biochem Biophys Res Commun* **2001**, 281, 249–58.
24 Broccardo, C., *et al. Cytogenet Cell Genet* **2001**, 92, 264–70.
25 Kaminski, W.E., *et al. Biochem Biophys Res Commun* **2000**, 273, 532–8.
26 Langmann, T., *et al. Biochem Biophys Res Commun* **1999**, 257, 29–33.
27 Lawn, R.M., *et al. Arterioscler Thromb Vasc Biol* **2001**, 21, 378–85.
28 Remaley, A.T., *et al. Proc Natl Acad Sci USA* **1999**, 96, 12685–90.
29 Bungert, S., *et al. J Biol Chem* **2001**, 24, 24.
30 Tanaka, A.R., *et al. Biochem Biophys Res Commun* **2001**, 283, 1019–25.
31 Fitzgerald, M.L., *et al. J Biol Chem* **2001**, 29, 29.
32 Neufeld, E.B., *et al. J Biol Chem* **2001**, 10, 10.
33 Hamon, Y., *et al. Nat Cell Biol* **2000**, 2, 399–406.
34 Takahashi, Y. and J.D. Smith. *Proc Natl Acad Sci USA* **1999**, 96, 11358–63.
35 Fielding, C.J. and P.E. Fielding. *J Lipid Res* **1995**, 36, 211–28.
36 Castro, G.R. and C.J. Fielding. *Biochemistry* **1988**, 27, 25–9.
37 Fielding, P.E., *et al. Biochemistry* **2000**, 39, 14113–20.
38 Haghpassand, M., *et al. J Clin Invest* **2001**, 108, 1315–20.
39 Brooks-Wilson, A., *et al. Nat Genet* **1999**, 22, 336–45.
40 Rust, S., *et al. Nat Genet* **1999**, 22, 352–355.
41 Bodzioch, M., *et al. Nat Genet* **1999**, 22, 347–351.
42 Lawn, R.M., *et al. J Clin Invest* **1999**, 104, R25–31.
43 Assmann, G., *et al.* In: Scriver, C.R., et al., eds. *The Metabolic Basis of Inherited Disease.* New York: McGraw-Hill **2001**, 2937–2960.
44 Walter, M., *et al. Biochem Biophys Res Commun* **1994**, 205, 850–6.
45 Remaley, A.T., *et al. Arterioscler Thromb Vasc Biol* **1997**, 17, 1813–21.
46 Rogler, G., *et al. Arterioscler Thromb Vasc Biol* **1995**, 15, 683–90.
47 Francis, G.A., *et al. J Clin Invest* **1995**, 96, 78–87.
48 Drobnik, W., *et al. Gastroenterology* **2001**, 120, 1203–11.
49 McNeish, J., *et al. Proc Natl Acad Sci USA* **2000**, 97, 4245–50.
50 Orso, E., *et al. Nat Genet* **2000**, 24, 192–196.
51 Wang, N., *et al. J Biol Chem* **2000**, 275, 33053–8.
52 Oram, J.F., *et al. J Biol Chem* **2000**, 275, 34508–11.
53 Yancey, P.G., *et al. Biochemistry* **1995**, 34, 7955–65.
54 Fidge, N.H. *J Lipid Res* **1999**, 40, 187–201.
55 Chambenoit, O., *et al. J Biol Chem* **2001**, 276, 9955–60.
56 Wang, N., *et al. J Biol Chem* **2001**, 17, 17.
57 Remaley, A.T., *et al. Biochem Biophys Res Commun* **2001**, 280, 818–23.
58 Luciani, M.F. and G. Chimini. *EMBO J* **1996**, 15, 226–35.
59 Ellis, R.E., *et al. Genetics* **1991**, 129, 79–94.
60 Wood, W., *et al. Development* **2000**, 127, 5245–52.
61 Tepper, A.D., *et al. J Cell Biol* **2000**, 150, 155–64.
62 Bevers, E.M., *et al. Biochim Biophys Acta* **1999**, 1439, 317–30.
63 Zwaal, R.F.A. and A.J. Schroit. *Blood* **1997**, 89, 1121–32.
64 Simons, K. and D. Toomre, *Nat Rev* **2000**, 1, 31–9.
65 Simons, K. and E. Ikonen. *Science* **2000**, 290, 1721–6.
66 Sprong, H., *et al. Nat Rev Mol Cell Biol* **2001**, 2, 504–13.
67 Sulston, J.E. and H.R. Horvitz. *Dev Biol* **1977**, 56, 110–56.

68 Ellis, R. E., *et al. Annu Rev Cell Biol* **1991**, 7, 663–98.

69 Hengartner, M. O. and H. R. Horvitz. *Curr Opin Genet Dev* **1994**, 4, 581–6.

70 Gumienny, T. L., *et al. Cell* **2001**, 107, 27–41.

71 Wu, Y., *et al., Dev Cell 2001* **2001**, 1, 491–502.

72 Zhou, Z., *et al., Dev Cell 2001* **2001**, 1, 477–89.

73 Reddien, P. W. and H. R. Horvitz. *Nat Cell Biol* **2000**, 2, 131–6.

74 Tosello-Trampont, A. C., *et al. J Biol Chem* **2001**, 276, 13797–802.

75 Albert, M., *et al., Nat Cell Biol* **2000**, 2, 899–906.

76 Wu, Y. and H. R. Horvitz. *Nature* **1998**, 329, 501–4.

77 Su, H. P., *et al. J Biol Chem* **2000**, 275, 9542–9.

78 Liu, Q. A. and M. O. Hengartner. *Curr Biol* **1999**, 9, 1347–50.

79 Liu, Q. and M. Hengartner. *Cell* **1998**, 93, 961–72.

80 Wu, Y. and R. H. Horvitz. *Cell* **1998**, 93, 951–60.

81 Culetto, E. and D. B. Sattelle. *Hum Mol Genet* **2000**, 9, 869–77.

82 Zhou, Z., *et al. Cell* **2001**, 104, 43–56.

83 Platt, N. and S. Gordon. *J Clin Invest* **2001**, 108, 649–54.

84 Febbraio, M., *et al. J Clin Invest* **2001**, 108, 785–91.

85 Krieger, M. *J Clin Invest* **2001**, 108, 793–797.

86 Krieger, M. and D. M. Stern. *J Clin Invest* **2001**, 108, 645–7.

87 Savill, J. and V. Fadok. *Nature* **2000**, 407, 784–8.

88 Savill, J., *et al. Immunol Today* **1993**, 14, 131–6.

89 Henson, P. M., *et al. Curr Biol* **2001**, 11, R795–805.

90 Parnaik, R., *et al. Curr Biol* **2000**, 10, 857–60.

7
Innate Immunity and Apoptosis: CD14-dependent Clearance of Apoptotic Cells

Christopher D. Gregory and Andrew Devitt

7.1
Introduction: CD14, A Multifunctional Molecule involved in Innate Immune Responses

7.1.1
Background

Discovered over a decade ago [1, 2], CD14 is a glycoprotein that appears to serve multiple receptor functions in relation to the innate immune system. Its prototypic and still its best-known function is as a receptor for the endotoxin of Gram-negative bacteria, lipopolysaccharide (LPS). Functional in both plasma-membrane-anchored and soluble forms, CD14 is renowned for its ability to activate severe pro-inflammatory responses following LPS binding and in so doing contribute to septic shock, a significant factor in post-surgical mortality [3, 4]. More recently – and surprisingly in view of its history as a pro-inflammatory receptor – CD14 has been implicated in apoptotic cell clearance [5], a normally anti-inflammatory process that represents the culmination of the apoptotic programme.

7.1.2
Molecular Structure and Distribution of CD14

CD14 is expressed strongly by most monocytes and some tissue macrophages, and weakly by granulocytes. In all these cells the molecule is attached to the outer leaflet of the plasma membrane by a glycosylphosphatidylinositol (GPI) anchor at its C-terminus. Soluble CD14 is also present in plasma at high levels (around 4 μg/ml [6]) and this appears to be produced by two independent mechanisms – one involving cleavage of the membrane-anchored form by protease or phospholipase D activity and the other by direct secretion [7]. Other cell lineages including B lymphocytes, epithelial cells, endothelial cells, trophoblast, hepatocytes and fibroblasts have also been reported to express CD14 [8–15]. Some studies are limited to the mRNA level, whereas others fail to discriminate between exogenously produced, adsorbed CD14 and endogenous protein. It is probable, however, that the distribution of membrane-anchored CD14 is more widespread than was pre-

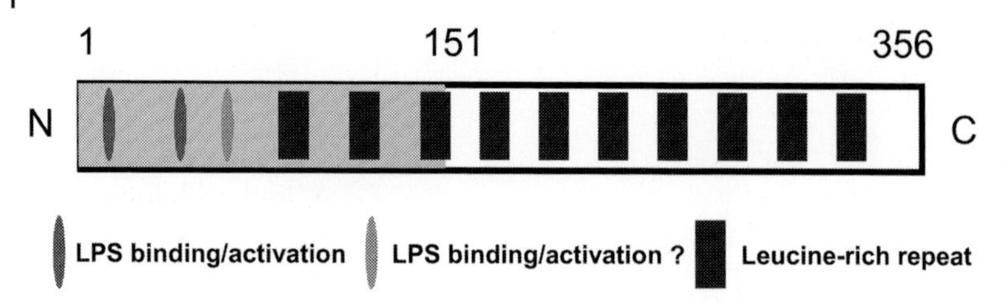

Fig. 7.1 Schematic representation of the CD14 protein sequence illustrating (1) the active region of the molecule for LPS responsiveness (grey shading), (2) the three putative LPS-binding/activation sites around positions 10, 40 and 60, and (3) the 10 leucine-rich repeats. The three LPS-binding/activation regions may be closely apposed in the tertiary structure of CD14.

viously thought, but the functional attributes of this distribution are not yet understood.

Human CD14 is a 356-amino-acid, 53–55-kDa protein that is glycosylated at *O*- and *N*-linked sites [16]. The C-terminal 21-amino-acid domain is hydrophobic and is involved in anchoring the molecule into the lipid of the membrane via a GPI anchor. The extracellular domain contains 10 leucine-rich repeats that are evolutionarily conserved in many unrelated proteins (Fig. 7.1). The function of these repeats in CD14 is unknown, though it seems likely that they are involved in interactions with other proteins [17], one contender being Toll-like receptor (TLR)-4 that is known to cooperate with CD14 to promote responsiveness to LPS [18–21]. Since the plasma-membrane-anchored form of CD14 contains no intracytoplasmic domain, it is widely thought that such cooperation of CD14 with integral, transmembrane proteins is required for signal transduction following ligand binding.

7.1.3
CD14 as a LPS Receptor that Signals LPS Responses

In both its soluble (sCD14) and membrane-anchored (mCD14) forms, CD14 has been shown to bind LPS and to generate LPS responsiveness – the activation of inflammatory signaling pathways. CD14–LPS binding is enhanced by the LPS-binding protein (LBP) which catalyzes transfer of LPS from micellar aggregates of the glycolipid to CD14 to form CD14–LPS complexes [22–24]. It has largely been assumed that mCD14 and sCD14 bind LPS in identical ways, although there are reasons to believe that this assumption may be incorrect [25]. Whether or not this is the case, mCD14 generates LPS responsiveness in many different cell types after CD14 gene transfer as well as in cells on which it is expressed constitutively. sCD14–LPS complexes are able to generate LPS responsiveness in a variety of cell types, notably endothelial cells, that are $mCD14^-$ [25]. The nature of the receptors that are responsible for such responses remains ill defined, although candidates include TLR-4 [18–21], CR3 [26, 27] and an uncharacterized 216-kDa cell-surface protein [28]. LPS responsiveness in both myeloid and endothelial cells involves signaling pathways that induce phosphorylation of the mitogen-activated protein (MAP) kinases ERK1, ERK2 and p38, and activate NF-κB translocation.

Mutational analyses have indicated that the N-terminal 151 or 152 amino acids are sufficient for CD14 to function as a LPS receptor that can signal LPS responsiveness [29, 30]. In other words, less than half of the molecule encompassing only three of its 10 leucine-rich repeats is involved in this function (Fig. 7.1). Considered from this perspective, it is perhaps not surprising that CD14 proves to be a multi-functional receptor. For mCD14, no function has yet been ascribed to the GPI anchor, as a chimeric mCD14 that contains the transmembrane domain and cytoplasmic tail of tissue factor in place of the GPI anchor is fully capable of generating LPS responsiveness [31]. Studies of deletion mutants of mCD14 indicated that LPS binding and responsiveness are dependent on amino acids at or around positions 10, 39 and 60 [32]. Detailed investigations of alanine substitution mutants confirmed the importance of the regions around amino acids 10 and 39, but demonstrated that mutants of amino acids 57–64 failed to affect LPS responsiveness of mCD14 [33]. By contrast, several lines of evidence point to this region being important for the function of sCD14 in binding, and generating responses, to LPS. Thus, following LPS binding, this region is protected by Asp-N protease digestion [34]. Also, this amino acid sequence is required for the binding to sCD14 of LPS and of the monoclonal antibody MEM-18 which blocks LPS responsiveness [35, 36]. Mutations around amino acid 10 cause loss of sCD14 responsiveness to LPS. Furthermore, a second monoclonal antibody, 3C10, maps to sCD14 between amino acids 7 and 14, and blocks LPS responsiveness, but not binding [35, 36]. Taken together, these results highlight three areas of CD14, clustered in the first fifth of the molecule, that are important for LPS responses (Fig. 7.1). Recent studies of charge reversal mutations support this view [37] and, along with epitope mapping studies, suggest that these three highly hydrophilic areas are closely apposed in the tertiary structure of CD14, together contributing to the LPS-binding site [32, 33, 37]. Further structure–function relationships will undoubtedly be clarified once the three-dimensional structure of CD14 has been solved.

7.1.4 CD14 Binds Multiple and Diverse Ligands

While the vast majority of current knowledge of CD14 is based upon its capacity to bind and respond to LPS, ligands of CD14 are not limited to this glycolipid. Indeed, over the past decade it has become clear that CD14 can interact directly with, and/or generate responses to, many and varied ligands of both microbial and non-microbial origin (Tab. 7.1). As well as ill-defined components, known bacterial structures that elicit a range of CD14-dependent responses such as NF-κB activation, nitric oxide (NO) production and secretion of tumor necrosis factor (TNF)-α, interleukin (IL)-1β, IL-6, IL-8 and IL-12, include lipoteichoic acid (LTA), peptidoglycan (PGN), lipoarabinomannan (LAM), rhamnose glucose polymers, uronic acid polymers, glycolipids and lipoproteins. Furthermore, numerous examples of proteins, lipids and phospholipids of viral, fungal, yeast or mammalian origin are known to interact directly or indirectly with CD14, in some cases to generate pro-inflammatory responses and, in others, to prevent such responses

(Tab. 7.1). Given the diversity of structures, carbohydrate, lipid or protein in nature, that can interact with CD14 – and examples of each have been shown directly to bind either to sCD14, mCD14 or both [24, 38–46] – the origins of these structures and the range of responses (pro- and anti-inflammatory) that can be generated via CD14-dependent mechanisms, it must be concluded that CD14 has evolved to perform functions which include, but which extend beyond, innate anti-microbial defense. In this regard, CD14 not only qualifies as a classical pattern recognition receptor (PRR) [47] (see Section 7.3.1), but also as a receptor with further functional attributes that allow it to interact with 'self' in addition to 'non-self' structures. Amongst the 'self' structures with which CD14 interacts are apoptotic cells.

7.2 Evidence that Apoptotic Cells Interact with CD14

7.2.1 61D3, a CD14 Monoclonal Antibody that Blocks Apoptotic Cell Clearance

The first evidence that CD14 is involved in the clearance of apoptotic cells emerged from experiments in which monoclonal antibodies were tested for their capacity to modulate the recognition and phagocytosis of apoptotic leukocytes by human monocyte-derived macrophages. These studies identified a monoclonal antibody, 61D3, which potently inhibited apoptotic cell clearance *in vitro*, both as whole IgG and as $F(ab')_2$ fragments [48]. This monoclonal antibody had been raised several years previously by immunizing mice with human monocytes [49], but its specificity remained undefined. Ultimately, transient expression cloning in eukaryotic cells successfully resulted in positive identification of the molecule, known to be expressed on the surface of monocytes and macrophages, carrying the epitope recognized by 61D3. The molecule proved to be CD14 [5]. Although studies with CD14 mutants have not yet been performed, the 61D3 monoclonal antibody appears to bind to the same region of CD14 as LPS. This conclusion is based upon (1) the observation that MEM-18, a monoclonal antibody known to block LPS binding to CD14, inhibits apoptotic cell clearance by macrophages to the same extent as 61D3 [5], and (2) the finding that 61D3 inhibits MEM-18 binding to sCD14 and mCD14 ([5] and our unpublished observations). 61D3 also inhibits LPS-induced TNF-α secretion by macrophages, lending further support to the probability that 61D3 interacts with the LPS-binding site of CD14 [5].

7.2.2 Exogenous Expression of CD14 in Non-myeloid Cells

Confirmation that mCD14 acts as a receptor in apoptotic cell clearance has been obtained *in vitro* by overexpressing exogenous CD14 in cells of non-myeloid origin such as COS-1. Similar to previous studies which demonstrated the 'conversion' of

Tab. 7.1 Diversity of structures (excepting LPS) known to bind CD14 and/or induce CD14-dependent cellular responses.

Structure	*CD14 (m or s)*	*Responses*	*References*
Bacterial envelope components from Gram-positive organisms.	mCD14, sCD14	NO, IL-8, IgM expression	47
Components of *Staphylococcus aureus*	sCD14, mCD14	cytokine production, adhesion	112, 113
Bacterial cell wall components (not LPS) of *Neisseria gonorrhoeae*	sCD14	IL-8	114
Streptococcal cell wall polysaccharides (Rhamnose glucose polymers)	mCD14, sCD14	TNF-α	38
Uronic acid polymers (including mannuronic acid from *Pseudomonas aeruginosa* and teichuronic acids from *Micrococcus luteus*)	mCD14, sCD14	TNF-α, IL-8	63, 65–67
Lipoteichoic acid (LTA)	mCD14	IL-12 p40, NO, iNOS, IL-8, IL-1β, NF-κB	98, 115–117
Peptidoglycan (PGN)	mCD14, sCD14	IL-1β, IL-6, MAP kinases, ATF1/CREB, AP-1, TNF-α, sIgM, NF-κB, IκB degradation	39, 42, 118–122
Synthetic muramyl dipeptide	mCD14	IL-6	39
Lipoarabinomannan (LAM)	mCD14, sCD14	NF-κB, IL-6, TNF-α, IL-1β, Ca^{2+}, migration, NO, IL-8, IgM	31, 123–127
Flavolipin	mCD14	NF-κB	128
Glycolipids from *Treponema maltophylum*	mCD14	TNF-α, NF-κB	129
Soluble tuberculosis factor (STF)	mCD14	NF-κB	127
Outer membrane lipoproteins Osp (outer surface protein) A and OspC of *Borrelia burgdorferi*	sCD14	NF-κB, IL-6, IL-8	41
Acylpolygalactoside (from *Klebsiella pneumoniae* membrane)	mCD14	modulation of CD14 and CR3	130
Group B *Streptococcus* (released soluble factor)	mCD14	TNF-α	131
WI-1 Yeast cell wall protein from *Blastomyces dermitidis*	mCD14	none measured	132
RSV fusion (F) protein	mCD14	IL-6	133
Hyphal fragments of *Aspergillus fumigatus*	not clear (plasma required)	TNF-α	134

Tab. 7.1 (continued)

Structure	*CD14 (m or s)*	*Responses*	*References*
Fucoidan	mCD14 (or sCD14 from serum)	TNF-α, IL-6	64
HSP60	mCD14	IL-6	69
HSP70	mCD14	TNF-α, IL-1β, IL-6, NF-κB, Ca^{2+}	68
Ceramide	mCD14	co-localization with CR3 in rafts	70
Phospholipids: PI, PC, PS, PE and phosphatidic acid	mCD14, sCD14	inhibition of LPS-induced IL-8 secretion and NF-κB activation, arachidonate metabolism	24, 40, 46, 135
Pulmonary surfactant proteins (SP-A and SP-D)	mCD14, sCD14	TNF-α inhibited	43, 44
Rat mannose-binding protein A	sCD14	none measured	45
IL-2	mCD14	IL-1β, IL-8	136
Endothelial cells (IL-1α, TNF-α or IFN-γ stimulated)	mCD14 (?)	none measured	137
Apoptotic cells	mCD14	phagocytosis, no TNF-α	5

an amateur phagocyte of apoptotic cells to one closer to a professional upon transfer of *CD36* cDNA [50], transfer of *CD14* cDNA to COS-1 cells substantially enhanced the capacity of these cells to bind and phagocytose apoptotic lymphocytes [5]. Notably, however, considering the massive amounts of CD14 expressed by COS-1 cells relative to that of macrophages, the levels of interaction of apoptotic cells with COS-1 never approached those observed with macrophages. Therefore, in order to function efficiently in apoptotic cell clearance, CD14 may require concordant overexpression of additional cooperating molecule(s) that are necessary for generating CD14-dependent responses to apoptotic cells.

7.2.3
Apoptotic Cell-associated Ligands of CD14

While it remains possible that sCD14 could modulate apoptotic cell clearance by amateur or professional phagocytes, evidence for the involvement of CD14 in the removal of apoptotic cells is currently limited to the membrane-anchored form of the molecule. Furthermore, in light of available evidence, it appears that apoptotic cells interact directly with mCD14. Thus, CD14-dependent interactions between apoptotic cells and phagocytes – be they macrophages or CD14 transfectants of

COS-1 cells – can occur in the absence of added plasma components that may otherwise have served to opsonize or 'bridge' apoptotic cell surface structures with CD14 [5]. It is conceivable, however, that CD14-dependent clearance of apoptotic cells involves binding intermediates that associate with the apoptotic cell or phagocyte surfaces since exhaustive removal of these factors has not been investigated. Furthermore, it remains possible that, while CD14-dependent apoptotic cell clearance can proceed in the absence of such factors, their presence may improve efficiency. It is now known that multiple proteins, in addition to integral membrane components, contribute to complexes at the interacting cell surfaces that enable engulfment of apoptotic cells via pathways involving phagocyte molecules that include the $\alpha_v\beta_3$ vitronectin receptor, CD36 and CD91 ([51, 52] and see Tab. 7.2). In view of the variety of molecules able to interact with CD14 (Tab. 7.1), it is tempting to speculate that, for optimal activity in mediating apoptotic cell clearance, CD14 may require molecules in addition to those that are integral to the plasma membrane of either the phagocyte or the apoptotic cell. In this context it is noteworthy that members of the collectin family have been shown to contribute to complexes at phagocyte surfaces that mediate apoptotic cell clearance [52, 53] and at least some members of this family, namely the pulmonary surfactant proteins, SP-A and SP-D and mannose-binding protein A are known to interact directly with CD14 [43–45].

Perhaps the most obvious potential CD14 ligands of apoptotic cells are phospholipids. Disruption of phospholipid distribution is the only well-defined feature of the apoptotic cell surface and phospholipids have proven capacity to bind directly to mCD14 [40], as well as sCD14 [24, 46]. These qualities argue strongly in favor of a role for phospholipids in CD14-dependent apoptotic cell clearance. Since phosphatidylserine (PS) becomes exposed on the outer plasma membrane leaflet during apoptosis and, unlike other phospholipids including phosphatidylinositol (PI), phosphatidylethanolamine (PE) and phosphatidylcholine (PC), is functional in apoptotic cell clearance [54], an intuitively obvious possibility is that PS interacts directly with macrophage CD14. Several lines of evidence, however, indicate that PS does not function as a preferential ligand for mCD14 in apoptotic cell clearance by human monocyte-derived macrophages *in vitro*. Firstly, although PS can interact with both mCD14 and sCD14, it does so with lower affinity than PI, PC or PE [40, 46]. Second, CD14-dependent clearance of apoptotic cells can be demonstrated in the absence of overt PS receptor activity [55]. Last, and by no means least, the dominant PS receptor for apoptotic cells (PS-R) has been cloned and bears no resemblance to CD14 [56]. Therefore, based on the available evidence it must be concluded that CD14 ligands other than PS predominate, at least in the case of human monocyte-derived macrophages, in clearance of apoptotic cells via mCD14.

In view of CD14's established lectin-like activity (Tab. 7.1), it is possible that glycan moieties at apoptotic cell surfaces may be functional as CD14 ligands. Although not yet well defined at the molecular level, it is clear that the glycocalyx becomes altered during apoptosis [57–62] and it is an attractive proposition that surface sugars could participate in the formation of molecular patterns that signal

Tab. 7.2 Molecules implicated in mammalian apoptotic cell clearance at interactive cell surfaces.

Location [a]	*Molecule*	*Reference* [b]
Phagocyte	lectins	57, 138
Phagocyte	$\alpha_v\beta_3$ integrin	139
Phagocyte	CD36	51
Phagocyte	PS receptor	56, 140
Phagocyte	asialoglycoprotein receptor	58
Phagocyte	CD68	141
Phagocyte	scavenger receptor A	142
Phagocyte [c]	ATP-binding cassette transporter (ABC)-1	143
Phagocyte	CD14	5
Phagocyte	CR3, CR4	144
Phagocyte	$\alpha_v\beta_5$ integrin	145
Phagocyte	scavenger receptor B1	146, 147
Phagocyte	PS	148
Phagocyte	MER	111
Phagocyte	CD91	52
Phagocyte	CD31	149
Apoptotic cell	sugars	57, 138
Apoptotic cell	PS	54
Apoptotic cell	ICAM-3	74
Apoptotic cell	oxidized structures	150
Apoptotic cell	CD31	149
Intermediate	thrombospondin	51
Intermediate	β_2-glycoprotein 1	151
Intermediate	C1q	107
Intermediate	C3bi	144
Intermediate	Gas6	152
Intermediate	C-reactive protein	153
Intermediate	mannose-binding lectin	52
Intermediate	surfactant protein-A	53
Intermediate	calreticulin	52
Intermediate	milk fat globule-EGF-factor 8	154

a) Molecules are categorized as (1) integral membrane structures of either phagocytes (including macrophages and amateur phagocytes), (2) integral membrane structures of apoptotic cells or (3) intermediate components including soluble factors that opsonize/bridge or peripheral membrane components.

b) Citations focus on the seminal studies.

c) Evidence from studies of the *C. elegans* homolog of ABC-1, CED-7, suggests that ABC-1 could act both on phagocyte and apoptotic cell surfaces [155].

apoptosis to the extracellular microenvironment. Polysaccharides that have been shown to induce CD14-dependent responses are currently limited to those of bacterial or algal origin [38, 63–67]. Therefore, there is no evidence to date to support the notion that CD14 may interact with apoptotic cell surface structures through its lectin-like activity. Other candidate molecules that have been shown to interact with CD14, but as yet have no known role in apoptotic cell clearance, are heat-shock proteins (HSP-60 and -70) [68, 69] and ceramide [70]. Intriguingly, these HSPs have been reported to become cell-surface associated during apoptosis [71, 72]. Ceramide has been proposed to emulate LPS in three-dimensional structure [73] and this provides a rationale for its potential to interact directly with CD14. Again, however, there is as yet no evidence that CD14-dependent clearance of apoptotic cells involves either HSPs or ceramide.

Circumstantial evidence suggests that the immunoglobulin superfamily (IgSF) member, intercellular adhesion molecule (ICAM)-3 (CD50), participates in the route to apoptotic cell clearance by human macrophages that requires CD14 [74]. ICAM-3 is expressed by all mature human leukocytes, but virtually nothing is known about its function, excepting its role on T cells in aiding interactions with antigen-presenting cells that promote T cell activation. During apoptosis, ICAM-3 becomes qualitatively altered such that it is unable to interact with its prototypic counter-receptor, LFA-1. Instead, it appears that ICAM-3 gains the capacity to interact with macrophage molecules that are functional in apoptotic cell clearance [74]. Thus, as a consequence of the apoptosis programme becoming activated, the function of ICAM-3 is radically changed from its pre-apoptotic role. That ICAM-3 gains the capacity to interact with CD14 as a result of the induction of apoptosis is implied from observations of anti-CD14 and anti-ICAM-3 blocking monoclonal antibodies which fail to produce additive inhibitory effects [74]. Since, in these experiments, the anti-CD14 monoclonal antibody blocked on the side of the macrophage and the anti-ICAM-3 monoclonal antibody blocked on the side of the apoptotic cells [74], one interpretation is that the monoclonal antibodies are blocking two arms of the same pathway. However, it is not yet clear whether ICAM-3, in its apoptotic form, can bind directly to CD14.

7.3
Mechanisms: Conceptualizing CD14's Role in Apoptotic Cell Clearance

7.3.1
CD14 as a PRR that Recognizes Apoptotic Cell-associated Molecular Patterns (ACAMPs)

The concept of pattern recognition was originally invoked to rationalize the capacity of the innate immune system to recognize conserved microbial structures which, on pathogens, have come to be known collectively as pathogen-associated molecular patterns (PAMPs) [75, 76]. PAMPs are recognized by innate immune receptors known as PRRs. It has been proposed that all macrophage receptors may be considered as PRRs [77], but CD14, by way of its capacity to recognize LPS, clearly falls into this category even when PRRs are viewed less generally [47]. Since CD14 is involved in binding not only PAMPs, but also 'self' components, its receptor function is wider than that originally envisaged of a classical PRR. Of relevance to its role in apoptotic cell clearance is the concept that CD14 interacts with a broader range of molecular patterns that includes ACAMPs [78, 79], otherwise known as 'eat me' signals [80]. The simplest model predicts that CD14 binds directly to ACAMPs (Fig. 7.2). However, since the CD14 ligands that function in apoptotic cell clearance remain elusive and taking account of the numerous soluble factors that contribute to the apoptotic cell clearance process, there is a strong possibility that CD14 associates with apoptotic cell surfaces through indirect means (Fig. 7.2).

Just as the ACAMPs that involve CD14 in the clearance of apoptotic cells are unknown, so too are the mechanisms that underlie their production, regardless of whether they interact directly or indirectly with CD14. Indeed, very little is yet known of the biochemical mechanisms that facilitate apoptotic cell clearance driven by any of the phagocyte receptors that are implicated in the process. Some experimental approaches have sought relationships between molecules controlling the self-destruction of the cell by apoptosis and the capacity to be phagocytosed by CD14-dependent (and -independent) mechanisms. For example, the anti-apoptosis molecule Bcl-2, although reported not to impinge upon the capacity of neutrophils to be phagocytosed despite promoting their survival in the absence of macrophages [81], prevented human lymphocytes from displaying features of apoptosis, including the capacity to be phagocytosed by CD14-dependent mechanisms [82]. While it is not yet known whether caspase activity is required for the surface changes that allow engulfment of apoptotic cells, recent evidence suggests that caspase-3, the effector caspase that is responsible for several of the hallmarks of apoptosis, is not required – either for CD14-dependent or -independent, mechanisms [83]. Intriguingly, caspase-3 activity is similarly superfluous for the externalization of PS that occurs during apoptosis [38, 84, 85], and in view of evidence that suggests that physiological cell death can occur in the absence of caspases [86–92] it will be important to understand the relationship between caspase activation in apoptosis and the requirements for apoptotic cell clearance. There is already clear evidence that the apoptosis-like demise of platelets leads to phagocytosis in the absence of caspase activation [93].

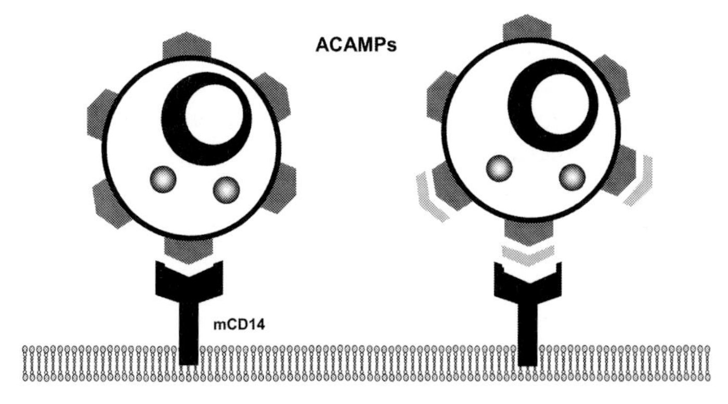

Fig. 7.2 Cartoon to illustrate modes of interaction between ACAMPs and macrophage mCD14. The scheme on the left shows the simplest scenario of direct interaction between CD14 and ACAMP, while on the right a simplistic view of indirect interaction involving bridging molecules or opsonins is shown.

7.3.2
Differential CD14 Signaling following Ligand Binding

Since CD14 can generate profoundly different responses to different ligands, it follows that ligand-dependent differences between signal transduction pathways downstream of receptor ligation must exist. In view of its GPI anchorage to the plasma membrane, it is probable that mCD14 cooperates with transmembrane co-receptors in order to couple extracellular ligand-binding events to intracytoplasmic signal transduction pathways [4] (although other mechanisms have also been proposed – reviewed in [94]). Candidate CD14 co-receptors in CD14-dependent pro-inflammatory responses are CR3 [26, 70], TLR-2 and TLR-4 [18–21, 67, 96], and MD-2 [70, 97, 98]. TLR-4 has been shown by genetic studies to be critical for physiological LPS responses [18, 21], whereas TLR-2 is involved in CD14-dependent responses to other microbial components such as peptidoglycan and lipoteichoic acid [99, 100] and including minor contaminants of LPS preparations [101]. Biochemical cross-linking and fluorescence resonance energy transfer microscopy have indicated that CR3, TLR-4 and MD-2 become physically closely associated with CD14 as a consequence of the presence of LPS [26, 27, 97]. Therefore, for responsiveness to LPS, CD14 forms one component of a multi-molecular receptor complex. The same seems likely to be true in the case of its role in responses to other ligands including those mediating apoptotic cell clearance. In this context it is noteworthy that CD14 has been shown to be capable of associating with CD36, a scavenger receptor implicated in apoptotic cell engulfment (Tab. 7.2) [70].

How could mCD14 elicit anti-inflammatory signaling following interaction (directly or indirectly) with ACAMPs? Three major scenarios are envisaged (Fig. 7.3). The first predicts that the ACAMPs determine the composition of a receptor complex around CD14 that differs from that which is determined by LPS. The hypothetical ACAMP-determined complex would then be expected to link up to an anti-inflammatory signaling pathway (Fig. 7.3 A), rather than the pro-inflammatory signal transduction pathway that is coupled to the LPS-determined receptor complex. In the second scenario (Fig. 7.3 B), CD14-interactive ACAMPs would, like LPS, elicit a default pro-inflammatory signaling response that would be suppressed by additional ACAMPs interacting with alternative phagocyte receptors. The implication here is that the complex nature of the apoptotic cell or body allows it to display multiple ligands that together coordinate the anti-inflammatory clearance process. In the third scenario, CD14 is envisaged as playing a tethering role only – perhaps because it is unable to associate with the signaling partners required for its pro-inflammatory function – and fails to activate any intracellular signals (Fig. 7.3 C). Again, this mechanism implies that additional ACAMP–receptor interactions are required to complement the tethering function of CD14 and generate anti-inflammatory phagocytosis.

All three of these scenarios accord well with available evidence, although that relating specifically to CD14 is currently limited. It has been proposed recently that the engulfment of apoptotic cells is a two-phase, 'tether-and-tickle', process composed (1) of mechanisms that facilitate cell–cell interactions, and (2) of events that drive anti-inflammatory signaling and engulfment [102]. It seems likely that the process is composed of three facets with separate molecular mechanisms: (1) tethering, (2) signal transduction culminating in engulfment and (3) signal transduction suppressing inflammatory responses. In this model, CD14 is currently thought to play a tethering role [5, 102, 103], but further work is required in order to determine whether it is restricted in this respect. An attractive working hypothesis is that CD14 acts to tether apoptotic cells to macrophages and, because it fails to interact preferentially with PS, allows PS-R ligation and subsequent signaling for anti-inflammation and engulfment, both functions recently claimed for the PS-R [56, 102, 104]. It remains possible, however, that CD14, through interaction with co-receptors, can also contribute to signal transduction in apoptotic cell clearance.

7.4
Conclusion: Relative Importance of CD14 in Apoptotic Cell Clearance

In view of the highly complex nature of the molecular interactions between apoptotic cells and phagocytes involving integral plasma-membrane elements of both cell types as well as non-integral 'intermediate' components (Tab. 7.2), it is impossible to judge the importance of one component in isolation. Superficially, the mechanisms responsible for the clearance of apoptotic cells appear redundant, reflecting the importance of efficient clearance, not least in militating against autoimmune disease pathogenesis. The multiplicity of mechanisms, however, may not represent redundancy, but rather individually essential mechanisms that (1) operate in different phagocytes including subsets of macrophages and (2) engage different phases of the apoptotic process. In the case of mCD14, the key to its signif-

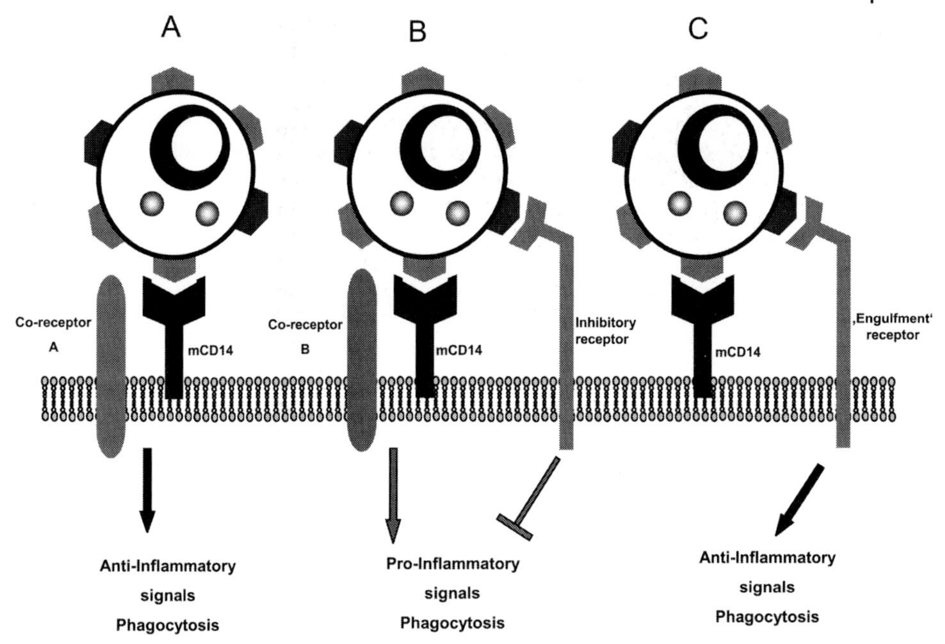

Fig. 7.3 Cartoon depicting theoretical models of signal generation following CD14 interacting (directly or indirectly) with ACAMPs that leads to anti-inflammatory phagocytosis of apoptotic cells. (A) In this scheme, GPI-anchored mCD14 cooperates with a transmembrane receptor to signal anti-inflammatory phagocytosis. The receptor complex would be predicted to be different from that involved in LPS responses. (B) Here, mCD14 is complexed with a co-receptor that generates pro-inflammatory signals (i.e. the receptor complex is identical to that invoked by LPS). These are switched off by an inhibitory receptor that engages other ACAMPs. (C) This represents mCD14's role simply as a signalling-inert tethering molecule. In this scenario, other ACAMPs engage additional receptors ('engulfment' receptors) that provide the signaling mechanisms for anti-inflammatory responses and phagocytosis. The PS-R would fit into the category of engulfment and anti-inflammatory signaling receptor. See text for details.

icance in apoptotic cell clearance lies in its restricted distribution. Thus, dendritic cells, which are functional in antigen presentation and immune activation following apoptotic cell engulfment, do not express CD14, in contrast to macrophages, which are immunosuppressive following apoptotic cell clearance [105, 106]. Therefore, one prediction based on the published literature would be that CD14 is involved in tolerogenic clearance, simply because of its cellular distribution [79]. In this respect it can be viewed as an important innate regulator of the adaptive immune response to apoptotic cell-associated antigens. Of course, this simple viewpoint could change radically should sCD14 also be shown to play a role in apoptotic cell clearance.

By far the majority of studies of the molecular mechanisms of apoptotic cell clearance, including those of CD14, have been carried out using *in vitro* models. Ultimately, only detailed *in vivo* studies can elucidate the relative significance of individual molecules in supporting the clearance of apoptotic cells and indeed how important the clearance process is in preventing disease development. Studies of single-knockout animals have begun [107–111] and extension of this approach to CD14, both alone and in combination with other components of the apoptotic cell clearance machinery, will help to provide a definitive appraisal of the role of CD14 – and of the apoptotic cell clearance process itself – in health and disease.

7.5 References

1 Goyert, S. M., E. Ferrero, *et al.* The CD14 monocyte differentiation antigen maps to a region encoding growth factors and receptors. *Science* **1988**, 239, 497–500.

2 Simmons, D. L., S. Tan, *et al.* Monocyte antigen CD14 is a phospholipid anchored membrane protein. *Blood* **1989**, 73, 284–9.

3 Wright, S. D., R. A. Ramos, *et al.* CD14, a receptor for complexes of lipopolysaccharide (LPS) and LPS binding protein. *Science* **1990**, 249, 1431–3.

4 Ulevitch, R. J. and P. S. Tobias. Receptor-dependent mechanisms of cell stimulation by bacterial endotoxin. *Annu Rev Immunol* **1995**, 13, 437–57.

5 Devitt, A., O. D. Moffatt, *et al.* Human CD14 mediates recognition and phagocytosis of apoptotic cells. *Nature* **1998**, 392, 505–509.

6 Grunwald, U., C. Kruger, *et al.* An enzyme-linked-immunosorbent-assay for the quantification of solubilized CD14 in biological-fluids. *J Immunol Methods* **1992**, 155, 225–32.

7 Stelter, F. Structure/function relationships of CD14. *Chemical Immunology* **2000**, 74, 25–41.

8 Gadd, S. Cluster report: CD14. In: W. Knapp, B. Dorken, W. R. Gilks, *et al.*, eds. *Leukocyte Typing IV White Cell Differentiation Antigens.* Oxford: Oxford University Press **1989**, 787–9.

9 Guilbert, L., S. A. Robertson, *et al.* The trophoblast as an integral component of a macrophage cytokine network. *Immunol Cell Biol* **1993**, 71, 49–57.

10 Ziegler-Heitbrock, H. W. L. and R. J. Ulevitch. CD14: cell surface receptor and differentiation marker. *Immunol Today* **1993**, 14, 121–5.

11 Fearns, C., V. V. Kravchenko, *et al.* Murine CD14 gene-expression *in vivo* – extramyeloid synthesis and regulation by lipopolysaccharide. *J Exp Med* **1995**, 181, 857–66.

12 Diamond, G., J. P. Russell, *et al.* Inducible expression of an antibiotic peptide gene in lipopolysaccharide-challenged tracheal epithelial cells. *Proc Natl Acad Sci USA* **1996**, 93, 5156–60.

13 Liu, S. B., L. S. Khemlani, *et al.* Expression of CD14 by hepatocytes: upregulation by cytokines during endotoxemia. *Infect Immun* **1998**, 66, 5089–98.

14 Su, G. L., K. Dorko, *et al.* CD14 expression and production by human hepatocytes. *J Hepatol* **1999**, 31, 435–42.

15 Jersmann, H. P. A., C. S. T. Hii, *et al.* Synthesis and surface expression of CD14 by human endothelial cells. *Infect Immun* **2001**, 69, 479–85.

16 Stelter, F., M. Pfister, *et al.* The myeloid differentiation antigen CD14 is *N*- and *O*-glycosylated – contribution of *N*-linked glycosylation to different soluble CD14 isoforms. *Eur J Biochem* **1996**, 236, 457–64.

17 Kobe, B. and J. Deisenhofer. The leucine-rich repeat – a versatile binding motif. *Trends Biochem Sci* **1994**, 19, 415–21.

18 Poltorak, A., X. L. He, *et al.* Defective LPS signaling in C3H/HeJ and C57BL/10ScCr mice: mutations in Tlr4 gene. *Science* **1998**, 282, 2085–8.

19 Chow, J. C., D. W. Young, *et al.* Toll-like receptor-4 mediates lipopolysaccharide-induced signal transduction. *J Biol Chem* **1999**, 274, 10689–92.

20 Hoshino, K., O. Takeuchi, *et al.* Cutting edge: Toll-like receptor 4 (TLR4)-deficient mice are hyporesponsive to lipopolysaccharide: evidence for TLR4 as the *Lps* gene product. *J Immunol* **1999**, 162, 3749–52.

21 Qureshi, S.T., L. Lariviere, *et al.* Endotoxin-tolerant mice have mutations in toll-like receptor 4 (Tlr4). *J Exp Med* **1999**, 189, 615–25.

22 Hailman, E., H.S. Lichenstein, *et al.* Lipopolysaccharide (LPS)-binding protein accelerates the binding of LPS to CD14. *J Exp Med* **1994**, 179, 269–77.

23 Tobias, P.S., K. Soldau, *et al.* Lipopolysaccharide-binding protein-mediated complexation of lipopolysaccharide with soluble CD14. *J Biol Chem* **1995**, 270, 10482–8.

24 Yu, B., E. Hailman, *et al.* Lipopolysaccharide binding protein and soluble CD14 catalyze exchange of phospholipids. *J Clin Invest* **1997**, 99, 315–24.

25 Tapping, R. I. and P. S. Tobias. Soluble CD14-mediated cellular responses to lipopolysaccharide. *Chemical Immunology* **2000**, 74, 108–21.

26 Zarewych, D.M., A.L. Kindzelskii, *et al.* LPS induces CD14 association with complement receptor type 3, which is reversed by neutrophil adhesion. *J Immunol* **1996**, 156, 430–3.

27 Jiang, Q.Q., S. Akashi, *et al.* Cutting edge: lipopolysaccharide induces physical proximity between CD14 and Toll-like receptor 4 (TLR4) prior to nuclear translocation of NF-kappa B. *J Immunol* **2000**, 165, 3541–4.

28 Vita, N., S. Lefort, *et al.* Detection and biochemical characteristics of the receptor for complexes of soluble CD14 and bacterial lipopolysaccharide. *J Immunol* **1997**, 158, 3457–62.

29 Juan, T.S.C., M.J. Kelley, *et al.* Soluble CD14 truncated at amino acid 152 binds lipopolysaccharide (LPS) and enables cellular-response to LPS. *J Biol Chem* **1995**, 270, 1382–7.

30 Viriyakosol, S. and T.N. Kirkland. The N-terminal half of membrane CD14 is a functional cellular lipopolysaccharide receptor. *Infect Immun* **1996**, 64, 653–6.

31 Pugin, J., V.V. Kravchenko, *et al.* Cell activation mediated by glycosylphosphatidylinositol-anchored or transmembrane forms of CD14. *Infect Immun* **1998**, 66, 1174–80.

32 Viriyakosol, S. and T.N. Kirkland. A region of human CD14 required for lipopolysaccharide-binding. *J Biol Chem* **1995**, 270, 361–8.

33 Stelter, F., M. Bernheiden, *et al.* Mutation of amino acids 39–44 of human CD14 abrogates binding of lipopolysaccharide and *Escherichia coli*. *Eur J Biochem* **1997**, 243, 100–9.

34 McGinley, M.D., L.O. Narhi, *et al.* CD14 – physical properties and identification of an exposed site that is protected by lipopolysaccharide. *J Biol Chem* **1995**, 270, 5213–8.

35 Juan, T.S.C., E. Hailman, *et al.* Identification of a lipopolysaccharide binding domain in CD14 between amino acids 57 and 64. *J Biol Chem* **1995**, 270, 5219–24.

36 Juan, T.S.C., E. Hailman, *et al.* Identification of a domain in soluble CD14 essential for lipopolysaccharide (LPS) signaling but not LPS binding. *J Biol Chem* **1995**, 270, 17237–42.

37 Cunningham, M.D., R.A. Shapiro, *et al.* CD14 employs hydrophilic regions to capture lipopolysaccharides. *J Immunol* **2000**, 164, 3255–63.

38 Soell, M., E. Lett, *et al.* Activation of human monocytes by streptococcal rhamnose glucose polymers is mediated by CD14 antigen, and mannan-binding protein inhibits TNF-alpha release. *J Immunol* **1995**, 154, 851–860.

39 Weidemann, B., J. Schletter, *et al.* Specific binding of soluble peptidoglycan and muramyldipeptide to CD14 on human monocytes. *Infect Immun* **1997**, 65, 858–64.

40 Wang, P.Y., R.L. Kitchens, *et al.* Phosphatidylinositides bind to plasma membrane CD14 and can prevent monocyte activation by bacterial lipopolysaccharide. *J Biol Chem* **1998**, 273, 24309–13.

41 Wooten, R.M., T.B. Morrison, *et al.* The role of CD14 in signaling mediated by outer membrane lipoproteins of *Borrelia burgdorferi*. *J Immunol* **1998**, 160, 5485–92.

42 Dziarski, R. and D. Gupta. Function of CD14 as a peptidoglycan receptor: differences and similarities with LPS. *J Endotoxin Res* **1999**, 5, 56–61.

43 Sano, H., H. Sohma, *et al.* Pulmonary surfactant protein A modulates the cellular response to smooth and rough lipopolysaccharides by interaction with CD14. *J Immunol* **1999**, 163, 387–95.

44 Sano, H., H. Chiba, *et al.* Surfactant proteins A and D bind CD14 by different mechanisms. *J Biol Chem* **2000**, 275, 22442–51.

45 Chiba, H., H. Sano, *et al.* Rat mannose-binding protein A binds CD14. *Infect Immun* **2001**, 69, 1587–92.

46 Sugiyama, T. and S.D. Wright. Soluble CD14 mediates efflux of phospholipids from cells. *J Immunol* **2001**, 166, 826–31.

47 Pugin, J., D. Heumann, *et al.* CD14 is a pattern recognition receptor. *Immunity* **1994**, 1, 509–16.

48 Flora, P.K. and C.D. Gregory. Recognition of apoptotic cells by human macrophages – inhibition by a monocyte/macrophage-specific monoclonal antibody. *Eur J Immunol* **1994**, 24, 2625–32.

49 Ugolini, V., G. Nunez, *et al.* Initial characterization of monoclonal antibodies against human monocytes. *Proc Natl Acad Sci USA* **1980**, 77, 6764–8.

50 Ren, Y., R.L. Silverstein, *et al.* CD36 gene transfer confers capacity for phagocytosis of cells undergoing apoptosis. *J Exp Med* **1995**, 181, 1857–62.

51 Savill, J., N. Hogg, *et al.* Thrombospondin cooperates with CD36 and the vitronectin receptor in macrophage recognition of neutrophils undergoing apoptosis. *J Clin Invest* **1992**, 90, 1513–22.

52 Ogden, C.A., A. deCathehneau, *et al.* C1q and mannose binding lectin engagement of cell surface calreticulin and CD91 initiates macropinocytosis and uptake of apoptotic cells. *J Exp Med* **2001**, 194, 781–95.

53 Schagat, T.L., J.A. Wofford, *et al.* Surfactant protein A enhances alveolar macrophage phagocytosis of apoptotic neutrophils. *J Immunol* **2001**, 166, 2727–33.

54 Fadok, V.A., D.R. Voelker, *et al.* Exposure of phosphatidylserine on the surface of apoptotic lymphocytes triggers specific recognition and removal by macrophages. *J Immunol* **1992**, 148, 2207–16.

55 Devitt, A., S. Pierce, *et al.* CD14-dependent clearance of apoptotic cells by human macrophages: the role of phosphatidylserine. *Cell Death Diff* (in press).

56 Fadok, V.A., D.L. Bratton, *et al.* A receptor for phosphatidylserine-specific clearance of apoptotic cells. *Nature* **2000**, 405, 85–90.

57 Duvall, E., A.H. Wyllie, *et al.* Macrophage recognition of cells undergoing programmed cell death (apoptosis). *Immunology* **1985**, 56, 351–8.

58 Dini, L., F. Autuori, *et al.* The clearance of apoptotic cells in the liver is mediated by the asialoglycoprotein receptor. *FEBS Lett* **1992**, 296, 174–8.

59 Hall, S.E., J.S. Savill, *et al.* Apoptotic neutrophils are phagocytosed by fibroblasts with participation of the fibroblast vitronectin receptor and involvement of a mannose/fucose-specific lectin. *J Immunol* **1994**, 153, 3218–27.

60 Falasca, L., A. Bergamini, *et al.* Human Kupffer cell recognition and phagocytosis of apoptotic peripheral blood lymphocytes. *Exp Cell Res* **1996**, 224, 152–62.

61 Russell, L., P. Waring, *et al.* Increased cell surface exposure of fucose residues is a late event in apoptosis. *Biochem Biophys Res Commun* **1998**, 250, 449–53.

62 Beaver, J.P. and C.L. Stoneman. Exposure of *N*-acetylglucosamine decreases early in dexamethasone-induced apoptosis in thymocytes, demonstrated by flow cytometry using wheat germ agglutinin and pokeweed mitogen. *Immunol Cell Biol* **1999**, 77, 224–35.

63 Espevik, T., M. Otterlei, *et al.* The involvement of CD14 in stimulation of cytokine production by uronic acid polymers. *Eur J Immunol* **1993**, 23, 255–61.

64 Cavaillon, J.M., C. Marie, *et al.* CD14–LPS receptor exhibits lectin-like properties. *J Endotoxin Res* **1996**, 3, 471–80.

65 Jahr, T.G., L. Ryan, *et al.* Induction of tumor necrosis factor production from monocytes stimulated with mannuronic acid polymers and involvement of lipopolysaccharide-binding protein, CD14, and bactericidal/permeability-increasing factor. *Infect Immun* **1997**, 65, 89–94.

66 Flo, T.H., L. Ryan, *et al.* Involvement of CD14 and beta 2-integrins in activating cells with soluble and particulate lipopolysaccharides and mannuronic acid polymers. *Infect Immun* **2000**, 68, 6770–6.

67 Yang, S.H., S. Sugawara, *et al.* *Micrococcus luteus* teichuronic acids activate human and murine monocytic cells in a CD14- and toll-like receptor 4-dependent manner. *Infect Immun* **2001**, 69, 2025–30.

68 Asea, A., S.K. Kraeft, *et al.* HSP70 stimulates cytokine production through a CD14-dependent pathway, demonstrating its dual role as a chaperone and cytokine. *Nature Med* **2000**, 6, 435–42.

69 Kol, A., A.H. Lichtman, *et al.* Cutting edge: heat shock protein (HSP) 60 activates the innate immune response: CD14 is an essential receptor for HSP60 activation of mononuclear cells. *J Immunol* **2000**, 164, 13–7.

70 Pfeiffer, A., A. Bottcher, *et al.* Lipopolysaccharide and ceramide docking to CD14 provokes ligand-specific receptor clustering in rafts. *Eur J Immunol* **2001**, 31, 3153–64.

71 Poccia, F., P. Piselli, *et al.* Heat-shock protein expression on the membrane of T cells undergoing apoptosis. *Immunology* **1996**, 88, 6–12.

72 Sapozhnikov, A.M., E.D. Ponomarev, *et al.* Spontaneous apoptosis and expression of cell surface heat-shock proteins in cultured EL-4 lymphoma cells. *Cell Prolif* **1999**, 32, 363–78.

73 Wright, S.D. and R.N. Kolesnick. Does endotoxin stimulate cells by mimicking ceramide. *Immunol Today* **1995**, 16, 297–302.

74 Moffatt, O.D., A. Devitt, *et al.* Macrophage recognition of ICAM-3 on apoptotic leukocytes. *J Immunol* **1999**, 162, 6800–6810.

75 Janeway, C.A. Approaching the asymptote – evolution and revolution in immunology. *Cold Spring Harbor Symp Quant Biol* **1989**, 54, 1–13.

76 Medzhitov, R. and C.A. Janeway. Innate immunity: the virtues of a nonclonal system of recognition. *Cell* **1997**, 91, 295–8.

77 Stahl, P.D. and R.A.B. Ezekowitz. The mannose receptor is a pattern recognition receptor involved in host defense. *Curr Opin Immunol* **1998**, 10, 50–5.

78 Franc, N.C., K. White, *et al.* Phagocytosis and development: back to the future. *Curr Opin Immunol* **1999**, 11, 47–52.

79 Gregory, C.D. CD14-dependent clearance of apoptotic cells: relevance to the immune system. *Curr Opin Immunol* **2000**, 12, 27–34.

80 Savill, J. and V. Fadok. Corpse clearance defines the meaning of cell death. *Nature* **2000**, 407, 784–8.

81 Lagasse, E. and I.L. Weissman. Bcl-2 inhibits apoptosis of neutrophils but not their engulfment by macrophages. *J Exp Med* **1994**, 179, 1047–52.

82 Flora, P.K., A. Devitt, *et al.* Bcl-2 delays macrophage engulfment of human B cells induced to undergo apoptosis. *Eur J Immunol* **1996**, 26, 2243–7.

83 Turner, C., A. Devitt, *et al.* Macrophage-mediated clearance of cells undergoing caspase-3-independent death. *Cell Death Diff* (in press).

84 Woo, M., R. Hakem, *et al.* Essential contribution of caspase 3 CPP32 to apoptosis and its associated nuclear changes. *Genes Dev* **1998**, 12, 806–19.

85 Zheng, T.S., S.F. Schlosser, *et al.* Caspase-3 controls both cytoplasmic and nuclear events associated with Fas-mediated apoptosis *in vivo*. *Proc Natl Acad Sci USA* **1998**, 95, 13618–23.

86 Xiang, J. L., D.T. Chao, *et al.* BAX-induced cell death may not require interleukin 1 beta-converting enzyme-like proteases. *Proc Natl Acad Sci USA* **1996**, 93, 14559–63.

87 McCarthy, N.J., M.K.B. Whyte, *et al.* Inhibition of Ced-3/ICE-related proteases does not prevent cell death induced by oncogenes, DNA damage, or the Bcl-2 homologue Bak. *J Cell Biol* **1997**, 136, 215–27.

88 Kawahara, A., Y. Ohsawa, *et al.* Caspase-independent cell killing by Fas-associated protein with death domain. *J Cell Biol* **1998**, 143, 1353–60.

89 Monney, L., I. Otter, *et al.* Defects in the ubiquitin pathway induce caspase-independent apoptosis blocked by Bcl-2. *J Biol Chem* **1998**, 273, 6121–31.

90 Quignon, F., F. De Bels, *et al.* PML induces a novel caspase-independent death process. *Nat Genet* **1998**, 20, 259–65.
91 Bojes, H. K., X. Feng, *et al.* Apoptosis in hematopoietic cells (FL5.12) caused by interleukin-3 withdrawal: relationship to caspase activity and the loss of glutathione. *Cell Death Different* **1999**, 6, 61–70.
92 Doerfler, P., K. A. Forbush, *et al.* Caspase enzyme activity is not essential for apoptosis during thymocyte development. *J Immunol* **2000**, 164, 4071–9.
93 Brown, S. B., M. C. H. Clarke, *et al.* Constitutive death of platelets leading to scavenger receptor-mediated phagocytosis – a caspase-independent cell clearance program. *J Biol Chem* **2000**, 275, 5987–96.
94 Gregory, C. D. and A. Devitt. CD14 and apoptosis. *Apoptosis* **1999**, 4, 11–20.
95 Yang, R.-B., M. R. Mark, *et al.* Toll-like receptor-2 mediates lipopolysaccharide-induced cellular signalling. *Nature* **1998**, 395, 284–8.
96 Yang, R. B., M. R. Mark, *et al.* Signaling events induced by lipopolysaccharide-activated toll-like receptor 2. *J Immunol* **1999**, 163, 639–43.
97 Correia, J. D., K. Soldau, *et al.* Lipopolysaccharide is in close proximity to each of the proteins in its membrane receptor complex – transfer from CD14 to TLR4 and MD-2. *J Biol Chem* **2001**, 276, 21129–35.
98 Schwandner, R., R. Dziarski, *et al.* Peptidoglycan- and lipoteichoic acid-induced cell activation is mediated by toll-like receptor 2. *J Biol Chem* **1999**, 274, 17406–9.
99 Dziarski, R., Q. L. Wang, *et al.* MD-2 enables toll-like receptor 2 (TLR2)-mediated responses to lipopolysaccharide and enhances TLR2-mediated responses to gram-positive and gram-negative bacteria and their cell wall components. *J Immunol* **2001**, 166, 1938–44.
100 Yoshimura, A., E. Lien, *et al.* Cutting edge: recognition of gram-positive bacterial cell wall components by the innate immune system occurs via toll-like receptor 2. *J Immunol* **1999**, 163, 1–5.
101 Hirschfeld, M., Y. Ma, *et al.* Cutting edge: repurification of lipopolysaccharide eliminates signaling through both human and murine toll-like receptor 2. *J Immunol* **2000**, 165, 618–622.
102 Hoffmann, P. R., A. M. deCathelineau, *et al.* Phosphatidylserine (PS) induces PS receptor-mediated macropinocytosis and promotes clearance of apoptotic cells. *J Cell Biol* **2001**, 155, 649–59.
103 Savill, J. Apoptosis – phagocytic docking without shocking. *Nature* **1998**, 392, 442–3.
104 Fadok, V. A., A. de Cathelineau, *et al.* Loss of phospholipid asymmetry and surface exposure of phosphatidylserine is required for phagocytosis of apoptotic cells by macrophages and fibroblasts. *J Biol Chem* **2001**, 276, 1071–7.
105 Albert, M. L., B. Sauter, *et al.* Dendritic cells acquire antigen from apoptotic cells and induce class I restricted CTLs. *Nature* **1998**, 392, 86–9.
106 Ronchetti, A., P. Rovere, *et al.* Immunogenicity of apoptotic cells *in vivo*: role of antigen load, antigen-presenting cells, and cytokines. *J Immunol* **1999**, 163, 130–6.
107 Botto, M., C. DellAgnola, *et al.* Homozygous C1q deficiency causes glomerulonephritis associated with multiple apoptotic bodies. *Nat Genet* **1998**, 19, 56–9.
108 Hamon, Y., C. Broccardo, *et al.* ABC1 promotes engulfment of apoptotic cells and transbilayer redistribution of phosphatidylserine. *Nat Cell Biol* **2000**, 2, 399–406.
109 Platt, N., H. Suzuki, *et al.* Apoptotic thymocyte clearance in scavenger receptor class A-deficient mice is apparently normal. *J Immunol* **2000**, 164, 4861–7.
110 Taylor, P. R., A. Carugati, *et al.* A hierarchical role for classical pathway complement proteins in the clearance of apoptotic cells *in vivo*. *J Exp Med* **2000**, 192, 359–66.
111 Scott, R. S., E. J. McMahon, *et al.* Phagocytosis and clearance of apoptotic cells is mediated by MER. *Nature* **2001**, 411, 207–11.
112 Kusunoki, T., E. Hailman, *et al.* Molecules from *Staphylococcus aureus* that bind CD14 and stimulate innate immune responses. *J Exp Med* **1995**, 182, 1673–82.

113 Kusunoki, T. and S. D. Wright. Chemical characteristics of *Staphylococcus aureus* molecules that have CD14-dependent cell-stimulating activity. *J Immunol* **1996**, 157, 5112–7.

114 Fichorova, R. N., A. O. Cronin, *et al.* Response to *Neisseria gonorrhoeae* by cervicovaginal epithelial cells occurs in the absence of toll-like receptor 4-mediated signaling. *J Immunol* **2002**, 168, 2424–32.

115 Cleveland, M. G., J. D. Gorham, *et al.* Lipoteichoic acid preparations of Gram-positive bacteria induce interleukin-12 through a CD14-dependent pathway. *Infect Immun* **1996**, 64, 1906–12.

116 Hattor, Y., K. Kasai, *et al.* Induction of NO synthesis by lipoteichoic acid from *Staphylococcus aureus* in J774 macrophages: involvement of a CD14-dependent pathway. *Biochem Biophys Res Commun* **1997**, 233, 375–9.

117 Sugawara, S., R. Arakaki, *et al.* Lipoteichoic acid acts as an antagonist and an agonist of lipopolysaccharide on human gingival fibroblasts and monocytes in a CD14-dependent manner. *Infect Immun* **1999**, 67, 1623–32.

118 Weidemann, B., H. Brade, *et al.* Soluble peptidoglycan-induced monokine production can be blocked by anti-CD14 monoclonal antibodies and by lipid A partial structures. *Infect Immun* **1994**, 62, 4709–15.

119 Gupta, D., Y. P. Jin, *et al.* Peptidoglycan induces transcription and secretion of TNF-alpha and activation of lyn, extracellular signal-regulated kinase, and rsk signal-transduction proteins in mouse macrophages. *J Immunol* **1995**, 155, 2620–30.

120 Gupta, D., T. N. Kirkland, *et al.* CD14 is a cell-activating receptor for bacterial peptidoglycan. *J Biol Chem* **1996**, 271, 23310–6.

121 Dziarski, R., R. I. Tapping, *et al.* Binding of bacterial peptidoglycan to CD14. *J Biol Chem* **1998**, 273, 8680–90.

122 Gupta, D., Q. L. Wang, *et al.* Bacterial peptidoglycan induces CD14-dependent activation of transcription factors CREB ATF and AP-1. *J Biol Chem* **1999**, 274, 14012–20.

123 Zhang, Y. H., M. Doerfler, *et al.* Mechanisms of stimulation of interleukin-1-beta and tumor necrosis factor-alpha by *Mycobacterium tuberculosis* components. *J Clin Invest* **1993**, 91, 2076–83.

124 Savedra, R., R. L. Delude, *et al.* Mycobacterial lipoarabinomannan recognition requires a receptor that shares components of the endotoxin signaling system. *J Immunol* **1996**, 157, 2549–54.

125 Bernardo, J., A. M. Billingslea, *et al.* Differential responses of human mononuclear phagocytes to mycobacterial lipoarabinomannans: role of CD14 and the mannose receptor. *Infect Immun* **1998**, 66, 28–35.

126 Means, T. K., E. Lien, *et al.* The CD14 ligands lipoarabinomannan and lipopolysaccharide differ in their requirement for Toll-like receptors. *J Immunol* **1999**, 163, 6748–55.

127 Medvedev, A. E., P. Henneke, *et al.* Induction of tolerance to lipopolysaccharide and mycobacterial components in Chinese hamster ovary/CD14 cells is not affected by overexpression of toll-like receptors 2 or 4. *J Immunol* **2001**, 167, 2257–67.

128 Gomi, K., K. Kawasaki, *et al.* Toll-like receptor 4–MD-2 complex mediates the signal transduction induced by flavolipin, an amino acid-containing lipid unique to *Flavobacterium meningosepticum. J Immunol* **2002**, 168, 2939–43.

129 Schroder, N. W. J., B. Opitz, *et al.* Involvement of lipopolysaccharide binding protein, CD14, and toll-like receptors in the initiation of innate immune responses by *Treponema* glycolipids. *J Immunol* **2000**, 165, 2683–93.

130 Hmama, Z., A. Mey, *et al.* CD14 and CD11b mediate serum-independent binding to human monocytes of an acylpolygalactoside isolated from *Klebsiella pneumoniae. Infect Immun* **1994**, 62, 1520–7.

131 Henneke, P., O. Takeuchi, *et al.* Novel engagement of CD14 and multiple toll-like receptors by group B streptococci. *J Immunol* **2001**, 167, 7069–76.

132 Newman, S. L., S. Chaturvedi, *et al.* The WI-1 antigen of blastomyces dermatitidis yeasts mediates binding to human macrophage CD11b/CD18 (CR3) and CD14. *J Immunol* **1995**, 154, 753–61.

133 Kurt-Jones, E.A., L. Popova, *et al.* Pattern recognition receptors TLR4 and CD14 mediate response to respiratory syncytial virus. *Nat Immunol* **2000**, 1, 398–401.

134 Wang, J.E., A. Warris, *et al.* Involvement of CD14 and toll-like receptors in activation of human monocytes by *Aspergillus fumigatus* hyphae. *Infect Immun* **2001**, 69, 2402–6.

135 Wang, P.Y. and R.S. Munford. CD14-dependent internalization and metabolism of extracellular phosphatidylinositol by monocytes. *J Biol Chem* **1999**, 274, 23235–41.

136 Bosco, M.C., I. EspinozaDelgado, *et al.* Functional role for the myeloid differentiation antigen CD14 in the activation of human monocytes by IL-2. *J Immunol* **1997**, 159, 2922–31.

137 Beekhuizen, H., I. Blokland, *et al.* CD14 contributes to the adherence of human monocytes to cytokine-stimulated endothelial cells. *J Immunol* **1991**, 147, 3761–7.

138 Morris, R. G., A. D. Hargreaves, *et al.* Hormone-induced cell-death. 2. Surface changes in thymocytes undergoing apoptosis. *Am J Pathol* **1984**, 115, 426–36.

139 Savill, J., I. Dransfield, *et al.* Vitronectin receptor-mediated phagocytosis of cells undergoing apoptosis. *Nature* **1990**, 343, 170–3.

140 Fadok, V.A., J.S. Savill, *et al.* Different populations of macrophages use either the vitronectin receptor or the phosphatidylserine receptor to recognize and remove apoptotic cells. *J Immunol* **1992**, 149, 4029–35.

141 Sambrano, G.R. and D. Steinberg. Recognition of oxidatively damaged and apoptotic cells by an oxidized low-density-lipoprotein receptor on mouse peritoneal macrophages: role of membrane phosphatidylserine. *Proc Natl Acad Sci USA* **1995**, 92, 1396–400.

142 Platt, N., H. Suzuki, *et al.* Role for the class A macrophage scavenger receptor in the phagocytosis of apoptotic thymocytes *in vitro*. *Proc Natl Acad Sci USA* **1996**, 93, 12456–60.

143 Luciani, M.-F. and G. Chimini. The ATP binding cassette transporter ABC1, is required for the engulfment of corpses generated by apoptotic cell death. *EMBO J* **1996**, 15, 226–35.

144 Mevorach, D., J.O. Mascarenhas, *et al.* Complement-dependent clearance of apoptotic cells by human macrophages. *J Exp Med* **1998**, 188, 2313–20.

145 Albert, M.L., S.F.A. Pearce, *et al.* Immature dendritic cells phagocytose apoptotic cells via alpha(v)beta(5) and CD36, and cross-present antigens to cytotoxic T lymphocytes. *J Exp Med* **1998**, 188, 1359–68.

146 Shiratsuchi, A., Y. Kawasaki, *et al.* Role of class B scavenger receptor type I in phagocytosis of apoptotic rat spermatogenic cells by Sertoli cells. *J Biol Chem* **1999**, 274, 5901–8.

147 Svensson, P.A., M.S.C. Johnson, *et al.* Scavenger receptor class B type I in the rat ovary: Possible role in high density lipoprotein cholesterol uptake and in the recognition of apoptotic granulosa cells. *Endocrinology* **1999**, 140, 2494–500.

148 Marguet, D., M.F. Luciani, *et al.* Engulfment of apoptotic cells involves the redistribution of membrane phosphatidylserine on phagocyte and prey. *Nat Cell Biol* **1999**, 1, 454–6.

149 Brown, S., I. Heinisch, *et al.* Apoptosis disables CD31-mediated cell detachment from phagocytes promoting binding and engulfment. *Nature* **2002**, 418, 200–3.

150 Chang, M.K., C. Bergmark, *et al.* Monoclonal antibodies against oxidized low-density lipoprotein bind to apoptotic cells and inhibit their phagocytosis by elicited macrophages: evidence that oxidation-specific epitopes mediate macrophage recognition. *Proc Natl Acad Sci USA* **1999**, 96, 6353–8.

151 Balasubramanian, K., J. Chandra, *et al.* Immune clearance of phosphatidylserine-expressing cells by phagocytes – the role of beta(2)-glycoprotein I in macrophage recognition. *J Biol Chem* **1997**, 272, 31113–7.

152 Ishimoto, Y., K. Ohashi, *et al.* Promotion of the uptake of PS liposomes and apoptotic cells by a product of growth arrest-specific gene, gas6. *J Biochem* **2000**, 127, 411–7.

153 Gershov, D., S. Kim, *et al.* C-reactive protein binds to apoptotic cells, protects the cells from assembly of the terminal complement components, and sustains an antiinflammatory innate immune response: Implications for systemic autoimmunity. *J Exp Med* **2000**, 192, 1353–63.

154 Hanayama, R., M. Tanaka, *et al.* Identification of a factor that links apoptotic cells to phagocytes. *Nature* **2002**, 417, 182–7.

155 Wu, Y.C. and H.R. Horvitz. The *C. elegans* cell corpse engulfment gene *ced-7* encodes a protein similar to ABC transporters. **1998**, *Cell* 93, 951–60.

Part 3
Autoimmunity Caused by Defective Execution of Apoptosis or Defective Clearance of Apoptotic Cells

8
Autoimmune Lymphoproliferative Syndromes (ALPS)

Frédéric Rieux-Laucat, Françoise Le Deist and Alain Fischer

8.1
Introduction

Studies on human and mouse natural mutants presenting with lymphoproliferative syndrome and autoimmunity (ALPS) have shed light on the role of Fas and Fas ligand (FasL) in the regulation of lymphocyte homeostasis. T cell expansion upon antigenic stimulation is followed by a contraction phase essentially driven by Fas. This is underscored by the observation that mice and humans lacking a functional Fas receptor develop a lymphoproliferative disease along with autoimmune manifestations. An identical phenotype is observed in mice carrying a defective FasL gene. Unraveling the genetic basis of the human pathology led to the identification of an apoptosis signaling defect, pointing to the crucial role of caspase-10 in the process of apoptosis induction. This chapter will discuss the main findings provided by the study of ALPS conditions.

8.2
Death Receptors and Signaling of Apoptosis

Apoptosis is a form of programmed cell death that can be triggered by specialized membrane-bound receptors belonging to the tumor necrosis factor (TNF) receptor (TNF-R)/nerve growth factor (NGF) receptor (NGF-R) superfamily. These 'death receptors' (DR) define a subfamily as they all contain cysteine-rich domains (CRDs) in their extracytoplasmic region and are characterized by the presence of a functional domain termed the 'death domain' (DD) within the cytoplasmic region [1]. Fas (also known as CD95 or Apo-1 and now as TNFRSF6) [2, 3] is a prototypical member with three extracellular CRDs and an 80-amino-acid intracellular DD.

Four other DD-containing receptors have been found on peripheral blood mononuclear cells, such as TNF-RI [4], TRAMP (TNF-R-related apoptosis mediating protein; DR3/wsl1/APO-3/LARD) [5–9], TRAIL (TNF-related apoptosis-inducing ligand)-R1 (DR4/APO-2) [10], TRAIL-R2 (DR5/Trick/Killer) [11–16] and DR6 [17]. The ligands that interact with these receptors, apart from the DR6 ligand that remains unknown, are structurally related molecules belonging to the TNF superfamily. FasL (CD95L) binds to Fas; TNF and lymphotoxin bind to TNF-RI; APO3 ligand (APO-3L; TWEAK) and TL1-A bind to DR3; and TRAIL [APO-2 ligand (APO-2L)] binds to DR4 and DR5 [18–22].

Decoy receptors are receptors that retain ligand-binding properties, but lack many of the functions of the corresponding receptors. Therefore, their *in vivo* function is debatable. Two of them lack a transmembrane domain and exist either as a soluble form, like DcR-3, the decoy receptor of FasL [23], or as a glycosylphosphatidylinositol (GPI) -linked receptor, like DcR-1 (TRAIL-R3) that binds TRAIL. A second decoy receptor for TRAIL, DcR-2 (TRAIL-R4), retains an incomplete DD, making it unable to trigger apoptosis, yet it has some signaling capabilities like NF-κB activation [12, 13, 23, 24].

The important feature of Fas and other DR signaling events is the recruitment of a signal transduction protein complex within minutes upon triggering by their cognate ligand or agonistic antibodies [25]. The formation of this 'death-inducing signaling complex' (DISC) is mediated by the specific conformation of the intracellular DD. This allows interaction with a cytoplasmic adapter protein called FADD (Fas-associated DD; also named MORT-1) by means of its DD to the clustered receptors' DD [26, 27]. FADD then recruits pro-enzymes, the pro-caspase-8 and -10 (also called Flice/MACH-1 and Flice-2, respectively) [28, 29, 30] via interactions of their mutual N-terminal 'death effector domain' (DED). It is thought that pro-caspase-8 is auto-processed in the vicinity of the DISC as a consequence of a high local concentration of the pro-enzyme. Studies with FADD and caspase-8 gene knockout (KO) mice as well as with human caspase-10 deficiencies established that these molecules are essential for apoptosis induction by Fas [31–34].

Caspases are cysteine proteases that cleave substrates after a specific aspartic residue. All caspases are synthesized as latent proenzymes made of an N-terminal caspase-recruiting domain (CARD), and a large (p20) and a small (p10) protease subunit. DEDs are specific examples of the homophilic interaction domains called CARDs. Processing of caspases results in proteolytic cleavage of the zymogene form, and association of the small and large protease domain in a heterotetramer complex forming the active caspase [35]. Activated caspase-8 and -10 can, in turn, activate other pro-caspases such as caspase-3, thereby initiating a proteolytic cascade that culminates in apoptosis. This pathway of apoptosis induction is independent of the apoptogenic activation of mitochondria and therefore cannot be inhibited by members of the Bcl-2 family. However, in some cells, referred to as type II cells (as opposed to type I cells using the former pathway), the amount of active caspase-8 generated is insufficient to activate the pro-caspase 3, but sufficient to cleave the BH3-only protein Bid. The truncated Bid (tBid) then activates the mitochondria and provokes the release of cytochrome *c*, a key element of a complex called the apoptosome and made of caspase-9, cytochrome *c*, ATP and Apaf-1. Activated caspase-9 can, in turn, cleave pro-caspase-3. It seems that thymocytes and peripheral T cells are identified as type I cells since their Fas-induced apoptosis is mitochondria independent, and cannot be inhibited by Bcl-2 and Bcl-x_L. Recent unraveling of Fas apoptosis execution in type I cells indicates that DISC assembly under receptor microaggregates is driven by caspase-8. Moreover, this process and the internalization of Fas-FasL complexes are actin dependent.

Other cytoplasmic proteins were reported to interact with Fas, such as Daxx [36], Rip [37] and FAF [38]. However, their roles in Fas-induced apoptosis remain controversial and may vary as a function of cell type or stage of differentiation.

The understanding of mechanisms underlying apoptosis induction upon triggering of Fas by its cognate ligand was recently modified by the work of Lenardo's group [39]. Previously, the 'post-ligand' model proposed that trimerization of Fas, and thus signaling of apoptosis, was triggered upon interaction with trimeric FasL. However, Siegel *et al.*, by using FRET technology, elegantly demonstrated that Fas molecules were previously trimerized through interactions of a N-terminal domain called PLAD (pre-ligand association domain). From a mechanistic point of view this means that ligand-receptor interaction may allow formation of receptor superclusters, a stoichiometry potentially required to induce apoptosis (as observed by using cross-linked anti-Fas antibodies). Alternatively, the FasL-Fas interaction could induce a change of the DD's conformation enabling further interactions with FADD and other downstream components. This may be the consequence of DD's cooperativity to self-associate and/or of the dissociation of a DD silencer (SODD) as described for TNF-RI [40].

Finally, execution of Fas apoptosis is controlled by a family of viral proteins called v-FLIPs (viral flice inhibitory proteins) and their cellular counterpart c-Flip (also called I-Flice, Cash, Casper or FLAME). They contain a DED similar to the corresponding domain in caspase-8 and -10 and FADD. These molecules are potent inhibitors of Fas-induced apoptosis by interacting with FADD and/or caspase-8, and may have an important role in the regulation of several DR-induced apoptotic events [41]. Resting naive T cells express little surface Fas, but T cell receptor (TCR) stimulation increases Fas expression and renders activated T cells progressively sensitive to Fas-induced apoptosis in the presence of interleukin (IL)-2. This has been attributed to a decrease in Flip that occurs after 3 days of T cell activation *in vitro* [42]. Additionally, it has been proposed that c-Flip may also connect Fas signaling to co-stimulation through the ERK pathway and activation of NF-κB [43]. It was reported, a decade ago, that antibodies to Fas might provide co-stimulation for human T cells *in vitro* [44]. This observation is supported by more recent studies showing that caspase-3 or -8 inhibitors can inhibit anti-CD3 T cell proliferation [45, 46]. Nevertheless, the physiological significance of these observations remains poorly understood since Fas deficiencies in patients or in *lpr* mice do not lead to immunodeficiency, but rather to uncontrolled lymphoproliferation (see below).

8.3 Clinical and Immunological Basis of ALPS

8.3.1 Definitions

More than 30 years ago, Canale and Smith reported a condition characterized by non-malignant lymphadenopathies associated with autoimmune features in children [47]. It turned out that these patients and a number of newly described ones have a genetic disorder caused by mutations of the Fas-encoding gene [48–50]. The syndrome was also named ALPS [48]. However, the genetic basis of this lymphoproliferation with autoimmunity was firstly identified in MRL *lpr* mice [51]. This natural mouse mutant strain was considered as a model of human lupus since these mice develop nephritis, hypergammaglobulinemia and antinuclear antibodies in addition to lymphadenopathy. These mice accumulate $CD4^-CD8^-$ TCR $\alpha\beta$ T cells in the periphery with age. This phenotype is likely the consequence of the CD8 downregulation on mature peripheral lymphocytes. The autoimmune features vary from strain to strain carrying the *lpr* mutation. Other natural mutants were subsequently described, i.e. lpr_{cg} and *gld* mice [52]. The *lpr* strain is characterized by an almost complete defect of Fas expression and consequently a complete defect of Fas-induced apoptosis as in the Fas-defective KO mouse [53]. The lpr_{cg} mutation allows the expression of a non-functional protein [51]. In the *gld* mouse, a missense mutation in the extracellular domain (ECD) of FasL abrogates its function [54]. The phenotype of these mice is very similar, with a variable time course of symptoms onset, the shortest one being observed in the Fas KO model [53].

At least five subtypes of ALPS have been described so far. ALPS 0 refers to complete Fas expression defect. ALPS Ia defines functional Fas deficiency (with slightly diminished or normal Fas expression). ALPS Ib is circumscribed to a FasL defect; however, this term may be inappropriate, as the phenotype of the unique patient described is dissimilar to other ALPS patients. ALPS II is used to describe a Fas-induced apoptosis defect in the absence of Fas mutation. Defects in the Fas signaling pathway were described in ALPS II. Finally, ALPS III designates patients presenting with ALPS symptoms but with a normal *in vitro* Fas-induced apoptosis. Nevertheless, as discussed below, four criteria characterize the ALPS condition: splenomegaly, lymphadenopathies, hypergammaglobulinemia and detection of TCR $\alpha\beta$ $CD4^-CD8^-$ T cells [double-negative (DN) T cells] in blood. With a few exceptions, ALPS patients present with at least three of these criteria.

8.3.2 Clinical Presentation

8.3.2.1 Lymphoproliferation

Lymphoproliferation is usually the most salient manifestation causing lymphadenopathy, splenomegaly and, in some cases, hepatomegaly. Onset of symptoms occurs in early childhood (0–5 years, around 6–12 months in most cases) [55, 56].

However, in some severe cases (ALPS 0 and in some ALPS Ia), massive lymphoproliferation is evident at birth, indicating a process that had started in the prenatal period [57]. Splenomegaly fluctuates in a given patient and is very variable from one patient to another one. It can be palpable at only 2 cm below the costal grill or, in contrast, be in the iliac fossa. In these later cases, splenectomy is often performed because of discomfort or hypersplenism. Blood lymphocytosis or increased adenopathies (in number and in size) are sometimes observed after splenectomy. Adenopathies are multifocal and their size fluctuates with time. A paradoxical decrease in lymph nodes has been observed during viral infection [47]. Hepatomegaly, when observed, is mild and is not associated with liver dysfunction. Lymphoproliferation may also involve the thymus, which is enlarged as visualized by computed tomography studies [58]. In severe cases, pulmonary infiltrate can be related to the lymphoproliferation [57]. Using a long-term follow-up of a number of patients, it has been possible to determine that there is a significant shrinkage of lymphoproliferation in a number of them over time [56].

8.3.2.2 Autoimmune Manifestations

Autoimmune manifestations are the second most frequent event in ALPS patients. They are present in about 70% of the patients [55, 56]. Age at onset varies considerably in contrast to the lymphoproliferative syndrome. It is therefore likely that, among reported patients, some who have not yet developed such manifestations will do so later on. For example, in one patient, the first ALPS manifestation was autoimmune thrombocytopenia and was observed at 18 years of age.

The most common autoimmune manifestations involve hematological lineages leading to anemia, thrombocytopenia and neutropenia, and are associated with the corresponding autoantibodies. Hemolytic anemia is the most frequent manifestation and has been found associated with dyserythropoiesis in two cases [59]. The magnitude of the cytopenia observed is variable from one patient to another; the hemolysis can be severe (Hb < 5mg/dl) as well as thrombocytopenia (< 10 000/µl), thus constituting a prognostic element. However, autoantibodies can be detected when the patient is clinically stable or even in case of the complete absence of clinical autoimmune manifestation [57, 60, 61]. Other autoimmune manifestations can be observed, such as glomerulonephritis, Guillain-Barre syndrome, uveitis, arthritis, hepatitis and diabetes [55, 56, 62]. Autoimmune manifestations involving the skin, including urticaria rashes and vasculitis, are common. An autoimmune basis is suspected in some cases of seizure, autism, ovarian failure and mucosal ulceration [63]. In none of the patients was typical lupus detected.

In addition to autoantibodies against blood cells, autoantibodies towards cardiolipin, smooth muscle and nuclear antigens are commonly detected as well as rheumatoid factor, but anti-DNA antibodies were never seen.

A striking observation is that autoimmunity appears to be always associated with autoantibodies, although direct pathogenic intervention of T cells in some of the autoimmune processes cannot be excluded.

8.3.2.3 Other Clinical Manifestations

Failure to thrive is a frequent symptom in children. Of note, splenectomy could reverse this in a number of cases. Fas mutations represent a significant risk factor for malignancy. A study performed on a large series of patients and relatives showed that the risk of non-Hodgkin's as well as of Hodgkin's lymphoma, in carriers of a heterozygous Fas mutation, was 14 and 51 times greater than expected, respectively [64]. In this series, the average age of lymphoma occurrence was 28 years. This observation is in accordance with the description of somatic Fas mutations in both children and adult leukemia and lymphomas [65–67]. Other malignant diseases have been reported – liver carcinoma in one patient (with hepatitis C infection), and multiple thyroid and breast adenomas together with basal cell carcinomas in another [50].

8.3.3 Laboratory Findings

8.3.3.1 Immunological Data

The lymphocyte count is variably increased, reflecting the intensity of the lymphoproliferative syndrome. Unusual T cells, DN T cells (TCR $\alpha\beta$ $CD4^-CD8^-$, $CD45RA^+$, $CD57^+$, $HLA\text{-}DR^+$) T cells are detected in excess in the blood of all patients with ALPS [55, 56]. This subset accounts for 1–60% of the blood T cell counts. Moreover, excess of $CD8^+$ T cells, TCR $\alpha\beta$ T cells and activated T cells was also frequently observed [55, 68]. Chronic lymphocyte activation was found, as demonstrated by the presence of high levels of HLA-DR expression on peripheral CD3 T cells and by the presence of high levels of serum activation markers such as soluble IL-2 receptor, soluble CD30 [61] and soluble FasL [69].

Polyclonal hyperimmunoglobulinemia G and A is a very frequent finding, while the level of serum IgM is usually reduced. However, in some cases hypogammaglobulinemia has been described [56]. Polyclonal B cell lymphocytosis with expansion of $CD5^+$ B cells was a characteristic finding [68].

A striking feature of ALPS Ia consists of overproduction of IL-10 as well as reduced IL-12 production by monocytes and increased IL-10 plasma levels [70]. It can be postulated to be a secondary regulatory event attempting to counterbalance the persistence and activation of autoimmune clones. This is in accordance with observations made in the mouse model where IL-10 was found to exacerbate the autoimmunity [71]. In addition, the T cells mostly have a T_h2 phenotype [72].

8.3.3.2 Pathological Findings

Lymph node paracortical areas are hyperplastic and contain many large lymphocytes with numerous mitotic figures. T cells as well as B cells accumulate in paracortical areas, while the overall architecture of lymphoid organs is preserved [57, 73]. Many of the cells express the Ki-67 antigen (indicative of active proliferation) and markers associated with cytotoxicity, such as perforin and CD57. A majority of the paracortical cells were DN cells TCR $\alpha\beta$ $CD3^+$ $CD45RO^-$ $CD45RA^+$. Spontaneous apoptosis is often seen. In addition, there is also excessive B lymphocyte accumulation with plasmocytosis in some patients, but not all. Most lymph nodes exhibited florid follicular hyperplasia. However, in some cases, follicular involution was seen.

In splenomegaly, expansion involves the red pulp and to a lesser extent the white pulp. Lymphoid cells in the red pulp are similar to the ones observed in paracortical areas of lymph nodes. Periarteriolar sheets are also enlarged.

DN T cells can also infiltrate the liver at the level of portal tracts and sinusoids. ALPS has to be considered as a possible differential diagnosis with lymphoma, especially involving T cells [74].

8.3.4 Treatment

Indications for treatment depend on the type and severity of the symptoms. In many patients, the clinical status does not require any treatment. Splenectomy, was often performed because of discomfort and hypersplenism, was also required, in some cases, because of protracted autoimmunity toward blood cells [55, 56]. In some patients autoimmunity tends to be severe, requiring aggressive immunosuppressive regimens including steroids and cyclophosphamide [57]. The antifolate drug Fansidar®, a combination of pyrimethamine and sulfadoxine, has been reported to be effective, especially on the lymphoproliferative manifestations [75]. In the same way, 6-mercaptopurine may be helpful (personal data).

In two severe cases, characterized by progression of lymphoproliferation in spite of chemotherapy including cyclophosphamide, vincristine and prednisone, bone marrow transplantation was performed from an unrelated donor in one case and from a haploidentical donor in the other one. In both cases, the bone marrow transplantation led to the correction of the Fas deficiency, and to the disappearance of clinical and biological manifestations [76, 77].

8.4 Genetic and Molecular Bases of ALPS

The recognition that *lpr* mice carry a mutation of the Fas gene was a major advance in the understanding of the role of the Fas molecule [51]. It has also pointed the way to the identification of the human counterpart. *lpr* mutant mice have an insertion of a transposon into intron 2 of the Fas gene that dramatically reduces the normal splicing of the Fas transcript [78]. A similar, albeit less severe, condition, lpr_{cg}, is associated with a missense mutation within the so-called DD-encoding part of the Fas gene [51]. Mutations of the FasL gene also result in lymphoproliferation [54]. Similarly, the human pathology is also diverse as at least five types can be described.

8.4.1 ALPS 0

ALPS 0 is a complete Fas expression deficiency. These cases are consequences of homozygous null mutations. Three cases of homozygous mutations were reported [48, 69, 79]. In one case, a stop codon in the ECD was predicted to lead to a complete expression defect. In contrast, in the two other cases mutations are localized in the intracellular domain (ICD), affecting the DD-encoding exon 9. These mutants should be expressed at the surface, according to truncation experiments performed on human Fas expressed in murine fibroblasts [2]. This defect can thus result from instability and premature elimination in the endoplasmic reticulum or retention due to the neo-peptide encoded in these mutants. Considering that the mutation was inherited from healthy parents (carrying the mutation at the heterozygous state), it was proposed that these mutations were recessive [79]. However, we were able to detect a partial Fas-induced apoptosis in cells from both parents of the patients we initially described (Le Deist *et al.*, unpublished observations). This is characteristic of a dominant effect exerted by the mutated protein with partial clinical penetrance as observed in many other cases of heterozygous mutations (see ALPS Ia). Another unpublished case of a family with an ALPS 0 patient supports this observation. In this family, the mother is healthy and carries the heterozygous mutation, and the father, with the same genotype, presented with symptoms of classical ALPS Ia. The proband, who received both mutated alleles, is presenting with typical ALPS 0. Thus, it can be suggested that most, if not all, mutations are dominant and that when homozygous they lead to a more severe phenotype – a classical observation in dominant diseases.

In a unique family, the patients are compound heterozygotes, with one mutation resulting in an amino acid substitution in the ECD [61]. It is unclear whether this modified Fas molecule has an impaired function. These patients exhibit a phenotype similar to the one of ALPS Ia.

8.4.2 ALPS Ia

More than 80 patients carrying a heterozygous Fas mutation have been so far reported in the literature [55–59, 61, 63, 68, 69, 73, 79–87]. A large spectrum of Fas gene mutations have been found associated with ALPS Ia, including mutations leading to truncated products or modified sequences (Fig. 8.1). The mutant Fas molecules exert a transdominant negative effect on Fas-mediated apoptosis as shown by transfection experiments [84, 88]. Moreover, mutant proteins can be detected at the membrane level (Rieux-Laucat *et al.*, unpublished observations) and, thus, random incorporation into Fas trimers of abnormal molecules likely accounts for it. These mutants result in reduced FADD binding and caspase recruitment, much greater than the 50% reduction predicted in a 1:1 non-cooperative interaction between Fas and FADD [88]. Therefore, this suggests that there is cooperation between Fas subunits in the recruitment of FADD, consistent with the

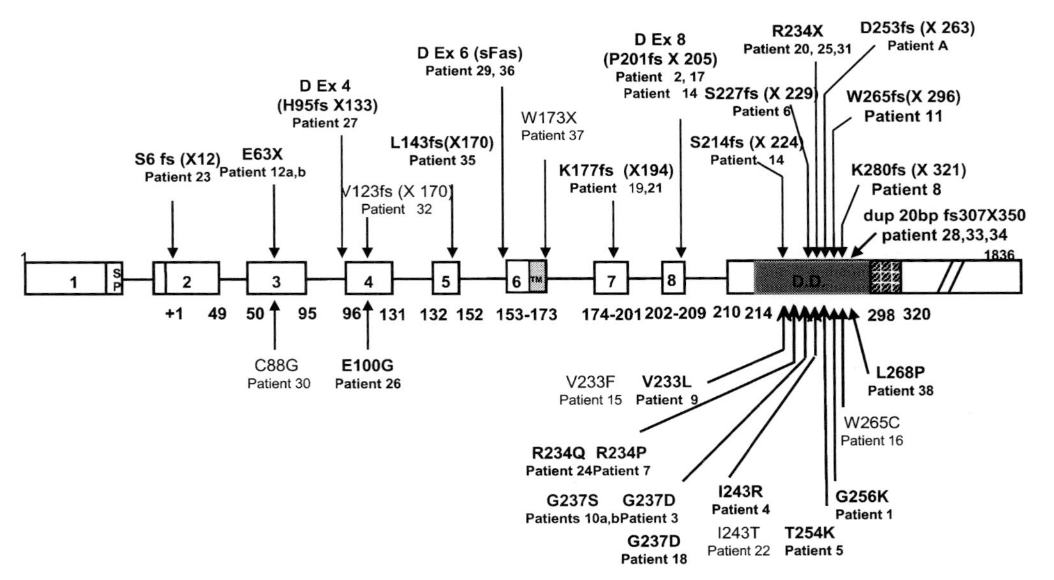

Fig. 8.1 Fas mutations characterized at our center are shown. Most of them are localized into the intracellular DD of Fas. Mutations in bold black are *de novo*, mutations in red have full clinical penetrance and those in blue have partial clinical penetrance.

presence of only one out of eight normal trimers in this setting. A majority of the identified mutations are located within the ICD. Sixty percent are localized within the DD. Nonsense and missense mutations are equally represented.

One-third of mutations alter the ECD of Fas and are usually associated with a less important Fas-induced apoptosis defect as compared with ICD mutations. ECD mutations can result in truncated products unable to anchor the membrane. These mutations should be considered as loss-of-function mutation. Although soluble forms of the Fas receptor were shown to inhibit *in vitro* Fas-induced apoptosis, no such results could be obtained with the supernatant of cells from patients with the ECD. According to the PLAD model, these mutants can be associated to the receptor complex through the N-terminal domain [39]. The identification of a mutation leading to a stop codon at position +12 (Rieux-Laucat et al., unpublished observations) suggests that the PLAD is restricted to the signal peptide and the very first residues of Fas. Alternatively it can be envisaged that such a short mutant would not be expressed, thus leading to haplo-insufficiency. This last hypothesis is supported by experiments performed on thymocytes from heterozygous Fas KO mice, which exhibit a reduced Fas-induced apoptosis [89]. Further experiments are needed to distinguish between these two possibilities. Missense mutations in the ECD result in expression of an abnormal Fas molecule, most likely incapable of interaction with FasL.

Of note, Fas deficiency is observed in cells from all carriers of heterozygous mutations. Thus, from a functional point of view, mutations are fully penetrant. In contrast, only 70% of the carriers of heterozygous Fas mutations develop clinical symptoms [56, 83, 84]. This clinical penetrance is highest for intracellular missense mutations, reaching 90%. Mutations leading to intracellular truncation have a clinical penetrance of roughly 75%. Finally, the clinical penetrance dropped to 30% for ECD mutations. Although these mutations lead to a lower Fas-induced apoptosis defect *in vitro* and a lower penetrance, there is no apparent correlation between the type of heterozygous mutations and the severity of the disease.

The partial clinical penetrance strongly suggests that a second event should be associated to Fas mutations in order to induce an overt ALPS Ia syndrome. It is likely that genetic rather than environmental factors, as in *lpr* mice, influence ALPS expression and account for variable penetrance of some of the mutations. So far, none of these gene modifiers have been identified. These gene products can be either directly involved in the apoptotic process, its regulation or in other events controlling activation, proliferation and survival of autoimmune clones. Importantly, Fas mutations on both alleles (either in homozygous or in compound heterozygous) are associated with full clinical penetrance. Similarly, in mice, symptoms develop only in homozygous animals, despite a potential defective Fas-induced apoptosis in heterozygous animals. More importantly, it was reported that double heterozygous lpr_{cg}/gld animals develop a mild lymphoproliferative disease [90]. These findings suggest that accumulation of defects in the Fas–FasL pathway may lead to development of the syndrome. In contrast, the finding of a slight, but statistically significant, increase of DN T cells in parents who do not carry a Fas mutation (and therefore have normal Fas-induced apoptosis) implies that the second event may be independent of the Fas pathway [68].

8.4.3
ALPS Ib

A unique case of a dominant FasL mutation has been described in a patient presenting with features of SLE along with chronic lymphoproliferation [91]. It was defined as ALPS Ib although the phenotype does not fulfill criteria of classical ALPS (DN T cells and splenomegaly were absent). The clinical manifestation could represent a bias as the search was performed in a cohort of patients with SLE. Moreover, there was no evidence of inheritance of this mutation event suggesting it could be a somatic event.

The similar phenotype of *lpr* and *gld* mice predicted an occurrence of FasL mutation in humans as frequent as Fas mutations. The absence of inherited FasL mutations in ALPS suggests several hypotheses. On the one hand, one may speculate that FasL is more important for human development than it is in mice. Thus, a FasL defect would not be compatible with life. On the other hand, a FasL defect might lead to a completely different phenotype, such as severe disease (related to potential extra-hematopoietic manifestations) masking the diagnosis of ALPS. Finally, an absence of the phenotype would then suggest the existence of an unforeseen ligand in humans.

8.4.4
ALPS II

Some patients present with all of the typical clinical and immunological features of ALPS along with abnormal lymphocyte Fas-mediated apoptosis *in vitro*. However, Fas molecule expression and sequence are normal in these patients. Recently, Wang *et al.* reported the occurrence in two families of caspase-10 mutations associated with ALPS [34]. These findings are extremely important as they demonstrate the key role of caspase-10 in Fas-mediated apoptosis. Mutated caspase-10 molecules impair auto-processing of the caspase-10 molecules and, more importantly, caspase-8 activation while DISC recruitment is normal. It is therefore strongly suggested that, besides caspase-8, the related caspase-10 is also incorporated into the DISC upon Fas engagement. More recent works confirmed the role of caspase-10 in Fas-induced apoptosis [92, 93].

A second interesting finding was that not only the Fas-mediated apoptosis pathway is impaired, but also apoptosis triggered by TNF-R1, DR3 and TRAIL receptors DR4 and DR5 (Fig. 8.2). Caspase-10 appears thus involved in the apoptotic cascade of all known receptors inducing lymphocyte apoptosis. Strikingly, unlike in ALPS Ia patients, dendritic cell accumulation was noted in lymphoid organs of caspase-10-deficient patients. The authors speculate that the defective death of T, B lymphocytes and dendritic cells in this setting could account for a possibly more severe autoimmune phenotype observed in these patients. Caspase-10-deficient patients were indeed reported to exhibit extremely severe autoimmune hemolytic anemia and thrombopenia, and optic neuritis and meningitis, respectively. As it has been previously reported that dendritic cell death by apoptosis is in-

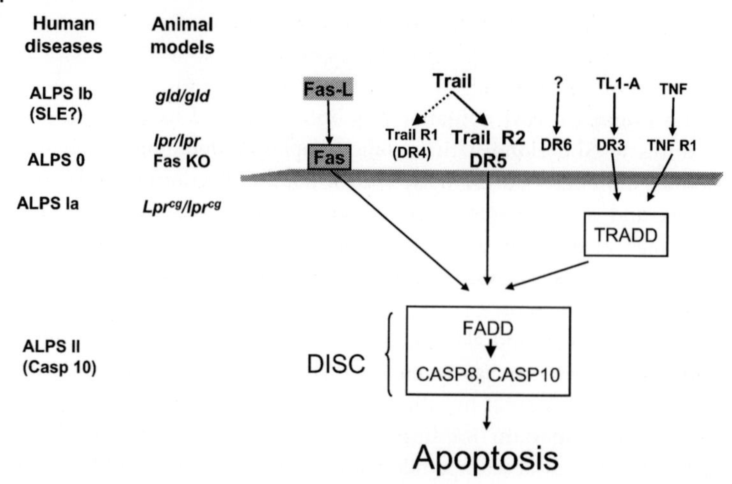

Fig. 8.2 Summary of known death ligand-receptor signaling defects.

volved in homeostasis of immune responses [94], abnormal persistence of dendritic cells might also contribute to the activation of autoimmune clones.

Interestingly, one of the mutations exerts a transdominant negative effect impairing Fas as well as other DR-mediated apoptosis, while the second requires expression on both alleles to induce a full-blown clinical picture. Detection of the latter mutation in the heterozygous state with a high frequency in the Danish population has disputed its involvement in the onset of ALPS II [95]. Nevertheless, description of a healthy individual carrying this homozygous mutation/polymorphism is still awaited. In any case, the dominant caspase-10 mutation has been validated and highlights the physiological role of this caspase in the cascade leading to apoptosis.

8.4.5 ALPS III

We and others have investigated a number of patients (over 30) who have presented with a clinical condition close to mild ALPS, associated with hypergammaglobulinemia and an excess of blood DN T cells (unpublished observation) [96]. Patients' lymphocytes exhibit a normal activation of the FasL-Fas pathway. No molecular defects have been found so far. Although not demonstrated, it is plausible that in these patients another lymphocyte apoptotic pathway is impaired. Recent descriptions of DR deficiencies in KO mouse models provide information about their potential role in lymphocyte homeostasis. Analyses of DR6-deficient mice showed enhanced $CD4^+$ T cell proliferation with a profound T_h2 polarization [97]. This phenotype appears independent of the apoptosis function of DR6, but rather implies connections with the c-Jun N-terminal kinase (JNK). Therefore, DR6 functions could be related to T_h2 attenuation through activation of JNK [97, 98]. This is of interest for ALPS studies as a bias toward T_h2 cytokine production was reported in some patients [72]. Another interesting observation was made in DR3-deficient mice [99]. These mice exhibit negative selection impairment in the thymus when crossed to the HY-transgenic model. Whereas anti-CD3-induced apoptosis of thymocytes is defective, pre-T cell checkpoint, positive selection and superantigen-induced negative selection are normal. This suggests that T cells with moderate but significant self-reactivity may reach the periphery in these mice. One may therefore consider that one of these receptors or both can be involved in ALPS III. They can also be implicated in ALPS Ia as a second genetic event participating in the triggering of the syndrome. If these hypotheses are correct, then Fas-independent apoptotic pathway(s) is (are) important mediator(s) of the control of lymphocyte homeostasis in humans. It would thus be important to determine which one.

8.5 Mechanism of Autoimmunity in ALPS

Autoimmunity is intrinsically a key feature of ALPS as it is in Fas-deficient murine mutants. It is present in nearly all patients and appears to be mostly antibody mediated. Therefore, the relationship between the defective FasL-Fas pathway and onset of autoimmunity is established. It appears that antigen-induced cell death (AICD) of autoimmune clones is more profoundly affected than that of clones specific for exogenous antigens. There is indeed no obvious expansion of T and/or B cell clones following infection or immunization. In contrast, a transient reduction in size of lymphoid organs has been noticed in ALPS patients during an infection [100]. This suggests that the FasL-Fas pathway is mostly involved in AICD during chronic exposure to autoantigens. Rathmell and Goodnow have proposed an elegant model to account for the role of the FasL-Fas pathway in the control of autoreactive T and B cells [101]. Chronically stimulated T cells downmodulate CD28 and therefore become more vulnerable to FasL-induced cell death according either to a suicide or a killing mechanism [102]. Fas-deficient T cells escape cell death, proliferate and can activate Fas-deficient B cells. In the physiological setting, FasL-expressing T cells kill normal autoimmune B cell clones, likely because chronic exposure to antigens no longer induces protecting signals from cell death. Fas-deficient B cells will thus escape this regulatory process and further proliferate. CD40 ligand-CD40 T-B interaction can induce a switch and further affinity maturation of Ig with autoimmune specificity. B cells produce autoantibodies in T cell zones where 'normal' autoimmune B cells, excluded from B cell zones, are programmed to die [101].

Therefore, FasL-Fas pathway deficiency creates a defect in peripheral tolerance. Central tolerance, at least of T cells, does not appear to be impaired as shown in *lpr* mice since negative selection normally occurs in *lpr* mice transgenic for a given TCR [103]. These results do not exclude a minor role of FasL-Fas in the thymus. Also, it would be interesting to know whether central tolerance is affected in the caspase-10-deficient setting.

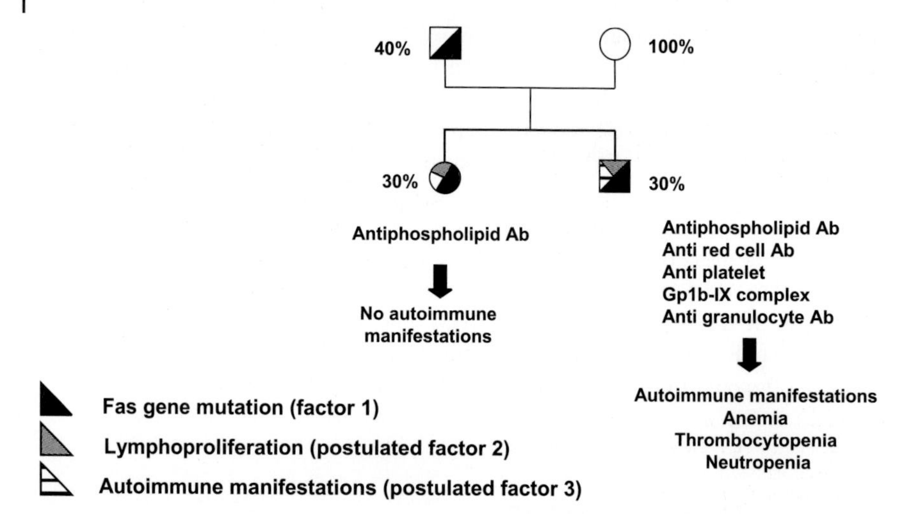

Fig. 8.3 Genetic events leading to lymphoproliferative syndrome and autoimmunity. (%: percentage of Fas-induced apoptosis).

Autoimmune features vary from strain to strain carrying the *lpr* mutation. These manifestations are moderate in C57BL6 or 129 mice [53, 89, 104]. This important finding shows that Fas mutations are a predisposing factor for the onset of autoimmunity, but that the latter requires other genetic susceptibility factors. Loci encoding some of them have been recently mapped [105]. Further genetic studies of mice should thus lead to the identification of other important gene products in the control of autoimmunity. Similarly in humans, autoimmune manifestations may be driven by a third genetic event. This is illustrated in a family with several mutant carriers (Fig. 8.3). The proband presented with lymphoproliferation and severe autoimmune hemolytic anemia, while his sister developed a lymphoproliferative syndrome without any autoimmune manifestation. Finally, the father who carries the mutation did not exhibit any clinical symptoms, although his lymphocytes displayed a Fas-induced apoptosis defect *in vitro*. This observation suggests that several modifier gene products might be involved in this case.

8.6 Conclusion

The recent unraveling of the molecular events leading to the ALPS syndromes in humans has brought a significant insight into the understanding of the function of the FasL–Fas pathway in lymphocyte homeostasis. More questions are now raised about the possible connections between apoptosis pathways, proliferation and autoimmunity, the three components of ALPS.

8.7 References

1 Ashkenazi A, Dixit VM. Death receptors: signaling and modulation. *Science* **1998**, 281, 1305–8.

2 Itoh N, Nagata S. A novel protein domain required for apoptosis. Mutational analysis of human Fas antigen. *J Biol Chem* **1993**, 268, 10932–7.

3 Oehm A, Behrmann I, Falk W, Pawlita M, Maier G, Klas C, Li-Weber M, Richards S, Dhein J, Trauth BC, Ponsting H, Krammer PH. Purification and molecular cloning of the APO-1 cell surface antigen, a member of the tumor necrosis factor/nerve growth factor receptor superfamily. *J Biol Chem* **1992**, 267, 10709–15.

4 Tartaglia LA, Ayres TM, Wong GHW, Goeddel DV. A novel domain within the 55 kd TNF receptor signals cell death. *Cell* **1993**, 74, 845–53.

5 Bodmer JL, Burns K, Schneider P, Hofmann K, Steiner V, Thome M, Bornand T, Hahne M, Schroter M, Becker K, Wilson A, French LE, Browning JL, MacDonald R, Tschopp J. TRAMP, a novel apoptosis-mediating receptor with sequence homology to TNFR1 and Fas (Apo-1/CD95). *Immunity* **1997**, 6, 79–89.

6 Chinnaiyan AM, O'Rourke K, Yu G, Lyons RH, Garg M, Duan DR, Xing L, Gentz R, Ni J, Dixit V. Signal transduction by DR3, a death domain-containing receptor related to TNF-R1 and CD95. *Science* **1996**, 274, 990–2.

7 Kitson J, Raven T, Jiang Y, Goeddel DV, Giles KM, Pun K, Grinham CJ, Brown R, Farrow SN. A death-domain-containing receptor that mediates apoptosis. *Nature* **1996**, 384, 372–5.

8 Marsters SA, Sheridan JP, Donahue CJ, Pitti RM, Gray CL, Goddard AD, Bauer KD, Ashkenazi A. Apo-3, a new member of the tumor necrosis factor receptor family, contains a death domain and activates apoptosis and NF-kappa B. *Curr Biol* **1996**, 6, 1669–76.

9 Screaton GR, Xu XN, Olsen AL, Cowper AE, Tan R, McMichael AJ, Bell JI. LARD: a new lymphoid-specific death domain containing receptor regulated by alternative pre-mRNA splicing. *Proc Natl Acad Sci USA* **1997**, 94, 4615–9.

10 Pan G, O'Rourke K, Chinnaiyan AM, Gentz R, Ebner R, Ni J, Dixit VM. The receptor for the cytotoxic ligand TRAIL. *Science* **1997**, 276, 111–3.

11 Chaudhary PM, Eby M, Jasmin A, Bookwalter A, Murray J, Hood L. Death receptor 5, a new member of the TNFR family, and DR4 induce FADD-dependent apoptosis and activate the NF-kappaB pathway. *Immunity* **1997**, 7, 821–30.

12 Pan G, Ni J, Wei YF, Yu G, Gentz R, Dixit VM. An antagonist decoy receptor and a death domain-containing receptor for TRAIL [see Comments]. *Science* **1997**, 277, 815–8.

13 Sheridan JP, Marsters SA, Pitti RM, Gurney A, Skubatch M, Baldwin D, Ramakrishnan L, Gray CL, Baker K, Wood WI, Goddard AD, Godowski P, Ashkenazi A. Control of TRAIL-induced apoptosis by a family of signaling and decoy receptors [see Comments]. *Science* **1997**, 277, 818–21.

14 Schneider P, Thome M, Burns K, Bodmer J, Hofmann K, Kataoka T, Holler N, Tschopp J. TRAIL Receptors 1 (DR4) and 2 (DR5) signal FADD-dependent Apoptosis and activate NF-κB. *Immunity* **1997**, 7, 831–6.

15 Screaton GR, Mongkolsapaya J, Xu XN, Cowper AE, McMichael AJ, Bell JI. TRICK2, a new alternatively spliced receptor that transduces the cytotoxic signal from TRAIL. *Curr Biol* **1997**, 7, 693–6.

16 Walczak H, Degli-Esposti MA, Johnson RS, Smolak PJ, Waugh JY, Boiani N, Timour MS, Gerhart MJ, Schooley KA, Smith CA, Goodwin RG, Rauch CT. TRAIL-R2: a novel apoptosis-mediating receptor for TRAIL. *EMBO J* **1997**, 16, 5386–97.

17 Pan G, Bauer JH, Haridas V, Wang S, Liu D, Yu G, Vincenz C, Aggarwal BB, Ni J, Dixit VM. Identification and functional characterization of DR6, a novel

death domain-containing TNF receptor. *FEBS Lett* **1998**, 431, 351–6.

18 Chicheporticheh Y, Bourdon PR, Xu H, Hsu YM, Scott H, Hession C, Garcia I, Browning JL. TWEAK, a new secreted ligand in the tumor necrosis factor family that weakly induces apoptosis. *J Biol Chem* **1997**, 272, 32401–10.

19 Marsters SA, Sheridan JP, Pitti RM, Brush J, Goddard A, Ashkenazi A. Identification of a ligand for the death-domain-containing receptor Apo3. *Curr Biol* **1998**, 8, 525–8.

20 Migone TS, Zhang J, Luo X, Zhuang L, Chen C, Hu B, Hong JS, Perry JW, Chen SF, Zhou JX, Cho YH, Ullrich S, Kanakaraj P, Carrell J, Boyd E, Olsen HS, Hu G, Pukac L, Liu D, Ni J, Kim S, Gentz R, Feng P, Moore PA, Ruben SM, Wei P. TL1A is a TNF-like ligand for DR3 and TR6/DcR3 and functions as a T cell costimulator. *Immunity* **2002**, 16, 479–92.

21 Suda T, Takahashi T, Golstein P, Nagata S. Molecular cloning and expression of the Fas ligand, a novel member of the tumor necrosis factor family. *Cell* **1993**, 75, 1169–78.

22 Wiley SR, Schooley K, Smolak PJ, Din WS, Huang C, Nicholl JK, Sutherland GR, Smith TD, Goodwin RG. Identification and characterization of a new member of the TNF family that induces apoptosis. *Immunity* **1995**, 3, 673–82.

23 Pitti RM, Marsters SA, Lawrence DA, Roy M, Kischkel FC, Dowd P, Huang A, Donahue CJ, Sherwood SW, Baldwin DT, Godowski PJ, Wood WI, Gurney AL, Hillan KJ, Cohen RL, Goddard AD, Botstein D, Ashkenazi A. Genomic amplification of a decoy receptor for Fas ligand in lung and colon cancer [In process citation]. *Nature* **1998**, 396, 699–703.

24 Marsters SA, Sheridan JP, Pitti RM, Huang A, Skubatch M, Baldwin D, Yuan J, Gurney A, Goddard AD, Godowski P, Ashkenazi A. A novel receptor for Apo2L/TRAIL contains a truncated death domain. *Curr Biol* **1997**, 7, 1003–6.

25 Kischkel FC, Hellbardt S, Behrmann I, Germer M, Pawlita M, Krammer PH, Peter ME. Cytotoxicity-dependent APO-1 (Fas/CD95)-associated proteins form a death-inducing signaling complex (DISC) with the receptor. *EMBO J* **1995**, 14, 5579–88.

26 Boldin M, Varfolomeev E, Pancer Z, Mett I, Camonis J, Wallach D. A novel protein that interacts with the death domain of fas/APO1 contains a sequence motif related to the death domain. *J Biol Chem* **1995**, 270, 7795–8.

27 Chinnaiyan A, O'Rourke K, Tewari M, Dixit V. FADD, a novel death domain-containing protein, interacts with the death domains of fas and initiates apoptosis. *Cell* **1995**, 81, 505–12.

28 Boldin MP, Goncharov TM, Goltsev YV, Wallach D. Involvement of MACH, a novel MORT1/FADD-interacting protease, in Fas/APO-1- and TNF Receptor-induced cell death. *Cell* **1996**, 85, 803–15.

29 Muzio M, Chinnaiyan AM, Kischkel FC, O'Rourke K, Shevchenko A, Ni J, Scaffidi C, Bretz JD, Zhang M, Gentz R, Mann M, Krammer PH, Peter ME, Dixit VM. FLICE, a novel FADD-homologous ICE/CED3-like Protease, is recruited to the CD95 (Fas/APO-1) death-inducing signaling complex. *Cell* **1996**, 85, 817–27.

30 Vincenz C, Dixit VM. Flice-2, an ICE/Ced-3 homologue, is proximally involved in CD95- and p55-mediated death signaling. *J Biol Chem* **1997**, 272, 6578–83.

31 Yeh WC, Pompa JL, McCurrach ME, Shu HB, Elia AJ, Shahinian A, Ng M, Wakeham A, Khoo W, Mitchell K, El-Deiry WS, Lowe SW, Goeddel DV, Mak TW. FADD: essential for embryo development and signaling from some, but not all, inducers of apoptosis. *Science* **1998**, 279, 1954–8.

32 Zhang J, Cado D, Chen A, Kabra NH, Winoto A. Fas-mediated apoptosis and activation-induced T-cell proliferation are defective in mice lacking FADD/Mort1. *Nature* **1998**, 392, 296–300.

33 Varfolomeev EE, Schuchmann M, Luria V, Chiannilkulchai N, Beckmann JS, Mett IL, Rebrikov D, Brodianski VM, Kemper OC, Kollet O, Lapidot T, Soffer D, Sobe T, Avraham KB, Goncharov T, Holtmann H, Lonai

P, Wallach D. Targeted disruption of the mouse Caspase 8 gene ablates cell death induction by the TNF receptors, Fas/Apo1, and DR3 and is lethal prenatally. *Immunity* **1998**, 9, 267–76.

34 Wang J, Zheng L, Lobito A, Chan FK, Dale J, Sneller M, Yao X, Puck JM, Straus SE, Lenardo MJ. Inherited human Caspase 10 mutations underlie defective lymphocyte and dendritic cell apoptosis in autoimmune lymphoproliferative syndrome type II. *Cell* **1999**, 98, 47–58.

35 Thornberry NA, Lazebnik Y. Caspases: enemies within. *Science* **1998**, 281, 1312–6.

36 Yang X, Khosravi-Far R, Baltimore D. Daxx, a novel Fas-binding protein that activates JNK and Apoptosis. *Cell* **1997**, 89, 1067–76.

37 Stanger B, Leder P, Lee T, Kim E, Seed B. RIP: a novel protein containing a death domain that interacts with fas/APO-1(CD95) in yeast and causes cell death. *Cell* **1995**, 81, 513–23.

38 Chu K, Niu X, Williams LT. A Fas-associated protein factor, FAF1, potentiates Fas-mediated apoptosis. *Proc Natl Acad Sci USA* **1995**, 11894–8.

39 Siegel RM, Frederiksen JK, Zacharias DA, Chan FK, Johnson M, Lynch D, Tsien RY, Lenardo MJ. Fas preassociation required for apoptosis signaling and dominant inhibition by pathogenic mutations. *Science* **2000**, 288, 2354–7.

40 Jiang Y, Woronicz JD, Liu W, Goeddel DV. Prevention of constitutive TNF receptor 1 signaling by silencer of death domains. *Science* **1999**, 283, 543–6.

41 Thome M, Tschopp J. Regulation of lymphocyte proliferation and death by FLIP. *Nat Rev Immunol* **2001**, 1, 50–8.

42 Refaeli Y, Van Parijs L, London CA, Tschopp J, Abbas AK. Biochemical mechanisms of IL-2-regulated Fas-mediated T cell apoptosis. *Immunity* **1998**, 8, 615–23.

43 Kataoka T, Budd RC, Holler N, Thome M, Martinon F, Irmler M, Burns K, Hahne M, Kennedy N, Kovacsovics M, Tschopp J. The caspase-8 inhibitor FLIP promotes activation of NF-kappaB and Erk signaling pathways. *Curr Biol* **2000**, 10, 640–8.

44 Alderson MR, Armitage RJ, Maraskovsky E, Tough TW, Roux E, Schooley K, Ramsdell F, Lynch DH. Fas transduces activation signals in normal human T lymphocytes. *J Exp Med* **1993**, 178, 2231–5.

45 Alam A, Cohen LY, Aouad S, Sekaly RP. Early activation of caspases during T lymphocyte stimulation results in selective substrate cleavage in nonapoptotic cells. *J Exp Med* **1999**, 190, 1879–90.

46 Kennedy NJ, Kataoka T, Tschopp J, Budd RC. Caspase activation is required for T cell proliferation. *J Exp Med* **1999**, 190, 1891–6.

47 Canale VC, Smith CH. Chronic lymphadenopathy simulating malignant lymphoma. *J Pediatr* **1967**, 70, 891–9.

48 Rieux-Laucat F, Le Deist F, Hivroz C, Roberts I, Debatin K, Fischer A, de Villartay J. Mutations in fas associated with human lymphoproliferative syndrome and autoimmunity. *Science* **1995**, 268, 1347–9.

49 Fisher GH, Rosenberg FJ, Straus SE, Dale JK, Middleton LA, Lin AY, Strober W, Lenardo MJ, Puck JM. Dominant interfering Fas gene mutations impair apoptosis in a human autoimmune lymphoproliferative syndrome. *Cell* **1995**, 81, 935–46.

50 Drappa J, Vaishnaw AK, Sullivan KE, Chu J-L, Elkon KB. Fas gene mutations in the Canale-Smith syndrome, an inherited lymphoproliferative disorder associated with autoimmunity. *N Engl J Med* **1996**, 335, 1643–9.

51 Watanabe-Fukunaga R, Brannan CI, Copeland NG, Jenkins NA, Nagata S. Lymphoproliferation disorder in mice explained by defect in Fas antigen that mediates apoptosis. *Nature* **1992**, 356, 314–7.

52 Nagata S, Suda T. Fas and Fas ligand: *lpr* and *gld* mutations. *Immunol Today* **1995**, 16, 39–43.

53 Adachi M, Suematsu S, Suda T, Watanabe D, Fukuyama H, Ogasawara J, Tanaka T, Yoshida N, Nagata S. Enhanced and accelerated lymphoproliferation in Fas-null mice. *Proc Natl Acad Sci USA* **1996**, 93, 2131–6.

54 Takahashi T, Tanaka M, Brannan CI, Jenkins NA, Copeland NG, Suda T, Nagata S. Generalized lymphoproliferative disease in mice, caused by a point mutation in the FAS ligand. *Cell* **1994**, 76, 969–76.
55 Sneller MC, Wang J, Dale JK, Strober W, Middelton LA, Choi YN, Fleisher TA, Lim MS, Jaffe ES, Puck JM, Lenardo MJ, Straus SE. Clinical, immunologic, and genetic features of an autoimmune lymphoproliferative syndrome associated with abnormal lymphocyte apoptosis. *Blood* **1997**, 89, 1341–8.
56 Rieux-Laucat F, Blachere S, Danielan S, De Villartay JP, Oleastro M, Solary E, Bader-Meunier B, Arkwright P, Pondare C, Bernaudin F, Chapel H, Nielsen S, Berrah M, Fischer A, Le Deist F. Lymphoproliferative syndrome with autoimmunity: a possible genetic basis for dominant expression of the clinical manifestations. *Blood* **1999**, 94, 2575–82.
57 Le Deist F, Emile JF, Rieux-Laucat F, Benkerrou M, Roberts I, Brousse N, Fischer A. Clinical, immunological, and pathological consequences of Fas-deficient conditions. *Lancet* **1996**, 348, 719–23.
58 Avila NA, Dwyer AJ, Dale JK, Lopatin UA, Sneller MC, Jaffe ES, Puck JM, Straus SE. Autoimmune lymphoproliferative syndrome: a syndrome associated with inherited genetic defects that impair lymphocytic apoptosis – CT and US features. *Radiology* **1999**, 212, 257–63.
59 Bader-Meunier B, Rieux-Laucat F, Croisille L, Yvart J, Mielot F, Dommergues JP, Ledeist F, Tchernia G. Dyserythropoiesis associated with a fas-deficient condition in childhood. *Br J Haematol* **2000**, 108, 300–4.
60 Carter LB, Procter JL, Dale JK, Straus SE, Cantilena CC. Description of serologic features in autoimmune lymphoproliferative syndrome. *Transfusion* **2000**, 40, 943–8.
61 Bettinardi A, Brugnoni D, Quiros-Roldan E, Malagoli A, La Grutta S, Correra A, Notarangelo LD. Missense mutations in the Fas gene resulting in autoimmune lymphoproliferative syndrome: a molecular and immunological analysis. *Blood* **1997**, 89, 902–9.
62 Pensati L, Costanzo A, Ianni A, Accapezzato D, Iorio R, Natoli G, Nisini R, Almerighi C, Balsano C, Vajro P, Vegnente A, Levrero M. Fas/Apo1 mutations and autoimmune lymphoproliferative syndrome in a patient with type 2 autoimmune hepatitis. *Gastroenterology* **1997**, 113, 1384–9.
63 Shenoy S, Arnold S, Chatila T. Response to steroid therapy in autism secondary to autoimmune lymphoproliferative syndrome. *J Pediatr* **2000**, 136, 682–7.
64 Straus SE, Jaffe ES, Puck JM, Dale JK, Elkon KB, Rosen-Wolff A, Peters AM, Sneller MC, Hallahan CW, Wang J, Fischer RE, Jackson CM, Lin AY, Baumler C, Siegert E, Marx A, Vaishnaw AK, Grodzicky T, Fleisher TA, Lenardo MJ. The development of lymphomas in families with autoimmune lymphoproliferative syndrome with germline Fas mutations and defective lymphocyte apoptosis. *Blood* **2001**, 98, 194–200.
65 Tamiya S, Etoh K, Suzushima H, Takatsuki K, Matsuoka M. Mutation of CD95 (Fas/Apo-1) gene in adult T-cell leukemia cells. *Blood* **1998**, 91, 3935–42.
66 Gronbaek K, Straten PT, Ralfkiaer E, Ahrenkiel V, Andersen MK, Hansen NE, Zeuthen J, Hou-Jensen K, Guldberg P. Somatic Fas mutations in non-Hodgkin's lymphoma: association with extranodal disease and autoimmunity. *Blood* **1998**, 92, 3018–24.
67 Beltinger C, Kurz E, Bohler T, Schrappe M, Ludwig WD, Debatin KM. CD95 (APO-1/Fas) mutations in childhood T-lineage acute lymphoblastic leukemia. *Blood* **1998**, 91, 3943–51.
68 Bleesing JJ, Brown MR, Straus SE, Dale JK, Siegel RM, Johnson M, Lenardo MJ, Puck JM, Fleisher TA. Immunophenotypic profiles in families with autoimmune lymphoproliferative syndrome. *Blood* **2001**, 98, 2466–73.
69 Kasahara Y, Wada T, Niida Y, Yachie A, Seki H, Ishida Y, Sakai T, Koizumi F, Koizumi S, Miyawaki T, Taniguchi N. Novel Fas (CD95/APO-1) mutations in

infants with a lymphoproliferative disorder. *Int Immunol* **1998**, 10, 195–202.

70 Lopatin U, Yao X, Williams RK, Bleesing JJ, Dale JK, Wong D, Teruya-Feldstein J, Fritz S, Morrow MR, Fuss I, Sneller MC, Raffeld M, Fleisher TA, Puck JM, Strober W, Jaffe ES, Straus SE. Increases in circulating and lymphoid tissue interleukin-10 in autoimmune lymphoproliferative syndrome are associated with disease expression. *Blood* **2001**, 97, 3161–70.

71 Watanabe N, Ikuta K, Nisitani S, Chiba T, Honjo T. Activation and differentiation of autoreactive B-1 cells by interleukin 10 induce autoimmune hemolytic anemia in Fas-deficient antierythrocyte immunoglobulin transgenic mice. *J Exp Med* **2002**, 196, 141–6.

72 Fuss IJ, Strober W, Dale JK, Fritz S, Pearlstein GR, Puck JM, Lenardo MJ, Straus SE. Characteristic T helper 2 T cell cytokine abnormalities in autoimmune lymphoproliferative syndrome, a syndrome marked by defective apoptosis and humoral autoimmunity. *J Immunol* **1997**, 158, 1912–8.

73 Lim MS, Straus SE, Dale JK, Fleisher TA, Stetler-Stevenson M, Strober W, Sneller MC, Puck JM, Lenardo MJ, Elenitoba-Johnson KS, Lin AY, Raffeld M, Jaffe ES. Pathological findings in human autoimmune lymphoproliferative syndrome. *Am J Pathol* **1998**, 153, 1541–50.

74 van der Werff ten Bosch J, Delabie J, Bohler T, Verschuere J, Thielemans K. Revision of the diagnosis of T-zone lymphoma in the father of a patient with autoimmune lymphoproliferative syndrome type II. *Br J Haematol* **1999**, 106, 1045–8.

75 van der Werff Ten Bosch J, Schotte P, Ferster A, Azzi N, Boehler T, Laurey G, Arola M, Demanet C, Beyaert R, Thielemans K, Otten J. Reversion of autoimmune lymphoproliferative syndrome with an antimalarial drug: preliminary results of a clinical cohort study and molecular observations. *Br J Haematol* **2002**, 117, 176–88.

76 Benkerrou M, Le Deist F, de Villartay J, Caillat-Zuckman S, Rieux-Laucat F, Jabado N, Cavazzana-Calvo M, Fischer A. Correction of fas (CD95) deficiency by haploidentical bone marrow transplantation. *Eur J Immunol* **1997**, 27, 2043–7.

77 Sleight BJ, Prasad VS, DeLaat C, Steele P, Ballard E, Arceci RJ, Sidman CL. Correction of autoimmune lymphoproliferative syndrome by bone marrow transplantation. *Bone Marrow Transplant* **1998**, 22, 375–80.

78 Chu JL, Drappa J, Parnassa A, Elkon KB. The defect in Fas mRNA expression in MRL/*lpr* mice is associated with insertion of the retrotransposon, ETn. *J Exp Med* **1993**, 178, 723–30.

79 van der Burg M, de Groot R, Comans-Bitter WM, den Hollander JC, Hooijkaas H, Neijens HJ, Berger RM, Oranje AP, Langerak AW, van Dongen JJ. Autoimmune lymphoproliferative syndrome (ALPS) in a child from consanguineous parents: a dominant or recessive disease? *Pediatr Res* **2000**, 47, 336–43.

80 Infante AJ, Britton HA, DeNapoli T, Middelton LA, Lenardo MJ, Jackson CE, Wang J, Fleisher T, Straus SE, Puck JM. The clinical spectrum in a large kindred with autoimmune lymphoproliferative syndrome caused by a Fas mutation that impairs lymphocyte apoptosis. *J Pediatr* **1998**, 133, 629–33.

81 Martin DA, Combadiere B, Hornung F, Jiang D, McFarland H, Siegel R, Trageser C, Wang J, Zheng L, Lenardo MJ. Molecular genetic studies in lymphocyte apoptosis and human autoimmunity. *Novartis Found Symp* **1998**, 215, 73–82; discussion 82–91.

82 Aspinall AI, Pinto A, Auer IA, Bridges P, Luider J, Dimnik L, Patel KD, Jorgenson K, Woodman RC. Identification of new Fas mutations in a patient with autoimmune lymphoproliferative syndrome (ALPS) and eosinophilia. *Blood Cells Mol Dis* **1999**, 25, 227–38.

83 Jackson CE, Fischer RE, Hsu AP, Anderson SM, Choi Y, Wang J, Dale JK, Fleisher TA, Middelton LA, Sneller MC, Lenardo MJ, Straus SE, Puck JM. Autoimmune lymphoproliferative syn-

drome with defective Fas: genotype influences penetrance. *Am J Hum Genet* **1999**, 64, 1002–14.

84 Vaishnaw AK, Orlinick JR, Chu JL, Krammer PH, Chao MV, Elkon KB. The molecular basis for apoptotic defects in patients with CD95 (Fas/Apo-1) mutations. *J Clin Invest* **1999**, 103, 355–63.

85 Arkwright PD, Rieux-Laucat F, Le Deist F, Stevens RF, Angus B, Cant AJ. Cytomegalovirus infection in infants with autoimmune lymphoproliferative syndrome (ALPS). *Clin Exp Immunol* **2000**, 121, 353–7.

86 Boulanger E, Rieux-Laucat F, Picard C, Legall M, Sigaux F, Clauvel JP, Oksenhendler E, Le Deist F, Meignin V. Diffuse large B-cell non-Hodgkin's lymphoma in a patient with autoimmune lymphoproliferative syndrome. *Br J Haematol* **2001**, 113, 432–4.

87 DeFranco S, Bonissoni S, Cerutti F, Bona G, Bottarel F, Cadario F, Brusco A, Loffredo G, Rabbone I, Corrias A, Pignata C, Ramenghi U, Dianzani U. Defective function of Fas in patients with type 1 diabetes associated with other autoimmune diseases. *Diabetes* **2001**, 50, 483–8.

88 Martin DA, Zheng L, Siegel RM, Huang B, Fisher GH, Wang J, Jackson CE, Puck JM, Dale J, Straus SE, Peter ME, Krammer PH, Fesik S, Lenardo MJ. Defective CD95/APO-1/Fas signal complex formation in the human autoimmune lymphoproliferative syndrome, type Ia. *Proc Natl Acad Sci USA* **1999**, 96, 4552–7.

89 Adachi M, Suematsu S, Kondo T, Ogasawara J, Tanaka T, Yoshida N, Nagata S. Targeted mutation in the Fas gene causes hyperplasia in peripheral lymphoid organs and liver. *Nat Genet* **1995**, 11, 294–300.

90 Kimura M, Ikeda H, Katagiri T, Matsuzawa A. Characterization of lymphoproliferation induced by interactions between *lprcg* and *gld* genes. *Cell Immunol* **1991**, 134, 359–69.

91 Wu JG, Wilson J, He J, Xiang LB, Schur PH, Mountz JD. Fas ligand mutation in a patient with systemic lupus erythematosus and lymphoproliferative disease. *J Clin Invest* **1996**, 98, 1107–13.

92 Wang J, Chun HJ, Wong W, Spencer DM, Lenardo MJ. Caspase-10 is an initiator caspase in death receptor signaling. *Proc Natl Acad Sci USA* **2001**, 98, 13884–8.

93 Kischkel FC, Lawrence DA, Tinel A, LeBlanc H, Virmani A, Schow P, Gazdar A, Blenis J, Arnott D, Ashkenazi A. Death receptor recruitment of endogenous caspase-10 and apoptosis initiation in the absence of caspase-8. *J Biol Chem* **2001**, 276, 46639–46.

94 Inguli E, Mondino A, Khoruts A, Jenkins MK. *In vivo* detection of dendritic cell antigen presentation to $CD4^+$ T cells. *J Exp Med* **1998**, 185, 2133–41.

95 Gronbaek K, Dalby T, Zeuthen J, Ralfkiaer E, Guidberg P. The V410I (G1228A) variant of the caspase-10 gene is a common polymorphism of the Danish population. *Blood* **2000**, 95, 2184–5.

96 Dianzani U, Bragardo M, DiFranco D, Alliaudi C, Scagni P, Buonfiglio D, Redoglia V, Bonissoni S, Correra A, Dianzani I, Ramenghi U. Deficiency of the Fas apoptosis pathway without Fas gene mutations in pediatric patients with autoimmunity/lymphoproliferation. *Blood* **1997**, 89, 2871–9.

97 Liu J, Na S, Glasebrook A, Fox N, Solenberg PJ, Zhang Q, Song HY, Yang DD. Enhanced $CD4^+$ T cell proliferation and T_h2 cytokine production in DR6-deficient mice. *Immunity* **2001**, 15, 23–34.

98 Zhao H, Yan M, Wang H, Erickson S, Grewal IS, Dixit VM. Impaired c-Jun amino terminal kinase activity and T cell differentiation in death receptor 6-deficient mice. *J Exp Med* **2001**, 194, 1441–8.

99 Wang EC, Thern A, Denzel A, Kitson J, Farrow SN, Owen MJ. DR3 regulates negative selection during thymocyte development. *Mol Cell Biol* **2001**, 21, 3451–61.

100 Straus S, Sneller M, Lenardo MJ, Puck JM, Strober W. The autoimmune lymphoproliferative syndrome. *Ann Intern Med* **1999**, 130, 591–601.

101 Rathmell JC, Cooke MP, Ho WY, Grein J, Townsend SE, Davis MM, Goodnow CC. CD95 (Fas)-dependent

elimination of self-reactive B cells upon interaction with CD4+ T cells. *Nature* **1995**, 376, 181–4.

102 Nagata S, Golstein P. The Fas death factor. *Science* **1995**, 267, 1449–55.

103 Singer GG, Abbas AK. The fas antigen is involved in peripheral but not thymic deletion of T lymphocytes in T cell receptor transgenic mice. *Immunity* **1994**, 1, 365–71.

104 Cohen PL, Eisenberg RA. *lpr* and *gld*: single gene models of systemic autoimmunity and lymphoproliferative disease. *Annu Rev Immunol* **1991**, 9, 243–69.

105 Vidal S, Kono DH, Theofilopoulos AN. Loci predisposing to autoimmunity in MRL-Fas *lpr* and C57BL/6-Fas *lpr* mice. *J Clin Invest* **1998**, 101, 696–702.

9
Infection and Inflammation as Cofactors for Autoimmunity in Systemic Lupus Erythematosus Patients

Hanns-Martin Lorenz

9.1
Introduction

Systemic lupus erythematosus (SLE) is a chronic autoimmune disease presenting with a highly variable clinical picture and course. There are cases with one major initial flare, which will be in remission for years after appropriate treatment. On the other hand, there are remittent courses with chronic disease activity requiring aggressive treatment and intensive observation. The reasons for this high variability in the clinical courses are unknown.

Infection in patients with SLE is a clinically and therapeutically difficult situation for several reasons, given the difficulty in the judgment of fever being a symptom of an ongoing infection (lowering immunosuppression, antibiotics) or a symptom of disease activity itself (more aggressive immunosuppression). Moreover, infections have been discussed for a long time as factors contributing to the induction of autoimmunity *per se* and possibly for the initiation of SLE flares. That certain bacteria and viruses can induce autoimmune phenomena has clearly been shown, both *in vitro* and *in vivo* in mice and in humans. In theory, infections could be involved in the pathogenesis of SLE by several pathways including dysregulation of apoptosis and promotion of immune responses against nuclear constituents altered during apoptosis. In clinical terms, in our own experience infections preceded SLE flares in certain cases, an observation which has been supported by various authors. On the other hand, this cannot be generalized for every SLE patient. Moreover, safety studies in SLE patients indicated that vaccination in SLE patients is save and effective in most cases, but may cause SLE flares in some patients. All of these points will be discussed within the chapter.

9.2
Infection and Autoimmunity

The fact that bacterial as well as viral infections can lead to autoimmune phenomena or manifest autoimmunopathies is known for a long time. Best examples are infections with *Borrelia burgdorferi*, *Chlamydia trachomatis* and enteropathic bacteria like *Salmonella, Yersinia* or *Campylobacter,* all known to be able to induce reactive or chronic arthritis, and various other clinical syndromes. A scleroderma-like disease develops as a consequence of a chronic infection with *B. burgdorferi*. Guillain-Barré syndrome (GBS) can develop after *Campylobacter* infection. It is important to note that – at least in Borreliosis – detection of bacterial antigens or DNA was not possible, indicating that the pathogenesis of the clinical symptoms is probably based on autoimmune reactions and not on defense mechanisms against replicating bacteria. In animal models, mycobacteria can prevent diabetes, but initiate a SLE-like disease in diabetes-prone non-obese diabetic mice [1] – again favoring those viral or bacterial infections as triggers for autoimmunity in an appropriate background and environment.

Infections with viruses like Hepatitis C and B/D or retroviruses have been linked to autoimmunity for a long time [2]. Retroviruses have been implicated in the pathogenesis of several autoimmunopathies including Sjögren's disease, primary biliary cirrhosis, type I diabetes and multiple sclerosis [3–5]. However, most of these data were obtained in murine models, where a specific infection of mice with retroviruses (ungulate caprine arthritis encephalitis virus, equine infectious anemia virus, Meadi-Visna virus) causes a well-defined autoimmunopathy. Data supporting a role for retroviruses in the pathogenesis of human autoimmune diseases are scarce. Retroviral particles could be detected in the salivary glands of patients with Sjögren's syndrome [6]. In an area heavily endemic for HTLV-1 infections in Japan, antibody positivity for HTLV-1 was significantly higher in Sjögren's syndrome patients than in the general population, a finding which was not true for the SLE patients [7]. On the other hand, viral load in the peripheral blood mononuclear cells was not necessarily high. It is important to note, however, that since Sjögren's syndrome patients usually present with high polyclonal γ-globulin levels, false positive reactivity against different viruses is always a concern. However, viral particles were detected by electron microscopy in biliary epithelium of patients with primary biliary cirrhosis [8] or the synovial fluid of patients with rheumatoid arthritis as well [9]. From plasma of patients with active SLE, but not of patients with Waldenström's disease, rheumatoid arthritis and myasthenia gravis, high molecular weight DNA and RNase-insensitive RNA could be isolated, which contained high levels of CpG motifs [10]. These data more directly link infection with autoimmune phenomena, but still are no prove that these viral particles play a direct pathogenetic role.

Hepatitis C-associated rheumatic complications are frequent and include mixed cryoglobulinemia, sicca symptoms, vasculitis, arthritis or fibromyalgia, development of antinuclear antibodies (ANA), rheumatoid factor, anticardiolipin antibodies, antithyroid antibodies or antiliver/kidney antibodies [11, 12]. Formation of immune complexes of viral particles and antiviral antibodies might occur and contribute to the pathogenesis of vasculitis, neuropathy or glomerulonephritis [12]. Case reports describe patients with Hepatitis C infection preceding or coincidently developing rheumatoid arthritis, SLE or polymyositis/dermatomyositis [12–15]. Another virus which needs to be considered in the diagnostic procedure at onset of autoimmune diseases is the Parvo B 19 virus. Parvo B 19 virus infection can cause fever, arthritis, malaise, rash, anemia, leukocytopenia, thrombocytopenia, presence of autoantibodies like ANA, hypocomplementemia, proteinuria or GBS. In one case, chronic SLE was triggered by Parvo B 19 infection [16–19]. The mechanisms responsible for the induction of autoimmunity in the course of these infections are unresolved. Molecular mimicry as one of the possible factors has clearly been demonstrated in herpes keratoconjunctivitis in mice: T lymphocytes reacting against the viral UL6 protein cross-react with a peptide derived from a corneal antigen [20]. Rheumatic fever is a good example for an autoimmune disease in humans pathophysiologically caused by molecular mimicry of streptococcal and cardiac myosin [21]. In autoimmune diabetes T cells recognize both a peptide derived from the autoantigen glutamic acid decarboxylase and a highly analogous peptide from Coxsackie P2-C protein [22]. These examples clearly show that molecular mimicry might centrally be involved in the pathogenesis of some autoimmune phenomena after certain infections.

Crohn's disease is another prototype of autoimmune disease in which infection and autoimmunity are closely connected also in pathogenetic terms. Murine models of this inflammatory bowel disease like T cell receptor (TCR) chain knockout mice or interleukin (IL)-2 knockout mice develop an inflammatory bowel disease resembling Crohn's disease. Under germ-free conditions disease does not develop [23, 24]. The role of the bacterial flora of the gut in the pathogenesis of human Crohn's disease is still unresolved. Another example is HLA-B27 transgenic rats, an animal model which develops a spondylarthropathy resembling ankylosing spondylitis or similar diseases as well an inflammatory bowel disease. Both immunopathies only develop if the animals are kept under non-germ-free conditions [25, 26]. These data show that infections might indeed be involved in the initiation and/or perpetuation of autoimmunity. Thus, it is intriguing to speculate that development of SLE is initiated by an infection as well and then leads to chronic inflammatory autoimmune responses due to yet unknown mechanisms. This hypothesis is certainly not only discussed for pathogenesis of SLE, but for other autoimmunopathies as well like Graves' disease [27], Sjögren's syndrome and many others as listed above.

9.3 Infection, Inflammation and SLE: Theory and Practical Aspects

9.3.1 Theoretical Considerations for the Pathogenesis of SLE

Based on these data and the arguments linking defective apoptosis as a central mechanism in the pathogenesis of autoimmunity it is intriguing to hypothesize that regulation of apoptotic cell death is altered in the course of infection and inflammation. A hallmark in autoimmune diseases is the presence of serum autoantibodies mostly directed against nuclear constituents, which are not readily accessible for immunocompetent cells under normal circumstances. The question how the sensitization against these antigens develops is unresolved. However, a central hypothesis claims that a dysregulation of programmed cell death is involved in the generation of autoantibodies and the pathogenesis of SLE (for review, see Chapter 15 or [28]).

For several reasons, this hypothesis is relevant and highly interesting especially in the context of SLE and infections. Rosen *et al.* described that after induction of apoptosis by Sindbis virus infection in HeLa cells viral antigens and autoantigens co-cluster exclusively in small surface blebs of apoptotic cells [29]. These blebs form antigenic structures of mixed viral and self origin, and could define a novel immune context. Thus, an immune response originally directed against the virus could easily react with apoptotic bodies. This could lead to autoantibody formation against intracellular constituents and initiation of autoimmune diseases. These data again support the hypothesis of an induction of (auto)immunity against nuclear constituents in the context of uncontrolled apoptosis during infections and inflammation.

During bacterial or viral challenges antigen-specific T lymphocytes become activated and expand within the adjacent lymph node or the inflamed tissue. From our own data, we demonstrate that regulation of apoptosis is altered in activated lymphocytes from patients with SLE. This is evidenced by data showing that activated lymphoblasts from SLE patients, but not mixed connective tissue disease/vasculitis patients, were hyporesponsive to γ_c chain cytokines leading to accelerated apoptosis if the cells were obtained during a phase with high serological activity markers [erythrocyte sedimentation rate, low C4, elevated double-stranded (ds) DNA antibody titers or tumor necrosis factor (TNF)-α levels] or T_h1 dominance in the serum [high IL-12, detectable interferon (IFN)-γ] [70]. *Vice versa*, 16 of 60 SLE patients had none of these serological parameters in the abnormal range as defined and response to γ_c chain cytokines was not different to ND control lymphoblasts in these SLE patients. In further support of this statement we showed in this paper that accelerated apoptosis can be specifically found in *in vivo* activated lymphocytes derived from SLE patients suffering from an infection with fever (as a clinical marker for systemic involvement of the infectious process). Similar results were not seen in cells from non-autoimmune patients or patients with non-SLE autoimmunopathies during episodes with infections and fever [70]. Moreover, especially activated lymphoblasts could serve as a source for autoantigens stemming from the nucleus. In our current work, we show that histones are present in the cytoplasm of activated lymphoblasts in very early phases of apoptosis. This was not true for quiescent lymphocytes (Gabler *et al.*, unpublished). Moreover, cytoplasmic histones were different from the nuclear histones in their phosphorylation and acetylation state, possibly qualifying these cytoplasmic proteins as a source for (originally nuclear) autoantigens.

How could infection, inflammation and the induction of apoptosis in activated human lymphoblasts contribute to provoke autoimmunity? An important consequence of apoptosis as opposed to necrosis is that the cellular membranes are preserved, until finally the apoptotic cell body is rapidly removed and degraded via phagocytosis without induction of an inflammatory response [30–32]. Thus, under physiological circumstances, cellular constituents are not released and therefore cannot activate immunocompetent cells ([30, 33] and Chapter 5). However, an increased rate of apoptosis could theoretically lead to an overflow of the phagocytic system with apoptotic cell bodies. Thus, intracellular constituents like apoptotic DNA fragments would be presented to, and recognized as non-self antigens by, immunocompetent cells, leading to autoantibody formation against intracellular molecules like dsDNA. In support of this scenario, Casciola-Rosen *et al.* [31] reported that after ultraviolet irradiation-induced *in vitro* apoptotic cell death of keratinocytes, most of the known SLE autoantigens are clustered within surface blebs of apoptotic cells. This generated high concentrations of known autoantigens within discrete subcellular packages, and could lead to autoantibody formation against intracellular constituents and initiation of autoimmune diseases. Furthermore, modifications of nuclear constituents occurring during apoptotic degradation like methylation, de/phosphorylation, citrullination, oxidative stress, activation of transglutaminases, acetylation (reviewed in [34]) or degradation by granzyme B [35] could change immunogenicity. Meanwhile, at least 39 proteins are known which are proteolytically cleaved during apoptosis and possibly modified in their antigenicity. Of these proteins, at least 17 could be detected as autoantigens or as components of complexes containing known autoantigens [34]. In our previous work, we found that stimulation of peripheral blood mononuclear cells with autologous apoptotic material leads to formation of histone-specific T cell clones [36]. Furthermore, these autoantigen-specific T cells could stimulate autologous B lymphocytes to production of dsDNA autoantibodies [36], again bringing increased onflow of apoptotic material in a context with induction of autoimmunity and production of autoantibodies (Fig. 9.1).

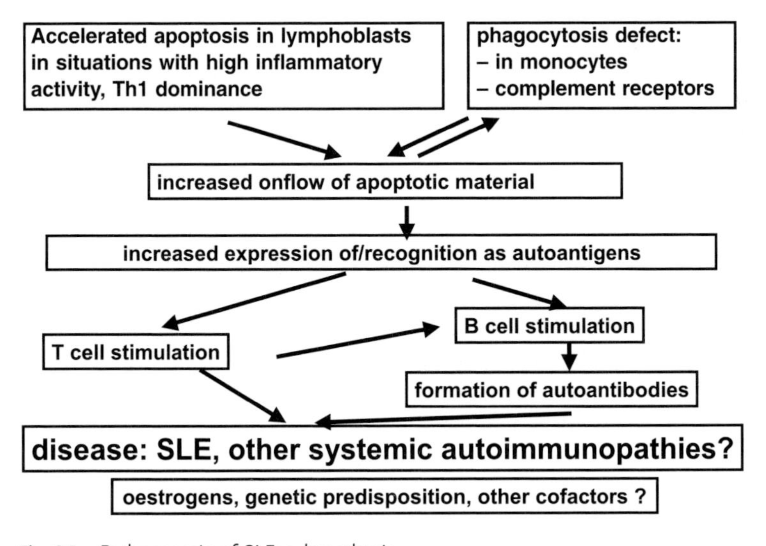

Fig. 9.1 Pathogenesis of SLE: a hypothesis.

All of these factors might occur during infections and inflammation in SLE patients. SLE lymphoblasts might be the source of apoptotic material as mentioned above [70, Gabler *et al.*, unpublished]. An increased onflow of apoptotic material, possibly combined with a failure to adequately clear apoptotic material ([37] and Chapter 11), might be especially deleterious in an highly activated tissue environment in the presence of activated antigen-presenting cells and pro-inflammatory cytokines, originally assembled to respond against an infectious agent. In this context, the data by Casciola-Rosen [35] are of special interest: granzyme B is a protease which is released from $CD8^+$ T lymphocytes or natural killer (NK) cells after challenge with infectious organisms [38, 39]. On the other hand, granzyme B specifically cleaves nuclear autoantigens [35]. Thus, one could speculate that during infections elevated granzyme B levels meet increased concentrations of apoptotic cells, leading to specific degradation of nuclear constituents which become accessible in the course of the degradation of apoptotic material. This could cause recognition of these specifically degraded proteins as autoantigens, being followed by an immune response mounted against these autoantigens. This chain of events could especially be promoted and facilitated if these (neo)autoantigens are exposed in a highly inflammatory environment (originally activated during the response against bacteria or virus).

Our hypothesis that infection and inflammation are causative factors or promoting cofactors in the induction of autoimmunity was again supported by the recent work of Berden and his group: in MRL/*lpr* mice nucleosomes were persistently present in the plasma and complexed into nucleosome immune complexes. Injection of bacterial lipopolysaccharide increased the release of nucleosomes due to an enhancement of apoptosis and a decrease in the clearance of apoptotic cells [40]. Several bacteria, viruses or parasites have been shown to induce apoptosis in mammalian cells [41, 42], so that a direct influence of certain bacteria or infectious agents on pathogenesis or on flares of SLE could be envisioned through these concepts.

In an alternative or complementary scenario, it is hypothesized that DNA-anti-DNA immune complex deposition contributes to the precipitation of the inflammatory response to the kidney glomerulum. The mechanisms responsible for the binding to the glomerular basement membrane are unresolved. Heparan sulfates might be the central ligand in the glomerulum responsible for immune complex binding [43]. On the other hand, Tabata *et al.* [44] investigated lupus-prone mice strains developing immune complex-dependent glomerulonephritis. It has earlier been shown that the envelope glycoprotein gp70 of a xenotropic endogenous retrovirus impacts the formation and deposition of immune complexes in the glomerula. Furthermore, after injection of anti-gp70-producing hybridoma cells or purified anti-gp70, IgG antibodies into syngeneic, non-autoimmune mice or SCID mice proliferative or wire loop-like glomerular lesions developed. These data favor a hypothesis in which immunological responses to viruses or bacteria at least influence, if not co-induce, autoimmune phenomena and help to precipitate organ tissue damage.

Bacteria carry dsDNA in their nuclei as well. Thus, in the context of development of SLE and infection, it is intriguing to speculate that immunization against bacterial DNA during infections leads to formation of anti-human dsDNA autoantibodies. This has been intensively studied by Pisetsky *et al.* Robertson *et al.* studied sera of eight patients with proven *Escherichia coli* bacteremia. Single-stranded bacterial DNA could be detected using ELISA techniques in five of these patients antibodies against [45]. The isotype distribution as well as the avidity as assessed by competition ELISA suggested that during the course of the infection some patients may produce antibodies with immunochemical properties similar to those arising in SLE. Furthermore, they found that sera from patients with SLE showed high levels of binding capacities to all bacterial DNA tested [46, 47]. This was true for sera from healthy subjects as well, but these sera showed lower reactivity and a different subclass distribution as compared to SLE sera [46]. On the other hand, in an animal model of SLE, the NZB/NZW mice, immunization with bacterial dsDNA from *E. coli*, complexed with bovine serum albumin in adjuvant, induced significant titers of anti-dsDNA autoantibodies, but decreased the amount of proteinuria and glomerular pathology as compared to control mice [48].

Bacterial DNA differs from vertebrate DNA in the frequency and methylation of CpG dinucleotides, with bacterial DNA being hypomethylated [49]. In SLE sera elevated levels of circulating plasma DNA were found, which is enriched in hypomethylated CpG motifs [10, 49]. On the other hand, genomic DNA is also hypomethylated in SLE. Nevertheless, CpG DNA has immunological properties: CpG DNA isolated from SLE sera and transfected into endothelial cells induced the expression of ICAM-1 on the protein level as well as IL-6, IL-8, TNF-α and IFN-γ on the mRNA level [50]. In similar experiments, DNA isolated from SLE sera contained CpG motifs and induced expression of HLA-DR and ICAM-1 on monocytes as well as mRNA expression of IL-12 and IFN-γ, and triggered proliferation of mononuclear cells [51]. CpG motifs can induce resistance to apoptosis, thereby possibly causing survival of autoreactive cells and favoring an autoimmune situation [52, 53]. Although it is still an open question whether bacterial DNA directly influences formation of anti-human dsDNA antibodies, it is an intriguing hypothesis. However, mechanisms leading to generation of autoantibodies against nuclear constituents other than DNA remain unclear in this scenario.

9.3.2 Practical Aspects in Human SLE

The practical experience in our lupus cohort tells us that in the majority of the patients infections do not precede SLE flares. On the other hand, in some patients and on some occasions, we saw a coincidence of infections and SLE disease activation. Of course, a causal relationship cannot be proven, but the coincidence was at least remarkable. Duffy *et al.* [54] studied 82 SLE patients admitted to their hospital in 5 years. By logistic regression analysis infection was significantly associated with SLE disease activity, but not with disease duration or prednisolone dosage. In a Korean SLE population, incidence of infection was associated with high SLE disease activity [55]. These data show that infections can influence SLE disease activity *in vivo* in humans and might indirectly argue for a scenario as has earlier been discussed.

This is again supported by the study of Nived *et al.* [56]. These authors found that in 102 SLE patients, disease activity was accompanied by an increase in the frequency of bacterial infections, occurring both before and after initiation of the SLE flare. In contrast, ter Borg *et al.* [57] prospectively followed 72 SLE patients for an average of 18.5 months. During this time 33 SLE flares and 31 infections were observed. Only one infection preceded an SLE flare within a period of 3 months. The authors concluded that significant infections were neither related to rises in levels of anti-dsDNA antibodies nor to the induction of exacerbations of SLE.

On the other hand, certain infections like Hepatitis B or HIV might decrease SLE activity: Lu *et al.* reported that in Taiwan, a hyperendemic area for Hepatitis B infection, SLE disease activity of patients with Hepatitis B infection was significantly lower as compared to the patients without the infection [58]. However, this study needs confirmation, since the number of Hepatitis B-infected SLE patients was only six versus 167 non-infected SLE patients. Moreover, several case reports state that SLE significantly improved or went into remission after infection with HIV [59–61]. *Vice versa*, in an HIV-infected patient, SLE developed after initiation of highly effective antiretroviral treatment [62]. These data again support the notion that infections indeed can influence disease activity in autoimmune disorders.

This is again supported by the clinical experience that vaccination in SLE – despite being safe and effective in the majority of the patients [63–66] – can lead to exacerbation of SLE or precipitation of autoimmune symptoms in rare cases [63, 67–69]. For security reasons patients with steroid dosages of more than 20 mg/day or on immunosuppressive drugs should not be immunized with live vaccines as proposed in the guidelines of the British Society of Rheumatology [63].

9.4 Conclusions

In summary, without any doubt certain infectious organisms can cause autoimmune phenomena or manifest autoimmunity, supporting the notion that infections and autoimmunity are affiliated. In theory, this can be extrapolated to a pathogenetic scenario in SLE, in which a dysregulation of apoptosis especially in activated lymphoblasts (after infections, during inflammatory responses) and/or an impaired phagocytosis of apoptotic material during certain phases of immune responses (like infections) play a central role. Secretion of granzyme B during infections within a highly inflammatory and activated environment might generate specifically cleaved and degraded nuclear antigens, leading to recognition as autoantigens, B and T lymphocyte activation, and precipitation of autoimmunity. The question whether bacterial DNA directly contributes as antigen for generation of autoantibodies against anti-human dsDNA remains open. Our clinical experience tells that in the majority of SLE patients overt infections do not lead to SLE flares. In some cases, however, this has been observed in our cohort and reported by other colleagues. This observation is supported by the fact that vaccination rarely can lead to exacerbation of SLE disease activity.

9.5 References

1 Baxter AG, Horsfall AC, Healey D, Ozegbe P, Day S, Williams DG, *et al.* Mycobacteria precipitate an SLE-like syndrome in diabetes-prone NOD mice. *Immunology* **1994**, 83, 227–31.

2 Obermayer-Straub P, Manns MP. Hepatitis C and D, retroviruses and autoimmune manifestations. *J Autoimmun* **2001**, 16, 275–85.

3 Herrmann M, Neidhart M, Gay S, Hagenhofer M, Kalden JR. Retrovirus-associated rheumatic syndromes. *Curr Opin Rheumatol* **1998**, 10, 347–54.

4 Nakagawa K, Harrison LC. The potential roles of endogenous retroviruses in autoimmunity. *Immunol Rev* **1996**, 152, 193–236.

5 Perron H, Seigneurin JM. Human retroviral sequences associated with extracellular particles in autoimmune diseases: epiphenomenon or possible role in aetiopathogenesis? *Microbes Infect* **1999**, 1, 309–22.

6 Mason AL, Xu L, Guo L, Garry RF. Retroviruses in autoimmune liver disease: genetic or environmental agents? *Arch Immunol Ther Exp* **1999**, 47, 289–97.

7 Terada K, Katamine S, Eguchi K, Moriuchi R, Kita M, Shimada H, *et al.* Prevalence of serum and salivary antibodies to HTLV-1 in Sjögren's syndrome. *Lancet* **1994**, 344, 1116–9.

8 Meilof JF, Smeenk RJ. Detection of retroviral antibodies in primary biliary cirrhosis. *Lancet* **1998**, 352, 739–40.

9 Neidhart M, Rethage J, Kuchen S, Kunzler P, Crowl RM, Billingham ME, *et al.* Retrotransposable L1 elements expressed in rheumatoid arthritis synovial tissue: association with genomic DNA hypomethylation and influence on gene expression. *Arthritis Rheum* **2000**, 43, 2634–47.

10 Krapf FE, Herrmann M, Leitmann W, Kalden JR. Are retroviruses involved in the pathogenesis of SLE? Evidence demonstrated by molecular analysis of nucleic acids from SLE patients' plasma. *Rheumatol Int* **1989**, 9, 115–21.

11 Buskila D. Hepatitis C-associated arthritis. *Curr Opin Rheumatol* **2000**, 12, 295–9.

12 McMurray RW, Elbourne K. Hepatitis C virus infection and autoimmunity. *Semin Arthritis Rheum* **1997**, 26, 689–701.

13 Rivera J, Garcia-Monforte A. Hepatitis C virus infection presenting as rheumatoid arthritis. Why not? *J Rheumatol* **1999**, 26, 2062–3.

14 Fiore G, Giacovazzo F, Giacovazzo M. HCV and dermatomyositis: report of 5 cases of dermatomyositis in patients with HCV infection. *Riv Eur Sci Med Farmacol* **1996**, 18, 197–201.

15 Ramos-Casals M, Font J, Garcia-Carrasco M, Cervera R, Jimenez S, Trejo O, *et al.* Hepatitis C virus infection mimicking systemic lupus erythematosus: study of hepatitis C virus infection in a series of 134 Spanish patients with systemic lupus erythematosus. *Arthritis Rheum* **2000**, 43, 2801–6.

16 Trapani S, Ermini M, Falcini F. Human parvovirus B19 infection: its relationship with systemic lupus erythematosus. *Semin Arthritis Rheum* **1999**, 28, 319–25.

17 Moore TL. Parvovirus-associated arthritis. *Curr Opin Rheumatol* **2000**, 12, 289–94.

18 Roblot P, Roblot F, Ramassamy A, Becq-Giraudon B. Lupus syndrome after parvovirus B19 infection. *Rev Rhum Engl Ed* **1997**, 64, 849–51.

19 Yamaoka Y, Isozaki E, Kagamihara Y, Matsubara S, Hirai S, Takagi K. A case of Guillain-Barre syndrome (GBS) following human parvovirus B19 infection. *Rinsho Shinkeigaku* **2000**, 40, 471–5.

20 Zhao ZS, Granucci F, Yeh L, Schaffer PA, Cantor H. Molecular mimicry by herpes simplex virus-type 1: autoimmune disease after viral infection. *Science* **1998**, 279, 1344–7.

21 Guilherme L, Cunha-Neto E, Tanaka AC, Dulphy N, Toubert A, Kalil J. Heart-directed autoimmunity: the case of rheumatic fever. *J Autoimmun* **2001**, 16, 363–7.

22 Kukreja A, Maclaren NK. Current cases in which epitope mimicry is considered as a component cause of autoimmune disease: immune-mediated (type 1) diabetes. *Cell Mol Life Sci* **2000**, 57, 534–41.

23 Mizoguchi A, Mizoguchi E, Saubermann LJ, Higaki K, Blumberg RS, Bhan AK. Limited CD4 T-cell diversity associated with colitis in T-cell receptor alpha mutant mice requires a T helper 2 environment. *Gastroenterology* **2000**, 119, 983–95.

24 Sadlack B, Merz H, Schorle H, Schimpl A, Feller AC, Horak I. Ulcerative colitis-like disease in mice with a disrupted interleukin-2 gene. *Cell* **1993**, 75, 253–61.

25 Taurog JD, Maika SD, Satumtira N, Dorris ML, McLean IL, Yanagisawa H, *et al.* Inflammatory disease in HLA-B27 transgenic rats. *Immunol Rev* **1999**, 169, 209–23.

26 Onderdonk AB, Richardson JA, Hammer RE, Taurog JD. Correlation of cecal microflora of HLA-B27 transgenic rats with inflammatory bowel disease. *Infect Immun* **1998**, 66, 6022–3.

27 Koh LD, Napolitano G, Singer DS, Molteni M, Scorza R, Shimojo N, *et al.* Graves' disease: a host defense mechanism gone awry. *Int Rev Immunol* **2000**, 19, 633–64.

28 Lorenz HM, Herrmann M, Winkler T, Gaipl U, Kalden JR. Role of apoptosis in autoimmunity. *Apoptosis* **2000**, 5, 443–9.

29 Rosen A, Casciola-Rosen L, Ahearn J. Novel packages of viral and self-antigens are generated during apoptosis. *J Exp Med* **1995**, 181, 1557–61.

30 Savill J, Fadok V, Henson P, Haslett C. Phagocyte recognition of cells undergoing apoptosis. *Immunol Today* **1993**, 14, 131–6.

31 Casciola-Rosen LA, Anhalt G, Rosen A. Autoantigens targeted in systemic lupus erythematosus are clustered in two populations of surface structures on apoptotic keratinocytes. *J Exp Med* **1994**, 179, 1317–30.

32 Voll R, Herrmann M, Roth E, Stach C, Kalden JR, Girkontaite I. Immunosuppressive effects of apoptotic cells. *Nature* **1997**, 390, 350–1.

33 Flora PK, Gregory CD. Recognition of apoptotic cells by human macrophages: inhibition by a monocyte/macrophage-specific monoclonal antibody. *Eur J Immunol* **1994**, 24, 2625–32.

34 Utz PJ, Anderson P. Posttranslational protein modifications, apoptosis, and the bypass of tolerance to autoantigens. *Arthritis Rheum* **1998**, 41, 1152–60.

35 Casciola-Rosen L, Andrade F, Ulanet D, Wong WB, Rosen A. Cleavage by granzyme B is strongly predictive of autoantigen status: implications for initiation of autoimmunity. *J Exp Med* **1999**, 190, 815–26.

36 Voll R, Roth E, Girkontaite I, Fehr H, Herrmann M, Lorenz H-M, *et al.* Histone-specific T_h0 and T_h1 clones derived from systemic lupus erythematosus patients induce double-stranded DNA antibody production. *Arthritis Rheum* **1997**, 40, 2162–9.

37 Herrmann M, Voll RE, Zoller OM, Hagenhofer M, Ponner BB, Kalden JR. Impaired phagocytosis of apoptotic material by monocyte-derived macrophages from patients with systemic lupus erythematosus. *Arthritis Rheum* **1998**, 41, 1241–50.

38 Lauw FN, Simpson AJ, Hack CE, Prins JM, Wolbink AM, van Deventer SJ, *et al.* Soluble granzymes are released during human endotoxemia and in patients with severe infection due to gram-negative bacteria. *J Infect Dis* **2000**, 182, 206–13.

39 Sandberg JK, Fast NM, Nixon DF. Functional heterogeneity of cytokines and cytolytic effector molecules in human CD8(+) T lymphocytes. *J Immunol* **2001**, 167, 181–7.

40 Licht R, van Bruggen MC, Oppers-Walgreen B, Rijke TP, Berden JH. Plasma levels of nucleosomes and nucleosome-autoantibody complexes in murine lupus: effects of disease progression and lipopolysacharide administration. *Arthritis Rheum* **2000**, 44, 1320–30.

41 Grassme H, Kirschnek S, Riethmueller J, Riehle A, von Kurthy G, Lang F, *et al.* CD95/CD95 ligand interactions

on epithelial cells in host defense to Pseudomonas aeruginosa. *Science* **2000**, 290, 527–30.

42 Zychlinsky A, Prevost MC, Sansonetti PJ. *Shigella flexneri* induces apoptosis in infected macrophages. *Nature* **1992**, 358, 167–9.

43 Raats CJ, van den Born J, Berden JH. Glomerular heparan sulfate alterations: mechanisms and relevance for proteinuria. *Kidney Int* **2000**, 57, 385–400.

44 Tabata N, Miyazawa M, Fujisawa R, Takei YA, Abe H, Hashimoto K. Establishment of monoclonal anti-retroviral gp70 autoantibodies from MRL/*lpr* lupus mice and induction of glomerular gp70 deposition and pathology by transfer into non-autoimmune mice. *J Virol* **2000**, 74, 4116–26.

45 Robertson CR, Pisetsky DS. Immunochemical properties of anti-DNA antibodies in the sera of patients with *Escherichia coli* bacteremia. *Int Arch Allergy Immunol* **1992**, 98, 311–6.

46 Wu ZQ, Drayton D, Pisetsky DS. Specificity and immunochemical properties of antibodies to bacterial DNA in sera of normal human subjects and patients with systemic lupus erythematosus (SLE). *Clin Exp Immunol* **1997**, 109, 27–31.

47 Pisetsky D, Drayton D, Wu ZQ. Specificity of antibodies to bacterial DNA in the sera of healthy human subjects and patients with systemic lupus erythematosus. *J Rheumatol* **1999**, 26, 1934–8.

48 Gilkeson GS, Ruiz P, Pippen AM, Alexander AL, Lefkowith JB, Pisetsky DS. Modulation of renal disease in autoimmune NZB/NZW mice by immunization with bacterial DNA. *J Exp Med* **1996**, 183, 1389–97.

49 Krieg AM. CpG DNA: a pathogenic factor in systemic lupus erythematosus? *J Clin Immunol* **1995**, 15, 284–92.

50 Miyata M, Ito O, Kobayashi H, Sasajima T, Ohira H, Suzuki S, *et al.* CpG-DNA derived from sera in systemic lupus erythematosus enhances ICAM-1 expression on endothelial cells. *Ann Rheum Dis* **2001**, 60, 685–9.

51 Sato Y, Miyata M, Sato Y, Nishimaki T, Kochi H, Kasukawa R. CpG motif-containing DNA fragments from sera of patients with systemic lupus erythematosus proliferate mononuclear cells *in vitro*. *J Rheumatol* **1999**, 26, 294–301.

52 Wang Z, Karras JG, Colarusso TP, Foote LC, Rothstein TL. Unmethylated CpG motifs protect murine B lymphocytes against Fas-mediated apoptosis. *Cell Immunol* **1997**, 180, 162–7.

53 Yi AK, Chang M, Peckham DW, Krieg AM, Ashman RF. CpG oligodeoxyribonucleotides rescue mature spleen B cells from spontaneous apoptosis and promote cell cycle entry. *J Immunol* **1998**, 160, 5898–906.

54 Duffy KN, Duffy CM, Gladman DD. Infection and disease activity in systemic lupus erythematosus: a review of hospitalized patients. *J Rheumatol* **1991**, 18, 1180–4.

55 Suh CH, Jeong YS, Park HC, Lee CH, Lee J, Song CH, *et al.* Risk factors for infection and role of C-reactive protein in Korean patients with systemic lupus erythematosus. *Clin Exp Rheumatol* **2001**, 19, 191–4.

56 Nived O, Sturfelt G, Wollheim F. Systemic lupus erythematosus and infection: a controlled and prospective study including an epidemiological group. *Q J Med* **1985**, 55, 271–87.

57 ter Borg EJ, Horst G, Hummel E, Limburg PC, Kallenberg CG. Rises in anti-double stranded DNA antibody levels prior to exacerbations of systemic lupus erythematosus are not merely due to polyclonal B cell activation. *Clin Immunol Immunopathol* **1991**, 59, 117–28.

58 Lu CL, Tsai ST, Chan CY, Hwang SJ, Tsai CY, Wu JC, *et al.* Hepatitis B infection and changes in interferon-alpha and -gamma production in patients with systemic lupus erythematosus in Taiwan. *J Gastroenterol Hepatol* **1997**, 12, 272–6.

59 Molina JF, Citera G, Rosler D, Cuellar ML, Molina J, Felipe O, *et al.* Coexistence of human immunodeficiency virus infection and systemic lupus erythematosus. *J Rheumatol* **1995**, 22, 347–50.

60 Byrd VM, Sergent JS. Suppression of systemic lupus erythematosus by the human immunodeficiency virus. *J Rheumatol* **1996**, 23, 1295–6.

61 Fox RA, Isenberg DA. *Arthritis Rheum* **1997**, 40, 1168–72.

62 Diri E, Lipsky PE, Berggren RE. Emergence of systemic lupus erythematosus after initiation of highly active antiretroviral therapy for human immunodeficiency virus infection. *J Rheumatol* **2000**, 27, 2711–4.

63 Ioannou Y, Isenberg DA. Immunisation of patients with systemic lupus erythematosus, the current state of play. *Lupus* **1999**, 8, 497–501.

64 Aron-Maor A, Shoenfeld Y. Vaccination and systemic lupus erythematosus: the bidirectional dilemmas. *Lupus* **2001**, 10, 237–40.

65 Abu-Shakra M, Zalmanson S, Neumann L, Flusser D, Sukenik S, Buskila D. Influenza virus vaccination of patients with systemic lupus erythematosus: effects on disease activity. *J Rheumatol* **2000**, 27, 1681–5.

66 Battafarano DF, Battafarano NJ, Larsen L, Dyer PD, Older SA, Muehlbauer S, *et al.* Antigen-specific antibody responses in lupus patients following immunization. *Arthritis Rheum* **1998**, 41, 1828–34.

67 Older SA, Battafarano DF, Enzenauer RJ, Krieg AM. Can immunization precipitate connective tissue disease? Report of five cases of systemic lupus erythematosus and review of the literature. *Semin Arthritis Rheum* **1999**, 29, 131–9.

68 Senecal JL, Bertrand C, Coutlee F. Severe exacerbation of systemic lupus erythematosus after hepatitis B vaccination and importance of pneumococcal vaccination in patients with autosplenectomy. *Arthritis Rheum* **1999**, 42, 1307–8.

69 Maillefert JF, Tavernier C, Sibilia J, Vignon E. Exacerbation of systemic lupus erythematosus after hepatitis B vaccination. *Arthritis Rheum* **2000**, 43, 468–9.

70 Lorenz H-M, Grünke M, Hieronymus T, Winkler S, Blank N, Rascu A, Wendler J, Geiler T, Kalden JR. Hyporesponsiveness to γc-chain cytokines in activated lymphocytes from patients with Systemic Lupus Erythematosus (SLE) leads to accelerated apoptosis. *Eur J Immunol*, **2002**, 32, 1253–1263.

10
Apoptosis in Rheumatoid Arthritis

Yuji Yamanishi and Gary S. Firestein

10.1
Introduction

Rheumatoid arthritis (RA) is a systemic inflammatory disease characterized by synovial hyperplasia and mononuclear cell infiltration into the synovium. The disease occurs in about 1% of adults worldwide, and has both genetic and environmental components. The primary manifestation of RA is inflammatory synovitis with hyperplasia of the synovium and invasion of articular structures. This process ultimately leads to the destruction of bone and cartilage, thereby causing significant disability. Although current management of RA has improved with the advent of cytokine antagonists and immunomodulators, a significant percentage of patients have debilitating signs and symptoms despite aggressive therapy.

The rheumatoid synovium is marked by dramatically increased cellularity in both the intimal lining and sublining regions (Fig. 10.1). Lining hyperplasia results from accumulation of both macrophage-like type A synoviocytes and fibroblast-like type B synoviocytes (FLS). The sublining region is infiltrated with mononuclear cells, primarily CD4 T cells and macrophages, along with lesser numbers of B cells and stromal cells. The remarkable accumulation of cells in RA synovium, which is reminiscent of a locally invasive tumor in the joint, results from an imbalance in the forces that increase and decrease cellularity (Fig. 10.2). Migration of peripheral blood cells or bone marrow cells into synovium and proliferation of FLS contributes to accumulation of synoviocytes and sublining mononuclear cells. On the other hand, apoptosis and egress of the cells out of the joint via the lymphatics or to synovial fluid diminish cellularity.

Increased proliferation of RA synovial cells, especially FLS, has been suggested by studies evaluating [^{3}H]thymidine incorporation and expression of proliferation markers in the intimal lining [1–3]. Although there is some disagreement, most of the studies indicate that the proliferation rate of RA synoviocytes is rather modest compared with the extent of hyperplasia [2, 3]. In addition, the number of apoptotic cells is relatively low in RA synovium and indicates that a prolonged lifespan of synoviocytes might play an important role. DNA fragmentation occurs in RA synovium, but cells showing typical morphological features of apoptosis are rare. These data suggest that defective apoptosis might participate in synovial hy-

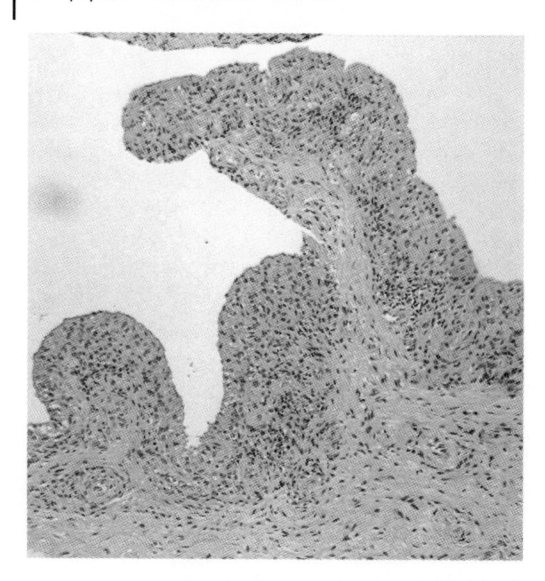

Fig. 10.1 Histology of RA synovial tissue, showing intimal lining hyperplasia and mononuclear cell infiltration into the synovial sublining.

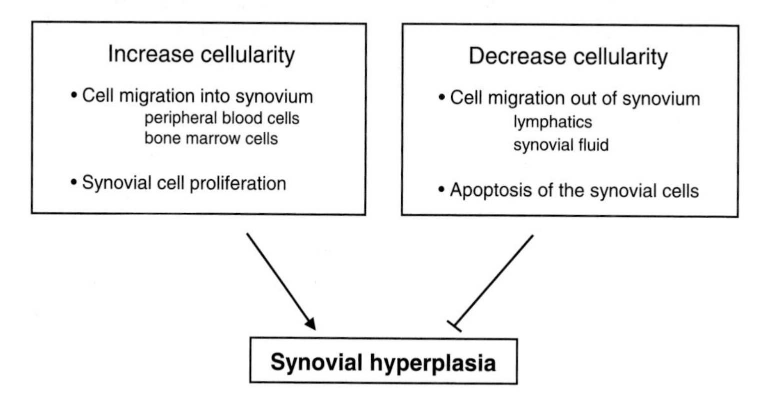

Fig. 10.2 Balance of cellularity in RA synovium. The remarkable accumulation of cells in RA synovium results from an imbalance in the forces that increase and decrease cellularity. Migration of peripheral blood cells or bone marrow cells into synovium and proliferation of synovial cells contribute to accumulation of synoviocytes and sublining mononuclear cells. On the other hand, apoptosis and egress of the cells out of the joint via the lymphatics or to synovial fluid diminish cellularity.

perplasia and pannus formation (Tab. 10.1). As a corollary, the induction of apoptosis in proliferating synoviocytes is a potential treatment for RA. In this chapter, we will review the mechanisms of apoptosis in RA and explore how defects in this process might exacerbate the disease.

Tab. 10.1 Factors that decrease apoptosis in RA synovium.

Anti-apoptotic genes	high Bcl-2 expression
	high sentrin expression
Tumor suppressor genes	low PTEN expression
	p53 mutations
Cytokines	TGF-β, IL-15, IL-4, IL-13, IL-2
Cell-cell interaction	T cells and FLS

10.2 Apoptosis in RA Synovium

The extent of apoptosis in RA synovium has been evaluated using several different methods, including *in situ* end-labeling (ISEL), terminal deoxynucleotidyl transferase-mediated dUTP-biotin nick end-labeling (TUNEL) and electron microscopic analysis [4, 5]. By and large, the methods that most precisely define apoptosis demonstrate that only a small percentage of cells in the hyperplastic intimal lining undergo apoptosis. DNA ladders of rheumatoid synovium can demonstrate a faint ladder pattern and indicate that at least some apoptosis occurs. Ultrastructural studies using electron microscopy on RA synovium indicate that only 3% of the synovial cells, mainly in sublining, are apoptotic [6]. TUNEL techniques are more variable (and less specific), but also suggest that a limited number of cells undergo programmed cell death [5, 7, 8]. Abundant single-strand DNA breaks have also been noted with ISEL indicating local DNA damage [4]. It is possible that synovial cells showing apoptosis are not detected using these techniques due to the rapidity of receptor-mediated scavenging processes. However, the varied results from TUNEL and other labeling techniques suggest that extensive oxidative DNA damage occurs in the joint without effective apoptosis (Tab. 10.1).

The relative lack of cell death in the lining could result from the expression of anti-apoptotic genes or alteration of tumor suppressor gene function. For instance, Matumoto *et al.* reported that Bcl-2 is mainly expressed in the RA FLS in the synovial lining layer, which might result in decreased apoptosis [6]. Sentrin is a novel ubiquitin-like protein that binds to the death domain of Fas/APO-1 and tumor necrosis factor receptor 1 (TNF-R1), and protects cells from both anti-Fas and TNF-induced cell death [9]. High expression of sentrin-1 mRNA was observed in RA synovium, predominantly in synovial fibroblasts of the lining layer and at sites of invasion by RA synovium into cartilage, while no sentrin mRNA was detected in normal synovium [10]. In addition, cultured RA FLS showed significantly higher (approximately 15-fold) sentrin mRNA expression than cultured osteoarthritis (OA) FLS and normal dermal fibroblasts [10]. Similarly, deficient PTEN gene expression in RA synoviocytes might suppress apoptosis resulting from DNA damage. PTEN is a tumor suppressor and mutations in the PTEN gene have been detected in various malignancies [11, 12]. In contrast to the p53 tumor suppressor, mutations of PTEN have not been detected in cultured RA FLS [13]. However, expression of PTEN mRNA is quite low in the lining layer of RA synovium, despite abundant expression in the sublining [13]. In addition, defective p53-mediated cell death might result from local synovial mutations of the p53 gene (see below).

The expression of some surface adhesion molecules might distinguish FLS subsets that are susceptible to cell death. FLS derived from RA synovium have been classified into intracellular adhesion molecule (ICAM)-1 positive or negative cells. ICAM-1^+ cells are at the G_0/G_1 phase and exhibit some apoptotic changes, whereas ICAM-1^- cells are viable and observed at the S to G_2/M phase [14]. In ICAM-1^+ cells, increased expression of p53 and $p21^{WAF}$ and suppression of cyclin-dependent protein kinase 6 activity were also observed. Fas expression is higher on ICAM-1^+ cells compared to ICAM-1^- cells, which might suggest participation of Fas in apoptosis of ICAM-1^+ cells.

Lymphoid aggregates in RA synovial tissues appear to be protected from both DNA damage and apoptosis [4]. Expression of Bcl-2 is markedly high in these regions, which undoubtedly contributes to this phenomenon [4, 15]. High levels of Bcl-2 were observed in both T and B cells in lymphoid follicles, although cells in germinal centers are Bcl-2^- [15, 16]. Bcl-2 expression in lymphocytes infiltrating into synovium is not specific for RA, and has been detected in other inflammatory or degenerative joint diseases like reactive arthritis and OA [15]. More than 90% of RA peripheral blood lymphocytes are Bcl-2^+, which is also similar to reactive arthritis and healthy control lymphocytes. Interestingly, Bcl-2 expression in RA synovial fluid lymphocytes is lower than in peripheral blood cells [15]. This is also not specific for RA, since similar results were observed in reactive arthritis patients.

10.3
Apoptosis in Synovial Fluid T Cells

Apoptosis of T cells in RA synovial fluid is quite rare, which contrasts with crystal-induced arthritis where many T cells are $TUNEL^+$ [17]. Surprisingly, RA synovial fluid T cells demonstrate spontaneous apoptosis when removed from inflammatory environment in the joint. Therefore, the defect in apoptosis observed *in vivo* is not intrinsic but depends on external influences [17]. The synovial fluid T cells are highly differentiated $CD45RB^{dull}$ $CD45RO^{bright}$ cells and express low Bcl-2, and high Bax and Fas, which is quite distinct from the phenotype of T cells infiltrating into synovium in terms of Bcl-2 expression. Since highly differentiated $CD45RB^{dull}$ cells from peripheral blood show marked susceptibility to apoptosis, resistance to apoptosis of synovial fluid T cells is probably due to the inflammatory environment in the joint. Cultured synovial fibroblasts provide some protection from T cell apoptosis *in vitro* and suggest that certain mesenchymal-derived trophic factors contribute to T cell survival.

The mechanisms promoting T cell survival in RA synovial fluid are still uncertain, but cytokines and soluble mediators could play a role. For instance, IL-2 is a potent inhibitor of T cell apoptosis [18], but is an unlikely explanation since IL-2 levels are extremely low in the RA joint [19]. Expression of IL-15, which has powerful anti-apoptotic effect through induction of Bcl-2 and the large splice variant of Bcl-x (Bcl-x_L) [17], is abundant in RA synovial fluid and could contribute [20]. Perhaps the most important reason for the longevity of synovial fluid T cells is the interaction between T cells and stromal cells [17]. When synovial fluid T cells are co-cultured with FLS, Bcl-x_L, but not Bcl-2, is increased, and apoptosis of T cells is inhibited. Integrin binding appears to participate in the protective effect of mesenchymal cells. Peptides containing the RGD (Arg–Gly–Asp) motif, which is important for many integrin-binding sites, prevents rescue from T cell apoptosis by synovial fibroblasts [17].

10.4
Regulation of Apoptosis by Cytokines

Regulation of apoptosis is complex, and involves many intracellular signaling molecules as well as secreted and membrane bound proteins. Several cytokines and growth factors that have been identified in rheumatoid synovial tissue and fluid have been implicated in this process, and can contribute to the balance between cell survival and death. The cytokine network in RA has been the object of intense investigation over the last decade. These studies have demonstrated that macrophage- and fibroblast-derived cytokines, including TNF-α, interleukin (IL)-1, IL-6, IL-15, IL-18 and granulocyte macrophage colony stimulating factor (GM-CSF), are abundantly expressed in RA synovium [21]. In addition, several anti-inflammatory cytokines, such as IL-10 and transforming growth factor (TGF)-β, are also produced in RA synovium and serve as counter-regulatory mechanisms [22]. Both pro-inflammatory and anti-inflammatory cytokines might play a role in apoptosis of RA cells.

TNF-α is especially interesting in light of recent data demonstrating a beneficial clinical effect of TNF inhibitors in RA. Various conflicting reports suggest that TNF-α can regulate apoptosis in FLS, although the mechanism and degree of modulation remains uncertain. TNF-α can, in some culture conditions, directly induce DNA fragmentation in cultured synoviocytes [4]. In another report, TNF-α alone did not cause apoptosis but instead sensitized cells to the effects of anti-Fas antibody [23, 24]. Fas and FADD (Fas-associated death domain protein) expression are not increased by TNF-α in FLS, but enhanced enzymatic activity of caspase-8 and -3 was observed. In addition, Fas-mediated apoptosis in TNF-treated FLS was almost completely inhibited by caspase-8- or -3-specific inhibitors, which suggests that these caspases are involved in the sensitization of FLS by TNF-α [23]. However, Ohshima *et al.* reported that TNF-α inhibited apoptosis of RA FLS in the presence of anti-Fas antibody, and TNF-α blockade by neutralizing anti-TNF-α antibody restored the apoptotic effect of anti-Fas antibody in FLS [25]. The differences reported by various investigators likely results from differences in patient material (e.g., stage of the disease) and culture conditions. Clearly, additional work is needed to define the role of TNF-α in synoviocyte apoptosis.

Several other cytokines, including IL-2, IL-4, IL-7, IL-9, IL-13 and IL-15, use the shared receptor signaling component of a common IL-2 receptor γ chain, and these cytokines can suppress apoptosis in some circumstances. The anti-apoptotic effects can be partially explained by induction of Bcl-2 and Bcl-x_L expression. In particular, IL-15 has a prominent anti-apoptotic effect among these cytokines and can prevent apoptosis of synovial fluid T cells [17]. In addition, IL-4 and IL-13, but not IL-10, protect RA FLS from nitric oxide-induced apoptosis [26]. The anti-apoptotic effects of IL-4 and IL-13 are blocked by phosphatidylinositol-3 kinase and protein kinase C inhibitors. However, the relative paucity of these cytokines in RA suggests that this mechanism plays a relatively minor role *in vivo*.

TGF-β is multifunctional regulatory polypeptide with 25 kDa, and its abundant expression has been reported in RA synovium and synovial fluid [27, 28]. TGF-β enhances the proliferation of FLS in a concentration-dependent manner. Stimulation of FLS with TGF-β also causes marked resistance to Fas-mediated apoptosis [29]. Anti-apoptotic effects by TGF-β are mediated by both decreased Fas antigen expression and increased Bcl-2 expression in FLS. Therefore, TGF-β favors synovial hyperplasia by both inducing proliferation of FLS and protecting against Fas-mediated apoptosis.

10.5 p53 Mutations in RA Synovium and Fibroblast-like Synoviocytes

10.5.1 The Role and Regulation of p53 Tumor Suppressor Protein

The tumor suppressor protein p53 plays a central role in cell cycle regulation, DNA replication, DNA repair, senescence and apoptosis (Tab. 10.2) [30, 31]. When DNA is damaged by genotoxic stress, p53 induces cell cycle arrest at G_1/S and G_2/M interphase through the transcriptional activation of several genes, including $p21^{WAF}$, or induce apoptosis [30–34]. These actions either allow repair of damaged nucleotides prior to DNA synthesis, or maintain DNA integrity by deleting cells with overwhelming DNA damage.

The p53 protein is normally detected only at very low levels but its expression is induced in response to various stimuli, such as DNA damage (by ionizing radiation, ultraviolet radiation and chemotherapeutic drugs, etc.), hypoxia, heat shock, viral infection, growth factor deprivation, and oncogene activation [35, 36]. MDM2 oncoprotein, which is a critical protein that suppresses p53 degradation, accumulates to high levels and contributes to the loss of cell cycle control in malignant tissue [37–39]. Since wild-type p53 protein has a relatively short half-life (less than 20 min), it is usually not detected in normal tissues by immunohistochemistry [40].

10.5.2
p53 Protein Expression in Inflammation

p53 protein expression is typically quite low in normal joint tissues and has been examined extensively in rodents and humans. In the presence of synovial inflammation, including murine collagen-induced arthritis or rat adjuvant arthritis, p53 expression increases dramatically [41, 42]. A similar phenomenon has been described in many human inflammatory diseases, including RA, ulcerative colitis, psoriasis and peptic ulcers [41]. These data suggest that increased p53 protein expression is a general phenomenon of inflammation and that the tumor suppressor might have a protective role (Tab. 10.2). Recent data in p53 knockout mice confirms this hypothesis, since the lack of p53 protein increases joint destruction and inflammation in murine collagen-induced arthritis. The mechanism of protection provided by p53 appears to be related to both increased apoptosis as well as suppression of pro-inflammatory cytokines and matrix metalloproteinases [42]. p53 gene therapy in rabbit IL-1-induced arthritis also increases synovial apoptosis and decreases the local inflammatory response [43]. Taken together, these studies suggest that p53 protein has a normal homeostatic control mechanism in inflammation and abnormalities in p53 function could enhance disease (Tab. 10.3).

As in the animal models, p53 protein expression is low in normal human synovium but is increased in rheumatoid synovium. This phenomenon is observed in both long-standing disease as well as early RA as determined by both immunohistochemistry and Western blot analysis [44]. Of interest, p53 protein expression is greater in RA synovium compared with other inflammatory joint diseases, includ-

Tab. 10.2 The roles of p53 tumor suppressor gene.

Cell cycle regulation
DNA replication
DNA repair
Senescence
Apoptosis
Anti-inflammatory

Tab. 10.3 p53 in RA.

Overexpressed in RA synovium
p53 mutations in RA synovium
G/A and T/C transition mutations
Dominant-negative p53 mutations
Inactivation of p53 protein increases proliferation and invasion of synoviocytes
p53 knockout mice with collagen-induced arthritis
increased severity of arthritis and joint destruction
increased synovial IL-1, IL-6 and collagenase-3
decreased synovial apoptosis

ing inflammatory OA and reactive arthritis [44]. This might be due to the intensity of the inflammatory response and oxidative stress in RA. In addition, p53 mutations could also contribute to high levels of the protein.

10.5.3
p53 Mutations in RA Synovial Tissue

Although there is some variability among reports, mutations of p53 in RA synovium and FLS have been identified independently by several investigators using different methods [45–48]. Initially, we detected p53 mutations in exons 5–10 using RNA mismatch detection assay and cDNA sequencing in the majority of RA synovium and FLS samples. Similar mutations were not found in OA synovium, or RA and OA skin samples [45]. Rème *et al.* focused on only exon 6 and confirmed p53 mutations at the genomic DNA level by the single-strand conformation polymorphism (SSCP) technique and sequence analysis in three of 20 RA synovium samples and one of four RA FLS lines [46]. Recently, a Japanese group also detected p53 mutations in exons 4–11 using a modified SSCP technique in nearly half of RA FLS lines [48]. Surprisingly, no p53 mutations were detected in RA FLS lines from German patients, but the same group confirmed the mutations in RA FLS from U.S. patients [47]. In general, approximately 5–30% of the cDNA pools from RA synovial cells have p53 mutation [45, 46, 48]. More recently, we have found a consistent mutation rate of 7–10% in the cDNA pool of RA synoviocytes (unpublished data). While there are some differences in the frequency of mutations, these are likely due to variations in disease activity, disease duration, genetic background, and methods used for detection.

10.5.4
Possible Mechanism of Occurrence of p53 Mutations

Most p53 mutations in RA synovial cells are single base pair G/A or T/C transition mutations; transversion mutations are rare [45, 48]. Transition mutations are characteristically due to oxidative stress, such as reactive oxygen species (ROS) and reactive nitrogen species (RNS) [49, 50]. ROS and RNS can induce DNA damage and mutations both *in vitro* and *in vivo* [51, 52]. Therefore, we proposed that inflammatory cells that are recruited to a diseased site produce ROS and RNS, and provide a mutagenic stress. This notion has been confirmed in many studies indicating elevated levels of ROS and RNS in the synovial microenvironment [53–55]. Chronic inflammation might be a risk factor of somatic gene mutation in other diseases, such as ulcerative colitis [56]. Therefore, oxidative stress could play a major role in the production of somatic mutations in many chronic inflammatory diseases.

Although most attention has been focused on p53, mutations in other genes have also been detected in RA tissues. The *hprt* gene, which codes for hypoxanthine guanine phosphoribosyltransferase, has been frequently used as a marker to detect gene mutation in human cells. *hprt* mutations were detected in T cells derived from RA synovial tissues and peripheral blood [57]. The frequency of *hprt* mutant T cells in synovial tissue was significantly greater than that in peripheral blood of the same patients. These data suggest that mutations in T cells are induced by the inflammatory environment as T cells traffic through synovial tissue into peripheral blood [57]. H-*ras* gene mutations in codons 13 and 14 were also reported in RA synovial tissues, but their pathogenetic role remains unknown [58].

10.5.5 Function of p53 Mutations in RA Synovial Cells

After identifying p53 mutations in RA synovium, the functional sequelae of this observation were investigated. Some of p53 mutations detected in RA synovium have dominant-negative effects, which suppress normal function of wild-type p53 and as observed in many neoplastic diseases [59]. Co-transfection of plasmids expressing wild-type p53 and mutant p53 genes found in RA synovial samples inhibit *bax* mRNA expression as well as *bax* promoter activity [59]. Since Bax is a critical inducer of apoptosis, inhibition of this gene could lead to decreased apoptosis and contribute to synovial hyperplasia. Other studies have shown that the loss of p53 function increases collagenase gene expression and promoter activity [60, 61]. As an alternative method to assess the function of p53 in RA synovium, synoviocytes were infected with the papilloma virus 18 E6, which accelerates degradation of wild-type p53 protein [62, 63]. E6 expression in FLS resulted in decreased apoptosis as well as increased anchorage-independent growth, proliferation and invasiveness into cartilage extract [62]. These findings were confirmed *in vivo* using a severe combined immunodeficiency (SCID) mouse RA model, where E6 not only increased cartilage invasion of RA synoviocytes but also converted non-invasive normal FLS to a rheumatoid phenotype [63].

10.6 Fas-Fas Ligand (FasL) Apoptotic Pathway in RA

10.6.1 Fas and FasL

Fas (Apo-1/CD95) is a type I transmembrane glycoprotein that belongs to the TNF-R/nerve growth factor receptor (NGF-R) family [64]. This molecule is expressed on various types of cells, including activated lymphocytes and certain transformed cells [64]. FasL (CD178) is 40-kD type II transmembrane protein that is homologous to TNF and its binding to Fas transmits an apoptotic signal to susceptible target cells [65]. One of the major roles of Fas-dependent apoptotic pathway is the termination of immune responses by deletion of activated T and B cells. In fact, mice that have mutations of Fas (*lpr*) or FasL (*gld*) develop a lupus-like disease associated with autoantibodies, lymphadenopathy, splenomegaly and glomerulonephritis due to interference of physiologic apoptosis [66, 67]. The Fas-dependent pathway is also involved in formation and regression of tumors [68]. Thus, Fas and FasL have an important role in the pathogenesis of some diseases through defective apoptosis.

10.6.2 Fas-FasL Expression in Synovium

Fas expression has been observed in RA synovial tissue by immunohistochemistry or flow cytometry [5, 7, 8]. The majority of Fas-expressing cells were FLS and fewer mononuclear cells were also Fas^{+}. Fas-expressing cells were located in the lining [7] or sublining layer [8]. As with Fas, many studies indicate that FasL expression is significantly increased in RA synovial tissue compared with OA or post-traumatic tissues [7, 8, 69], although one report indicates a relative deficiency of FasL expression in RA joint tissue [70]. When present, FasL is predominantly expressed by infiltrating mononuclear cells, but only small number of cells (less than 5%) co-express both Fas and FasL [7, 8]. Interestingly, $FasL^{+}$ mononuclear cells are occasionally found in contact with Fas-expressing synoviocytes.

10.6.3 Fas-FasL Expression in Synovial Fluid Cells

Most of RA synovial fluid T and B cell express Fas and the proportion of Fas-expressing mononuclear cells in RA synovial fluid is significantly higher than peripheral blood from the same patients [70]. Fas expression is induced in T cells through TCR crosslinking or by IL-2 or interferon (IFN)-γ exposure [71, 72]. However, since concentrations of IL-2 or IFN-γ in the joints are relatively low [19, 21], another mechanism for Fas induction in RA synovial fluid T cells, such as cell-cell interactions, probably exists. FasL expression in RA synovial fluid lymphocytes was also confirmed by RT-PCR, but not by flow cytometry [73]. This discrepancy might be due to rapid cleavage of FasL to soluble FasL (sFasL). FasL mRNA expression was not observed in peripheral blood mononuclear cells from RA patients whose synovial fluid lymphocytes express FasL [73].

10.6.4 sFas

An alternatively spliced form of Fas mRNA has been found in peripheral blood mononuclear cells from normal individuals, and angioimmunoblastic lymphadenopathy and SLE patients [74]. This molecule lacks the transmembrane domain of Fas (membrane Fas) and is secreted as a soluble form of Fas (sFas). sFas inhibits Fas-dependent apoptosis in peripheral blood mononuclear cells *in vitro*, and alters lymphocyte development and proliferation in response to self antigen *in vivo* by neutralization of anti-Fas antibody [74]. Clinically, elevated serum concentrations of sFas were reported in patients with SLE, juvenile rheumatoid arthritis and T cell acute lymphoblastic leukemia [74, 75].

The concentration of sFas in the synovial fluid from RA patients is significantly higher than that from OA patients [76]. However, there are no significant differences of serum concentration of sFas among RA, OA and healthy controls. In RA patients, the concentration of sFas in synovial fluid correlates with serum markers of disease activity like erythrocyte sedimentation rate and C-reactive protein, and inflammatory molecules in synovial fluid, such as IL-2 receptor, IL-6 and ICAM-1 [76]. The source of sFas production is probably synoviocytes and infiltrating mononuclear cells. Since sFas exhibits inhibitory effects on Fas-dependent apoptosis, increased sFas in synovial fluid might prevent apoptosis of synoviocytes in the RA joints.

10.6.5 sFasL

sFasL is the 26-kDa soluble form of FasL, which is cleaved from 40-kDa membrane FasL by the action of a metalloproteinase [77, 78]. sFasL has been detected in the serum of patients with various diseases, e.g. large granular lymphocytic leukemia and natural killer cell lymphoma [77]. Since most of those patients with high levels of serum sFasL were neutropenic and had hepatic dysfunction, sFasL might have caused systemic tissue damage when it released into the circulation [79]. However, its actual pathogenic role remains unknown.

The concentrations of sFasL in RA synovial fluid are significantly higher than in OA [73]. On the other hand, sFasL was not detected in RA and OA serum samples from the same patients. The levels of sFasL in synovial fluid are significantly higher in severe RA patients compared to mild RA patients. Since sFasL inhibits the membrane FasL-induced killing of peripheral blood T cells [80] and RA synovial cells are resistant to apoptosis by human sFasL [73], sFasL might suppress Fas-mediated apoptosis in RA synovial cells.

10.6.6 Fas-Mediated Apoptosis

Approximately 10–30% of Fas-expressing cells in RA synovium show DNA fragmentation, suggesting that Fas-mediated apoptosis could play a role in RA synovium [7, 8]. Cultured RA FLS may be more sensitive to anti-Fas IgM monoclonal antibody-mediated apoptosis compared with OA FLS, although this difference is controversial [5, 73]. Since Fas-mediated apoptosis of RA FLS is associated with activation of c-Jun N-terminal kinase (JNK) and activator protein 1 (AP-1) [81], JNK/AP-1 signaling participates in this process. In another study, the effects of TNF and basic fibroblast growth factor (bFGF), both of which can induce synovial cell proliferation, on the Fas-mediated apoptosis were examined. Fas-mediated apoptosis by anti-Fas antibody was increased in TNF-treated RA FLS, but only weakly induced in bFGF-treated FLS [24]. Thus, Fas-induced apoptosis of RA synoviocytes might be differentially regulated by TNF-α and bFGF.

More than half of T cells infiltrating RA synovium die after treatment with anti-Fas antibody in an *in vitro* model, whereas no effect was observed in synovial T cells from OA tissue or peripheral blood mononuclear cells from RA and OA patients [69]. RA synovial fluid mononuclear cells are also susceptible to apoptosis by anti-Fas antibody, although peripheral blood mononuclear cells from the same patients are resistant [69]. Fas^+ cells among the $CD3^+$ T lymphocytes from RA synovium and synovial fluid increased in number compared with those from the peripheral blood of RA patients or those from blood and synovium of OA patients. The main phenotype of T cells susceptible to anti-Fas antibody is the CD45 RO^+ population [69].

10.7
Therapeutic Target of Molecules for Inducing Apoptosis in Synovial Tissues

10.7.1
Fas-FasL and Related Molecules

As the Fas-FasL system has the ability to induce apoptosis of many key cells involved in inflammatory synovitis, considerable effort has been expended to explore Fas-FasL-directed therapies in RA models. Therapeutic benefit of anti-Fas monoclonal antibody was observed in human T cell leukemia virus type 1 (HTLV-1) tax transgenic mice, which develop chronic erosive arthritis with pannus formation and lymphocyte infiltration [82]. Intra-articular administration of anti-Fas monoclonal antibody improved paw swelling and histological features of arthritis within 2 days, and these effects persisted for at least 9 days [83]. Many apoptotic cells in the synovium were detected by TUNEL and electron microscopic analysis [83]. In this study, no serious side effects were observed by intra-articular administration of anti-Fas monoclonal antibody, although lethal liver damage has been reported in mice after intraperitoneal administration [84]. Similarly, gene therapy using an adenoviral construct encoding FasL was effective in murine collagen-induced arthritis [85].

As an alternative to anti-Fas antibodies or direct gene therapy, cell therapy approaches using transfectants expressing the human FasL gene were tested on engrafted human RA synovium in SCID mice [86]. The local injection of human FasL-expressing cells eliminated synoviocytes and mononuclear cells in engrafted human synovium of SCID mice. Of interest, cytotoxicity of human FasL transfectants against RA but not OA synoviocytes was observed even though similar levels of Fas expression in RA were confirmed [86]. Neutralizing monoclonal antibodies against Fas and FasL inhibited human FasL transfectants-induced cytotoxicity in synoviocytes, indicating that the cytotoxic effect was through the Fas-FasL apoptotic pathway.

FADD plays a critical role in Fas-mediated apoptosis of RA synoviocytes. FADD binds to Fas through interactions with the death domains and leads to the activation of the caspase cascade. The therapeutic utility of FADD gene for RA was evaluated using a FADD-expressing adenoviral vector (Ad-FADD) [87]. Infection of RA FLS with Ad-FADD increases FADD gene expression and causes apoptosis in dose-dependent manner. When Ad-FADD is locally injected into RA synovium engrafted in SCID mice, synoviocytes and mononuclear cells infiltration decrease through induction of apoptosis [87]. Ad-FADD-induced apoptosis only occurs in synovium, but not in cartilage, which is engrafted along with synovium. Therefore, cartilage is protected from apoptosis while the synovium itself is targeted.

10.7.2
Nuclear Factor (NF)-κB and Related Molecules

NF-κB is a ubiquitous transcription factor that plays an important role in inflammatory gene transcription. It is activated by wide range of pro-inflammatory signals and cellular stresses, such as TNF-α, IL-1, ionizing radiation, lipopolysaccharide and H_2O_2. NF-κB resides in the cytoplasm in an inactive form through its association with inhibitory proteins called inhibitors of nuclear factor κB (IκB). IκB family members, including IκBα, IκBβ and IκBε, regulate the DNA binding and subcellular localization of NF-κB proteins by masking a nuclear localization signal. NF-κB activation occurs through signal-induced degradation of IκB in the cytoplasm, allowing the translocation of NF-κB to the nucleus. Immunohistochemistry and electromobility shift assays show that the transcription factor is highly activated in rheumatoid synovium as well as animal models of arthritis [88].

The ability of NF-κB to regulate apoptosis in arthritis was examined by using an animal model of RA [89]. NF-κB is activated in the synovium of rats with streptococcal cell wall (SCW) arthritis. Suppression of NF-κB by either proteasome inhibitors, which inhibit NF-κB activation by preventing IκBα degradation, or intra-articular adenoviral gene transfer of the IκBα super-repressor increase apoptosis in the synovium of arthritic rats and suppress clinical arthritis [89]. These findings suggest that activation of NF-κB protects the cells in the synovium against apoptosis, and suppression of NF-κB could induce apoptosis in synovial cells. Similar results were observed in murine collagen-induced arthritis where a T cell-specific NF-κB inhibitor decreased joint inflammation [90]. In addition, intra-articular gene therapy with a dominant negative IKKβ construct decreases synovial NF-κB activation and inflammation in the rat adjuvant arthritis model [91].

Other gene therapy approaches also confirm the role of NF-κB as an attractive apoptosis-regulating target for RA. IκBα dominant-negative adenovirus (Ad-IκB-DN) was used to determine whether inhibition of NF-κB nuclear translocation leads to increased apoptosis of RA synoviocytes [92]. Apoptosis was induced in RA synoviocytes treated with Ad-IκB-DN plus TNF-α *in vitro*, while no apoptosis was induced in RA cells infected with Ad-IκB-DN alone. To examine *in vivo* effects of Ad-IκB-DN, the SCID mice model of RA was used 4 weeks after transplantation of RA FLS into knee joints of the mice. Mice were treated with Ad-IκB-DN intra-articularly followed by intra-peritoneal TNF-α administration. Extensive apoptosis was detected in engrafted FLS of SCID mice treated with Ad-IκB-DN plus TNF-α, but not in those treated with control vector plus TNF-α.

TNF-α-mediated activation of NF-κB also induces expression of the anti-apoptosis protein X-linked inhibitor of apoptosis (XIAP). Increased XIAP gene expression in RA FLS is inhibited by Ad-IκB-DN infection, which blocks nuclear translocation of NF-κB. Because XIAP plays a role in TNF-α-induced apoptosis of RA cells, XIAP antisense expressing adenovirus vector (Ad-XIAP-AS) was used to evaluate the importance of this pathway [92]. After TNF-α exposure, significant apoptosis was observed in the cells transfected with Ad-XIAP-AS but not in the cells transfected with control vector. Therefore, XIAP, which is induced by NF-κB, may be a critical inhibitor of apoptosis in activated RA synoviocytes.

10.7.3 p53 Tumor Suppressor Gene

In light of data suggesting defective p53 function in RA synovium, the therapeutic utility of p53 gene in RA was examined in a rabbit model with a wild-type p53 expressing adenovirus vector (Ad-p53). Infection with Ad-p53-induced apoptosis in FLS through overexpression of wild-type p53 in a dose-dependent manner, although FLS from RA and OA patients were equally susceptible to p53-mediated apoptosis [43]. Intra-articular injection of Ad-p53 into the joints of rabbit with IL-1-induced arthritis caused rapid induction of apoptosis (less than 24 h) throughout the lining resulting in a significant reduction of cellularity. However, the cytotoxic effect by Ad-p53 was not observed in articular cartilage [43].

10.7.4 Proteasome

Proteasome function also can regulate programmed cell death [93]. Apoptosis of cultured FLS is induced in a dose-dependent manner by LLL-CHO (*Z*-Leu-Leu-Leu-aldehyde), a potent proteasome inhibitor, through the activation of caspase cascade, including caspase-3 [94]. Pretreatment of FLS with TNF-α enhances apoptosis by LLL-CHO, while pretreatment with TGF-β suppresses cell death [94]. The regulation of proteasome inhibitor-related apoptosis by these cytokines is similar to Fas-mediated apoptosis [24, 94].

10.8 Conclusion

While the extent and mechanisms of apoptosis remain somewhat uncertain in RA, it is clear that cell death is overwhelmed by the forces of cell accumulation (proliferation, recruitment and retention). Understanding the relevant molecular processes that regulate apoptosis in RA will hopefully identify targets that can be targeted with therapies to delete pathogenic cells in the synovium. *In vivo* studies suggest that this approach has great potential, although balancing the benefit with toxicity is a serious concern. In the upcoming years, apoptosis-directed treatments might be feasible in RA and other immune-mediated human diseases.

10.9 References

1 Qu, Z., Garcia, C.H., O'Rourke, L.M., Planck, S.R., Kohli, M., Rosenbaum, J.T. *Arthritis Rheum* **1994**, 37, 212–20.

2 Lalor, P.A., Mapp, P.I., Hall, P.A., Revell, P.A. *Rheumatol Int* **1987**, 7, 183–6.

3 Konttinen, Y.T., Nykanen, P., Nordstrom, D., Saari, H., Sandelin, J., Santavirta, S., Kouri, T. *J Rheumatol* **1989**, 16, 339–45.

4 Firestein, G.S., Yeo, M., Zvaifler, N.J. *J Clin Invest* **1995**, 96, 1631–8.

5 Nakajima, T., Aono, H., Hasunuma, T., Yamamoto, K., Shirai, T., Hirohata, K., Nishioka, K. *Arthritis Rheum* **1995**, 38, 485–91.

6 Matsumoto, S., Muller-Ladner, U., Gay, R. E., Nishioka, K., Gay, S. *J Rheumatol* **1996**, 23, 1345–52.

7 Asahara, H., Hasumuna, T., Kobata, T., Yagita, H., Okumura, K., Inoue, H., Gay, S., Sumida, T., Nishioka, K. *Clin Immunol Immunopathol* **1996**, 81, 27–34.

8 Chou, C. T., Yang, J. S., Lee, M.R. *J Pathol* **2001**, 193, 110–6.

9 Okura, T., Gong, L., Kamitani, T., Wada, T., Okura, I., Wei, C.F., Chang, H.M., Yeh, E.T. *J Immunol* **1996**, 157, 4277–81.

10 Franz, J.K., Pap, T., Hummel, K.M., Nawrath, M., Aicher, W.K., Shigeyama, Y., Muller-Ladner, U., Gay, R.E., Gay, S. *Arthritis Rheum* **2000**, 43, 599–607.

11 Li, J., Yen, C., Liaw, D., Podsypanina, K., Bose, S., Wang, S.I., Puc, J., Miliaresis, C., Rodgers, L., *et al. Science* **1997**, 275, 1943–7.

12 Steck, P.A., Pershouse, M.A., Jasser, S.A., Yung, W.K., Lin, H., Ligon, A.H., Langford, L.A., Baumgard, M.L., Hattier, T., *et al. Nat Genet* **1997**, 15, 356–62.

13 Pap, T., Franz, J.K., Hummel, K.M., Jeisy, E., Gay, R., Gay, S. *Arthritis Res* **1999**, 2, 59–64.

14 Tanaka, Y., Nomi, M., Fujii, K., Hubscher, S., Maruo, A., Matsumoto, S., Awazu, Y., Saito, K., Eto, S., *et al. Arthritis Rheum* **2000**, 43, 2513–22.

15 Isomaki, P., Soderstrom, K.O., Punnonen, J., Roivainen, A., Luukkainen, R., Merilahti-Palo, R., Nikkari, S., Lassila, O., Toivanen, P. *Br J Rheumatol* **1996**, 35, 611–9.

16 Zdichavsky, M., Schorpp, C., Nickels, A., Koch, B., Pfreundschuh, M., Gause, A. *Rheumatol Int* **1996**, 16, 151–7.

17 Salmon, M., Scheel-Toellner, D., Huissoon, A.P., Pilling, D., Shamsadeen, N., Hyde, H., D'Angeac, A.D., Bacon, P.A., Emery, P., *et al. J Clin Invest* **1997**, 99, 439–46.

18 Nieto, M.A., Gonzalez, A., Lopez-Rivas, A., Diaz-Espada, F., Gambon, F. *J Immunol* **1990**, 145, 1364–8.

19 Firestein, G.S., Xu, W.D., Townsend, K., Broide, D., Alvaro-Garcia, J., Glasebrook, A., Zvaifler, N.J. *J Exp Med* **1988**, 168, 1573–86.

20 McInnes, I.B., al-Mughales, J., Field, M., Leung, B.P., Huang, F.P., Dixon, R., Sturrock, R.D., Wilkinson, P.C., Liew, F.Y. *Nat Med* **1996**, 2, 175–82.

21 Firestein, G.S., Alvaro-Garcia, J.M., Maki, R. *J Immunol* **1990**, 144, 3347–53.

22 Miossec, P., van den Berg, W. *Arthritis Rheum* **1997**, 40, 2105–15.

23 Kobayashi, T., Okamoto, K., Kobata, T., Hasunuma, T., Sumida, T., Nishioka, K. *Arthritis Rheum* **1999**, 42, 519–26.

24 Kobayashi, T., Okamoto, K., Kobata, T., Hasunuma, T., Kato, T., Hamada, H., Nishioka, K. *Arthritis Rheum* **2000**, 43, 1106–14.

25 Ohshima, S., Mima, T., Sasai, M., Nishioka, K., Shimizu, M., Murata, N., Yoshikawa, H., Nakanishi, K., Suemura, M., *et al. Cytokine* **2000**, 12, 281–8.

26 Relic, B., Guicheux, J., Mezin, F., Lubberts, E., Togninalli, D., Garcia, I., van den Berg, W.B., Guerne, P.A. *J Immunol* **2001**, 166, 2775–82.

27 Fava, R., Olsen, N., Keski-Oja, J., Moses, H., Pincus, T. *J Exp Med* **1989**, 169, 291–6.

28 Lotz, M., Kekow, J., Carson, D.A. *J Immunol* **1990**, 144, 4189–94.

29 Kawakami, A., Eguchi, K., Matsuoka, N., Tsuboi, M., Kawabe, Y., Aoyagi, T.,

Nagataki, S. *Arthritis Rheum* **1996**, 39, 1267–76.

30 Ko, L.J., Prives, C. *Genes Dev* **1996**, 10, 1054–72.

31 Levine, A.J. *Cell* **1997**, 88, 323–31.

32 Di Leonardo, A., Linke, S.P., Clarkin, K., Wahl, G.M. *Genes Dev* **1994**, 8, 2540–51.

33 Agarwal, M.L., Agarwal, A., Taylor, W.R., Stark, G.R. *Proc Natl Acad Sci USA* **1995**, 92, 8493–7.

34 Taylor, W.R., Stark, G.R. *Oncogene* **2001**, 20, 1803–15.

35 Maltzman, W., Czyzyk, L. *Mol Cell Biol* **1984**, 4, 1689–94.

36 Kastan, M.B., Onyekwere, O., Sidransky, D., Vogelstein, B., Craig, R.W. *Cancer Res* **1991**, 51, 6304–11.

37 Haupt, Y., Maya, R., Kazaz, A., Oren, M. *Nature* **1997**, 387, 296–9.

38 Midgley, C.A., Lane, D.P. *Oncogene* **1997**, 15, 1179–89.

39 Buschmann, T., Minamoto, T., Wagle, N., Fuchs, S.Y., Adler, V., Mai, M., Ronai, Z. *J Mol Biol* **2000**, 295, 1009–21.

40 Rogel, A., Popliker, M., Webb, C.G., Oren, M. *Mol Cell Biol* **1985**, 5, 2851–5.

41 Tak, P.P., Klapwijk, M.S., Broersen, S.F., van de Geest, D.A., Overbeek, M., Firestein, G.S. *Arthritis Res* **2000**, 2, 229–35.

42 Yamanishi, Y., Boyle, D.L., Pinkoski, M.J., Mahboubi, A., Lin, T., Han, Z., Zvaifler, N.J., Green, D.R., Firestein, G.S. *Am J Pathol*, **2002**, 160, 123–30.

43 Yao, Q., Wang, S., Glorioso, J.C., Evans, C.H., Robbins, P.D., Ghivizzani, S.C., Oligino, T.J. *Mol Ther* **2001**, 3, 901–10.

44 Tak, P.P., Smeets, T.J., Boyle, D.L., Kraan, M.C., Shi, Y., Zhuang, S., Zvaifler, N.J., Breedveld, F.C., Firestein, G.S. *Arthritis Rheum* **1999**, 42, 948–53.

45 Firestein, G.S., Echeverri, F., Yeo, M., Zvaifler, N.J., Green, D.R. *Proc Natl Acad Sci USA* **1997**, 94, 10895–900.

46 Reme, T., Travaglio, A., Gueydon, E., Adla, L., Jorgensen, C., Sany, J. *Clin Exp Immunol* **1998**, 111, 353–8.

47 Kullmann, F., Judex, M., Neudecker, I., Lechner, S., Justen, H.P., Green, D. R., Wessinghage, D., Firestein, G.S., Gay, S., *et al. Arthritis Rheum* **1999**, 42, 1594–1600.

48 Inazuka, M., Tahira, T., Horiuchi, T., Harashima, S., Sawabe, T., Kondo, M., Miyahara, H., Hayashi, K. *Rheumatology* **2000**, 39, 262–6.

49 Wink, D.A., Kasprzak, K.S., Maragos, C.M., Elespuru, R.K., Misra, M., Dunams, T.M., Cebula, T.A., Koch, W.H., Andrews, A.W., *et al. Science* **1991**, 254, 1001–3.

50 Nguyen, T., Brunson, D., Crespi, C.L., Penman, B.W., Wishnok, J.S., Tannenbaum, S.R. *Proc Natl Acad Sci USA* **1992**, 89, 3030–4.

51 Driscoll, K.E., Carter, J.M., Howard, B.W., Hassenbein, D.G., Pepelko, W., Baggs, R.B., Oberdorster, G. *Toxicol Appl Pharmacol* **1996**, 136, 372–80.

52 Zhuang, J.C., Lin, C., Lin, D., Wogan, G.N. *Proc Natl Acad Sci USA* **1998**, 95, 8286–91.

53 Bashir, S., Harris, G., Denman, M.A., Blake, D.R., Winyard, P.G. *Ann Rheum Dis* **1993**, 52, 659–66.

54 Maurice, M.M., Nakamura, H., Gringhuis, S., Okamoto, T., Yoshida, S., Kullmann, F., Lechner, S., van der Voort, E.A., Leow, A., *et al. Arthritis Rheum* **1999**, 42, 2430–9.

55 McInnes, I.B., Leung, B.P., Field, M., Wei, X.Q., Huang, F.P., Sturrock, R.D., Kinninmonth, A., Weidner, J., Mumford, R., *et al. J Exp Med* **1996**, 184, 1519–24.

56 Brentnall, T.A., Crispin, D.A., Rabinovitch, P.S., Haggitt, R.C., Rubin, C.E., Stevens, A.C., Burmer, G.C. *Gastroenterology* **1994**, 107, 369–78.

57 Cannons, J.L., Karsh, J., Birnboim, H.C., Goldstein, R. *Arthritis Rheum* **1998**, 41, 1772–82.

58 Roivainen, A., Jalava, J., Pirila, L., Yli-Jama, T., Tiusanen, H., Toivanen, P. *Arthritis Rheum* **1997**, 40, 1636–43.

59 Han, Z., Boyle, D.L., Shi, Y., Green, D.R., Firestein, G.S. *Arthritis Rheum* **1999**, 42, 1088–92.

60 Sun, Y., Wenger, L., Rutter, J.L., Brinckerhoff, C.E., Cheung, H.S. *J Biol Chem* **1999**, 274, 11535–40.

61 Sun, Y., Cheung, J.M., Martel-Pelletier, J., Pelletier, J.P., Wenger, L., Alt-

man, R.D., Howell, D.S., Cheung, H.S. *J Biol Chem* **2000**, 275, 11327–32.
62 Aupperle, K.R., Boyle, D.L., Hendrix, M., Seftor, E.A., Zvaifler, N.J., Barbosa, M., Firestein, G.S. *Am J Pathol* **1998**, 152, 1091–8.
63 Pap, T., Aupperle, K.R., Gay, S., Firestein, G.S., Gay, R.E. *Arthritis Rheum* **2001**, 44, 676–81.
64 Itoh, N., Yonehara, S., Ishii, A., Yonehara, M., Mizushima, S., Sameshima, M., Hase, A., Seto, Y., Nagata, S. *Cell* **1991**, 66, 233–43.
65 Suda, T., Takahashi, T., Golstein, P., Nagata, S. *Cell* **1993**, 75, 1169–78.
66 Watson, M.L., Rao, J.K., Gilkeson, G.S., Ruiz, P., Eicher, E.M., Pisetsky, D.S., Matsuzawa, A., Rochelle, J.M., Seldin, M.F. *J Exp Med* **1992**, 176, 1645–56.
67 Takahashi, T., Tanaka, M., Brannan, C.I., Jenkins, N.A., Copeland, N.G., Suda, T., Nagata, S. *Cell* **1994**, 76, 969–76.
68 Nagata, S. *Adv Immunol* **1994**, 57, 129–44.
69 Hoa, T.T., Hasunuma, T., Aono, H., Masuko, K., Kobata, T., Yamamoto, K., Sumida, T., Nishioka, K. *J Rheumatol* **1996**, 23, 1332–7.
70 Cantwell, M.J., Hua, T., Zvaifler, N.J., Kipps, T.J. *Arthritis Rheum* **1997**, 40, 1644–52.
71 Brunner, T., Mogil, R.J., LaFace, D., Yoo, N.J., Mahboubi, A., Echeverri, F., Martin, S.J., Force, W.R., Lynch, D.H., *et al. Nature* **1995**, 373, 441–44.
72 Tucek-Szabo, C.L., Andjelic, S., Lacy, E., Elkon, K.B., Nikolic-Zugic, J. *J Immunol* **1996**, 156, 192–200.
73 Hashimoto, H., Tanaka, M., Suda, T., Tomita, T., Hayashida, K., Takeuchi, E., Kaneko, M., Takano, H., Nagata, S., *et al. Arthritis Rheum* **1998**, 41, 657–62.
74 Cheng, J., Zhou, T., Liu, C., Shapiro, J.P., Brauer, M.J., Kiefer, M.C., Barr, P.J., Mountz, J.D. *Science* **1994**, 263, 1759–62.
75 Knipping, E., Krammer, P.H., Onel, K.B., Lehman, T.J., Mysler, E., Elkon, K.B. *Arthritis Rheum* **1995**, 38, 1735–7.
76 Hasunuma, T., Kayagaki, N., Asahara, H., Motokawa, S., Kobata, T., Yagita, H., Aono, H., Sumida, T., Okumura, K., *et al. Arthritis Rheum* **1997**, 40, 80–6.
77 Tanaka, M., Suda, T., Takahashi, T., Nagata, S. *Embo J* **1995**, 14, 1129–35.
78 Kayagaki, N., Kawasaki, A., Ebata, T., Ohmoto, H., Ikeda, S., Inoue, S., Yoshino, K., Okumura, K., Yagita, H. *J Exp Med* **1995**, 182, 1777–83.
79 Tanaka, M., Suda, T., Haze, K., Nakamura, N., Sato, K., Kimura, F., Motoyoshi, K., Mizuki, M., Tagawa, S., *et al. Nat Med* **1996**, 2, 317–22.
80 Suda, T., Hashimoto, H., Tanaka, M., Ochi, T., Nagata, S. *J Exp Med* **1997**, 186, 2045–50.
81 Okamoto, K., Fujisawa, K., Hasunuma, T., Kobata, T., Sumida, T., Nishioka, K. *Arthritis Rheum* **1997**, 40, 919–26.
82 Iwakura, Y., Tosu, M., Yoshida, E., Takiguchi, M., Sato, K., Kitajima, I., Nishioka, K., Yamamoto, K., Takeda, T., *et al. Science* **1991**, 253, 1026–8.
83 Fujisawa, K., Asahara, H., Okamoto, K., Aono, H., Hasunuma, T., Kobata, T., Iwakura, Y., Yonehara, S., Sumida, T., *et al. J Clin Invest* **1996**, 98, 271–8.
84 Ogasawara, J., Watanabe-Fukunaga, R., Adachi, M., Matsuzawa, A., Kasugai, T., Kitamura, Y., Itoh, N., Suda, T., Nagata, S. *Nature* **1993**, 364, 806–9.
85 Yao, Q., Glorioso, J.C., Evans, C.H., Robbins, P.D., Kovesdi, I., Oligino, T.J., Ghivizzani, S.C. *J Gene Med* **2000**, 2, 210–9.
86 Okamoto, K., Asahara, H., Kobayashi, T., Matsuno, H., Hasunuma, T., Kobata, T., Sumida, T., Nishioka, K. *Gene Ther* **1998**, 5, 331–8.
87 Kobayashi, T., Okamoto, K., Kobata, T., Hasunuma, T., Kato, T., Hamada, H., Nishioka, K. *Gene Ther* **2000**, 7, 527–33.
88 Han, Z., Boyle, D.L., Manning, A.M., Firestein, G.S. *Autoimmunity* **1998**, 28, 197–208.
89 Miagkov, A.V., Kovalenko, D.V., Brown, C.E., Didsbury, J.R., Cogswell, J.P., Stimpson, S.A., Baldwin, A.S., Makarov, S.S. *Proc Natl Acad Sci USA* **1998**, 95, 13859–64.
90 Gerlag, D.M., Ransone, L., Tak, P.P., Han, Z., Palanki, M., Barbosa, M.S., Boyle, D., Manning, A.M., Firestein, G.S. *J Immunol* **2000**, 165, 1652–8.

91 Tak, P. P., Gerlag, D. M., Aupperle, K. R., van de Geest, D. A., Overbeek, M., Bennett, B. L., Boyle, D. L., Manning, A. M., Firestein, G. S. *Arthritis Rheum* **2001**, 44, 1897–907.

92 Zhang, H. G., Huang, N., Liu, D., Bilbao, L., Zhang, X., Yang, P., Zhou, T., Curiel, D. T., Mountz, J. D. *Arthritis Rheum* **2000**, 43, 1094–105.

93 Cui, H., Matsui, K., Omura, S., Schauer, S. L., Matulka, R. A., Sonenshein, G. E., Ju, S. T. *Proc Natl Acad Sci USA* **1997**, 94, 7515–20.

94 Kawakami, A., Nakashima, T., Sakai, H., Hida, A., Urayama, S., Yamasaki, S., Nakamura, H., Ida, H., Ichinose, Y., *et al. Arthritis Rheum* **1999**, 42, 2440–8.

11
Systemic Lupus Erythematosus

Thomas D. Beyer, Susanne Kuenkele, Udo S. Gaipl, Wasilis Kolowos,
Reinhard E. Voll, Irith Baumann, Joachim R. Kalden and Martin Herrmann

11.1
Introduction

Systemic lupus erythematosus (SLE) is a heterogeneous systemic autoimmune disease displaying a broad variety of symptoms and involving multiple organ systems. The age- and sex-adjusted prevalence rate for SLE in January 1993 was approximately 1.22 per 1000. The incidence rate of a Rochester SLE cohort 1980–1992 was 5.56 per 100000 and tripled since 1950 due to improved recognition of early stages of the disease. Due to new therapeutic approaches, the survival of SLE patients has improved over the past 4 decades, but it is still worse than in the general population. Furthermore, although most of the patients experience remissions [1], SLE remains incurable. The diagnosis is usually based on the American Rheumatism Association criteria for the classification of SLE [2]. The etiopathogenesis of SLE is multifactorial. Multiple genetic as well as environmental factors, including viruses and other infectious agents, drugs, chemicals and food, are discussed to contribute to the disease [1]. Profound immune alterations lead to activation of autoreactive T cells and consecutively to the generation of various autoantibodies. The presence of autoantibodies against double-stranded (ds) DNA and other nuclear antigens is a characteristic feature of SLE. In this chapter, we discuss a mechanism which might contribute to the induction and maintenance of autoimmunity in SLE.

11.2
Involvement of B Cells in the Development of SLE

Autoantibodies to dsDNA are a hallmark of SLE [3]. The establishment of anti-DNA hybridomas from patients with SLE and the analysis of their immunoglobulin variable genes have contributed important information on possible induction mechanisms [4–7]. Initially, it was reported that even normal individuals are able to express anti-dsDNA antibodies with a similar specificity as SLE patients. However, recently published studies clearly indicate that the anti-dsDNA antibodies detected in healthy individuals are mainly of the IgM isotype, are characterized by a wide cross-reactivity and are usually encoded by gene segments in the germline configuration. In contrast, high-affinity anti-dsDNA antibodies isolated from SLE patients had often undergone isotype switch (IgG or IgA) and affinity maturation [4, 6, 8, 9]. Both processes are dependent on T cell help, and take place in the germinal centers of lymph nodes and spleen. It is of special interest that there was a prominent exchange towards the amino acids arginine and asparagine, which are particularly important for dsDNA binding by electrostatic interaction. The bias towards arginine and asparagine had been generated by the usage of a particular reading frame of the D elements, by frameshifts in the V_κ–J_κ junction or by somatic mutations. These data show that the IgG anti-dsDNA antibody response bears all the characteristics of an antigen-driven, T cell-dependent immune response. The analysis of V_H gene usage of monoclonal anti-dsDNA revealed that IgG anti-DNA preferentially use members of the V_H3 or V_H4 gene family [10]. The κ and λ light chains were represented normally. In contrast, no V_H restriction was observed for anti-dsDNA of the IgM isotype.

The nature of the epitopes recognized by anti-dsDNA is still a matter of controversial discussion. Employing synthetic oligonucleotides, we showed that SLE patient-derived anti-dsDNA preferentially selected DNA sequences that were predicted not to form B-DNA structures. Instead, DNA molecules selected from a random library contained significantly more adenosine triplets, suggesting that SLE derived anti-dsDNA preferentially bind DNA [11].

11.3 Involvement of T Cells in the Development of SLE

It is well established that affinity maturation, isotype switch and memory formation are T cell-dependent processes which take place in germinal centers of secondary lymphoid organs. The isolation of autoreactive T cells from SLE patients recognizing high-mobility group proteins and histones [12, 13], and the efficiency of the T cell-specific immune-suppressive cyclosporin A in treatment, suggest that DNA-protein complexes like nucleosomes [14–16] may represent the source for autoantigens in SLE [17]. In the immune reaction against nucleosomes, both histones and DNA may provide T and B cell epitopes, respectively.

Ex vivo immunization experiments revealed that increased concentrations and/or an abnormal presentation of nuclear antigens are able to stimulate autoreactive peripheral T cells. Autologous apoptotic cells or isolated histones induced *in vitro* T cell proliferation of $CD4^+$ T cell clones isolated from peripheral blood of both non-Hodgkin's disease and patients with SLE [13]. Furthermore, co-incubation of the histone-specific T cells with autologous B cells induced the production of anti-dsDNA antibodies *in vitro* [18–20]. However, the nature of the antigens or peptides stimulating the proliferation of these autoreactive T cells remained elusive. Experiments with thymectomized [21], T cell receptor-depleted [22] or anti-CD4 antibody-treated mice [23] confirmed the contribution of T cells in the etiopathogenesis of murine SLE.

11.4
Genetic Factors for the Development of SLE

The association with the haplotype C2Q0, HLA-A10, B18, Dw2, BfS, suggests a possible linkage of SLE to immune response genes [24]. The deficiency of C1q is strongly associated with the development of SLE in humans [25–27]. Furthermore, low levels of C1q due to complement consumption and acquired C1q 'deficiency' secondary to anti-C1q autoantibodies are associated with severe nephritis in human SLE [28]. In addition, deficiencies of other components of the classical complement pathway are associated with disease susceptibility [29, 30].

The FcγRIIa polymorphism affects the immune complex (IC) metabolism and influences the clinical manifestations and course of SLE. FcγRIIa-H131 binds much better to complexed IgG2 and IgG3 than FcγRIIa-R131. This might be linked to variability in IC handling and therefore be related to the pathogenesis of the disease. However, it did not represent a genetic risk factor for the occurrence of the disease [31].

11.5
Animal Models for the Immunogenicity of Dying Cells

Tumor cells undergoing apoptosis display a low but detectable immunogenicity [32], which was enhanced in IL-10 knockout mice [33]. Apoptotic cells opsonized with antiphospholipid antibodies resumed their immunogenicity since they were efficiently processed by murine immature dendritic cells (DCs) *in vivo* [34, 35]. It was shown that high numbers of apoptotic cells, mimicking impaired clearance function, are able to trigger DC maturation. Under these circumstances, the presentation of intracellular antigens from apoptotic cells triggers an immune response, even in the absence of exogenous 'danger' signals [36, 37]. The cross-activation of cytotoxic T lymphocytes was dependent on the rate of apoptosis and on the accumulation of late apoptotic cells [37].

In recent years, several mouse models with impaired clearance of either apoptotic cells or chromatin have been reported. Mice with targeted deletion of the C1qa [38] displayed an impaired clearance, an accumulation of apoptotic cells in the glomeruli and a lupus-like disease [39]. Targeted deletions of serum amyloid P resulted in impaired solubilization and degradation of chromatin, inefficient clearance of apoptotic cells, and development of antinuclear antibodies. The animals displayed a SLE-like disease including severe glomerulonephritis [38, 40]. In MER knockout mice, the detection of uningested apoptotic cells in the thymus is associated with antinuclear autoantibodies [41] and investigation of DNase I-deficient mice suggested that lack or reduction of DNase I is a critical factor in the initiation of human SLE [42].

11.6 Phagocytosis of Apoptotic Cells

In tissues, apoptotic cells are usually engulfed by macrophages in the early phase of apoptotic cell death. This process induces neither inflammation nor an immune response [32, 43, 44]. In the thymus, the bone marrow, and the germinal centers of lymph nodes and spleen, specialized phagocytes that clear dying cells are referred to as tingible body macrophages. This clearance is so efficient that only few apoptotic cells can be detected in the thymus, although it functions as a mass grave for futile or dangerous thymocytes [45]. The instant removal of apoptotic cells offers the needed space for the heavily proliferating thymocytes or centroblasts, and is a prerequisite for a proper function of thymus, bone marrow and lymph nodes. The early engulfment of apoptotic cells is necessary to avoid cells entering the late stages of apoptosis. During late stages of apoptosis cells often cannot maintain their plasma membrane integrity, get secondary necrotic and release intracellular contents [46–48]. In the case of the germinal center, released cytoplasmic compounds have the potency to serve as autoantigens for the affinity maturation of B cells. Therefore, the instant removal of apoptotic cells is of major importance to avoid exposure of maturating B cells to nuclear autoantigens, which would otherwise provide a survival signal for autoreactive B cells. Cytoplasmic blebs of apoptotic cells and chromatin have been demonstrated to bind C1q [49] and to activate the classical complement cascade [50]. Consecutively, nuclear material of apoptotic cells opsonized by iC3b, C3d and C3dg can be immobilized by follicular DCs (FDCs) via CR2/CD21. The latter are known to trap complement-binding immune complexes, which under physiological conditions serve as antigens for affinity maturation. In the presence of uningested, complement-binding apoptotic material, FDCs may serve as autoantigen repositories.

11.7 Reduced Phagocytosis of Apoptotic Cells in SLE Patients Challenges T Cell Tolerance

11.7.1 Increased Apoptosis in SLE Patients?

Several studies have demonstrated an increased apoptosis of *in vitro* cultured blood mononuclear cells of SLE patients [51–56]. However, similar data were obtained in patients with systemic vasculitis, mixed connective tissue disease [52] and rheumatoid arthritis [56]. This leads to the conclusion that an increased apoptosis is not specific for SLE. The increased rate of apoptosis of SLE patients' lymphocytes was shown to be normalized by addition of IL-2. This points out that the increased rate of apoptosis rather reflects an increased number of activated lymphocytes and not a SLE-specific genetic defect [52]. However, lymphocytes of SLE patients with bacterial infections do have increased spontaneous apoptosis within their peripheral blood mononuclear cells [52]. This could be responsible for the clinical observation that a flare of SLE can follow an infectious episode [57, 58].

11.7.2
Reduced Phagocytosis of Apoptotic Cells by *In Vitro* Generated Macrophages from SLE Patients

It has been known for a long time that macrophages from patients with SLE have an impaired phagocytic activity for yeast and bacteria [59–62]. In addition, *in vitro* differentiated macrophages from a subgroup of SLE patients show a significantly reduced phagocytosis of apoptotic cells (Fig. 11.1). Impaired clearance functions for dying cells may explain accumulation of apoptotic cells and, subsequently, of secondary necrotic cells in various tissues of SLE patients [55, 63, 64]. In contrast to early apoptotic cells, in the late phase of apoptosis when the cells enter the stage of secondary necrosis, the cells lose their membrane integrity and cytoplasmic compounds are released. This may explain the increased levels of DNA and nucleosomes reported in the circulation of SLE patients [55, 65–67]. These facts led to the hypothesis that the clearance defect may be a major primary event in the etiopathogenesis of a subgroup of SLE patients [38, 55, 63, 64] (Fig. 11.2).

11.7.3
Hypothesis: Accumulation of Secondary Necrotic Cells Challenges T Cell Tolerance in SLE

In Sindbis virus-infected cells undergoing apoptosis, cellular proteins change their subcellular location and are clustered in blebs at the surface near sites of increased generation of reactive oxygen species [68, 69]. Thereby, an immune response initiated against viral antigens may spread to adjoining autoantigens. This 'epitope-spreading' was observed in animal models of BK virus infection [16].

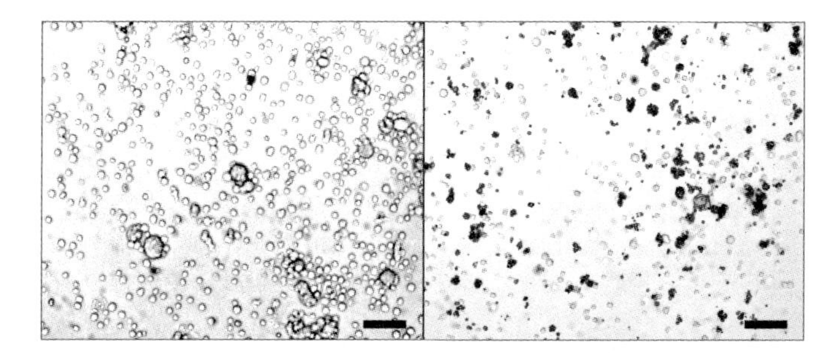

Fig. 11.1 *In vitro* phagocytosis is reduced in a subgroup of SLE patients. After 15 days of *in vitro* culture and differentiation macrophages isolated from a subgroup of SLE patients showed a significantly reduced phagocytosis of apoptotic cells (right column) as compared to the cells isolated from normal healthy donors (left column). Scale bars represent 100 μm.

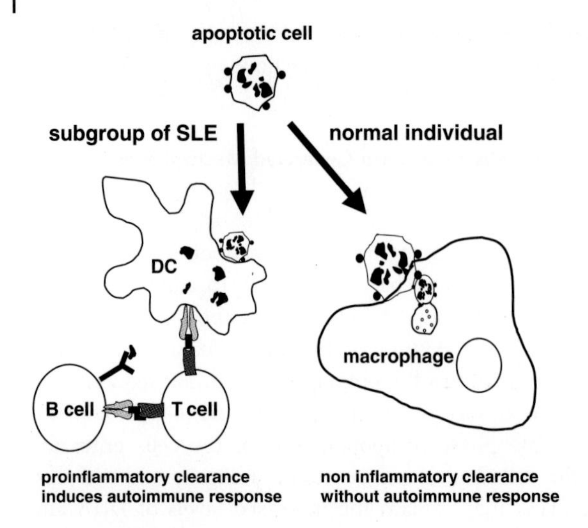

Fig. 11.2 Hypothesis: reduced clearance of apoptotic cells by SLE macrophages may result in uptake of dying cells by DCs. The uptake and degradation of early apoptotic cells is usually fast, efficient and nonphlogistic. In a subgroup of SLE patients with a reduced phagocytosis capacity the early apoptotic cells are insufficiently cleared. Therefore, apoptosis progresses and the cells enter the stage of secondary necrosis. The uningested apoptotic material including modified autoantigens is taken up by DCs. Cryptic auto-epitopes get dominant if processing alters the epitope hierarchy of antigen presentation.

During necrotic as well as apoptotic cell death, proteins are cleaved and otherwise modified [69–72]. At least 39 proteins are known which are proteolytically cleaved during execution of apoptosis and thereby probably modified in their immunogenicity. Many of these modified proteins are components of complex particles, frequently targeted by autoantibodies of SLE patients [73]. In particular, proteolytic cleavage of cellular proteins by the action of granzyme B was shown to modify most targets of systemic humoral autoimmune responses [72]. Therefore, cleavability by granzyme B was suggested to be the unifying feature of autoantigens. Since proteolytic cleavage changes the epitope hierarchy for antigen presentation, apoptosis may render cryptic epitopes immunodominant and lead to antigen presentation of epitopes to which the immune system has not achieved tolerance [74].

Several surface receptors and soluble adaptor molecules are involved in the nonphlogistic clearance of apoptotic cells (Fig. 11.3). Opsonization with antiphospholipid antibodies, which are frequently found in SLE patients, directs apoptotic cells into a pro-inflammatory pathway. Phagocytosis of opsonized apoptotic cells resulted in massive tumor necrosis factor-α secretion, putatively enhancing their immunogenicity [75].

Dendritic cells need to be activated to prime naïve T cells. Necrotic or virus-infected cells, but not apoptotic ones, have been described to serve as 'natural adjuvants'. Furthermore, it was suggested that stressed autologous cells might initiate

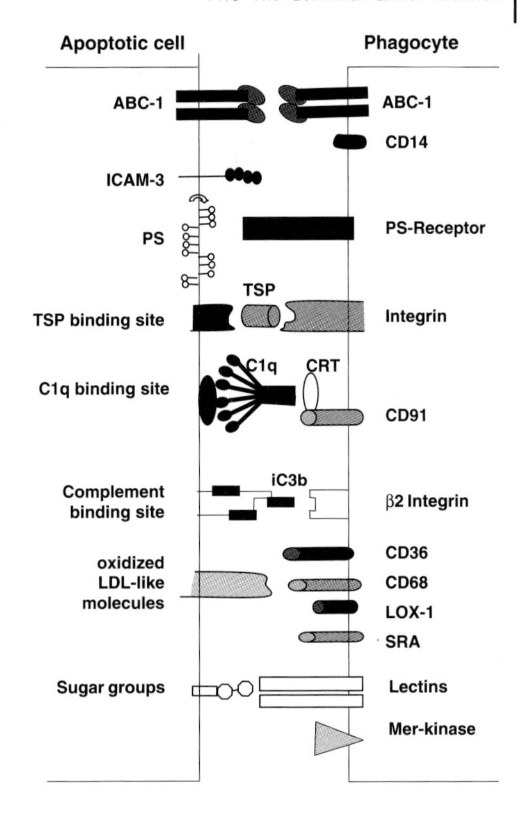

Fig. 11.3 Receptors for the phagocytosis of apoptotic cells. This schema illustrates surface receptors and soluble adaptor molecules involved in the non-phlogistic phagocytosis of apoptotic cells. Abbreviations: ABC, ATP binding cassette; ICAM, intercellular adhesion molecule; LDL, low-density lipoproteins; TSP, thrombospondine

some forms of autoimmunity [76]. To investigate antigen cross-presentation, mice were immunized with increasing numbers of syngeneic apoptotic tumor cells. Only in mice that had been injected with high amounts of dying cells – mimicking insufficient clearance – DCs did mature in the absence of exogenous 'danger' signals. These cells presented intracellular antigens from the apoptotic tumor cells and elicited tumor-specific cytotoxic T lymphocytes responses [36] (Fig. 11.4).

11.8 The Germinal Center Reaction

Surface immunoglobulin-negative B cells proliferate as large centroblasts in the dark zone of germinal centers. After few divisions they stop cycling, express surface immunoglobulin and enter the light zone as centrocytes. Here the latter come in contact with immune complexes immobilized by FDCs via the complement receptor 2 (CR2/CD21) [77, 78]. The contact with the antigens of the immune complexes provides an essential short-term survival signal for the centrocytes. The latter subsequently can re-enter the dark zone or leave the germinal

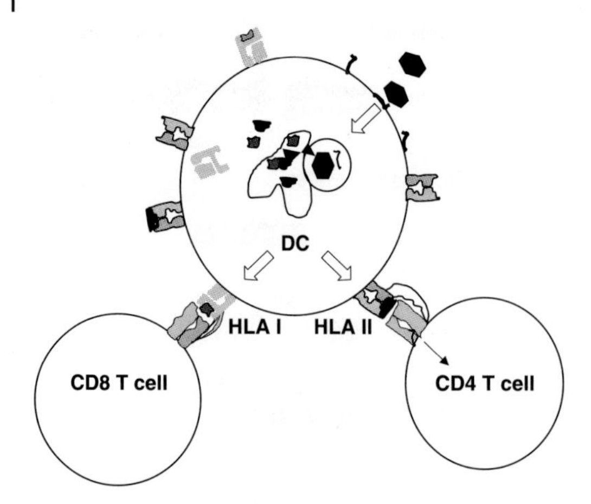

Fig. 11.4 The processing of autoantigens from apoptotic cells by DCs results in an activation of autoreactive cytotoxic CD8$^+$ T cells (cross-priming). High numbers of late apoptotic cells, disintegrated apoptotic nuclei and chromatin generate an amount of intracellular antigens. Endocytosis of these antigens leads to DC maturation and to the presentation of intracellular antigens from apoptotic cells via HLA class I and class II to cytotoxic CD8$^+$ T lymphocytes and CD4$^+$ T cells, respectively. Thus the DCs may trigger an immune response, even in the absence of exogeneous 'danger' signals. The presentation of extracellular antigens via HLA I molecules is referred to as cross priming.

center and migrate into the mantle zone. Interaction with CD4$^+$ T lymphocytes in the mantle zone leads to a long-term survival signal [79]. Both signals, i.e. from their antigen on FDCs and from T helper cells, are required to rescue the germinal center B cells from apoptotic cell death. This is a prerequisite for the generation of antibody-secreting plasma cells and memory B cells.

If there is a lack of antigen or if somatic mutations of the B cell receptor lead to a reduced affinity for the antigen, centrocytes do not receive survival signals and undergo apoptosis. The apoptotic cells are instantly removed by specialized phagocytes referred to as tingible body macrophages. These are CD68$^+$ cells and have a typical morphology. Due to their high phagocytic activity they often contain multiple apoptotic nuclei and reach diameters of up to 60 μm. The extremely fast and efficient uptake and degradation of early apoptotic cells is very important as it eliminates apoptotic cell-derived autoantigens released if late apoptotic cells enter the stage of secondary necrosis. Surface blebs of apoptotic cells are able to activate the classical complement pathway independent of antibodies [49, 50, 80]. Therefore, uningested apoptotic material including potential autoantigens may become coated with C3d and hence bind to CR2/CD21 on FDCs, which may then provide survival signals for autoreactive B cells (germinal center reaction reviewed in [81]) (see Fig. 11.5).

Protein-free B-form DNA *per se* is a poor immunogen and does not bear T cell epitopes, which are required to initiate T cell help for DNA-specific B cells. However, DNA released from apoptotic or necrotic cells is complexed with proteins that had been modified during the death program. Therefore, DNA-specific B

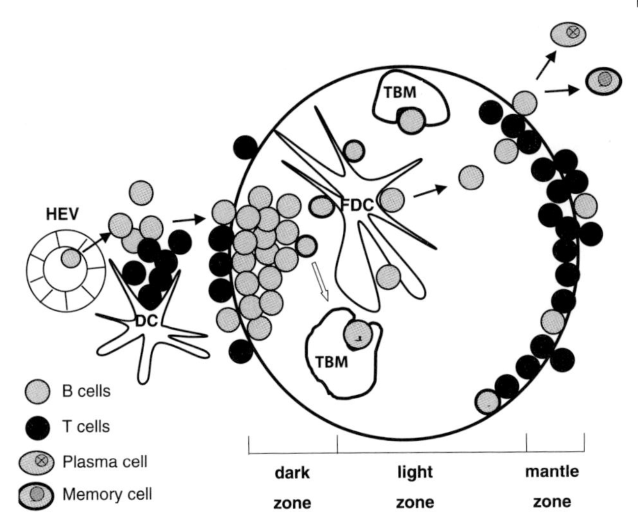

Fig. 11.5 Germinal center reaction. Unspecific activated B cells migrate from the high endothelia venules (HEV) into the lymph node. Here they get in contact with helper T cells, become stimulated and start proliferation. As centroblasts, these proliferating B cells compose the dark zone of the germinal center. After a few divisions they stop cycling, maturate to centrocytes expressing surface immunoglobulin and enter the light zone. Here they come in contact with immune complexes immobilized by FDCs and thereby obtain a short-term survival signal. Consecutively, they can re-enter the dark zone or leave the germinal center and migrate into the mantle zone. Interaction with $CD4^+$ T lymphocytes in the mantle zone leads to a second long-term survival signal, required to rescue the B cells from apoptotic cell death. Consecutively these cells can differentiate into plasma cells or memory B cells. If there is a lack of antigen or if somatic mutations of the B cell receptor lead to a reduced affinity for the antigen, centrocytes do not receive survival signals and undergo apoptosis. The apoptotic cells are instantly removed by specialized phagocytes referred to as tingible body macrophages (TBM).

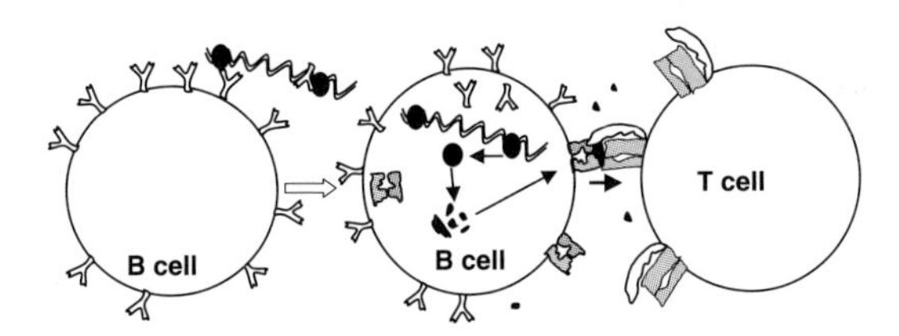

Fig. 11.6 Nucleosomes released from apoptotic cells contain B and T cell epitopes. DNA released from late apoptotic or necrotic cells is complexed with proteins modified during the death program. Therefore, DNA-specific B cells can recognize and internalize these protein-DNA complexes. Furthermore these cells may process and present the DNA-associated altered self-proteins to T cells.

cells can recognize and internalize protein-DNA complexes, process and present DNA-associated altered self-proteins, activate CD4-positive T cells, and thereby receive T cell help [82, 83] (Fig. 11.6).

11.9
Reduced Phagocytosis of Apoptotic Cells in SLE Patients Challenges B Cell Tolerance

11.9.1
The Number of Tingible Body Macrophages is Reduced in the Germinal Centers of SLE Patients

The presence of $CD68^+$ tingible body macrophages is a characteristic feature of germinal centers. Histological analyses of lymph nodes of patients with benign follicular hyperplasia indicate numerous large tingible body macrophages distributed in between centrocytes and centroblasts. Apoptotic cells with pyknotic nuclei representing condensed chromatin can almost exclusively be observed in the cytoplasm of these tingible body macrophages. In the germinal centers of patients with malign follicular lymphoma only very few small tingible body macrophages were found. However, only very few apoptotic cells were present, they had been all ingested by CD68 macrophages, were observed.

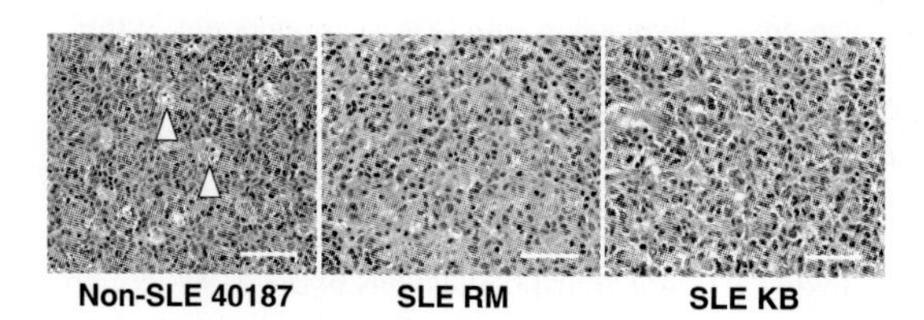

Fig. 11.7 Tingible body macrophages are reduced in the germinal centers of a subgroup of SLE patients (hematoxylin and eosin staining of lymph node sections). Lymph nodes from patients with SLE or from patients with non-malignant lymphadenopathy (non-SLE control) were formalin fixed and paraffin embedded. Sections stained with hematoxylin and eosin show germinal centers with numerous tingible body macrophages (marked with an arrow) in non-SLE controls, whereas tingible body macrophages were virtually absent in a subgroup of patients with SLE. The scale bars represent 100 μm.

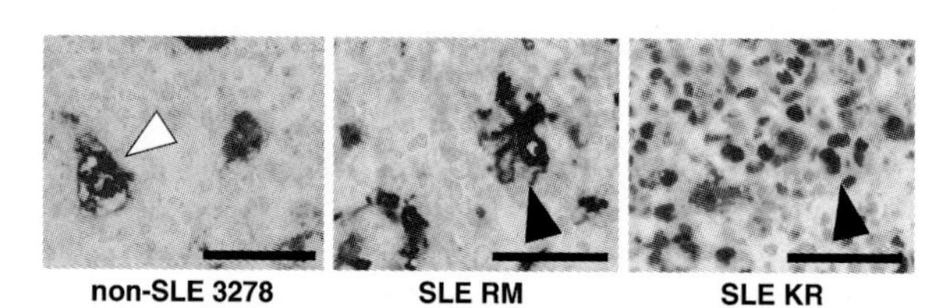

Fig. 11.8 Tingible body macrophages are reduced in the germinal centers of a subgroup of SLE patients (TUNEL staining of lymph node sections). Numerous tingible body macrophages could be detected in the germinal centers of lymph nodes of patients with benign follicular hyperplasia (left column). Apoptotic cells with pyknotic nuclei can almost exclusively be observed in the cytoplasm of these specialized macrophages (marked with a white arrow head). In contrast, in the germinal center of a subgroup of SLE (middle and right column), a significantly reduced number of tingible body macrophages and uningested apoptotic nuclei (marked with black arrow heads) were detected. The scale bars represent 100 μm.

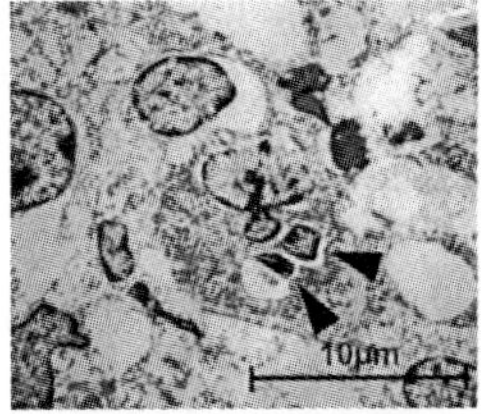

Fig. 11.9 Lymph node sections were analyzed by transmission electron microscopy. Apoptotic nuclei ingested by tingible body macrophages of non-SLE patient 5153 are marked with black arrow heads. Uningested apoptotic nuclei in SLE patient RM are marked with a white arrow head.

In contrast, the lymph nodes of a subgroup of SLE patients contained a significantly reduced number of tingible body macrophages, detected by hematoxylin and eosin staining (Fig. 11.7). In addition, although the count of the $CD68^+$ macrophages was near to normal, these cells showed an obviously different morphology than in controls. They were rather small, lacked the typical shape of tingible body macrophages and rarely contained intracellular apoptotic cell material. In some biopsies uningested apoptotic nuclei were to be found outside the $CD68^+$ macrophages (Figs. 11.8 and 11.9). However, the total number of apoptotic nuclei was not increased in the germinal centers of the lymph nodes of the SLE patients. The appearances of the germinal centers are compatible with the explanation that the phagocytic activity for apoptotic cells is reduced in a subgroup of SLE patients.

11.9.2
Accumulation of Apoptotic Cell-Derived Nuclear Fragments in the Germinal Centers of SLE Patients

As in healthy individuals, in the germinal centers of patients with benign follicular hyperplasia apoptotic material was located exclusively within tingible body macrophages and no free apoptotic cells could be detected. In the germinal centers of patients with follicular lymphoma markedly less and smaller tingible body macrophages were found. The total amount of apoptotic cells was rather low and all apoptotic material had been ingested by the tingible body macrophages. The lack of apoptotic material might be a consequence of the inability of these malignant cells to physiologically execute apoptosis.

In contrast, in a subgroup of SLE patients with reduced numbers of tingible body macrophages multiple non-engulfed apoptotic cells and cellular fragments were to be detected. In addition, disintegrated apoptotic nuclei and chromatin

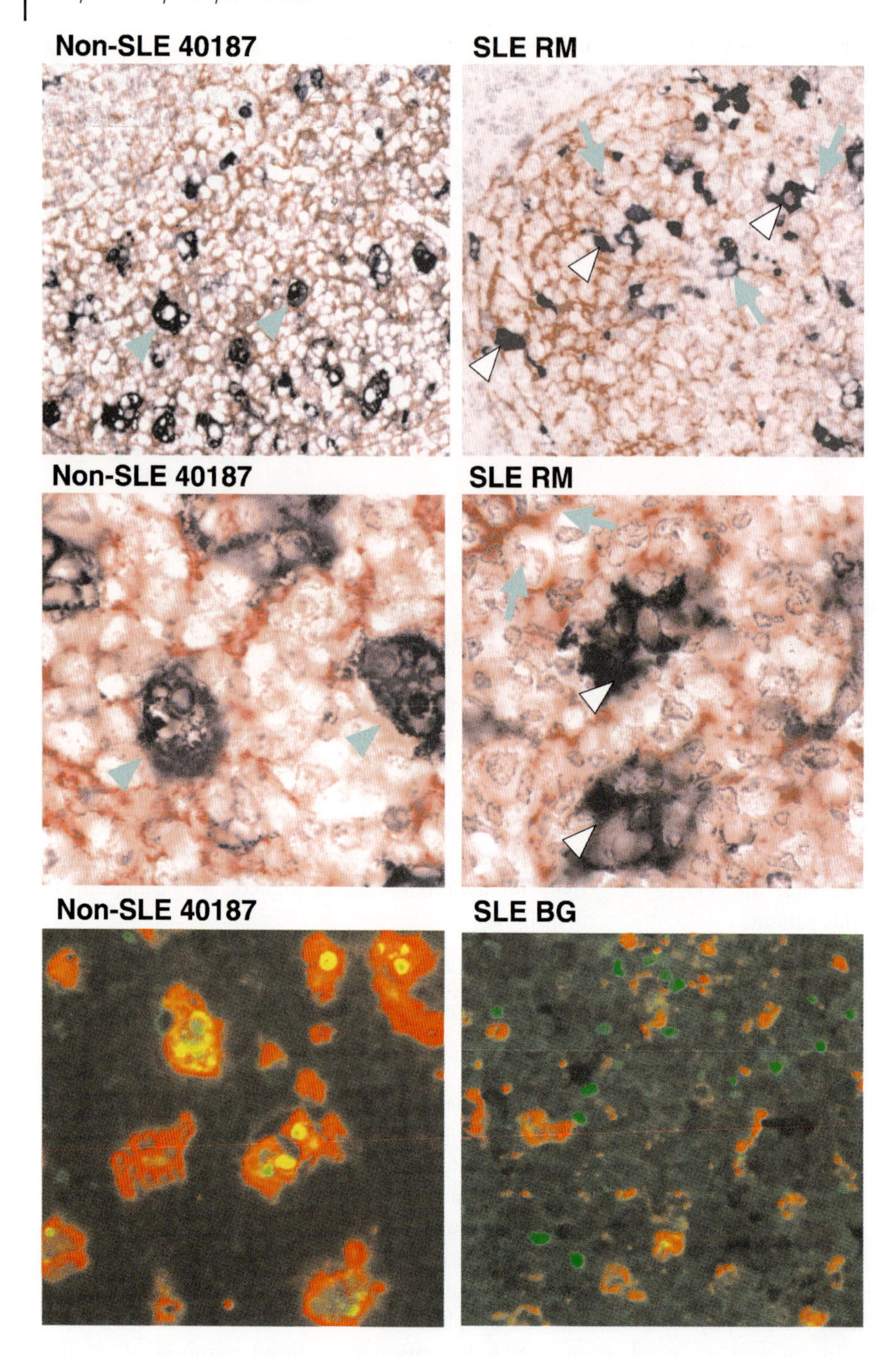
Non-SLE 40187
SLE RM
Non-SLE 40187
SLE RM
Non-SLE 40187
SLE BG

were observed to be attached to the surfaces of FDCs (Fig. 11.10, upper rows) which form a characteristic network within the germinal center. Confocal microscopy revealed that the remnants of the apoptotic nuclei have not been internalized by the FDCs (Fig. 11.10, lower row). Furthermore, the network of FDCs by itself appears to be unaffected in these patients. This observation suggests that in the germinal centers of some SLE patients apoptotic nuclei were not taken up by macrophages or by FDCs. The apoptotic cells accumulate in the germinal centers since they are not adequately cleared in the early phase of apoptosis. Therefore, apoptosis progresses and the cells enter secondary necrosis. In this stage of cell death, membrane integrity cannot be maintained and intracellular autoantigens including dsDNA are released.

Due to the low number of lymph nodes investigated, no correlation of the phagocytotic activity with any clinical feature could be established. However, the phenotype does not correlate with disease activity and treatment making an epiphenomenon unlikely.

11.9.3
Hypothesis: Accumulation of Apoptotic Dell-Derived Nuclear Fragments in the Germinal Centers Challenges B Cell Tolerance in SLE

In the germinal center of SLE patients, due to impaired phagocytic capacity, the apoptotic B cells are not adequately cleared in the early phase of apoptosis. Therefore, apoptosis progresses and the cells enter secondary necrosis. Subsequently, the membrane integrity is lost and intracellular potential autoantigens get accessible [55, 63, 64]. In this state, the classical complement cascade is activated, resulting in deposition of C3b on the surfaces of disintegrated apoptotic cells and nuclear debris [49, 50]. Via C3b and its fragments, this opsonized material binds to

Fig. 11.10 Tingible body macrophages are reduced in the germinal centers of a subgroup of SLE patients. TUNEL/CD21 staining of germinal centers analyzed by light microscopy (upper and middle rows). Characteristic lymph node sections from patients with non-malignant follicular hyperplasia (column 1) and of SLE patients (column 2) were double stained for TUNEL (dark blue) and FDCs (CR2/CD21 displayed in red). Apoptotic nuclei ingested by tingible body macrophages are marked with green arrow heads. Uningested apoptotic nuclear material putatively associated with FDCs are marked with white arrow heads. Nuclear material decorating the surfaces of FDCs are marked with green arrows. Low and high power magnifications are shown in panel 1 and 2, respectively. TUNEL/CD68 staining of lymph nodes analyzed by confocal laser microscopy (lower row). Characteristic lymph node sections of a non-SLE patient with reactive follicular hyperplasia (column 1) and of a SLE patient (column 2) were stained with TUNEL (FITC, green) and monoclonal anti-CD68 antibody (Cy5, red). The sections were analyzed by confocal laser microscopy. Co-localization of TUNEL-positive nuclear material and the macrophage cytoplasmic antigen CD68 is displayed in yellow. Whereas there germinal centers (GC) of all non-SLE controls showed large amounts of TUNEL-positive apoptotic nuclei taken up by giant tingible body macrophages, the germinal centers of some SLE patients contained many apparently free apoptotic nuclei, seemingly ignored by the small germinal center macrophages.

CR2/CD21 on the FDCs. The latter may then provide survival signals for those autoreactive B cells that had been generated incidentally by somatic mutations [81]. Thus an important initial control mechanism of B cell tolerance is circumvented under these conditions. Consecutively, B cell tolerance relies now only on the presence of a functional T cell tolerance for nuclear autoantigens. However, as discussed in the previous section, an impaired clearance also challenges the T cell tolerance.

11.10 Conclusion

Increased apoptosis and/or reduced clearance of dying cells by macrophages provoke accumulation of cellular fragments in various tissues. This may lead to the uptake of modified autoantigens from apoptotic nuclei or chromatin by DCs. The latter present altered self-epitopes to naïve T cells. Thus, potentially autoreactive T cells are activated and may now provide T cell help for B cells that present peptides processed from apoptotic prey.

In the germinal centers of secondary lymph organs, impaired phagocytic removal of early apoptotic cells may cause accumulation of secondary necrotic cells and debris. The latter bind complement and can therefore be trapped on the surfaces of FDCs. B cells may contact with nuclear autoantigens that had been released during late stages of apoptotic cell death and are immobilized by FDCs. Consecutively, B cells that had gained specificity for nuclear autoantigens during random somatic mutations can receive a short-term survival signal.

After migration into the mantle zone these autoreactive B cells may finally be activated by autoreactive $CD4^{+}$ T helper cells. B cells then differentiate into memory or plasma cells. The latter produce those pathogenic nuclear autoantibodies thought to be responsible for the tissue destruction and the pathogenesis of SLE.

11.11 References

1 Uramoto KM, Michet CJ, Jr, Thumboo J, Sunku J, WM OF, Gabriel SE. Trends in the incidence and mortality of systemic lupus erythematosus, 1950–1992. *Arthritis Rheum* **1999**, 42, 46–50.

2 Tan EM, Cohen AS, Fries JF, Masi AT, McShane DJ, Rothfield NF, *et al.* The 1982 revised criteria for the classification of systemic lupus erythematosus. *Arthritis Rheum* **1982**, 25, 1271–7.

3 Tan EM, Schur PH, Carr PI, Kunkel HG. Desoxyribonucleic acid (DNA) and antibodies to DNA in the serum of patients with systemic lupus erythematosus. *J Clin Invest* **1966**, 45, 1732–8.

4 Winkler TH, Jahn S, Kalden JR. IgG human monoclonal anti-DNA autoantibodies from patients with systemic lupus erythematosus. *Clin Exp Immunol* **1991**, 85, 379–85.

5 van Es JH, Gmelig Meyling FH, van de Akker WR, Aanstoot H, Derksen RH, Logtenberg T. Somatic mutations in the variable regions of a human IgG anti-double-stranded DNA autoantibody suggest a role for antigen in the induc-

tion of systemic lupus erythematosus. *J Exp Med* **1991**, 173, 461–70.

6 Winkler TH, Fehr H, Kalden JR. Analysis of immunoglobulin variable region genes from human IgG anti-DNA hybridomas. *Eur J Immunol* **1992**, 22, 1719–28.

7 Winkler TH, Kalden JR. Origin of anti-DNA autoantibodies in SLE [Editorial]. *Lupus* **1994**, 3, 75–6.

8 Radic MZ, Mackle J, Erikson J, Mol C, Anderson WF, Weigert M. Residues that mediate DNA binding of autoimmune antibodies. *J Immunol* **1993**, 150, 4966–77.

9 Radic MZ, Weigert M. Genetic and structural evidence for antigen selection of anti-DNA antibodies. *Annu Rev Immunol* **1994**, 12, 487–520.

10 Isenberg DA, Ehrenstein MR, Longhurst C, Kalsi JK. The origin, sequence, structure, and consequences of developing anti-DNA antibodies. A human perspective. *Arthritis Rheum* **1994**, 37, 169–80.

11 Herrmann M, Winkler TH, Fehr H, Kalden JR. Preferential recognition of specific DNA motifs by anti-double-stranded DNA autoantibodies. *Eur J Immunol* **1995**, 25, 1897–904.

12 Desai-Mehta A, Mao C, Rajagopalan S, Robinson T, Datta SK. Structure and specificity of T cell receptors expressed by potentially pathogenic anti-DNA autoantibody-inducing T cells in human lupus. *J Clin Invest* **1995**, 95, 531–41.

13 Voll RE, Roth EA, Girkontaite I, Fehr H, Herrmann M, Lorenz HM, *et al.* Histone-specific T_h0 and T_h1 clones derived from systemic lupus erythematosus patients induce double-stranded DNA antibody production. *Arthritis Rheum* **1997**, 40, 2162–71.

14 Rekvig OP, Fredriksen K, Brannsether B, Moens U, Sundsfjord A, Traavik T. Antibodies to eukaryotic, including autologous, native DNA are produced during BK virus infection, but not after immunization with non-infectious BK DNA. Scand *J Immunol* **1992**, 36, 487–95.

15 Rekvig OP, Andreassen K, Moens U. Antibodies to DNA – towards an understanding of their origin and pathophysiological impact in systemic lupus erythematosus [Editorial]. *Scand J Rheumatol* **1998**, 27, 1–6.

16 Andreassen K, Bredholt G, Moens U, Bendiksen S, Kauric G, Rekvig OP. T cell lines specific for polyomavirus T-antigen recognize T-antigen complexed with nucleosomes: a molecular basis for anti-DNA antibody production. *Eur J Immunol* **1999**, 29, 2715–28.

17 Berden JH, Licht R, van Bruggen MC, Tax WJ. Role of nucleosomes for induction and glomerular binding of autoantibodies in lupus nephritis. *Curr Opin Nephrol Hypertens* **1999**, 8, 299–306.

18 Theocharis S, Sfikakis PP, Lipnick RN, Klipple GL, Steinberg AD, Tsokos GC. Characterization of *in vivo* mutated T cell clones from patients with systemic lupus erythematosus. *Clin Immunol Immunopathol* **1995**, 74, 135–42.

19 Shivakumar S, Tsokos GC, Datta SK. T cell receptor alpha/beta expressing double-negative ($CD4^-/CD8^-$) and $CD4^+$ T helper cells in humans augment the production of pathogenic anti-DNA autoantibodies associated with lupus nephritis. *J Immunol* **1989**, 143, 103–12.

20 Murakami M, Kumagai S, Sugita M, Iwai K, Imura H. *In vitro* induction of IgG anti-DNA antibody from high density B cells of systemic lupus erythematosus patients by an HLA DR-restricted T cell clone. *Clin Exp Immunol* **1992**, 90, 245–50.

21 Steinberg AD, Roths JB, Murphy ED, Steinberg RT, Raveche ES. Effects of thymectomy or androgen administration upon the autoimmune disease of MRL/Mp-*lpr*/*lpr* mice. *J Immunol* **1980**, 125, 871–3.

22 Mihara M, Ohsugi Y, Saito K, Miyai T, Togashi M, Ono S, *et al.* Immunologic abnormality in NZB/NZW F_1 mice. Thymus-independent occurrence of B cell abnormality and requirement for T cells in the development of autoimmune disease, as evidenced by an analysis of the athymic nude individuals. *J Immunol* **1988**, 141, 85–90.

23 Wofsy D, Seaman WE. Reversal of advanced murine lupus in NZB/NZW F_1 mice by treatment with monoclonal antibody to L3T4. *J Immunol* **1987**, 138, 3247–53.

24 Schur PH. Genetics of complement deficiencies associated with lupus-like syndromes. *Arthritis Rheum* **1978**, 21 (5 suppl), S153–60.

25 Petry F. Molecular basis of hereditary C1q deficiency. *Immunobiology* **1998**, 199, 286–94.

26 Morgan BP, Walport MJ. Complement deficiency and disease. *Immunol Today* **1991**, 12, 301–6.

27 Hartung K, Baur MP, Coldewey R, Fricke M, Kalden JR, Lakomek HJ, *et al.* Major histocompatibility complex haplotypes and complement C4 alleles in systemic lupus erythematosus. Results of a multicenter study. *J Clin Invest* **1992**, 90, 1346–51.

28 Moroni G, Trendelenburg M, Del Papa N, Quaglini S, Raschi E, Panzeri P, *et al.* Anti-C1q antibodies may help in diagnosing a renal flare in lupus nephritis. *Am J Kidney Dis* **2001**, 37, 490–8.

29 Agnello V, De Bracco MM, Kunkel HG. Hereditary C2 deficiency with some manifestations of systemic lupus erythematosus. *J Immunol* **1972**, 108, 837–40.

30 Hauptmann G, Grosshans E, Heid E, Mayer S, Basset A. [Acute lupus erythematosus with total absence of the C4 fraction of complement]. *Nouv Presse Med* **1974**, 3, 881–2.

31 Manger K, Repp R, Spriewald BM, Rascu A, Geiger A, Wassmuth R, *et al.* Fcgamma receptor IIa polymorphism in Caucasian patients with systemic lupus erythematosus: association with clinical symptoms. *Arthritis Rheum* **1998**, 41, 1181–9.

32 Ponner BB, Stach C, Zoller O, Hagenhofer M, Voll R, Kalden JR, *et al.* Induction of apoptosis reduces immunogenicity of human T-cell lines in mice. *Scand J Immunol* **1998**, 47, 343–7.

33 Ronchetti A, Rovere P, Iezzi G, Galati G, Heltai S, Protti MP, *et al.* Immunogenicity of apoptotic cells *in vivo*: role of antigen load, antigen-presenting cells, and cytokines. *J Immunol* **1999**, 163, 130–6.

34 Rovere P, Manfredi AA, Vallinoto C, Zimmermann VS, Fascio U, Balestrieri G, *et al.* Dendritic cells preferentially internalize apoptotic cells opsonized by anti-beta2-glycoprotein I antibodies. *J Autoimmun* **1998**, 11, 403–11.

35 Rovere P, Sabbadini MG, Vallinoto C, Fascio U, Recigno M, Crosti M, *et al.* Dendritic cell presentation of antigens from apoptotic cells in a proinflammatory context: role of opsonizing anti-beta2-glycoprotein I antibodies. *Arthritis Rheum* **1999**, 42, 1412–20.

36 Rovere P, Vallinoto C, Bondanza A, Crosti MC, Rescigno M, Ricciardi-Castagnoli P, *et al.* Bystander apoptosis triggers dendritic cell maturation and antigen-presenting function. *J Immunol* **1998**, 161, 4467–71.

37 Rovere P, Sabbadini MG, Vallinoto C, Fascio U, Zimmermann VS, Bondanza A, *et al.* Delayed clearance of apoptotic lymphoma cells allows cross-presentation of intracellular antigens by mature dendritic cells. *J Leuk Biol* **1999**, 66, 345–9.

38 Botto M, Dell'Agnola C, Bygrave AE, Thompson EM, Cook HT, Petry F, *et al.* Homozygous C1q deficiency causes glomerulonephritis associated with multiple apoptotic bodies [see Comments]. *Nat Genet* **1998**, 19, 56–9.

39 Taylor PR, Carugati A, Fadok VA, Cook HT, Andrews M, Carroll MC, *et al.* A hierarchical role for classical pathway complement proteins in the clearance of apoptotic cells *in vivo* [In process citation]. *J Exp Med* **2000**, 192, 359–66.

40 Bickerstaff MC, Botto M, Hutchinson WL, Herbert J, Tennent GA, Bybee A, *et al.* Serum amyloid P component controls chromatin degradation and prevents antinuclear autoimmunity [see Comments]. *Nat Med* **1999**, 5, 694–7.

41 Scott RS, McMahon EJ, Pop SM, Reap EA, Caricchio R, Cohen PL, *et al.* Phagocytosis and clearance of apoptotic cells is mediated by MER. *Nature* **2001**, 411, 207–11.

42 Napirei M, Karsunky H, Zevnik B, Stephan H, Mannherz HG, Moroy T. Features of systemic lupus erythematosus in DNase1-deficient mice. *Nat Genet* **2000**, 25, 177–181.

43 Voll RE, Herrmann M, Roth EA, Stach C, Kalden JR, Girkontaite I. Immunosuppressive effects of apoptotic cells. *Nature* **1997**, 390, 350–1.

44 Fadok VA, Bratton DL, Konowal A, Freed PW, Westcott JY, Henson PM. Macrophages that have ingested apoptot-

ic cells *in vitro* inhibit proinflammatory cytokine production through autocrine/paracrine mechanisms involving TGF-beta, PGE2, and PAF. *J Clin Invest* **1998**, 101, 890–8.
45 Surh CD, Sprent J. T-cell apoptosis detected *in situ* during positive and negative selection in the thymus [see Comments]. *Nature* **1994**, 372, 100–3.
46 Haslett C. Granulocyte apoptosis and its role in the resolution and control of lung inflammation. *Am J Respir Crit Care Med* **1999**, 160 (5 pt 2), S5–11.
47 Vermes I, Haanen C, Richel DJ, Schaafsma MR, Kalsbeek-Batenburg E, Reutelingsperger CP. Apoptosis and secondary necrosis of lymphocytes in culture. *Acta Haematol* **1997**, 98, 8–13.
48 Louagie H, Cornelissen M, Philippe J, Vral A, Thierens H, De Ridder L. Flow cytometric scoring of apoptosis compared to electron microscopy in gamma irradiated lymphocytes. *Cell Biol Int* **1998**, 22, 277–83.
49 Korb LC, Ahearn JM. C1q binds directly and specifically to surface blebs of apoptotic human keratinocytes: complement deficiency and systemic lupus erythematosus revisited. *J Immunol* **1997**, 158, 4525–8.
50 Mevorach D, Mascarenhas JO, Gershov D, Elkon KB. Complement-dependent clearance of apoptotic cells by human macrophages [In process citation]. *J Exp Med* **1998**, 188, 2313–20.
51 Emlen W, Niebur J, Kadera R. Accelerated *in vitro* apoptosis of lymphocytes from patients with systemic lupus erythematosus. *J Immunol* **1994**, 152, 3685–92.
52 Lorenz HM, Grunke M, Hieronymus T, Herrmann M, Kuhnel A, Manger B, *et al.* *In vitro* apoptosis and expression of apoptosis-related molecules in lymphocytes from patients with systemic lupus erythematosus and other autoimmune diseases [see Comments]. *Arthritis Rheum* **1997**, 40, 306–17.
53 Kovacs B, Liossis SN, Dennis GJ, Tsokos GC. Increased expression of functional Fas-ligand in activated T cells from patients with systemic lupus erythematosus [In process citation]. *Autoimmunity* **1997**, 25, 213–21.
54 Chan EY, Ko SC, Lau CS. Increased rate of apoptosis and decreased expression of bcl-2 protein in peripheral blood lymphocytes from patients with active systemic lupus erythematosus. *Asian Pac J Allergy Immunol* **1997**, 15, 3–7.
55 Perniok A, Wedekind F, Herrmann M, Specker C, Schneider M. High levels of circulating early apoptic peripheral blood mononuclear cells in systemic lupus erythematosus. *Lupus* **1998**, 7, 113–8.
56 Courtney PA, Crockard AD, Williamson K, McConnell J, Kennedy RJ, Bell AL. Lymphocyte apoptosis in systemic lupus erythematosus: relationships with Fas expression, serum soluble Fas and disease activity. *Lupus* **1999**, 8, 508–13.
57 Krieg AM. CpG DNA: a pathogenic factor in systemic lupus erythematosus? *J Clin Immunol* **1995**, 15, 284–92.
58 Tsai YT, Chiang BL, Kao YF, Hsieh KH. Detection of Epstein–Barr virus and cytomegalovirus genome in white blood cells from patients with juvenile rheumatoid arthritis and childhood systemic lupus erythematosus. *Int Arch Allergy Immunol* **1995**, 106, 235–40.
59 Svensson B. Monocyte *in vitro* function in systemic lupus erythematosus (SLE). I. A clinical and immunological study. *Scand J Rheumatol Suppl* **1980**, 31, 29–41.
60 Svensson B. Occurrence of deficient monocyte yeast cell phagocytosis in presence of rheumatic sera. *Scand J Rheumatol Suppl* **1980**, 31, 21–7.
61 Hurst NP, Nuki G, Wallington T. Evidence for intrinsic cellular defects of 'complement' receptor-mediated phagocytosis in patients with systemic lupus erythematosus (SLE). *Clin Exp Immunol* **1984**, 55, 303–12.
62 Salmon JE, Kimberly RP, Gibofsky A, Fotino M. Defective mononuclear phagocyte function in systemic lupus erythematosus: dissociation of Fc receptor-ligand binding and internalization. *J Immunol* **1984**, 133, 2525–31.
63 Herrmann M, Zoller OM, Hagenhofer M, Voll R, Kalden JR. What triggers anti-dsDNA antibodies? *Mol Biol Rep* **1996**, 23, 265–7.
64 Herrmann M, Voll RE, Zoller OM, Hagenhofer M, Ponner BB, Kalden JR. Impaired phagocytosis of apoptotic

cell material by monocyte-derived macrophages from patients with systemic lupus erythematosus. *Arthritis Rheum* **1998**, 41, 1241–50.

65 Raptis L, Menard HA. Quantitation and characterization of plasma DNA in normals and patients with systemic lupus erythematosus. *J Clin Invest* **1980**, 66, 1391–9.

66 McCoubrey-Hoyer A, Okarma TB, Holman HR. Partial purification and characterization of plasma DNA and its relation to disease activity in systemic lupus erythematosus. *Am J Med* **1984**, 77, 23–34.

67 Steinman CR. Circulating DNA in systemic lupus erythematosus. Isolation and characterization. *J Clin Invest* **1984**, 73, 832–41.

68 Casciola-Rosen LA, Anhalt G, Rosen A. Autoantigens targeted in systemic lupus erythematosus are clustered in two populations of surface structures on apoptotic keratinocytes [see Comments]. *J Exp Med* **1994**, 179, 1317–30.

69 Rosen A, Casciola-Rosen L, Ahearn J. Novel packages of viral and self-antigens are generated during apoptosis. *J Exp Med* **1995**, 181, 1557–61.

70 Casiano CA, Ochs RL, Tan EM. Distinct cleavage products of nuclear proteins in apoptosis and necrosis revealed by autoantibody probes. *Cell Death Different* **1998**, 5, 183–90.

71 Rosen A, Casciola-Rosen L. Macromolecular substrates for the ICE-like proteases during apoptosis. *J Cell Biochem* **1997**, 64, 50–4.

72 Casciola-Rosen L, Andrade F, Ulanet D, Wong WB, Rosen A. Cleavage by granzyme B is strongly predictive of autoantigen status: implications for initiation of autoimmunity. *J Exp Med* **1999**, 190, 815–26.

73 Utz PJ, Hottelet M, van Venrooij WJ, Anderson P. Association of phosphorylated serine/arginine (SR) splicing factors with the U1-small ribonucleoprotein (snRNP) autoantigen complex accompanies apoptotic cell death. *J Exp Med* **1998**, 187, 547–60.

74 Moudgil KD, Sercarz EE, Grewal IS. Modulation of the immunogenicity of antigenic determinants by their flanking residues. *Immunol Today* **1998**, 19, 217–20.

75 Manfredi AA, Rovere P, Galati G, Heltai S, Bozzolo E, Soldini L, *et al.* Apoptotic cell clearance in systemic lupus erythematosus. I. Opsonization by antiphospholipid antibodies. *Arthritis Rheum* **1998**, 41, 205–14.

76 Gallucci S, Lolkema M, Matzinger P. Natural adjuvants: endogenous activators of dendritic cells. *Nat Med* **1999**, 5, 1249–55.

77 Carroll MC. The role of complement and complement receptors in induction and regulation of immunity. *Annu Rev Immunol* **1998**, 16, 545–68.

78 Camacho SA, Kosco-Vilbois MH, Berek C. The dynamic structure of the germinal center. *Immunol Today* **1998**, 19, 511–4.

79 Shokat KM, Goodnow CC. Antigen-induced B-cell death and elimination during germinal-centre immune responses. *Nature* **1995**, 375, 334–8.

80 Gaipl US, Kuenkele S, Voll RE, Beyer TD, Kolowos W, Heyder P, *et al.* Complement binding is an early feature of necrotic and a rather late event during apoptotic cell death. *Cell Death Different* **2001**, 8, 327–34.

81 Carroll MC. The role of complement in B cell activation and tolerance. *Adv Immunol* **2000**, 74, 61–88.

82 Rekvig O. Polyoma induced autoimmunity to DNA, experimental systems and clinical observations in human SLE. *Lupus* **1997**, 6, 325–6.

83 Moens U, Seternes OM, Hey AW, Silsand Y, Traavik T, Johansen B, *et al. In vivo* expression of a single viral DNA-binding protein generates systemic lupus erythematosus-related autoimmunity to double-stranded DNA and histones. *Proc Natl Acad Sci USA* **1995**, 92, 12393–7.

Part 4
Immunogenicity of Apoptotic Cells

12
Dendritic Cells Pulsed with Apoptotic Tumor Cells as Vaccines

Lars Jenne and Birthe Sauter

12.1
Introduction

Apoptotic cell death has long been regarded as a programmed cellular auto-destruction solely designed to remove potentially antigenic cellular particles without causing inflammation or uncontrolled liberation of antigen by damaging tissue. Although the relevance of apoptosis in tissue homeostasis as well as its morphological criteria strongly support this opinion, concerns about the immunomodulatory properties of apoptotic debris especially in the context of tumors or HIV infection were raised and challenged the paradigm that cellular debris left over from apoptotic cells is just removed by macrophages without functional consequences. It recently became apparent that dendritic cells (DCs) phagocytose apoptotic cell debris and whole apoptotic cells. As DCs constitute a specialized system of antigen-presenting cells (APCs) that are initiators and modulators of the immune response against microbial, tumor and self-antigens, this notion implied that antigen from apoptotic cells might be presented and thus play a role in the maintenance of peripheral tolerance as well as in the introduction of antiviral or even antitumoral immunity. Recent research data suggest that the efficient presentation of antigen derived from apoptotic cells can be explored for the *ex vivo* loading of DCs with apoptotic tumor cells in order to generate an effective cellular vaccine aiming at the induction of antitumoral immunity. The development of techniques to generate clinical grade DCs has facilitated DC-based immunotherapeutic approaches. Numerous phase-I and -II clinical trials have been introduced in order to evaluate the efficacy of tumor antigen-pulsed DCs in cancer patients. Although antitumoral T cell responses have been induced in clinical trials, DC-based therapy is in its infancy. Utilizing apoptotic tumor cells as tumor antigen donors in *ex vivo* antigen loading approaches might facilitate the loading of DCs with whole tumor cells leading to the MHC-restricted presentation of a broad array of tumoral antigens. In this chapter we will discuss the basis of DC loading, apoptotic cell utilization and sum up the experimental data available for different apoptotic tumor cells.

12.2
Immature DCs

12.2.1
Immature DCs Capture Antigen

DCs are virtually found in all organs. They exist as interstitial DCs in peripheral tissues, Langerhans and dermal dendritic cells in the epidermis and dermis, respectively, as interdigitating cells in lymph nodes and as circulating APCs in the blood stream [1]. This general presence makes DCs ideal sentinels to spot immunological alterations at various sites. While located in peripheral tissues, DCs are considered to be immature. This developmental stage is characterized by a high capacity to capture and process antigens.

DCs can acquire exogenous antigens by numerous mechanisms such as macropinocytosis, receptor-mediated endocytosis, phagocytosis and via specific receptors. Receptors include Fcγ [2] and Fcε receptors [3] for acquisition of immune complexes, the lectin receptors macrophage mannose receptor (MMR) [4] and DEC-205 [5] for acquisition of glycosylated proteins, and DC-SIGN, a receptor that mediates the uptake of HIV [6]. DCs also acquire antigens through receptor-independent mechanisms including macropinocytosis and phagocytosis [4]. Furthermore, DCs can acquire antigens through direct infection by pathogens, which introduce or synthesize proteins in the DC cytoplasm. DCs express Toll-like receptors (TLRs) which have specificity for conserved molecular patterns shared by groups of pathogens including lipopolysaccharides of bacteria and CpG-containing bacterial DNA. These receptors enable DCs to 'sense' microbes, acquire pathogen-derived antigens [7] and simultaneously undergo maturation.

12.2.2
Phagocytosis of Apoptotic Cells

Apoptosis in cells is defined by a series of unique cellular changes which include the compaction of nuclear chromatin into dense masses that move to the edge of the intact nuclear envelope associated with extensive DNA degradation through the activation of endonucleases; fragmentation of these chromatin masses and condensation of the cytoplasm with shrinkage of the cell; and finally fragmentation of the cell into pieces called apoptotic bodies and blebs which are still enclosed by intact cell membrane and contain the remaining antigen of the cell [8]. Clearance of intact apoptotic cells by phagocytes protects surrounding tissues from intracellular factors and reduces the likelihood of tissue damage caused by inappropriate autoimmune responses. The uptake of apoptotic cells can be performed by various cell types including microvascular endothelial cells, monocytes, macrophages and DCs. Fig. 12.1 shows an electron microscope picture of a DC after co-cultivation with apoptotic macrophages, leading to the phagocytosis of apoptotic bodies. Multiple ligands and receptors have been implicated in the recognition and uptake of apoptotic cells, yet attempts to assign function to discrete

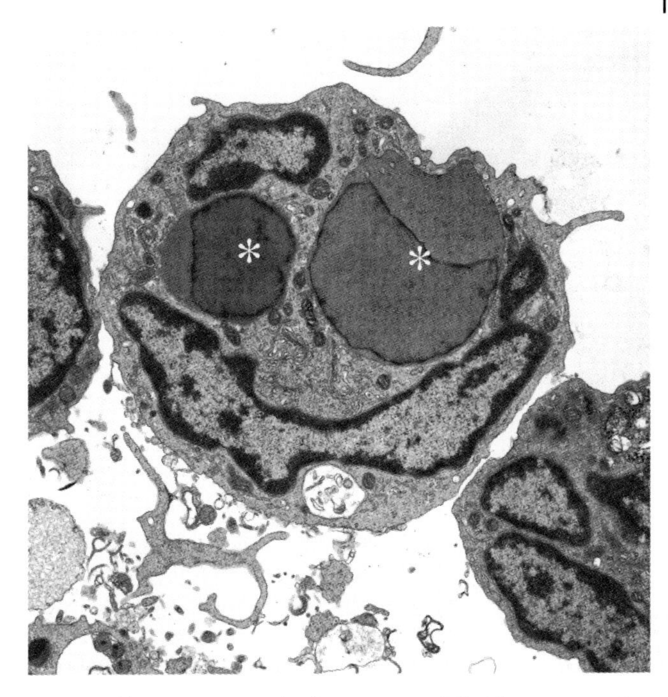

Fig. 12.1 Electron micrograph of an immature DC with two apoptotic bodies (*) derived from apoptotic macrophages within its cytoplasm.

receptors have proven difficult. During apoptosis, the asymmetry of plasma membrane phospholipids is lost, which exposes phosphatidylserine (PS) externally. PS seems to be one of the well-characterized 'eat me' flags. Recently, Fadok *et al.* characterized a new PS receptor (PSR) critical in mediating the uptake of apoptotic cells [9]. This group further found that individual or multiple engagements of the AC receptors CD36, the $\alpha_v\beta_3$ integrin and $\alpha_v\beta_5$ integrin, CD14, and CD68 caused tethering of erythrocytes (their 'model food'), but little internalization. In contrast, engagement of the PSR alone through PS-coated erythrocytes induced neither tethering nor uptake. However, ligation of both PSR and other receptors converted the adhesion mediated by the latter to ingestion. Hoffmann *et al.* [10] recently proposed that ligation of PSR on phagocytes delivers a 'tickle' signal which stimulates the internalization of apoptotic cells, including bystander cells, that are 'tethered' through other recognition receptors. PSR-independent pathways clearly exist: defense collagens such as the collectins [surfactant protein (SP)-A; mannose-binding lectin (MBL)] and the component C1q coat apoptotic cells via their globular heads and initiate uptake via interacting with phagocyte receptors through their collagenous tail groups. Other possible important bridging molecules include thrombospondin, lactadherin, iC3b and β_2-glycoprotein I. Phagocytosis of apoptotic cells by phagocytes (i.e. macrophages) via the 'tether and tickle' mechanism leads to the secretion of immunosuppressive cytokines suppressing any potential autoimmune responses. DCs preferentially phagocytose apoptotic cells via the $a_v\beta_5$ integrin [11]. While binding to CD36 seems to influence DC responsiveness towards anti-inflammatory signals [12], binding to the $a_v\beta_5$ integrin might direct the cytoskeletal changes necessary for the phagocyte surface to envelope apoptotic cells through recruitment of the CrkII–DOCK180–Rac1 signaling complex and possibly route dead cells towards a specialized antigen-presentation pathway [13]. The phagocytosis of murine apoptotic cells was facilitated by the binding of anti-β_2-glycoprotein I antibodies indicating a pathophysiological role of antiphospholipid antibodies in autoimmune diseases [14]. In contrast, the acute-phase protein PTX3, which binds to dying cells, inhibits the phagocytosis of apoptotic or necrotic cells and thus seems to play a protective role in the induction of autoimmune reactions in inflamed tissue [15].

12.2.3
Processing of Apoptotic Cells

As DCs continuously shuttle the material derived from their intracellular processing to regional lymph nodes [16] it is of interest to know whether apoptotic material is also presented. Following internalization, exogenous antigen is conserved in endosomes but gains access to the MHC I antigen-presentation machinery restricted to cytoplasmic antigen. Apart from the possible leakage of smaller-sized antigen through the endosomal membrane into the cytosol, DCs were shown to have a unique membrane transport pathway linking the lumen of endocytic compartments and the cytosol [17]. Furthermore, processing of the antigen within the endosomes and subsequent 'regurgitation' might lead to the MHC I-restricted antigen presentation on the same or neighboring cells, and this might be an additional mechanism [18]. Although *in vitro* studies have suggested that both transporter associated with antigen processing (TAP)-dependent and -independent pathways exist for cross-presentation, *in vivo* data demonstrate an absolute requirement for a functional TAP system for cross-presentation of exogenous antigen [19].

Most likely *in vivo* the antigenic material that gets access to the antigen presentation machinery of a DC is derived from both apoptotic and necrotic cells. One common denominator of both cells is heat shock protein (HSP), which can be found enclosed in apoptotic cells and apoptotic bodies and blebs derived from such cells [20] as well as being released by necrotic cell material [21]. HSPs are highly conserved chaperones designed for the cellular transport and folding of proteins. They are taken up by DCs, presumably by receptor-mediated phagocytosis (CD91 seems to be the molecule involved) [21, 22], and both HSP70 [23] and HSP96 [24] induce a maturation of immature DCs, indicating a physiological role for HSP-dependent antigen presentation. This concept is further supported by the *in vivo* induction of protective immunity and cytotoxic T lymphocytes (CTLs) by using HSPs derived from virally infected or tumor cells for murine vaccination studies [25].

12.3
Mature DCs

12.3.1
Mature DCs Present Antigen and Prime T Lymphocytes

When exposed to pathogenic and inflammatory stimuli, the DCs migrate to draining lymph nodes and undergo maturation. When mature, DCs downregulate their ability to acquire antigens, but increase their T cell-stimulatory capacity through (1) upregulation of co-stimulatory molecules and MHC molecules, (2) increased expression and durability of peptide–MHC complexes, (3) development of immunoproteasomes, and (4) the capacity to synthesize cytokines such as interleukin (IL)-12, IL-15 and IL-18 [26]. Several factors induce DC maturation including microorganisms (bacteria, viruses, yeasts), CD40 ligand on activated T cells, cytokines (e.g. tumor necrosis factor (TNF)-α, IL-1β), bacterial and viral products and nucleotides. A complete shift in the chemokine receptor expression profile enables their entry into regional lymph node.

T cells recognize antigens presented as small peptides in the grooves of human leukocyte antigen (HLA) molecules on the surfaces of APCs. $CD4^+$ T cells detect peptides derived from exogenously acquired antigens and which are processed within specialized HLA class II-rich compartments of immature DCs. During maturation, the peptides are bound to HLA molecules and transferred to the cell surface with co-stimulatory molecules for activation of $CD4^+$ T cells [27]. The complexes remain there for several days to allow interactions with naïve $CD4^+$ T cells. Following recognition of antigen on HLA molecules, $CD4^+$ T cells can differentiate into T helper 1 type (T_h1) or T helper 2 type (T_h2) cells, which are distinguished by the production of interferon (IFN)-γ and IL-4, respectively. Compared to other APCs, DCs are up to 1000-fold more efficient in activating resting T cells [28]. $CD4^+$ T cells, in turn, are essential for licensing the DCs to activate $CD8^+$ T cells and to maintain their memory.

$CD8^+$ T cells mediate protective immunity to intracellular pathogens and tumors, through production of cytokines such as IFN-γ and TNF-α, and/or direct cytolytic mechanisms. Typically, $CD8^+$ T cells are primed by antigens that are presented on HLA class I molecules and which have been acquired 'endogenously', e.g. all proteins synthesized within the DC's cytoplasm including viral antigens when the DC is infected by viruses. These proteins are cleaved by enzymatic units called proteosomes into peptides, transported into the endoplasmic reticulum via transporter molecules and loaded onto HLA class I molecules for export to the cell surface. Maturing DCs upregulate a subfraction of proteosomes termed immunoproteosomes which may enhance antigen processing, as well as HLA class I synthesis. Furthermore, endogenously synthesized antigens can be presented on HLA class II molecules, allowing for the simultaneous class I- and class II-restricted presentation of antigens brought into DCs by viruses or vector systems.

The priming of $CD8^+$ T cells, including their expansion, development into effector cells and maintenance as memory cells, also requires $CD4^+$ helper cells and the cytokine IL-15 [29], which is produced by DCs following T cell contact. In animal models, DCs bearing viral or tumoral antigens introduced as peptides, proteins, DNA, RNA, or within viral vectors or dead cells, induce protective and therapeutic CTL responses and can reverse tolerance in some cases (reviewed in [26]).

12.3.2
Cross-priming of $CD8^+$ T Cells

It has now become evident that exogenously acquired antigens can also be presented on HLA class I molecules to $CD8^+$ T cells. This phenomenon was first described in 1976 by Bevan [30] and refers to the capacity of APCs to acquire antigens from other cells *in vivo*, resulting in the generation of an antigen-specific $CD8^+$ T cell response. Bevan showed that mice immunized with cells that express foreign minor histocompatibility antigens mount an antigen-specific response that was restricted to self class I, demonstrating the importance of the exogenous pathway in the induction of $CD8^+$ T cell responses. Bevan termed the MHC I-restricted presentation of antigens from cells by APC cross-presentation. Subsequently studies confirmed that $CD8^+$ T cell responses could be generated to antigens expressed in peripheral autologous tissue that lacked APC functions, such as viral antigens or H-Y self-antigens and tumor antigens [31–33]. The activation of antigen-specific $CD8^+$ CTL is now commonly referred to as cross-priming, a mechanism considered to be essential for the induction of immunity against pathogens that fail to infect APCs directly [e.g. Epstein-Barr virus (EBV)] or to tumors which themselves present antigen extremely inefficiently (reviewed in [34]). The determination of the cells involved in the cross-presentation was the next key step in the attempts to decipher the process of priming $CD8^+$ T lymphocytes against exogenous antigen. First of all, the relevance of bone marrow-derived APCs was demonstrated [33, 35]. Together with growing concern about the *in vivo* relevance of DCs in inducing and regulating cellular immunity, data showing their capacity to cross-present cellular antigen emerged. *In vivo* studies in animals have confirmed the immunogenicity of apoptotic tumor cell-pulsed DCs but not macrophages in cross-priming [36].

12.3.3
Cross-Priming versus Cross-Tolerance

Although the existence of DC lineages specifically designed to induce tolerance has been postulated [37], the current data generally support a concept based on the maturation status of a DC in order to explain the relevance of DCs for the induction of immunity and the maintenance of peripheral tolerance. Both immature and mature DCs can present antigen in a MHC-restricted fashion. Yet, the context of this presentation is substantially different. Immature DC express only a limited amount of co-stimulatory molecules. T cell activation heavily depends on the context of antigen presentation as the activation threshold is substantially lowered when co-stimulatory molecules are triggered [38]. A lack of co-stimulation not only fails to activate T cell proliferation but instead induces an anergic state and thus tolerance [39] or even antigen-specific regulatory T cells which can hamper T cell proliferation [40]. These mechanisms of regulation are thought to play a major role in the induction and maintenance of peripheral tolerance as in fact bone marrow-derived APCs, and not the peripheral tissue itself, are responsible for the tolerization of antigen-specific CTL [41]. Studies of the interaction of cross-presenting DCs with $CD8^+$ T cells have revealed a substantial role for DCs in the regulation of peripheral tolerance by cross-presenting antigen from apoptotic cells [42, 43]. Furthermore, there is experimental evidence now for an amendment of the second signal theory, which reduces the balance of cross-tolerance versus cross-priming to the sole expression of co-stimulatory molecules and the amount of antigen on the DCs. It became clear that $CD4^+$ helper T cells play a major role in inducing an activation stage of DCs, licensing them to prime CTL to become activated [44]. This activation status of the DCs seems to be more than just the further upregulation of co-stimulatory molecules and instead a third signal of yet undefined nature active at the DC–$CD4^+$ T cell interface [43]. Physiologically, the cross-presentation of cellular antigen derived from apoptotic cells seems to play a role for the induction of autoimmune diseases such as lupus erythematodes [45] and in the induction of immunity against antigens structurally related tumoral antigens, e.g. in paraneoplastic cerebral degeneration [46]. It is less clear which role can be attributed to cross-presenting DC *in vivo* in the induction of antiviral or antitumoral immunity. As apoptosis in tumor cells occurs not only physiologically but, instead, is induced by various treatment modalities [47, 48], experimental determination is clearly of great interest. Studying the interaction of apoptotic tumor cells with APCs is a prerequisite to solve this problem.

12.4 Immunomodulating Properties of Apoptotic Cells

In general, tumor cells are thought to have the capacity to block DC functions by secreting regulatory factors such as IL-10, transforming growth factor (TGF)-β, prostaglandin E_2 or other, yet undefined molecules [49]. We will concentrate here on the effects described for apoptotic tumor cells. The dualistic function of antigen presentation by immature DCs participating in the maintenance of peripheral tolerance and the inhibition of antigen-specific $CD8^+$ T cells through the generation of suppressor or regulatory $CD8^+$ T cells *in vivo* [50] versus the immunogenic presentation of antigen by mature DC is the basis for intense studies determining the maturational capacities of apoptotic tumor cells. Consensus now seems to be reached for the non-maturing effects of apoptotic cells derived from both tumor cell lines and tumor cells up to a certain DC:tumor cell ratio of about 1:1 [51]. Alterations of the chemokine system, responsible for DC migration and T cell interaction, have been reported in the murine system by two groups. An upregulation of CCR7 with functional consequences was observed after co-cultivating mouse DC with apoptotic fibrosarcoma cells [52], and apoptotic BL6-10 melanoma cells induced the synthesis of chemokines such as MIP-1α, MIP-1β and MIP-2 together with the expression of CCR7 and the downregulation of CCR2 and CCR5, receptors that modulate responsiveness to MIP-1α, MIP-1β and RANTES [53]. These experiments await further analysis in the human system.

Concerning the *ex vivo* maturation of DCs, one report showed that beyond the proinflammatory cytokines used by most researchers for the maturation of antigen-loaded DCs, CD40 ligand plus IFN-γ adds to the cross-presentation of tumor antigens derived from apoptotic cells [54]. This report thus supports the opinion that in order to achieve cross-priming DCs have to mature completely while antigen presentation with immature or half mature DCs, as used in another report [43], leads to cross-tolerance.

12.5 Antigen Loading

Although the injection of DCs into tumors had some effect related to intra-tumoral apoptotic murine breast cancer cells [55] most researchers will prefer to load DC *ex vivo*. *Ex vivo* manipulation has certain advantages: it allows for the control of DC quality (i.e. maturation status, DC subset) and expression level of desired antigens. Furthermore, injection of the prepared DCs can be performed at anatomical sites of interest (i.e. lymph nodes or tumors). Recently gained knowledge of tumor-associated antigens (TAAs) has opened a myriad of possibilities to manipulate DCs for maximal antigen-specific presentation. In principle it is desirable to aim for the parallel presentation of HLA class I- and class II-restricted antigens, as the absence of $CD4^+$ helper cells affects the generation of long-term $CD8^+$ T cell memory [56] and $CD4^+$ helper T cells are essential for an antitumor immune responses [57]. The cloning of TAAs was followed by the determination of immunodominant peptide sequences suitable for MHC I-restricted presentation by a given HLA molecule. The syntheses of large quantities of 8- to 10-amino-acid long peptides that fit into the HLA class I groove is technically rather easy and several companies now provide clinical grade peptides. In a process termed 'pulsing' peptides are added to DC populations generated *ex vivo*. These peptides replace others, which are bound to class I molecules with lower affinities. However, there are certain caveats with this approach: (1) the longevity of these HLA-peptide complexes *in vivo* is unknown, (2) the affinity of peptides for their various HLA molecules varies, (3) competition between peptides may affect immunogenicity, (4) epitopes that activate $CD4^+$ T cells have yet to be identified in detail and (5) the approach is inherently tailored for individuals as it is dependent upon the HLA type. Nevertheless, some approaches have been developed to address these concerns. One is the development of heteroclitic epitopes where the affinity of the peptides derived from TAAs for HLA molecules is improved by exchanging amino acids [58]. In order to vaccinate patients with multiple TAA-derived peptides (which is desirable to avoid the development of HLA^- tumor cells) it is probably wise to pool aliquots of DCs, which have been pulsed with individual peptides to avoid competition between peptides for HLA binding. Recently, HLA class II-binding peptides have been defined for some TAAs as well, making it feasible to load DCs with epitopes that activate both $CD4^+$ and $CD8^+$ T cells. The technique of peptide pulsing has been shown to induce peptide-specific CTL in healthy subjects [59] and melanoma patients [60]. Furthermore, clinical studies with peptide-pulsed DCs were applied against a broad variety of tumors (for review, see [61]) with partial effectiveness.

In contrast to peptide pulsing, using whole tumor cell preparations for DC loading avoids the need for detailed tumor analysis and individual HLA typing, as it is assumed that all tumoral antigens, including as yet undefined TAAs and rare mutations, will be presented on MHC class I and class II molecules by autologous DCs. The disadvantage of this approach includes the uncertainty regarding the induction of autoimmunity [62] and the necessity to obtain a sufficient number of autologous tumor cells by invasive procedures. Furthermore, tumor metastases may have a different antigen profile than the one expressed by primary cells or the cells obtained for antigen loading. It is a well-known phenomenon that tumors can 'escape' detection via loss of HLA molecules or TAAs. When allogeneic cell lines are used for the loading of DCs, not all tumoral antigens of the autologous tumor might be expressed and cell lines tend to have an unstable gene expression, resulting in a possible loss of relevant tumoral antigens.

The preparations used for antigen loading are usually mechanically or thermally disrupted tumor cells. Necrotic tumor cell material has the capacity to induce DC maturation when given to immature DCs [51], but we feel that it is desirable to induce further maturation with a standard stimulus prior to clinical use. Although clinical trials have been performed using DCs loaded with tumor cell lysates [63–65], little is known about the efficacy of antigen loading and the antigen concentrations required to achieve antigen presentation. *In vitro*, at least, there is debate as to whether tumor lysates can be efficient sources of antigens for DCs [66, 67]. Furthermore, when comparing the loading efficacy of necrotic cellular material with further processed lysates (generated by removing cellular debris) in the generation of an anti-EBV immunity, conflicting data were presented [66, 68]. However, soluble antigen such as tumor-derived proteins is taken up by rather inefficient pinocytosis, while cellular antigen is taken up by receptor-mediated phagocytosis, presumably leading to a more efficient cross-presentation [67]. With regard to the phagocytosis mechanisms mentioned in Section 12.2.2, uptake and processing of antigen derived from apoptotic cells is assumed to be very efficient. The efficiency of cross-presentation of viral antigens from apoptotic cells, at least *in vitro*, is high: one apoptotic cell fed to 100 DCs is sufficient to elicit an antiviral $CD8^+$ T cell response. Furthermore, the antiviral CTL responses induced are equivalent to those elicited by direct infection of DC with virus or peptide pulsing with nanomolar concentrations of relevant viral antigen [34]. For tumor antigens it has been shown that cross-presentation of melanoma-derived TAAs is less effective for single TAA than peptide pulsing, but the overall efficiency of killing tumor cells is better with cross-primed CTL [69]. An augmentation of this cross-presentation can be achieved in a myeloma *in vitro* system by coating the apoptotic tumoral cells with an antitumoral monoclonal antibody [70]. We will summarize the available data concerning the usage of apoptotic tumor cells for the loading of DCs in the following section.

12.6 Apoptotic Tumor Cells for Loading

The first report that apoptotic cells/bodies are potential sources for antigen presentation stems from Bellone *et al.* [71] who demonstrated that macrophages are capable of phagocytosing lymphoma cells and presenting the resulting antigen to specific T lymphocytes. Thereafter, the efficient MHC I-restricted presentation of viral antigen derived from influenza-infected and thus apoptotic macrophages was demonstrated to be a feature restricted to DCs [72], while macrophages were incapable to stimulate T cells. This work inspired many researchers to search for the cross-presentation of tumoral antigen derived from apoptotic tumor cells. Russo *et al.* [73] were the first to demonstrate the cross-presentation of MAGE-3 antigen derived from DCs transduced with a retroviral vector containing recombinant MAGE-3. The first report on the cross-presentation of tumoral antigen derived from genuine apoptotic tumor cells was done with melanoma cells by our group [69] and we found that in contrast to the efficient cross-presentation of viral antigen, TAAs derived from theses melanoma cells did not reach the antigen levels achieved by peptide pulsing. Yet, the CTL primed with cross-presenting DCs were far more capable of lysing melanoma cells than CTLs generated by peptide-pulsed DCs. The research work following these initial sparks is listed in Tab. 12.1 and summarized in the following section. It is not clear or proven whether the use of apoptotic instead of other tumor cell preparations has advantages. One recent report describes similar *in vivo* capacities for necrotic or apoptotic B16 melanoma cells loaded on DCs [74], while apoptotic cell loading was advantageous over tumor lysates in other, human systems [75, 76]. Whenever clinicians initiate DC-based trials we recommend testing the loading efficacy of the given tumor. In the following subsections we will summarize date available for different tumors. For an overview about clinical studies involving DCs we recommend the following reviews: [61, 77, 78].

12.6.1 Melanoma

Melanoma is an attractive target for immunotherapy given that several TAAs have been identified and an anti-TAA T cell response can be detected readily in melanoma patients. The first clinical trial using DCs was conducted in stage IV melanoma patients using immature monocyte-derived DCs loaded with TAA-derived peptides or tumor lysates delivered intranodally [63]. This trial received criticism for the use of fetal calf serum (FCS) in DC generation, the simultaneous loading of

Tab. 12.1 Apoptotic tumor cells to load DCs

Tumor cells	*Loading method*	*Interaction with DCs*	*Results*	*Reference*
Melanoma	immature DCs loaded, matured with pro-inflammatory cytokines, compared with peptide pulsing	no maturation induced in immature DCs; maturation of loaded DC unproblematic	lysis of HLA-matched melanoma cells by using HLA-miss-matched melanoma cells as antigen donor; killing of tumor cells by cross-primed T cells better than by CTLs, generated with peptide-pulsed DCs	69
	loading with apoptotic melanoma cells only	not determined	shows the generation of melanoma-specific CTLs by cross-presenting CD40 ligand-matured DCs	79
	necrotic or apoptotic uveal melanoma cells	increased proliferation induced by DCs loaded with both methods (no DC maturation induced artificially, negative control missing)	comparing apoptotic and necrotic uveal melanoma cells as loading agents; only DCs loaded with apoptotic melanoma cells elicited an antimelanoma CTL response	76
	apoptotic or necrotic BL6-10 mouse melanoma cells versus TRP2 peptide loading	maturation induced by both cell preparations	efficient clearance of melanoma lung metastasis by melanoma cell-loaded DCs, but not TRP2 peptide-pulsed DCs	80
Solid tumors	apoptotic colorectal tumor cells	maturational effect of apoptotic tumor cells not determined; further maturation by monocyte-conditioned medium	mixed *in vitro* response: two patients with activated T lymphocytes, two patients with suppressed tumor-specific T cell response	86
Prostate cancer	immature DCs loaded with apoptotic PC3 prostate carcinoma cells	not determined	killing of DU145 prostate carcinoma cells	108

Tab. 12.1 (continued)

Tumor cells	*Loading method*	*Interaction with DCs*	*Results*	*Reference*
Hematological malignancies	apoptotic RMA (T cell lymphoma) cell for loading of murine DCs	not determined	immunization with DCs, but not macrophages, pulsed with apoptotic cells primes tumor-specific CTLs and confers protection against a tumor challenge (two out of five mice)	36
	murine bone marrow-derived DCs loaded with apoptotic B lymphocytic leukemia cell line	maturation induced by adding the apoptotic, retrovirally transduced cells	equally prolonged survival of mice when pretumor exposure vaccination with apoptotic leukemia cells or puified hCD4 protein-loaded DCs	94
	C2B8 (anit-CD20 antibody)-treated, apoptotic Daudi lymphoma cells	phagocytosis induces features of maturation	*In vitro* re-stimulation of T cells autologous to the loaded DCs induces efficient anti-Daudi cell antigen-specific CTLs responses	95
RCC	apoptotic RCC cells or RCC lysate	not determined	repetitive re-stimulation yielded tumor-specific $CD8^+$ T cell clones (both loading methods equal)	102
SCCHN	immature DCs loaded with apoptotic PCI-13 or lysates of the same line	no maturation induced by the uptake of apoptotic SCC cells; further maturation induced by pro-inflammatory cytokines	tumor cell-specific responses *ex vivo* with CTLs from healthy donors; apoptotic SCCHN cells superior to lysate from the very sample cells	75
	bone marrow-derived murine DCs loaded with apoptotic KLN205 cells	not determined	only combination with IL-2 resulted in suppressed tumor growth	103

multiple peptides competing for a given HLA molecule onto one batch of DC and the application of immature DCs, but it demonstrated that the route of delivery was safe and led to delayed-type hypersensitivity (DTH) and some clinical responses. Since then other groups have delivered mature monocyte- or $CD34^+$ HPC derived DCs subcutaneously and intravenously. Although clinical responses were partially reported and TAA-specific T cells were induced or boosted by these peptide-pulsed DCs, alternatives to antigen loading are clearly needed to improve this strategy. After the initial demonstration of the feasibility of using apoptotic melanoma cells for the antigen loading of DCs [69] the data were supported by others [79] and expanded to uveal melanoma [76]. With striking similarity, these reports demonstrate that apoptotic melanoma cells do not inhibit the artificial maturation of the loaded DCs necessary to obtain mature DCs. Comparing necrotic uveal melanoma cells in their loading capacity with apoptotic melanoma cells, the latter report suggested the superior loading capacity of apoptotic melanoma cells [76]. Furthermore, in a mouse model both apoptotic and necrotic BL6-10 melanoma cells were more efficient than peptide-pulsed DCs in the eradication of melanoma lung metastasis [80], supporting the *in vitro* data mentioned above [69].

12.6.2
Solid Tumors: Breast Cancer, Ovarian Cancer, Colorectal Cancer and Lung Cancer

The therapeutic support of DC-based immunotherapies has been probed in plenty of solid tumor. Unlike melanoma or renal cell carcinoma, most of these tumors were not characterized as immunogenic, a fact that somewhat limits the expectations of DC-based vaccines. Furthermore, the wide array of TAAs described for melanoma are not present in most of these tumors, a circumstance making the loading with tumor cell preparations desirable. On the other hand, tumor material for DC loading will be readily available as most patients undergo surgery. For breast and ovarian cancer overexpressed self-antigens like the HER-2/*neu* proto-oncogene or the MUC1 gene, a heavily glycosylated protein which is expressed in ductal epithelial cells of the normal breast, have been targeted by DC-based vaccination trials in a pilot study demonstrating the feasibility of DC therapy in heavily pretreated patients [81]. Lysates from ovarian cancer cells have been used successfully to load autologous DCs [82], but no apoptotic breast or ovarian cancer cells have been applied so far.

For colorectal cancer, the carcinoembryonic antigen (CEA) is an overexpressed self-antigen which is partially also found on breast cancer cells as well as colorectal and lung cancer and might provide yet another potential target for immunotherapies. To generate an *in vitro* CTL antitumoral activity, DCs have been successfully loaded with RNA [83], peptides derived from the immunodominant CTL epitope and a recombinant avipoxvirus [84]. Furthermore, initial clinical trials suggest a role for antigen-loaded DCs in the treatment of such malignancies [85]. Apoptotic cancer cells of gastrointestinal cancers were used in one study to activate autologous CTLs *in vitro* [86], but mixed results were reported as for some patients a reduced IFN-γ synthesis of T cells in response to autologous tumor cells was found.

DC-based vaccination studies have been probed for other tumors. For example, lysates derived from a pancreatic tumor cells have been used to load DC [87] for pancreatic cancer and solid tumors in pediatric patients (Wilm's tumor, Ewing's tumor, osteosarcoma, fibrosarcoma were targeted) [65], but so far, no apoptotic cells have been applied to the best of our knowledge.

12.6.3 Prostate Cancer

Antigens that are targeted in prostate cancer are differentiation antigens such as prostatic acid phosphatase, prostate-specific antigen (PSA) and prostate-specific membrane antigen. HLA-A*0201-restricted peptides are available for all these antigens and have been used in a number of DC-based trials [88]. Different routes of DC administration have been explored with prostate cancer patients. Comparing intravenous, intradermal and intralymphatic injection of DCs, comparable results in the induction of antiprostate cancer CTLs were achieved [89]. The potential of apoptotic prostate cancer cells has been probed in one report. Apoptotic prostate cancer lines were used to load DC and demonstrated the generation of CTL specific for yet another prostate cancer cell line after several rounds of re-stimulation [90]. Immature DCs were used for this *in vitro* study and no maturation status of the DCs was determined, thus further evaluation is recommended before clinical trials are set up.

12.6.4 Hematological Malignancies

Multiple myeloma [91] and B cell lymphoma [92] have been targeted with DC-based immunotherapies with limited success so far. Furthermore, unloaded DCs derived from patients with acute myelomatous leukemia (AML) were reported to stimulate a lytic response of autologous lymphocytes against leukemia cells [93]. Loading of DCs with whole leukemia cells might be an attractive option as not many relevant TAAs have been described for hematological malignancies. Some reports support the utilization of apoptotic leukemia cells for the loading. Although apoptotic RMA cells (a mouse T cell lymphoma) were 20-fold less efficient in the induction of antitumoral cellular immunity than viable, non-proliferating cells of the same lineage, DC pulsed with apoptotic RMA cells were rather efficient in the induction of an antitumoral immunity (protection of two out of five mice protected from subsequent challenge with RMA cells) [36]. In another mouse model, the loading efficacy of B cell lymphoma cells retrovirally transfected with hCD4 was probed and compared with the efficacy of hCD4 protein. Comparable anti-B cell lymphoma T cell activity and antitumoral protections were reported [94]. *In vitro,* treatment of human Daudi lymphoma cells with the C2B8 anti-CD20 antibody induced apoptotic cell death followed by phagocytosis of DCs [95]. After maturation, the DCs were capable of cross-priming CTLs against Daudi cell-associated antigens. This observation might indicate that the beneficial effect of an anti-CD20 antibody treatment (Rituximab), as applied for B cell non-Hodgkin's lymphomas, might not only be related to tumor cell killing but also to the generation of an antilymphoma immunity after cross-presentation of lymphoma antigens. In one recent study, the efficacy of apoptotic myeloma cells as loading agents for DC was demonstrated [70]. Interestingly, this study demonstrated that coating of the apoptotic tumor cells with an antitumoral antibody increased the cross-presentation of tumoral antigen.

12.6.5 Neurological Tumors

The feasibility of DC-based vaccination as additional therapy for patients with tumor material-loaded DCs has been proven for different tumors (reviewed in [96]). DCs fused with glioma cells [97] or DCs loaded with astrocytoma cell preparations [98] were used *in vivo*. Furthermore, nine patients with glioblastoma multiforme or anaplastic astrocytoma were treated with immature monocyte-derived DCs pulsed with peptides eluted from autologous brain tumor cells. While some tumor-specific cytotoxicity was observed, two subjects had evidence of $CD8^+$ T cell infiltration in areas of the tumor [99]. Thus, whenever glioblastoma cells are recovered by surgical procedures they might be used to load autologous DC. Yet, apoptotic glioma, glioblastoma or astrocytoma cells have not been used so far to the best of our knowledge.

12.6.6 Renal Cell Carcinoma (RCC)

Based on the occurrence of spontaneous remissions and the partial efficacy of non-specific immunoactivators such as IL-2, RCC is considered to be amenable to treatment with immunotherapy. Renal cell tumor lysates have been used in two trials and some cellular responses to tumor lysates were observed *in vitro* following the vaccinations [100, 101]. The cross-presentation efficacy of lysates generated from RCC cells and apoptotic RCC cells was compared in one report [102]. Several rounds of re-stimulation yielded comparable levels of antitumoral $CD8^+$ T cell clones. Thus, apoptotic RCC cells might be an option for the loading of autologous DCs for vaccination strategies or the *ex vivo* generation of anti-RCC T cells before their adoptive transfer.

12.6.7 Squamous Cell Carcinoma of the Head and Neck (SCCHN)

SCCHN are considered to be poorly immunogenic and immunosuppressive tumors. Restoration of a patient immunity against this tumor might be an option to support other antitumoral strategies. With only a few well-characterized TAAs (including CASP-8 and SART-1), whole tumor cell preparations have to be considered as antigen-loading agents for DC-based vaccination strategies. In a murine model, apoptotic cells of the poorly immunogenic SCC line KLN 205 were used to load DC *ex vivo*. Tumor growth of the same cell line was suppressed when intraperitoneal IL-2 was added [103]. Furthermore, splenic T cells of the treated mice produced larger amounts of IFN-γ when stimulated with KLN 205 tumor cells. In a human *in vitro* system, the activation of anti-SCCHN T cells was demonstrated when apoptotic PCI-13 SCCHN cells were used, but not when lysate of the same line was added to immature DC [75]. Thus, apoptotic SCCHN cells have to be considered as loading agent for DC-based vaccination strategies for these kinds of tumors.

12.7 Concluding Remarks

Several areas of investigation will be required to optimize DC-based immunotherapy. It will be important to optimize DCs preparations, in terms of subsets and maturity. Furthermore, antigen loading and monitoring of the resulting MHC-peptide complex formation will be as important as the determination of the route and timing of administration.

It is our strong opinion that DCs need to be given as mature cells given recent data showing that immature DCs induce immunoregulatory T cells *in vivo* [50] and fail to prime T cells to TAAs [104]. DC immunizations have been delivered intradermally, subcutaneously, intravenously, intranodally or into lymphatic vessels. Since optimal presentation of antigens in secondary lymph nodes is crucial for both the initiation and maintenance of T cell immunity, the route of administration must be carefully considered as this will affect DC migration to lymphoid tissue and thus most likely their immunogenicity.

Loading of DCs with apoptotic tumor cells for immunotherapeutic approaches seems to combine the efficient antigen expression, probably physiologically designed to support peripheral T cell tolerance, with the *ex vivo* maturation of DCs leading to an optimal antigen presentation and immunogenicity. When using antigen-loaded DCs, side effects have been mostly low grade, ranging from local reactions, fevers, myalgias, development of autoantibodies but without autoimmune disease (antinuclear antibodies, anti-DNA, anti-thyroid-stimulating hormone) and vitiligo to transfusion-like reactions with rigors. Anaphylactic reactions in anecdotal reports have been linked to the use of FCS in DC culture systems [105].

Loading of DCs with apoptotic tumor cells has yet to be directly compared *in vitro* and *in vivo* to other approaches that allow exploitation of the whole antigenic spectrum of a tumor such as fusion [106] and RNA transfection [107]. The latter approach is particularly exciting and attractive as RNA from minimal tumor samples can be amplified to provide a seamless unlimited supply of individualized antigen.

Yet, it has to be kept in mind that therapeutic protocols based on DCs loaded with whole tumor cell preparations inherit a certain risk of breaking the peripheral tolerance against presumably important autologous antigens presented with similar efficacy as the desired tumoral antigens. So far, no such side effects have been observed in multiple clinical trials. This fact might be due to the sustained and permanent *in vivo* tolerization of T cells against such antigens by powerful mechanisms [62].

12.8 References

1 R.M. Steinman, *Annu Rev Immunol* **1991**, 9, 271–96.

2 F. Sallusto, A. Lanzavecchia, *J Exp Med* **1994**, 179, 1109–18.

3 D. Maurer, E. Fiebiger, B. Reininger, C. Ebner, P. Petzelbauer, G.P. Shi, H.A. Chapman, G. Stingl, *J Immunol* **1998**, 161, 2731–9.

4 F. Sallusto, M. Cella, C. Danieli, A. Lanzavecchia, *J Exp Med* **1995**, 182, 389–400.

5 W. Jiang, W. J. Swiggard, C. Heufler, M. Peng, A. Mirza, R.M. Steinman, M.C. Nussenzweig, *Nature* **1995**, 375, 151–5.

6 T.B. Geijtenbeek, D.S. Kwon, R. Torensma, S.J. van Vliet, G.C. van Duijnhoven, J. Middel, I.L. Cornelissen, H.S. Nottet, V.N. Kewal-Ramani, D. R. Littman, C.G. Figdor, Y. van Kooyk, *Cell* **2000**, 100, 587–97.

7 B. Pulendran, K. Palucka, J. Banchereau, *Science* **2001**, 293, 253–6.

8 J.F. Kerr, C.M. Winterford, B.V. Harmon, *Cancer* **1994**, 73, 2013–26.

9 V.A. Fadok, D.L. Bratton, D.M. Rose, A. Pearson, R.A. Ezekewitz, P.M. Henson, *Nature* **2000**, 405, 85–90.

10 P.R. Hoffmann, A.M. deCathelineau, C.A. Ogden, Y. Leverrier, D.L. Bratton, D.L. Daleke, A.J. Ridley, V.A. Fadok, P.M. Henson, *J Cell Biol* **2001**, 155, 649–60.

11 M.L. Albert, S.F. Pearce, L.M. Francisco, B. Sauter, P. Roy, R.L. Silverstein, N. Bhardwaj, *J Exp Med* **1998**, 188, 1359–68.

12 B.C. Urban, N. Willcox, D.J. Roberts, *Proc Natl Acad Sci USA* **2001**, 98, 8750–5.

13 M.L. Albert, J.I. Kim, R.B. Birge, *Nat Cell Biol* **2000**, 2, 899–905.

14 P. Rovere, *J Autoimmun* **1998**, 11, 403–11.

15 P. Rovere, G. Peri, F. Fazzini, B. Bottazzi, A. Doni, A. Bondanza, V.S. Zimmermann, C. Garlanda, U. Fascio, M. G. Sabbadini, C. Rugarli, A. Mantovani, A.A. Manfredi, *Blood* **2000**, 96, 4300–6.

16 F.P. Huang, N. Platt, M. Wykes, J.R. Major, T.J. Powell, C.D. Jenkins, G.G. MacPherson, *J Exp Med* **2000**, 191, 435–44.

17 A. Rodriguez, A. Regnault, M. Kleijmeer, P. Ricciardi-Castagnoli, S. Amigorena, *Nat Cell Biol* **1999**, 1, 362–8.

18 J.W. Yewdell, C.C. Norbury, J.R. Bennink, *Adv Immunol* **1999**, 73, 1–77.

19 A.Y. Huang, A.T. Bruce, D.M. Pardoll, H. I. Levitsky, *Immunity* **1996**, 4, 349–55.

20 H. Feng, Y. Zeng, L. Whitesell, E. Katsanis, *Blood* **2001**, 97, 3505–12.

21 S. Basu, R.J. Binder, R. Suto, K.M. Anderson, P.K. Srivastava, *Int Immunol* **2000**, 12, 1539–46.

22 R.J. Binder, D.K. Han, P.K. Srivastava, *Nat Immunol* **2000**, 1, 151–5.

23 S. Todryk, A.A. Melcher, N. Hardwick, E. Linardakis, A. Bateman, M.P. Colombo, A. Stoppacciaro, R.G. Vile, *J Immunol* **1999**, 163, 1398–408.

24 H. Singh-Jasuja, H.U. Scherer, N. Hilf, D. Arnold-Schild, H.G. Rammensee, R.E. Toes, H. Schild, *Eur J Immunol* **2000**, 30, 2211–5.

25 P.K. Srivastava, A. Menoret, S. Basu, R.J. Binder, K.L. McQuade, *Immunity* **1998**, 8, 657–65.

26 J. Banchereau, F. Briere, C. Caux, J. Davoust, S. Lebecque, Y.J. Liu, B. Pulendran, K. Palucka, *Annu Rev Immunol* **2000**, 18, 767–811.

27 S.J. Turley, K. Inaba, W.S. Garrett, M. Ebersold, J. Unternaehrer, R.M. Steinman, I. Mellman, *Science* **2000**, 288, 522–7.
28 N. Bhardwaj, J.W. Young, A.J. Nisanian, J. Baggers, R.M. Steinman, *J Exp Med* **1993**, 178, 633–42.
29 J.S. Kuniyoshi, C.J. Kuniyoshi, A.M. Lim, F.Y. Wang, E.R. Bade, R. Lau, E.K. Thomas, J.S. Weber, *Cell Immunol* **1999**, 193, 48–58.
30 M.J. Bevan, *J Exp Med* **1976**, 143, 1283–8.
31 R.D. Gordon, E. Simpson, L.E. Samelson, *J Exp Med* **1975**, 142, 1108–20.
32 L.R. Gooding, C.B. Edwards, *J Immunol* **1980**, 124, 1258–62.
33 A.Y. Huang, P. Golumbek, M. Ahmadzadeh, E. Jaffee, D. Pardoll, H. Levitsky, *Science* **1994**, 264, 961–5.
34 M. Larsson, J.F. Fonteneau, N. Bhardwaj, *Trends Immunol* **2001**, 22, 141–8.
35 L.J. Sigal, S. Crotty, R. Andino, K.L. Rock, *Nature* **1999**, 398, 77–80.
36 A. Ronchetti, P. Rovere, G. Iezzi, G. Galati, S. Heltai, M.P. Protti, M.P. Garancini, A.A. Manfredi, C. Rugarli, M. Bellone, *J Immunol* **1999**, 163, 130–6.
37 S.D. Reid, G. Penna, L. Adorini, *Curr Opin Immunol* **2000**, 12, 114–21.
38 A. Lanzavecchia, F. Sallusto, *Cell* **2001**, 106, 263–6.
39 C. Kurts, H. Kosaka, F. R. Carbone, J.F. Miller, W.R. Heath, *J Exp Med* **1997**, 186, 239–45.
40 H. Jonuleit, E. Schmitt, G. Schuler, J. Knop, A.H. Enk, *J Exp Med* **2000**, 192, 1213–22.
41 W.R. Heath, C. Kurts, J.F. Miller, F.R. Carbone, *J Exp Med* **1998**, 187, 1549–53.
42 R.M. Steinman, S. Turley, I. Mellman, K. Inaba, *J Exp Med* **2000**, 191, 411–6.
43 M.L. Albert, M. Jegathesan, R.B. Darnell, *Nat Immunol* **2001**, 2, 1010–7.
44 A. Lanzavecchia, *Nature* **1998**, 393, 413–4.
45 M. Herrmann, R.E. Voll, J.R. Kalden, *Immunol Today* **2000**, 21, 424–6.
46 M.L. Albert, *Nat Med* **1998**, 4, 1321–4.
47 I. Herr, D. Wilhelm, T. Bohler, P. Angel, K.M. Debatin, *EMBO J* **1997**, 16, 6200–8.
48 E.A. Reap, K. Roof, K. Maynor, M. Borrero, J. Booker, P.L. Cohen, *Proc Natl Acad Sci USA* **1997**, 94, 5750–5.
49 S.M. Kiertscher, J. Luo, S.M. Dubinett, M.D. Roth, *J Immunol* **2000**, 164, 1269–76.
50 M.V. Dhodapkar, R.M. Steinman, J. Krasovsky, C. Munz, N. Bhardwaj, *J Exp Med* **2001**, 193, 233–8.
51 B. Sauter, M.L. Albert, L. Francisco, M. Larsson, S. Somersan, N. Bhardwaj, *J Exp Med* **2000**, 191, 423–34.
52 M. Hirao, N. Onai, K. Hiroishi, S.C. Watkins, K. Matsushima, P.D. Robbins, M.T. Lotze, H. Tahara, *Cancer Res* **2000**, 60, 2209–17.
53 Z. Chen, T. Moyana, A. Saxena, R. Warrington, Z. Jia, J. Xiang, *Int J Cancer* **2001**, 93, 539–48.
54 T.K. Hoffmann, N. Meidenbauer, J. Muller-Berghaus, W.J. Storkus, T.L. Whiteside, *J Immunother* **2001**, 24, 162–71.
55 K.A. Candido, K. Shimizu, J.C. McLaughlin, R. Kunkel, J.A. Fuller, B.G. Redman, E.K. Thomas, B.J. Nickoloff, J.J. Mule, *Cancer Res* **2001**, 61, 228–36.
56 A.J. Zajac, J.N. Blattman, K. Murali-Krishna, D.J. Sourdive, M. Suresh, J.D. Altman, R. Ahmed, *J Exp Med* **1998**, 188, 2205–13.
57 R.E. Toes, F. Ossendorp, R. Offringa, C.J. Melief, *J Exp Med* **1999**, 189, 753–6.
58 J.S. Serody, E.J. Collins, R.M. Tisch, J.J. Kuhns, J.A. Frelinger, *J Immunol* **2000**, 164, 4961–7.
59 M.V. Dhodapkar, R.M. Steinman, M. Sapp, H. Desai, C. Fossella, J. Krasovsky, S.M. Donahoe, P.R. Dunbar, V. Cerundolo, D.F. Nixon, N. Bhardwaj, *J Clin Invest* **1999**, 104, 173–80.
60 B. Schuler-Thurner, D. Dieckmann, P. Keikavoussi, A. Bender, C. Maczek, H. Jonuleit, C. Roder, I. Haendle, W. Leisgang, R. Dunbar, V. Cerundolo, D.P. von Den, J. Knop, E.B. Brocker, A. Enk, E. Kampgen, G. Schuler, *J Immunol* **2000**, 165, 3492–6.
61 L. Jenne, N. Bhardwaj, Perspectives of DC based immunotherapies. In: V.T. DeVita, S. Hellman, eds. *Principles and Practice of Oncology*, 6th edn. Baltimore, MD: Lippincott Williams & Wilkins **2001**, 1–15.
62 E. Gilboa, *Nat Immunol* **2001**, 2, 789–92.

63 F.O. Nestle, S. Alijagic, M. Gilliet, Y. Sun, S. Grabbe, R. Dummer, G. Burg, D. Schadendorf, *Nat Med* **1998**, 4, 328–32.

64 M. Thurnher, C. Rieser, L. Holtl, C. Papesh, R. Ramoner, G. Bartsch, *Urol Int* **1998**, 61, 67–71.

65 J. Geiger, R. Hutchinson, L. Hohenkirk, E. McKenna, A. Chang, J. Mule, *Lancet* **2000**, 356, 1163–5.

66 G. Ferlazzo, C. Semino, G.M. Spaggiari, M. Meta, M. C. Mingari, G. Melioli, *Int Immunol* **2000**, 12, 1741–7.

67 M. Li, G.M. Davey, R.M. Sutherland, C. Kurts, A.M. Lew, C. Hirst, F.R. Carbone, W.R. Heath, *J Immunol* **2001**, 166, 6099–103.

68 W. Herr, E. Ranieri, W. Olson, H. Zarour, L. Gesualdo, W.J. Storkus, *Blood* **2000**, 96, 1857–64.

69 L. Jenne, J.F. Arrighi, H. Jonuleit, J.H. Saurat, C. Hauser, *Cancer Res* **2000**, 60, 4446–52.

70 K.M. Dhodapkar, J. Krasovsky, B. Williamson, M.V. Dhodapkar, *J Exp Med* **2002**, 195, 125–33.

71 M. Bellone, G. Iezzi, P. Rovere, G. Galati, A. Ronchetti, M.P. Protti, J. Davoust, C. Rugarli, A.A. Manfredi, *J Immunol* **1997**, 159, 5391–9.

72 M.L. Albert, B. Sauter, N. Bhardwaj, *Nature* **1998**, 392, 86–9.

73 V. Russo, S. Tanzarella, P. Dalerba, D. Rigatti, P. Rovere, A. Villa, C. Bordignon, C. Traversari, *Proc Natl Acad Sci USA* **2000**, 97, 2185–90.

74 Y. Kotera, K. Shimizu, J.J. Mule, *Cancer Res* **2001**, 61, 8105–9.

75 T.K. Hoffmann, N. Meidenbauer, G. Dworacki, H. Kanaya, T.L. Whiteside, *Cancer Res* **2000**, 60, 3542–9.

76 M. Shaif-Muthana, C. McIntyre, K. Sisley, I. Rennie, A. Murray, *Cancer Res* **2000**, 60, 6441–7.

77 J. Banchereau, B. Schuler-Thurner, A.K. Palucka, G. Schuler, *Cell* **2001**, 106, 271–4.

78 R.M. Steinman, M. Dhodapkar, *Int J Cancer* **2001**, 94, 459–73.

79 F. Berard, P. Blanco, J. Davoust, E.M. Neidhart-Berard, M. Nouri-Shirazi, N. Taquet, D. Rimoldi, J.C. Cerottini, J. Banchereau, A. K. Palucka, *J Exp Med* **2000**, 192, 1535–44.

80 Z. Chen, T. Moyana, A. Saxena, R. Warrington, Z. Jia, J. Xiang, *Int J Cancer* **2001**, 93, 539–48.

81 P. Brossart, S. Wirths, G. Stuhler, V. L. Reichardt, L. Kanz, W. Brugger, *Blood* **2000**, 96, 3102–8.

82 A.D. Santin, P.L. Hermonat, A. Ravaggi, S. Bellone, S. Pecorelli, M.J. Cannon, G.P. Parham, *Am J Obstet Gynecol* **2000**, 183, 601–9.

83 S.K. Nair, D. Boczkowski, M. Morse, R.I. Cumming, H.K. Lyerly, E. Gilboa, *Nat Biotechnol* **1998**, 16, 364–9.

84 K.Y. Tsang, M. Zhu, J. Even, J. Gulley, P. Arlen, J. Schlom, *Cancer Res* **2001**, 61, 7568–76.

85 M.A. Morse, Y. Deng, D. Coleman, S. Hull, E. Kitrell-Fisher, S. Nair, J. Schlom, M.E. Ryback, H.K. Lyerly, *Clin Cancer Res* **1999**, 5, 1331–8.

86 A. Galetto, M. Contarini, A. Sapino, P. Cassoni, E. Consalvo, S. Forno, C. Pezzi, V. Barnaba, A. Mussa, L. Matera, *J Surg Res* **2001**, 100, 32–8.

87 M. Schnurr, P. Galambos, C. Scholz, F. Then, M. Dauer, S. Endres, A. Eigler, *Cancer Res* **2001**, 61, 6445–50.

88 B.A. Tjoa, G.P. Murphy, *Immunol Lett* **2000**, 74, 87–93.

89 L. Fong, D. Brockstedt, C. Benike, L. Wu, E.G. Engleman, *J Immunol* **2001**, 166, 4254–9.

90 M. Nouri-Shirazi, J. Banchereau, D. Bell, S. Burkeholder, E.T. Kraus, J. Davoust, K.A. Palucka, *J Immunol* **2000**, 165, 3797–803.

91 V.L. Reichardt, C.Y. Okada, A. Liso, C.J. Benike, K.E. Stockerl-Goldstein, E.G. Engleman, K.G. Blume, R. Levy, *Blood* **1999**, 93, 2411–9.

92 F.J. Hsu, *Blood* **1997**, 89, 3129–35.

93 B.A. Choudhury, J.C. Liang, E.K. Thomas, L. Flores-Romo, Q.S. Xie, K. Agusala, S. Sutaria, I. Sinha, R.E. Champlin, D.F. Claxton, *Blood* **1999**, 93, 780–6.

94 S. Paczesny, S. Beranger, J.L. Salzmann, D. Klatzmann, B.M. Colombo, *Cancer Res* **2001**, 61, 2386–9.

95 N. Selenko, O. Maidic, S. Draxier, A. Berer, U. Jager, W. Knapp, J. Stockl, *Leukemia* **2001**, 15, 1619–26.

96 A. Soling, N.G. Rainov, *Mol Med* **2001**, 7, 659–67.

97 T. Kikuchi, Y. Akasaki, M. Irie, S. Homma, T. Abe, T. Ohno, *Cancer Immunol Immunother* **2001**, 50, 337–44.
98 S. Yoshida, K. Morii, M. Watanabe, T. Saito, K. Yamamoto, R. Tanaka, *Cancer Immunol Immunother* **2001**, 50, 321–7.
99 J.S. Yu, C.J. Wheeler, P.M. Zeltzer, H. Ying, D.N. Finger, P.K. Lee, W.H. Yong, F. Incardona, R.C. Thompson, M.S. Riedinger, W. Zhang, R.M. Prins, K.L. Black, *Cancer Res* **2001**, 61, 842–7.
100 L. Holtl, C. Rieser, C. Papesh, R. Ramoner, G. Bartsch, M. Thurnher, *Lancet* **1998**, 352, 1358.
101 C. Rieser, R. Ramoner, L. Holtl, H. Rogatsch, C. Papesh, A. Stenzl, G. Bartsch, M. Thurnher, *Urol Int* **1999**, 63, 151–9.
102 T. Kurokawa, M. Oelke, A. Mackensen, *Int J Cancer* **2001**, 91, 749–56.
103 Y.I. Son, R.B. Mailliard, S.C. Watkins, M.T. Lotze, *Laryngoscope* **2001**, 111, 1472–8.
104 H. Jonuleit, A. Giesecke-Tuettenberg, T. Tuting, B. Thurner-Schuler, T.B. Stuge, L. Paragnik, A. Kandemir, P.P. Lee, G. Schuler, J. Knop, A.H. Enk, *Int J Cancer* **2001**, 93, 243–51.
105 A. Mackensen, R. Drager, M. Schlesier, R. Mertelsmann, A. Lindemann, *Cancer Immunol Immunother* **2000**, 49, 152–6.
106 J. Gong, D. Chen, M. Kashiwaba, D. Kufe, *Nat Med* **1997**, 3, 558–61.
107 D. Boczkowski, S.K. Nair, D. Snyder, E. Gilboa, *J Exp Med* **1996**, 184, 465–72.
108 M. Nouri-Shirazi, J. Banchereau, D. Bell, S. Burkeholder, E.T. Kraus, J. Davoust, K.A. Palucka, *J Immunol* **2000**, 165, 3797–803.

13
The Immune Response against Apoptotic Cells

Angelo A. Manfredi

13.1
Introduction: Immune Appraisal of Dying Cells

T cells normally ignore antigens expressed in peripheral tissues [1]. Kurts *et al.* demonstrated in a seminal paper that this rule applies to living tissues only, i.e. those tissues in which few scattered cells die at any given time. Adjacent phagocytes swiftly engulf cell corpses, thus sequestering their antigens from immune recognition. The inoculation of exogenous cytotoxic cells specific for peripheral antigens causes overwhelming apoptosis, an event that jeopardizes the ability of local scavenger phagocytes to clear cell corpses [2]. As a consequence, T cells specific for poorly expressed antigens actively proliferate in draining lymph nodes. Zhang *et al.* confirmed that the synchronized death of β cells in pancreatic islets induced by streptozotocin determines the priming of diabetogenic $CD8^+$ T cells, specific for β cell autoantigens [3, 4].

The inoculation of apoptotic cells, which mimics massive death of tissue cells, induces autoantibodies recognizing preferentially nuclear antigens or anionic phospholipids [5]. Autoantibodies are normally transient – they can be detected only in a narrow chronological window after the administration of dead cells. In some individuals, however, they persist and clinical manifestations develop. The causes are largely unknown, but genetic influences are implicated. For example, in NZW/NZW F_1 mice, the injection of dying thymocytes causes the development of a vast array of autoantibodies, including anti-double-stranded DNA, anti-β_2-glycoprotein I and antinucleosome antibodies. Injected animals undergo a dramatically accelerated lupus syndrome, with early death due to renal involvement (Manfredi *et al.*, unpublished). Conversely, mice that chronically fail to clear apoptotic cells almost invariably develop autoimmunity [6–9]. The injection of dying cells or their products in these 'clearance-defective' animals further accelerates and exacerbates autoimmunity [7, 9].

These findings support the concept that defective or deregulated disposal of cell corpses underlies systemic autoimmunity [10, 11] (see also other chapters in this book). The observation that autoantibodies preferentially recognize cell constituents that are preferentially cleaved, redistributed and clustered or otherwise post-translationally modified during apoptosis further strengthens this contention [12–

14]. Cells killed by cytotoxic T cell-assisted apoptosis are preferential sources of autoantigens [15]. This result is intriguing, since cytotoxic T cells are crucial for eradication of intracellular infection. Therefore, apoptosis of infected cells during infections could represent a trigger to autoimmunity (see also later).

Cells undergoing post-apoptotic necrosis are also apparently more immunogenic [16]. A defective clearance of apoptotic cells cause accumulation of uncleared dying cells, which develop features of advanced apoptosis *in vivo* [2, 17]. Autoimmune systemic lupus erythematosus (SLE) patients fail to properly clear autologous apoptotic cells [18–20] and products released from uncleared apoptotic cells accumulate in tissues [21–27]. The defective clearance of autoantigens and possibly of endogenous adjuvant moieties (see below) contributes to the origin of the vast array of autoantibodies recognizing cell death-associated antigens and to the tissue damage (in particular of kidney involvement) [26].

Why uncleared cell corpses should cause autoimmunity is poorly understood. Elicited autoantibodies are usually affinity-matured IgG. Isotype switching and affinity maturation of immunoglobulins require active help from T cells [28]. Activated helper T cells, in association with native antigen on follicular dendritic cells (DCs), indeed select in the germinal centers B cells that express high-affinity antigen receptors. Eventually, this leads to affinity maturation and to the generation of memory B cells [29]. In the case of autoantibodies preferentially recognizing dying cell epitopes, antigen-presenting phagocytes must have access to dying cells at the periphery, internalize and process them, and migrate in draining lymph nodes. Once there, phagocytes present their antigens either directly or indirectly [30] and activate autoreactive T cells, which in turn control the clonal expansion and the affinity maturation of autoreactive B cells in germinal centers.

Evidence accumulated in the last years implicating antigen-presenting phagocytes in the disposal of dying cells [30–38]. DCs that constitutively transport apoptotic intestinal epithelial cells to mesenteric lymph nodes *in vivo* have been identified [39]. This event is observed with relative ease in healthy rodents. Therefore, uptake of dying cells by unconventional phagocytes, which are endowed with the ability to professionally present antigens, occurs physiologically. DCs are the most potent antigen-presenting cells (APCs) [40]. They activate, after active processing of apoptotic cells, autoreactive T cells in the lymph node. The outcome of antigen presentation varies, ranging from establishment and maintenance of peripheral tolerance towards intracellular antigens to induction of productive immune responses [40–43]. Of course, productive immunity is, in most cases, the outcome during infections.

13.2
Death of the Host, Death of the Pathogen

Intracellular pathogens that elicit protective immune responses activate cytotoxic T lymphocytes (CTLs). Neither the pathogen nor infected cells in peripheral tissues are *per se* able to prime $CD8^+$ CTLs. Professional APCs derived from bone marrow precursors are required [44, 45]. They take up microbial antigens in infected tissues, process them for presentation in MHC class I- and class II-associated epitopes, and migrate to lymph nodes where they meet and activate T lymphocytes.

APCs that efficiently prime MHC class I-restricted $CD8^+$ T cells specific for exogenous antigens express a functional transporter associated with antigen processing (TAP) system. In the cytosol of infected cells, proteasomes generate antigenic peptides (or their precursors). TAP transports cytosolic peptides into the endoplasmic reticulum, where they associate with nascent MHC class I molecules [46]. Recent studies identified myeloid $CD8^+$ antigen-presenting DCs as the bone marrow-derived DCs implicated in MHC class I-restricted presentation of exogenous antigens *in vivo* [47]. These data suggest that the presence of the pathogen in the cytosol of infected APCs and the ensuing generation of antigenic peptides upon processing are limiting steps for the initiation of antimicrobial immunity.

One could therefore speculate that pathogens, to escape immune recognition and generation of protective immune responses, should simply avoid infecting DCs. However, this is not the case. DCs productively activate ('prime') pathogen-specific $CD8^+$ T cells in a TAP-dependent manner, even when they are not infected *in vivo* [44]. This implies, however, that uninfected DC somehow internalize and process the pathogenic antigens. Infected cells often die. They may represent an important source of antigens for $CD8^+$ T cell 'cross-priming' [48].

DC efficiently internalize dying infected cells and cross-present their antigens to MHC class I- and class II-restricted T cells specific for microbial antigens of the pathogen. Suitable phagocytic substrates are dying cells infected by influenza virus [49], cytomegalovirus [50, 51], canarypox virus [52, 53], Epstein-Barr virus [54], vaccinia virus [55] and *Salmonella typhimurium* [56]. The process is also independent of the lineage of infected cells. Immature DCs internalize and process infected dying monocytes [32, 49], macrophages [56], fibroblasts [50], transformed infected B cells [54] or even dying infected DC themselves [52, 53].

Larsson *et al.* provided a semiquantitative estimate of the efficiency by which a single viral epitope contained in infected dying cells is cross-presented. At a dead cell:mature DC 1:1 ratio, T cell activation was comparable to that induced by 10 nM of the synthetic peptide sequence [55]. Dead cells contain a plethora of antigens. This result, which is probably indicative of the efficiency by which any given processed epitope is presented, indicates that the cross-presentation is indeed strikingly efficient.

13.3
Infection, Cell Death and DC Maturation

DCs process antigens, present their fragments bound to MHC molecules and initiate immunity. However, the uptake of antigens and the initiation of productive immune responses are distinct events [40]. DCs at different stages of maturation carry out these functions. Mature DCs are better at inducing immunity. They express stable MHC–peptide complexes, high levels of membrane receptors involved in T cell binding, co-stimulation and activation. Mature DCs synthesize cytokines that promote T cell proliferation and differentiation, and undergo a tightly regulated reorganization of chemokines and chemokine receptor expression, which increases the movement into lymphoid organs and lymphatic vessels [57].

In contrast, immature DCs preferentially home to peripheral tissues, where they capture soluble and particulate antigens via dedicated receptors, and concentrate them into intracellular processing compartments. Immature DC sense pathogens via Toll-like receptors (TLRs) [58, 59]. Activation of TLRs upon recognition of microbial structures determines DC maturation. This event is a crucial step in the initiation of protective immune responses against the 'infectious non-self' [60].

Antigens derived from the processing of internalized dying cells initiate immune responses (see above). This implies that DCs received the signal to mature and migrate to lymphoid secondary organs. However, when DC phagocytose dying infected cells, direct recognition of pathogen molecules via TLRs does not take place. Other signals are possibly involved.

(1) *Cytokines.* Bystander cells belonging to the innate immune system secrete, upon pathogen recognition, a variety of pro-inflammatory cytokines, including tumor necrosis factor-α and interleukin (IL)-1β. These signals in turn can activate DCs that phagocytosed infected dying cells [61].

(2) *Heat shock proteins* (HSPs). Infection *per se* is a stressful event and stressed cells express HSPs. Purified HSPs trigger DC maturation via interaction with diverse membrane receptors (CD91 for HSP70, etc. [62]). Uningested infected cells possibly release soluble HSPs, which collaborate with pro-inflammatory cytokines in promoting DC maturation [63, 64]. Of interest, even HSPs contained in dying cells influence the outcome of the cross-presentation after uptake and processing by DCs, facilitating active cross-priming of CD4 and CD8 T cells [65]. HSPs also signal 'stressful' cell death to macrophages, limiting the release of immunosuppressive factors while inducing soluble factors that increase tumor immunogenicity [66].

(3) *Endogenous adjuvants.* Infections often cause the synchronized death of tissue cells. Uningested dying cells possibly release, besides HSPs, other activators of the antigen-presenting pathways. Most are still uncharacterized [67–69]. Double-stranded DNA fragments [70] and proteins loosely associated to the linker DNA (Rovere-Querini *et al.*, unpublished) are involved. Finally, natural anticoagulants, which interfere with the phosphatidylserine-dependent immunosuppressive clearance (see [71] and elsewhere in this book), interfere with the *in situ* immunogenicity of dying cells (Bondanza *et al.*, unpublished).

(4) *Adjuvant effects of uncleared cells.* Dying cells *per se* are poorly immunogenic. However, high numbers (5–10 per each DC) of apoptotic cells promote the autocrine production of pro-inflammatory cytokines and the maturation of DCs [34]. DCs fail to clear all dying cells in these systems. The latter undergo post-apoptotic necrosis and late plasma membrane failure [35], with release of endogenous adjuvants. Conversely, caspase inhibitors, which prevent apoptotic but not necrotic death of infected cells, block DC maturation [52] and the generation/release of intracellular adjuvants [67].

(5) *Adjuvant signals expressed by dying cells.* The nature of the dying cell, and in particular the expression on the membrane of signals that elicit the maturation of the phagocytosing DCs, like the CD40 ligand, possibly influence the final outcome of cross-presentation [37]. Accordingly, the local availability of CD40 ligand promoted the cross-priming of antineoplastic $CD4^+$ T cells as a consequence of apoptotic tumor cell processing [72].

All together, these results suggest that the death via apoptosis of infected cells may represent a privileged event, providing the sentinels of the immune system, the DCs, with both the pathogen antigens and the signal to initiate immune responses. In support of this hypothesis, most successful intracellular pathogens, and viruses in particular, developed complex and elegant strategies to prevent apoptosis induction in the infected cell [73]. It is tempting to speculate that the highly regulated, finely tuned cross-presentation pathway represents an evolutionary conserved template for acquired immunity against pathogens [48, 74, 75].

13.4 Infections, Cross-Presentation and Autoimmunity

The studies summarized above indicate that cross-presentation of antigens derived from the processing of dying infected cells contributes to the initiation of protective antimicrobial immune responses. However, infected dying cells are a reservoir of self-antigens. Endogenous and microbial antigens are likely to be presented in a similar fashion. Central and peripheral tolerance contributes to limit autoimmunity. However, it is not surprising that both acute and chronically infected patients develop autoantibodies [76]. Furthermore, different lines of evidence implicate infectious events as triggers or at least as facilitators in the development of autoimmune diseases [77, 78].

The autoimmune responses that infected patients develop are usually transient. Furthermore, they are likely to target cryptic epitopes derived from the processing of intracellular antigens [37]. Therefore, they do not necessarily cause overt clinical features. For example, Propato *et al.* recently showed that apoptotic cell death *in vivo* is related to the cross priming of vinculin-specific CTLs in HIV patients [37]. In other cases, the autoimmune response enforces a vicious self-maintaining circle. For example, the injection of apoptotic neutrophils causes the development of antiproteinase 3 (PR3) antibodies, a hallmark of systemic small vessels vasculitis [79]. In turn, anti-PR3 antibodies have been convincingly implicated in the maintenance of chronic inflammation and in vascular damage of vasculitis patients [80–82]. A similar role has convincingly been demonstrated for antiphospholipid antibodies [83–85], anti-Ro/SSA antibodies [86, 87] and antinucleosome antibodies [25] (for a detailed revision, see Chapter 5).

Opsonizing antibodies facilitate the access of internalized apoptotic cells into MHC class II compartments, the MHC class II-restricted presentation of relevant epitopes and the activation of $CD4^+$ helper T cells [22, 23, 85]. Cross-presentation by mature DCs is not sufficient to initiate immunity. In the absence of proper activation of T helper cells this event possibly determines tolerance towards antigens expressed by dying cells [43]. I propose that the ability of opsonizing autoantibodies, once established to promote the activation of T helper cells, is a crucial event in the shift from transient low-affinity autoimmune responses to chronic autoimmune diseases.

13.5 Cross-Presentation and Tumor Immunity

Autoimmune responses are probably common. However, several factors contribute to quench autoimmunity. The control of immune responses elicited by dying tumor cells is apparently less stringent. Tumor cells frequently die by apoptosis as a result of the unbalance between pro- and anti-apoptotic factors or as the consequence of antineoplastic treatments [88–90]. Scavenger phagocytes interspersed in the tumor masses or, more often, surrounding cells phagocytose the corpses.

DNA is horizontally transferred from apoptotic cells to recipient cells after phagocytosis [91–93]. When phagocytes lack the p53 'guardian', the DNA from the dying cell is propagated. The pathway is particularly efficient for the horizontal transfer of genes that confer to recipient cells a selective advantage. Therefore, the phagocytosis of apoptotic tumor cells contributes to genetic instability and diversity within tumors [91].

When cells with an intact p53 system phagocytose apoptotic tumor cells, the transfer of oncogenes is not detectable. In contrast, intracellular antigens of dying tumor cells survive the phagocytic process and are 'cross-presented' to tumor-specific T cells [31, 34, 37, 49, 65, 72, 94–111; reviewed in 1, 36, 38]. High numbers of lymphoma cells dying by apoptosis *in vivo* recruit in normal mice a long-lasting antineoplastic immune response, endowed with memory and specificity [96]. Agents that interfere with the immunosuppressive clearance of apoptotic lymphoma cells substantially enhance the immunogenicity of the tumor (unpublished). Several groups are actively involved in determining whether tumor cell death contributes to tumor immunity in neoplastic patients. Recently, Mercader *et al.* showed in patients with prostate tumors undergoing androgen ablative therapy that local apoptosis correlates with prominent $CD4^+$ T cell infiltration of tumor sites. Infiltrating T cells have features compatible with a local oligoclonal response [112]. Cell death at the tumor site represents an interesting source of antigens for immunotherapeutic approaches [105].

The access to APCs influences the immunogenicity of uningested apoptotic cells *in vivo*, since they compete with more efficient or more represented phagocytes. This step is bypassed challenging DCs *in vitro* with dying tumor cells. Accordingly, in diverse systems DC loaded with apoptotic tumor cells initiate immune responses against the antigens expressed by dying cells. DCs that had phagocytosed dying tumor cells are therefore an attractive target to achieve antineoplastic immunization [113, 114] (Tab. 13.1).

Tab. 13.1 Apoptotic cell-based cancer vaccines

Vaccine	*Species (strain)*	*Indication*	*Conclusions and references*
Irradiated DC-tumor cell conjugates	mouse (C57BL/6)	CTL induction, vaccination, cure, memory	physical interaction between DCs and irradiated tumor cells determines immunogenicity [94]
Killing of apoptosis-resistant/sensitive colon carcinoma cells *in vivo*	rat (BD-IX)	vaccination, proliferation of tumor-specific T cells	apoptosis favors tumor immunogenicity [95, 99]
Macrophages or DCs that phagocytosed apoptotic lymphoma cells	mouse (C57BL/6)	CTL induction, vaccination	DCs but no macrophages that phagocytosed apoptotic lymphoma cells are efficient vaccines [96]
Apoptotic or necrotic lymphoma cells	mouse (C57BL/6 wild-type or IL-10 knockout)	CTL induction, vaccination, memory, CD4/CD8 dependency	apoptotic lymphoma cells are immunogenic; the cytokine balance controls immunogenicity [96]
Thioglycollate-elicited peritoneal phagocytes that phagocytosed apoptotic colon carcinoma cells	rat (BD-IX)	tumor regression, CTL induction	APCs that phagocytosed apoptotic tumor cells are efficient vaccines [97]
HSP^+ or HSP^- apoptotic leukemia cells	mouse (BALB/c)	growth prevention, specificity	immunogenicity depends on stress before apoptosis [65]
Syngeneic DCs plus HSP^+ or HSP^- apoptotic leukemia cells	mouse (BALB/c)	growth prevention, specificity	immunogenicity depends on stress before apoptosis [65]
DCs plus apoptotic leukemia cells	mouse (DBA2)	CTL induction, immunization	DCs loaded with apoptotic tumor cells are efficient vaccines [104]
Macrophages conditioned with debris of apoptotic or necrotic colon carcinoma	mouse (C57BL/6)	immunization, cure	macrophages receive from dying tumor cells stress signals [66]
DCs plus purified apoptotic or necrotic melanoma cells	mouse (C57BL/6)	immunization, cure	no effect of HSP expression; effective priming with both apoptotic and necrotic cells [115]
DCs plus apoptotic squamous cell carcinoma $\pm$ IL-2	mouse (DBA2/j)	growth prevention, CTL induction	effective anti-tumor response requires IL-2 [107]

Tab. 13.1 (continued)

Vaccine	*Species (strain)*	*Indication*	*Conclusions and references*
DCs that phagocytosed apoptotic or necrotic melanoma cells	mouse (C57BL/6)	DC migration, T_h1 induction, eradication of lung metastasis	DCs that phagocytosed killed melanoma cells are efficient vaccines [108]
Apoptotic or necrotic fibrosarcoma cells plus CD40 ligand-transduced tumors	mouse (C57BL/6)	CD4 T_h1 induction, regression of brain metastasis, priming of T cells that mediate tumor regression	apoptotic but not necrotic tumor cells are efficient vaccines in the presence of CD40 ligand [72]
DCs that phagocytosed apoptotic fibrosarcoma or mastocytoma cells plus an oncogene expressing cytosolic HSP73	mouse (C57BL/6 and BALB/c)	CTL induction	HSP73 increases the immunogenicity of apoptotic tumor cells phagocytosed by DCs [112]
Apoptotic lymphoma cells, DCs that phagocytosed apoptotic lymphoma cells	mouse (C57BL/6)	CTL induction, epitope recognition	tumor epitopes are lost during apoptosis [110]

Other less well-characterized factors also contribute to increase (or quench) tumor immunogenicity. Feng *et al.* reported that the expression of HSPs enhances the immunogenicity of apoptotic leukemia cells [65]. This is in keeping with the observation that DCs that phagocytosed HSP-expressing apoptotic leukemia cells trigger a more efficient antineoplastic immune response [65].

Several studies have addressed the relative ability of tumor cells killed by diverse approaches to elicit the maturation of phagocytosing DC and to induce antineoplastic responses *in vivo*. This has been evaluated in mice vaccinated with dying tumor cells or with DC that phagocytosed *in vitro* dying tumor cells (Tab. 13.1). Results have been obtained in different experimental settings: both the susceptibility to cell death and elicited immune responses are likely to depend *in vivo* on features of each given cancer. Furthermore, often populations of apoptotic or necrotic cells are heavily cross-contaminated, due to the dynamic nature of the cell death program [115]. However, even if methodological details may differ, these studies agree on several issues.

(1) Dying tumor cells are efficiently processed after uptake by antigen presenting phagocytes. The efficiency of this event is not substantially different when tumor cells are killed by apoptosis or necrosis.
(2) Apoptotic tumor cells are normally poorly immunogenic.
(3) When apoptotic cells outnumber phagocytes, or when proper opsonins or adjuvant signals are provided, their immunogenicity is rescued *in vivo*.

It is intriguing that, to the best of our knowledge, autoimmune responses elicited by these approaches do not represent a major obstacle to antineoplastic immune therapy. This may indicate that the censorship mechanisms that limit the initiation and the maintenance of autoimmune responses are not apparently quenching antitumor autoimmunity.

13.6 Conclusions

The phagocytosis of dying cells prevents the initiation of immune responses against intracellular antigens in higher organisms. This finding has profound implications with respect to immune homeostasis and may allow designing therapies: (1) to interfere with the maintenance of autoimmune response or with the initiation of graft rejection and (2) to exploit the immune response against growing cancers. A more sophisticated understanding of the molecular events involved is thus necessary.

13.7 Acknowledgments

The author is supported by the Ministero della Sanità and by the AIRC.

13.8
References

1 Heath, W.R. and F.R. Carbone. Cross-presentation, dendritic cells, tolerance and immunity. *Annu Rev Immunol* **2001**, 19, 47–64.

2 Fadok, V.A., P.P. McDonald, D.L. Bratton and P.M. Henson. Regulation of macrophage cytokine production by phagocytosis of apoptotic and post-apoptotic cells. *Biochem Soc Trans* **1998**, 26, 653–6.

3 Trudeau, J.D., J.P. Dutz, E. Arany, D.J. Hill, W.E. Fieldus and D.T. Finegood. Neonatal beta-cell apoptosis: a trigger for autoimmune diabetes? *Diabetes* **2000**, 49, 1–7.

4 Zhang, Y., B. O'Brien, J. Trudeau, R. Tan, P. Santamaria and J.P. Dutz. *In situ* beta cell death promotes priming of diabetogenic CD8 T lymphocytes. *J Immunol* **2002**, 168, 1466–72.

5 Mevorach, D., J.L. Zhou, X. Song and K.B. Elkon. Systemic exposure to irradiated apoptotic cells induces autoantibody production. *J Exp Med* **1998**, 188, 387–92.

6 Botto, M., C. Dell'Agnola, A.E. Bygrave, E.M. Thompson, H.T. Cook, F. Petry, M. Loos, P.P. Pandolfi and M.J. Walport. Homozygous C1q deficiency causes glomerulonephritis associated with multiple apoptotic bodies. *Nat Genet* **1998**, 19, 56–9.

7 Bickerstaff, M.C., M. Botto, W.L. Hutchinson, J. Herbert, G.A. Tennent, A. Bybee, D.A. Mitchell, H.T. Cook, P.J. Butler, M.J. Walport and M.B. Pepys. Serum amyloid P component controls chromatin degradation and prevents antinuclear autoimmunity. *Nat Med* **1999**, 5, 694–7.

8 Lu, Q., M. Gore, Q. Zhang, T. Camenisch, S. Boast, F. Casagranda, C. Lai, M.K. Skinner, R. Klein, G.K. Matsushima, H.S. Earp, S.P. Goff and G. Lemke. Tyro-3 family receptors are essential regulators of mammalian spermatogenesis. *Nature* **1999**, 398, 723–8.

9 Scott, R.S., E.J. McMahon, S.M. Pop, E.A. Reap, R. Caricchio, P.L. Cohen, H.S. Earp and G.K. Matsushima. Phagocytosis and clearance of apoptotic cells is mediated by MER. *Nature* **2001**, 411, 207–11.

10 Lorenz, H.M., M. Herrmann, T. Winkler, U. Gaipl and J.R. Kalden. Role of apoptosis in autoimmunity. *Apoptosis* **2000**, 5, 443–9.

11 Rosen, A. and L. Casciola-Rosen. Clearing the way to mechanisms of autoimmunity. *Nat Med* **2001**, 7, 664–5.

12 Rosen, A. and L. Casciola-Rosen. Autoantigens as substrates for apoptotic proteases: implications for the pathogenesis of systemic autoimmune disease. *Cell Death Different* **1999**, 6, 6–12.

13 Utz, P.J., T.J. Gensler and P. Anderson. Death, autoantigen modifications and tolerance. *Arthritis Res* **2000**, 2, 101–14.

14 Rodenburg, R.J., J.M. Raats, G.J. Pruijn and W.J. van Venrooij. Cell death: a trigger of autoimmunity? *BioEssays* **2000**, 22, 627–36.

15 Casciola-Rosen, L., F. Andrade, D. Ulanet, W.B. Wong and A. Rosen. Cleavage by granzyme B is strongly predictive of autoantigen status: implications for initiation of autoimmunity. *J Exp Med* **1999**, 190, 815–26.

16 Wu, X., C. Molinaro, N. Johnson and C.A. Casiano. Secondary necrosis is a source of proteolytically modified forms of specific intracellular autoantigens: implications for systemic autoimmunity. *Arthritis Rheum* **2001**, 44, 2642–52.

17 Marguet, D., M.F. Luciani, A. Moynault, P. Williamson and G. Chimini. Engulfment of apoptotic cells involves the redistribution of membrane phosphatidylserine on phagocyte and prey. *Nat Cell Biol* **1999**, 1, 454–6.

18 Herrmann, M., O.M. Zoller, M. Hagenhofer, R. Voll and J.R. Kalden. What triggers anti-dsDNA antibodies? *Mol Biol Rep* **1996**, 23, 265–7.

19 Herrmann, M., R.E. Voll, O.M. Zoller, M. Hagenhofer, B.B. Ponner and J.R. Kalden. Impaired phagocytosis of apoptotic cell material by monocyte-derived macrophages from patients with

systemic lupus erythematosus. *Arthritis Rheum* **1998**, 41, 1241–50.

20 Baumann, I., W. Kolowos, R.E. Voll, B. Manger, U. Gaipl, W.L. Neuhuber, T. Kirchner, J.R. Kalden and M. Herrmann. Impaired uptake of apoptotic cells into tingible body macrophages in germinal centers of patients with systemic lupus erythematosus. *Arthritis Rheum* **2002**, 46, 191–201.

21 Rumore, P.M. and C.R. Steinman. Endogenous circulating DNA in systemic lupus erythematosus. Occurrence as multimeric complexes bound to histone. *J Clin Invest* **1990**, 86, 69–74.

22 Rovere, P., F. Fazzini, M.G. Sabbadini and AA. Manfredi. Apoptosis and systemic autoimmunity: the dendritic cell connection. *Eur J Histochem* **2000**, 44, 229–36.

23 Rovere, P., M.G. Sabbadini, A. Bondanza, V.S. Zimmermann, F. Fazzini, C. Rugarli and A.A. Manfredi. Remnants of suicidal cells fostering systemic autoaggression: apoptosis in the origin and maintenance of autoimmunity. *Arthritis Rheum* **2000**, 43, 1663–72.

24 Williams, R.C., Jr, C.C. Malone, C. Meyers, P. Decker and S. Muller. Detection of nucleosome particles in serum and plasma from patients with systemic lupus erythematosus using monoclonal antibody 4H7. *J Rheumatol* **2001**, 28, 81–94.

25 Amoura, Z., S. Koutouzov and J.C. Piette. The role of nucleosomes in lupus. *Curr Opin Rheumatol* **2000**, 12, 369–73.

26 Berden, J.H., R. Licht, M.C. van Bruggen and W.J. Tax. Role of nucleosomes for induction and glomerular binding of autoantibodies in lupus nephritis. *Curr Opin Nephrol Hypertens* **1999**, 8, 299–306.

27 Licht, R., M.C. van Bruggen, B. Oppers-Walgreen, T.P. Rijke and J.H. Berden. Plasma levels of nucleosomes and nucleosome-autoantibody complexes in murine lupus: effects of disease progression and lipopolysaccharide administration. *Arthritis Rheum* **2001**, 44, 1320–30.

28 Parker, D.C. T cell-dependent B cell activation. *Annu Rev Immunol* **1993**, 11, 331–60.

29 Bachmann, M.F. The role of germinal centers for antiviral B cell responses. *Immunol Res* **1998**, 17, 329–44.

30 Inaba, K., S. Turley, F. Yamaide, T. Iyoda, K. Mahnke, M. Inaba, M. Pack, M. Subklewe, B. Sauter, D. Sheff, M. Albert, N. Bhardwaj, I. Mellman and R.M. Steinman. Efficient presentation of phagocytosed cellular fragments on the major histocompatibility complex class II products of dendritic cells. *J Exp Med* **1998**, 188, 2163–73.

31 Bellone, M., G. Iezzi, P. Rovere, G. Galati, A. Ronchetti, M.P. Protti, J. Davoust, C. Rugarli and A.A. Manfredi. Processing of engulfed apoptotic bodies yields T cell epitopes. *J Immunol* **1997**, 159, 5391–9.

32 Albert, M.L., S.F. Pearce, L.M. Francisco, B. Sauter, P. Roy, R.L. Silverstein and N. Bhardwaj. Immature dendritic cells phagocytose apoptotic cells via alpha-beta5 and CD36 and cross-present antigens to cytotoxic T lymphocytes. *J Exp Med* **1998**, 188, 1359–68.

33 Albert, M.L., B. Sauter and N. Bhardwaj. Dendritic cells acquire antigen from apoptotic cells and induce class I-restricted CTL. *Nature* **1998**, 392, 86–9.

34 Rovere, P., C. Vallinoto, A. Bondanza, M.C. Crosti, M. Rescigno, P. Ricciardi-Castagnoli, C. Rugarli and A.A. Manfredi. Bystander apoptosis triggers dendritic cell maturation and antigen-presenting function. *J Immunol* **1998**, 161, 4467–71.

35 Rovere, P., M.G. Sabbadini, C. Vallinoto, U. Fascio, V. S. Zimmermann, A. Bondanza, P. Ricciardi-Castagnoli and A.A. Manfredi. Delayed clearance of apoptotic lymphoma cells allows cross-presentation of intracellular antigens by mature dendritic cells. *J Leuk Biol* **1999**, 66, 345–9.

36 Bellone, M. Apoptosis, cross-presentation and the fate of the antigen specific immune response. *Apoptosis* **2000**, 5, 307–14.

37 Propato, A., G. Cutrona, V. Francavilla, M. Ulivi, E. Schiaffella, O. Landt, R. Dunbar, V. Cerundolo, M. Ferrarini and V. Barnaba. Apoptotic cells overexpress vinculin and induce vinculin-spe-

cific cytotoxic T-cell cross-priming. *Nat Med* **2001**, 7, 807–13.

38 Larsson, M., J. F. Fonteneaou and N. Bhardwaj. Dendritic cells resurrect antigens from dead cells. *Trends Immunol* **2001**, 22, 142–8.

39 Huang, F. P., N. Platt, M. Wykes, J. R. Major, T. J. Powell, C. D. Jenkins and G. G. MacPherson. A discrete subpopulation of dendritic cells transports apoptotic intestinal epithelial cells to T cell areas of mesenteric lymph nodes. *J Exp Med* **2000**, 191, 435–44.

40 Steinman, R. M. and M. C. Nussenzweig. Inaugural Article: Avoiding horror autotoxicus: the importance of dendritic cells in peripheral T cell tolerance. *Proc Natl Acad Sci USA* **2002**, 99, 351–8.

41 Steinman, R. M., S. Turley, I. Mellman and K. Inaba. The induction of tolerance by dendritic cells that have captured apoptotic cells. *J Exp Med* **2000**, 191, 411–6.

42 Hawiger, D., K. Inaba, Y. Dorsett, M. Guo, K. Mahnke, M. Rivera, J. V. Ravetch, R. M. Steinman and M. C. Nussenzweig. Dendritic cells induce peripheral T cell unresponsiveness under steady state conditions *in vivo*. *J Exp Med* **2001**, 194, 769–79.

43 Albert, M. L., M. Jegathesan and R. B. Darnell. Dendritic cell maturation is required for the cross-tolerization of $CD8^+$ T cells. *Nat Immunol* **2001**, 2, 1010–7.

44 Sigal, L. J., S. Crotty, R. Andino and K. L. Rock. Cytotoxic T-cell immunity to virus-infected non-haematopoietic cells requires presentation of exogenous antigen. *Nature* **1999**, 398, 77–80.

45 Sigal, L. J. and K. L. Rock. Bone marrow-derived antigen-presenting cells are required for the generation of cytotoxic T lymphocyte responses to viruses and use transporter associated with antigen presentation (TAP)-dependent and -independent pathways of antigen presentation. *J Exp Med* **2000**, 192, 1143–50.

46 Cresswell, P., N. Bangia, T. Dick and G. Diedrich. The nature of the MHC class I peptide loading complex. *Immunol Rev* 172, 21–8.

47 den Haan, J. M., S. M. Lehar and M. J. Bevan. $CD8^+$ but not $CD8^-$ dendritic cells cross-prime cytotoxic T cells *in vivo*. *J Exp Med* **2000**, 192, 1685–96.

48 den Haan, J. M. and M. J. Bevan. Antigen presentation to $CD8^+$ T cells: cross-priming in infectious diseases. *Curr Opin Immunol* **2001**, 13, 437–41.

49 Albert, M. L., B. Sauter and N. Bhardwaj. Dendritic cells acquire antigen from apoptotic cells and induce class I-restricted CTL. *Nature* **1998**, 392, 86–9.

50 Arrode, G., C. Boccaccio, J. Lule, S. Allart, N. Moinard, J. P. Abastado, A. Alam and C. Davrinche. Incoming human cytomegalovirus pp65 (UL83) contained in apoptotic infected fibroblasts is cross-presented to $CD8^+$ T cells by dendritic cells. *J Virol* **2000**, 74, 10018–24.

51 Arrode, G., C. Boccaccio, J. P. Abastado and C. Davrinche. Cross-presentation of human cytomegalovirus pp65 (UL83) to $CD8^+$ T cells is regulated by virus-induced, soluble-mediator-dependent maturation of dendritic cells. *J Virol* **2002**, 76, 142–50.

52 Ignatius, R., M. Marovich, E. Mehlhop, L. Villamide, K. Mahnke, W. I. Cox, F. Isdell, S. S. Frankel, J. R. Mascola, R. M. Steinman and M. Pope. Canarypox virus-induced maturation of dendritic cells is mediated by apoptotic cell death and tumor necrosis factor alpha secretion. *J Virol* **2000**, 74, 11329–38.

53 Motta, I., F. Andre, A. Lim, J. Tartaglia, W. I. Cox, L. Zitvogel, E. Angevin and P. Kourilsky. Cross-presentation by dendritic cells of tumor antigen expressed in apoptotic recombinant canarypox virus-infected dendritic cells. *J Immunol* **2001**, 167, 1795–802.

54 Subklewe, M., C. Paludan, M. L. Tsang, K. Mahnke, R. M. Steinman and C. Munz. Dendritic cells cross-present latency gene products from Epstein-Barr virus-transformed B cells and expand tumor-reactive $CD8^+$ killer T cells. *J Exp Med* **2001**, 193, 405–11.

55 Larsson, M., J. F. Fonteneau, S. Somersan, C. Sanders, K. Bickham, E. K. Thomas, K. Mahnke and N. Bhardwaj. Efficiency of cross presentation of vaccinia virus-derived antigens by human dendritic cells. *Eur J Immunol* **2001**, 31, 3432–42.

56 Yrlid, U. and M.J. Wick. *Salmonella*-induced apoptosis of infected macrophages results in presentation of a bacteria-encoded antigen after uptake by bystander dendritic cells. *J Exp Med* **2000**, 191, 613–24.

57 Allavena, P., A. Sica, A. Vecchi, M. Locati, S. Sozzani and A. Mantovani. The chemokine receptor switch paradigm and dendritic cell migration: its significance in tumor tissues. *Immunol Rev* **2000**, 177, 141–9.

58 Rescigno, M., F. Granucci and P. Ricciardi-Castagnoli. Molecular events of bacterial-induced maturation of dendritic cells. *J Clin Immunol* **2000**, 20, 161–6.

59 Pulendran, B., K. Palucka and J. Banchereau. Sensing pathogens and tuning immune responses. *Science* **2001**, 293, 253–6.

60 Janeway, C.A., Jr. The immune system evolved to discriminate infectious nonself from noninfectious self. *Immunol Today* **1992**, 13, 11–6.

61 Roake, J.A., A.S. Rao, P.J. Morris, C.P. Larsen, D.F. Hankins and J.M. Austyn. Dendritic cell loss from nonlymphoid tissues after systemic administration of lipopolysaccharide, tumor necrosis factor and interleukin 1. *J Exp Med* **1995**, 181, 2237–47.

62 Basu, S., R.J. Binder, T. Ramalingam and P.K. Srivastava. CD91 is a common receptor for heat shock proteins gp96, hsp90, hsp70 and calreticulin. *Immunity* **2001**, 14, 303–13.

63 Basu, S., R.J. Binder, R. Suto, K.M. Anderson and P.K. Srivastava. Necrotic but not apoptotic cell death releases heat shock proteins, which deliver a partial maturation signal to dendritic cells and activate the NF-kappa B pathway. *Int Immunol* **2000**, 12, 1539–46.

64 Somersan, S., M. Larsson, J.F. Fonteneau, S. Basu, P. Srivastava and N. Bhardwaj. Primary tumor tissue lysates are enriched in heat shock proteins and induce the maturation of human dendritic cells. *J Immunol* **2001**, 167, 4844–52.

65 Feng, H., Y. Zeng, L. Whitesell and E. Katsanis. Stressed apoptotic tumor cells express heat shock proteins and elicit tumor-specific immunity. *Blood* **2001**, 97, 3505–12.

66 Gough, M.J., A.A. Melcher, A. Ahmed, M.R. Crittenden, D.S. Riddle, E. Linardakis, A.N. Ruchatz, L.M. Emiliusen and R.G. Vile. Macrophages orchestrate the immune response to tumor cell death. *Cancer Res* **2001**, 61, 7240–7.

67 Shi, Y., W. Zheng and K.L. Rock. Cell injury releases endogenous adjuvants that stimulate cytotoxic T cell responses. *Proc Natl Acad Sci USA* **2000**, 97, 14590–5.

68 Gallucci, S. and P. Matzinger. Danger signals: SOS to the immune system. *Curr Opin Immunol* **2001**, 13, 114–9.

69 Shi, Y. and K.L. Rock. Cell death releases endogenous adjuvants that selectively enhance immune surveillance of particulate antigens. *Eur J Immunol* **2002**, 32, 155–62.

70 Ishii, K.J., K. Suzuki, C. Coban, F. Takeshita, Y. Itoh, H. Matoba, L.D. Kohn and D.M. Klinman. Genomic DNA released by dying cells induces the maturation of APC. *J Immunol* **2001**, 167, 2602–7.

71 Henson, P.M., D.L. Bratton and V.A. Fadok. The phosphatidylserine receptor: a crucial molecular switch? *Nat Rev Mol Cell Biol* **2001**, 2, 627–33.

72 Fujita, N., H. Kagamu, H. Yoshizawa, K. Itoh, H. Kuriyama, N. Matsumoto, T. Ishiguro, J. Tanaka, E. Suzuki, H. Hamada and F. Gejyo. CD40 ligand promotes priming of fully potent antitumor $CD4^+$ T cells in draining lymph nodes in the presence of apoptotic tumor cells. *J Immunol* **2001**, 167, 5678–88.

73 Barber, G.N. Host defense, viruses and apoptosis. *Cell Death Different* **2001**, 8, 113–26.

74 Lopes, M.F., C.G. Freire-de-Lima and G.A. DosReis. The macrophage haunted by cell ghosts: a pathogen grows. *Immunol Today* **2000**, 21, 489–94.

75 DosReis, G.A. and M.A. Barcinski. Apoptosis and parasitism: from the parasite to the host immune response. *Adv Parasitol* **2001**, 49, 133–61.

76 Fairweather, D., Z. Kaya, G.R. Shellam, C.M. Lawson and N.R. Rose. From infection to autoimmunity. *J Autoimmun* **2001**, 16, 175–86.

77 Pisetsky, D.S. Immune responses to DNA in normal and aberrant immunity. *Immunol Res* **2000**, 22, 119–26.

78 Ronnblom, L. and G.V. Alm. An etiopathogenic role for the type I IFN system in SLE. *Trends Immunol* **2001**, 22, 427–31.

79 Patry, Y.C., D.C. Trewick, M. Gregoire, M.A. Audrain, A.M. Moreau, J.Y. Muller, K. Meflah and V.L. Esnault. Rats injected with syngenic rat apoptotic neutrophils develop antineutrophil cytoplasmic antibodies. *J Am Soc Nephrol* **2001**, 12, 1764–8.

80 Moosig, F., E. Csernok, G. Kumanovics and W.L. Gross. Opsonization of apoptotic neutrophils by anti-neutrophil cytoplasmic antibodies (ANCA) leads to enhanced uptake by macrophages and increased release of tumour necrosis factor-alpha (TNF-alpha). *Clin Exp Immunol* **2000**, 122, 499–503.

81 Harper, L. and C.O. Savage. Pathogenesis of ANCA-associated systemic vasculitis. *J Pathol* **2000**, 190, 349–59.

82 van der Geld, Y.M., P.C. Limburg and C.G. Kallenberg. Proteinase 3, Wegener's autoantigen: from gene to antigen. *J Leuk Biol* **2001**, 69, 177–90.

83 Manfredi, A.A., P. Rovere, G. Galati, S. Heltai, E. Bozzolo, L. Soldini, J. Davoust, G. Balestrieri, A. Tincani and M.G. Sabbadini. Apoptotic cell clearance in systemic lupus erythematosus. I. Opsonization by antiphospholipid antibodies. *Arthritis Rheum* **1998**, 41, 205–14.

84 Manfredi, A.A., P. Rovere, S. Heltai, G. Galati, G. Nebbia, A. Tincani, G. Balestrieri and M.G. Sabbadini. Apoptotic cell clearance in Systemic Lupus Erythematosus: II. Role for the beta$_2$-glycoprotein I. *Arthritis Rheum* **1998**, 41, 215–23.

85 Rovere P., M.G. Sabbadini, C. Vallinoto, U. Fascio, M. Rescigno, M. Crosti, P. Ricciardi Castagnoli, G. Balestrieri, A. Tincani and A.A. Manfredi. Dendritic cell presentation of antigens from apoptotic cells in a pro-inflammatory context: role of opsonizing anti-beta$_2$ glycoprotein I antibodies. *Arthritis Rheum* **1999**, 42, 1412–20.

86 Miranda, M.E., C.E. Tseng, W. Rashbaum, R.L. Ochs, C.A. Casiano, F. Di Donato, E.K. Chan and J.P. Buyon. Accessibility of SSA/Ro and SSB/La antigens to maternal autoantibodies in apoptotic human fetal cardiac myocytes. *J Immunol* **1998**, 161, 5061–9.

87 Miranda-Carus, M.E., A.D. Askanase, R.M. Clancy, F. Di Donato, T.M. Chou, M.R. Libera, E.K. Chan and J.P. Buyon. Anti-SSA/Ro and anti-SSB/La autoantibodies bind the surface of apoptotic fetal cardiocytes and promote secretion of TNF-alpha by macrophages. *J Immunol* **2000**, 165, 5345–51.

88 Pandey, S., B. Smith, P.R. Walker and M. Sikorska. Caspase-dependent and independent cell death in rat hepatoma 5123tc cells. *Apoptosis* **2000**, 5, 265–75.

89 Fernandez-Luna, J.L. Bcr-Abl and inhibition of apoptosis in chronic myelogenous leukemia cells. *Apoptosis* **2000**, 5, 315–8.

90 Nuydens, R., G. Dispersyn, K.G. van Den, M. de Jong, R. Connors, F. Ramaekers, M. Borgers and H. Geerts. Bcl-2 protects neuronal cells against taxol-induced apoptosis by inducing multinucleation. *Apoptosis* **2000**, 5, 335–43.

91 Bergsmedh, A., A. Szeles, M. Henriksson, A. Bratt, M.J. Folkman, A.L. Spetz and L. Holmgren. Horizontal transfer of oncogenes by uptake of apoptotic bodies. *Proc Natl Acad Sci USA* **2001**, 98, 6407–11.

92 Holmgren, L., A. Szeles, E. Rajnavolgyi, J. Folkman, G. Klein, I. Ernberg and K.I. Falk. Horizontal transfer of DNA by the uptake of apoptotic bodies. *Blood* **1999**, 93, 3956–63.

93 Spetz, A.L., B.K. Patterson, K. Lore, J. Andersson and L. Holmgren. Functional gene transfer of HIV DNA by an HIV receptor-independent mechanism. *J Immunol* **1999**, 63, 736–42.

94 Celluzzi, C.M. and L.D. Falo, Jr. Physical interaction between dendritic cells and tumor cells results in an immunogen that induces protective and therapeutic tumor rejection. *J Immunol* **1998**, 160, 3081–5.

95 Bonnotte, B., N. Favre, M. Moutet, A. Fromentin, E. Solary, M. Martin and F. Martin. Bcl-2-mediated inhibition of apoptosis prevents immunogenicity and restores tumorigenicity of spontaneously regressive tumors. *J Immunol* **1998**, 161, 1433–8.

96 Ronchetti, A., P. Rovere, G. Iezzi, G. Galati, S. Heltai, M.P. Protti, M.P. Garancini, A.A. Manfredi, C. Rugarli and M. Bellone. Immunogenicity of apoptotic cells *in vivo*: role of antigen load, antigen-presenting cells and cytokines. *J Immunol* **1999**, 163, 130–6.

97 Henry, F., O. Boisteau, L. Bretaudeau, B. Lieubeau, K. Meflah and M. Gregoire. Antigen-presenting cells that phagocytose apoptotic tumor-derived cells are potent tumor vaccines. *Cancer Res* **1999**, 59, 3329–32.

98 Russo, V., S. Tanzarella, P. Dalerba, D. Rigatti, P. Rovere, A. Villa, C. Bordignon and C. Traversari. Dendritic cells acquire the MAGE-3 human tumor antigen from apoptotic cells and induce a class I-restricted T cell response. *Proc Natl Acad Sci USA* **2000**, 97, 2185–90.

99 Bonnotte, B., N. Favre, M. Moutet, A. Fromentin, E. Solary, M. Martin and F. Martin. Role of tumor cell apoptosis in tumor antigen migration to the draining lymph nodes. *J Immunol* **2000**, 164, 1995–2000.

100 Shaif-Muthana, M., C. McIntyre, K. Sisley, I. Rennie and A. Murray. Dead or alive: immunogenicity of human melanoma cells when presented by dendritic cells. *Cancer Res* **2000**, 60, 6441–7.

101 Nouri-Shirazi, M., J. Banchereau, D. Bell, S. Burkeholder, E.T. Kraus, J. Davoust and K.A. Palucka. Dendritic cells capture killed tumor cells and present their antigens to elicit tumor-specific immune responses. *J Immunol* **2000**, 165, 3797–803.

102 Jenne, L., J.F. Arrighi, H. Jonuleit, J.H. Saurat and C. Hauser. Dendritic cells containing apoptotic melanoma cells prime human $CD8^+$ T cells for efficient tumor cell lysis. *Cancer Res* **2000**, 60, 4446–52.

103 Hoffmann, T.K., N. Meidenbauer, G. Dworacki, H. Kanaya and T.L. Whiteside. Generation of tumor-specific T-lymphocytes by cross-priming with human dendritic cells ingesting apoptotic tumor cells. *Cancer Res* **2000**, 60, 3542–9.

104 Paczesny, S., S. Beranger, J.L. Salzmann, D. Klatzmann and B.M. Colombo. Protection of mice against leukemia after vaccination with bone marrow- derived dendritic cells loaded with apoptotic leukemia cells. *Cancer Res* **2001**, 61, 2386–9.

105 Candido, K.A., K. Shimizu, J.C. McLaughlin, R. Kunkel, J.A. Fuller, B.G. Redman, E.K. Thomas, B.J. Nickoloff and J.J. Mule. Local administration of dendritic cells inhibits established breast tumor growth: implications for apoptosis-inducing agents. *Cancer Res* **2001**, 61, 228–36.

106 Hoffmann, T.K., N. Meidenbauer, J. Muller-Berghaus, W.J. Storkus and T.L. Whiteside. Proinflammatory cytokines and CD40 ligand enhance cross-presentation and cross-priming capability of human dendritic cells internalizing apoptotic cancer cells. *J Immunother* **2001**, 24, 162–71.

107 Son, Y.I., R.B. Mailliard, S.C. Watkins and M.T. Lotze. Dendritic cells pulsed with apoptotic squamous cell carcinoma have anti-tumor effects when combined with interleukin-2. *Laryngoscope* **2001**, 111, 1472–8.

108 Chen, Z., T. Moyana, A. Saxena, R. Warrington, Z. Jia and J. Xiang. Efficient antitumor immunity derived from maturation of dendritic cells that had phagocytosed apoptotic/necrotic tumor cells. *Int J Cancer* **2001**, 93, 539–48.

109 Galetto, A., M. Contarini, A. Sapino, P. Cassoni, E. Consalvo, S. Forno, C. Pezzi, V. Barnaba, A. Mussa and L. Matera. *Ex vivo* host response to gastrointestinal cancer cells presented by autologous dendritic cells. *J Surg Res* **2001**, 100, 32–8.

110 Castiglioni, P., A. Martin-Fontecha, G. Milan, V. Tomajer, F. Magni, J. Michaelsson, C. Rugarli, A. Rosato and M. Bellone. Apoptosis-dependent subversion of the T lymphocyte epitope hierarchy in lymphoma cells. *Cancer Res* **2002**, in press.

111 Kammerer, R., D. Stober, P. Riedl, C. Oehninger, R. Schirmbeck and J. Reimann. Noncovalent association with stress protein facilitates cross-priming of $CD8^+$ T cells to tumor cell antigens by dendritic cells. *J Immunol* **2002**, 168, 108–17.

112 Mercader, M., B.K. Bodner, M.T. Moser, P.S. Kwon, E.S. Park, R.G. Manecke, T.M. Ellis, E.M. Wojcik, D. Yang, R.C. Flanigan, W.B. Waters, W.M. Kast and E.D. Kwon. T cell infiltration of the prostate induced by androgen withdrawal in patients with prostate cancer. *Proc Natl Acad Sci USA* **2001**, 98, 14565–70.

113 Whiteside, T.L. Immunobiology and immunotherapy of head and neck cancer. *Curr Oncol Rep* **2001**, 3, 46–55.

114 Banchereau, J., B. Schuler-Thurner, A.K. Palucka and G. Schuler. Dendritic cells as vectors for therapy. *Cell* **2001**, 106, 271–4.

115 Kotera, Y., K. Shimizu and J.J. Mule. Comparative analysis of necrotic and apoptotic tumor cells as a source of antigen(s) in dendritic cell-based immunization. *Cancer Res* **2001**, 61, 8105–9.

Part 5
"Altered Self" in Dying Cells

14
Autoantigens as Substrates for Apoptotic Proteases: Implications for the Pathogenesis of Systemic Autoimmune Disease

Antony Rosen and Livia Casciola-Rosen

14.1
Autoantibodies: Probes of the Perturbed State

The cellular antigens targeted by the high-titer autoantibody response in systemic autoimmune diseases are a diverse group of molecules that are ubiquitously expressed and that share no obvious features in terms of subcellular distribution, protein structure or function [1]. In spite of such extraordinary diversity, the specificity of the autoimmune response is remarkably predictive of disease phenotype, such that specific autoantibodies have become clinically useful diagnostically and prognostically [2]. For example, antibodies to nucleosomes are strongly associated with the systemic lupus erythematosus (SLE) phenotype [3], while antibodies to topoisomerase I are associated with diffuse scleroderma [4]. Similarly, antibodies recognizing components of the centromere (e.g. CENP-B) are associated with the limited form of scleroderma [5] and are predictive of digit loss in this disease [6]. Although the targeted antigens in the different diseases do not share features that are readily apparent, there is a growing consensus that the highly specific humoral immune response to these molecules is T cell dependent and that flares of disease result when this primed immune system is rechallenged with self-antigen (reviewed in [7–9]). Thus, the autoantigens in systemic autoimmune diseases likely satisfied the stringent criteria for initiation of a primary immune response during disease development. Since initiation of an adaptive immune response requires that a unique molecular structure (not previously generated during development of immune tolerance) is presented to the immune system in a pro-immune context, we have proposed that the association of particular autoantibodies with specific phenotypes reflects the unique modification of autoantigen structure during a pro-immune initiating event in the target tissue [10].

Since T cell tolerance is only induced to dominant determinants in autoantigens (which are generated and presented at suprathreshold concentrations during natural processing of whole protein antigens), a potential exists for T cell autoreactivity directed against 'cryptic' determinants (which are generated at subthreshold concentrations during normal antigen processing (reviewed in [11, 12]). Such T cells recognizing the 'cryptic' self never encounter their antigen during natural antigen presentation and are therefore not tolerized. Indeed, several experimental systems have now clearly demonstrated the existence of such T cells (reviewed in [13, 14]). The structure of such antigens can be changed in a variety of ways, including novel autoantigen cleavage [15] and altered autoantigen processing induced by high-affinity ligand binding (e.g. to an antibody or receptor molecule [16–18]). In turn, these changes can alter the hierarchy of epitopes which are efficiently loaded onto MHC class II molecules, resulting in presentation of previously cryptic epitopes [17, 18]. We have proposed that autoantigens in systemic autoimmune diseases are all structurally altered during disease initiation, thus allowing efficient loading of previously 'cryptic' epitopes onto class II and the activation of autoreactive T cells [19–21]. We and others have therefore used high titer autoantibodies as probes of cell biology and biochemistry of autoantigens during different clinically relevant perturbed states, to search for those circumstances in which autoantigens become clustered, concentrated and structurally modified [19–29]. This chapter highlights the modifications of autoantigen structure that occur during different forms of cell death (particularly apoptosis), and focuses attention on unique forms of apoptotic death in distinct microenvironments as initiating and propagating events in systemic autoimmune diseases. The numerous current gaps in knowledge make parts of this chapter necessarily speculative. However, these gaps raise important questions about the normal immune consequences of different forms of apoptosis in tissues, and about how defects in the signaling, execution and clearance phases of the apoptotic process are susceptibility factors for the development of systemic autoimmune disease.

14.2
Lupus Autoantigens undergo a Striking Redistribution during Apoptosis, Becoming Clustered and Concentrated in the Surface Blebs of Apoptotic Cells

Studies to delineate the potential perturbed states that might initiate systemic autoimmune diseases have been focused by a striking clinical observation in SLE: that ultraviolet (UV) irradiation has a marked propensity to induce flares of both systemic and skin disease in lupus patients (reviewed in [30]). In this regard, the epidermis appears to be an important target of the immunopathologic response in lupus and constitutes an appropriate *in vitro* model with which to address the effects of flare-inducing stimuli (e.g. UVB) [31]. Earlier studies demonstrated that intracellular autoantigens can be stained at the exterior surface of keratinocytes incubated *in vitro* for 20–24 h after irradiation with UVB, although the mechanism of this 'redistribution' of antigens was not determined [32, 33]. To determine how this phenomenon might arise, we studied the subcellular distribution of lupus autoantigens at increasing times after UVB irradiation, in both intact and permeabilized cells [19, 34]. Our initial studies made several observations. (1) UVB-irradiated keratinocytes undergo apoptosis, beginning a few hours after irradiation. Apoptotic keratinocytes manifest the classic morphologic hallmarks of this process, including prominent surface blebbing (early event) and nuclear condensation and fragmentation into apoptotic bodies (later event). (2) Lupus autoantigens,

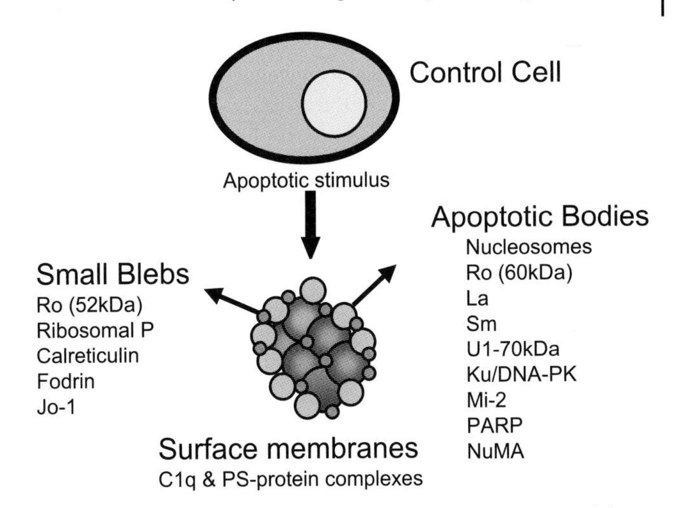

Fig. 14.1 Autoantigens cluster in unique subcellular structures in cells undergoing apoptosis. Although autoantigens are not restricted to any specific subcellular compartment in normal cells, they become clustered and concentrated within small surface blebs and apoptotic bodies in cells dying by apoptosis induced by various stimuli. Small blebs are enriched in ribosomal autoantigens and those found within the ER lumen, while apoptotic bodies are enriched in nuclear autoantigens. Additionally, PS rapidly redistributes early in apoptosis from the inner surface of the plasma membrane bilayer and appears at the external membrane surface.

which are not restricted to any specific subcellular compartment in control cells, are strikingly redistributed in apoptotic cells, such that they become clustered and concentrated within small surface blebs and apoptotic bodies (Fig. 14.1). Thus, small surface blebs [which contain fragmented rough endoplasmic reticulum (ER); see Fig. 14.1] are highly enriched in 52-kDa Ro, ribosomal autoantigens, as well as those autoantigens found within the ER lumen (e.g. calreticulin). This marked enrichment of autoantigens in small surface blebs is accompanied by a concomitant depletion from the cytosol. Nuclear autoantigens also undergo a striking redistribution and concentration during apoptosis (Fig. 14.1). Thus, 60-kDa Ro, La, the snRNPs, Ku and poly(ADP-ribose)polymerase (PARP), which normally have a diffuse nuclear distribution, initially become concentrated as a rim around the condensing chromatin in early apoptosis. As apoptosis progresses and the nucleus becomes fragmented into multiple membrane-bound apoptotic bodies, nuclear autoantigens remain rimmed around the condensed chromatin. Interestingly, the surface of these apoptotic surface blebs also has the capacity to concentrate potential autoantigens. For example, phosphatidylserine (PS) becomes concentrated at this site early in the apoptotic process [35–41]. PS, normally restricted to the inner surface of the plasma membrane bilayer, becomes rapidly redistributed early in apoptosis and appears at the external membrane surface. This generates a procoagulant external cell surface which has the capacity to bind several autoantigenic PS-binding proteins, including β_2-glycoprotein I and Annexin V

[35–39]. The demonstration that these phospholipid-binding proteins decorate the surface of apoptotic cells strongly suggests that the immunogenic phospholipid–protein complexes form at the surface of apoptotic cells *in vivo* [39]. This is further underscored by the observation that IgG purified from the plasma of patients with antiphospholipid syndrome binds to the surface of apoptotic cells and inhibits its procoagulant activity [37]. Recent data has also demonstrated that dual-specificity autoantibodies recognizing both PS and double-stranded DNA also recognize their cognate antigens at the surface of some apoptotic cells [42]. This observation is of great interest as it demonstrates for the first time that anti-DNA B cells can recognize apoptotic surface blebs and may play an important role in providing B cells with access to nuclear antigens [42].

It is also noteworthy that the surface blebs of apoptotic keratinocytes bind C1q [43, 44], whose collagen-like domains are the frequent (around 47%) target of a high-titer autoantibody response in patients with SLE [45]. The exact binding partner for C1q at the apoptotic surface is not yet known, but recent studies suggest that C1q binding to apoptotic cells may have a critical function in the non-inflammatory clearance of apoptotic cells *in vitro* [46] and in some microenvironments *in vivo* [47]. Interestingly, C1q deficiency is strongly associated with the development of SLE in both humans and mice (reviewed in [48]). In the C1q-null mouse, Walport *et al.* noted a marked accumulation of apoptotic cells in the kidney, suggesting that clearance of apoptotic cells is defective in this microenvironment [47] and focusing attention on clearance of apoptotic cells as an important potential defect underlying the development of systemic autoimmunity. In this regard, it is relevant that C1q also requires the binding of the acute phase reactant C-reactive protein (CRP) to the apoptotic cell surface for generating the anti-inflammatory consequences to apoptotic cells [46].

14.3 Susceptibility to Efficient Cleavage by Caspases unifies a Subgroup of Systemic Disease Autoantigens

Proteolysis plays an important mechanistic role in the apoptotic pathway, accomplished through the specific cleavage of a limited number of downstream substrates (reviewed in [49, 50]). This apoptosis-specific proteolysis is catalyzed by a unique family of cysteine proteases (called caspases, for cysteine proteases that cleave after aspartic acid) that have an absolute requirement for aspartic acid in the substrate P_1 position. Since initiation of the primary immune response requires that non-tolerized structure be generated, we were intrigued by the observation that the first proteolytic victims of the caspases discovered in apoptosis were PARP and lamins [51, 52], both of which are autoantigens targeted in systemic autoimmune diseases [53, 54]. We therefore addressed whether other autoantigens were similarly cleaved by caspases during apoptosis. Using western blotting of lysates of control and apoptotic cells, we detected a group of more than 20 autoantigens (including U1-70kDa, DNA-PK_{CS}, NuMA, topoisomerase I, NOR-90, fodrin, hnRNP C1/C2, PMS2, SRP72, La, CENP-C and Mi-2) that were recognized by a high-titer autoantibody response and which were cleaved early in apoptosis [21, 23, 24, 26, 28, 55–59]. Cleavage was prevented by caspase inhibitors, implicating the involvement of a caspase either directly or upstream of these cleavage events.

While susceptibility to efficient cleavage by a caspase is a frequent feature of autoantigens, it is not a universal feature of all autoantigens. For example, caspase-mediated proteolytic cleavage of components of the nucleosome, the frequently targeted Ro particle or the major scleroderma autoantigen CENP-B has not been observed [21, 26]. Furthermore, susceptibility to caspase-mediated cleavage does not appear to be specific to autoantigens. This imperfect, albeit striking correlation of susceptibility to caspase cleavage and status as an autoantigen requires that caution be exercised in ascribing a mechanistic role for caspase cleavage in defining molecules as autoantigens, since it suggests that additional (partially overlapping) properties might be relevant. For example, (1) caspases may have evolved to cleave a specific regulatory motif in proteins; the association with autoantigen status may be with the presence of this protein structure rather than with cleavage during apoptosis; (2) such structure may also be the target of additional proteases during specific forms of apoptosis in unique tissues/microenvironments [e.g. granzyme B (GrB); see below]; and (3) such a regulatory motif may be subject to additional post-translational modifications either during apoptosis or during other physiologic states relevant to disease propagation (e.g. phosphorylation, glutathiolation, transglutamination, citrullination, or formation of novel protein-protein or protein-nucleic acid complexes). In this regard, it is of interest that numerous autoantigens are phosphorylated during a variety of physiologic perturbations (including apoptosis) [29]. Furthermore, in several cases, recognition by autoantibodies is dependent on the phosphorylation state of the antigens. Thus, antibodies that preferentially recognize either the phosphorylated or dephosphorylated states of the large subunit of RNA polymerase II or SR proteins have been defined [60, 61]. Understanding what role changes in the structure of autoantigens during different physiologic states may play in altering the immunogenicity of self molecules is a priority, as it may provide important insights into the initiating and propagating states in systemic autoimmunity.

Autoantigen clustering and cleavage occurs in almost all forms of apoptosis described to date, which often occur in actively anti-inflammatory and non-immune contexts [62–65]. The marked frequency of apoptosis in normal development and homeostasis, coupled with the infrequency of systemic autoimmunity in the population, strongly suggests that only a very restricted subset of apoptotic events (e.g. those occurring in a pro-immune setting; see below) in individuals that are genetically predisposed to generation of novel autoantigen structure (e.g. from abnormalities in the clearance and degradation of apoptotic material in tissues; see below) will initiate a self-sustaining autoimmune response.

14.4
Novel Autoantigen Fragments are produced during Cytotoxic Lymphocyte-induced Target Cell Apoptosis

One form of apoptosis that frequently occurs in a pro-immune setting is the death of virally infected target cells, induced by cytotoxic T lymphocytes (CTLs) and natural killer (NK) cells. Cytotoxic lymphocytes use several pathways to induce target cell apoptosis, including Fas ligation and granule exocytosis (reviewed in [66]). GrB, a serine protease found in the cytoplasmic granules of CTLs and NK cells, has a similar substrate specificity to the caspases, in its near-absolute requirement for aspartic acid in the substrate P_1 position [67]. (The substrate residues immediately upstream of the cleavage site are termed P_1, P_2, P_3 and P_4 as distance increases from the scissile bond; previous studies have shown that the specificity of GrB for its substrates resides in the P_1–P_4 residues). Although GrB also has a similar specificity in the P_2–P_4 substrate positions to the upstream activating (group III) caspases (which prefer Ileu/Val in P_4 and Glu in P_3 [68, 69]), there are some amino acids in the P_2 and P_3 positions that are preferred exclusively by GrB and not tolerated by the caspases (e.g. proline in P_2, and glycine or serine in P_3, see Fig. 14.2). Recent studies have shown that GrB plays an important role in inducing apoptotic changes in target cells during granule exocytosis induced cytotoxicity [70–72], partly by catalyzing the cleavage and activation of sev-

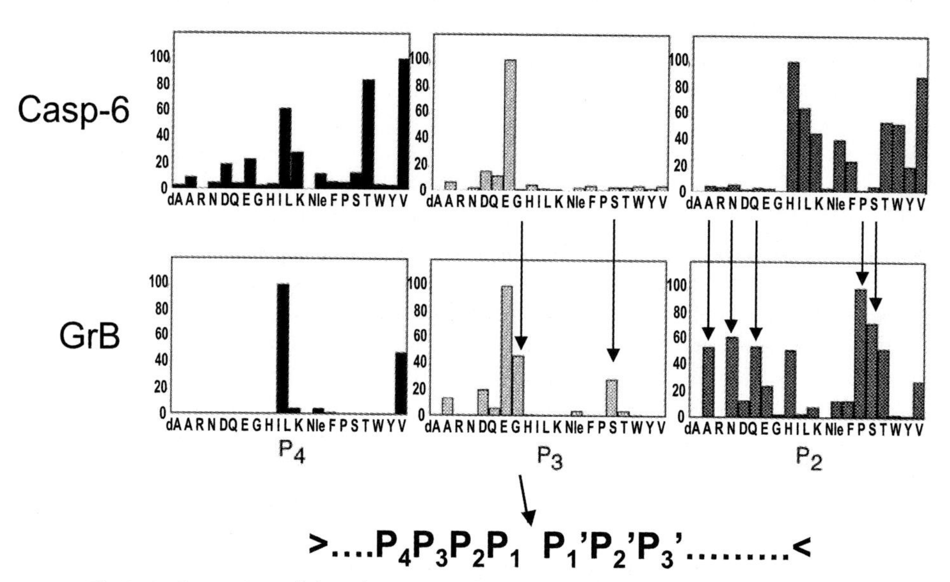

Fig. 14.2 Comparison of the substrate specificities of caspase-6 and GrB. The specificities of the two proteases shown were determined by Thornberry *et al.* using P_2, P_3 and P_4 scanning combinatorial libraries, and plots constructed from data in reference [68]. Although GrB and caspase-6 have similar specificities in the P_2–P_4 substrate positions, some amino acids are exclusively preferred by GrB in the P_2 and P_3 positions, and are not tolerated by caspase-6 (marked by vertical arrows).

Tab. 14.1 Comparison of the distinguishing features of caspases and GrB.

Caspases	***GrB***
Cysteine proteases	Serine protease
Near absolute requirement for Asp in the P_1 position	Near absolute requirement for Asp in the P_1 position
Highly fastidious proteases	Highly fastidious protease
Cleaves macromolecules as efficiently as tetrapeptides	Cleaves macromolecules much more efficiently than tetrapeptides
Ubiquitous expression	Very limited expression
Restricted set of intracellular substrates, whose cleavage is responsible for the apoptotic phenotype	Restricted set of intracellular substrates, many of which overlap with those of caspases

eral caspases (reviewed in [73]). The distinguishing features of caspases and GrB are compared in Tab. 14.1. GrB also initiates caspase-independent pathways which contribute to target cell death through direct targeting of several downstream caspase substrates that mediate critical components of the apoptotic phenotype. Thus, GrB directly cleaves Bid at a site close to that utilized by caspase-8 and recruits the downstream mitochondrial pathway [74–78]. Similarly, GrB cleaves ICAD allowing activation of the caspase-activated DNase, thereby generating the internucleosomal DNA degradation that is characteristic of caspase-mediated apoptotic death [79]. We have demonstrated that the majority of autoantigens targeted across the spectrum of systemic autoimmunity are directly and efficiently cleaved by GrB, both *in vitro* and in cells undergoing granule-induced cytotoxicity [20, 22, 59, 80]. GrB-mediated cleavage of these substrates generates unique fragments not generated during any other form of apoptosis studied to date. Interestingly, efficient cleavage by GrB (with the generation of distinct fragments) has also been observed for those systemic disease autoantigens that are not cleaved by caspases. These molecules include the scleroderma autoantigens CENP-B, fibrillarin and B23, PMS1 targeted in myositis, and the type 3 muscarinic receptor targeted in Sjögren's syndrome [20, 59, 80]. The recent demonstration that GrB cleaves subunit III of the glutamate receptor (GluR3) (an autoantigen in Rasmussen's encephalitis) is also of significance [81]. Interestingly, the GrB cleavage site is located within the epitope targeted by antibodies in this disease. In GluR3, the GrB cleavage site faces the outside of the plasma membrane and cleavage efficiency is markedly influenced by the glycosylation state of the receptor [81]. Thus, when GluR3 is fully glycosylated, it is relatively resistant to cleavage by GrB; the deglycosylated or non-glycosylated form of GluR3 is efficiently cleaved. The authors have proposed that generation of an 'under'-glycosylated form of GluR3 during inflammatory states may allow GluR3 cleavage by GrB and generation of previously cryptic peptide fragments [81]. Demonstrating that such circumstances occur *in vivo* will provide important evidence to support such a mechanism.

The striking susceptibility of many autoantigens to cleavage by GrB, together with their clustering at the same site in apoptotic cells, focuses attention on granule-induced apoptosis as a potential initiator of autoimmunity in the susceptible host. It will be extremely important to evaluate whether the immunogenicity of these novel forms of autoantigen is indeed increased over native forms of these molecules, and to demonstrate the presence of the relevant cleavage products *in vivo* during disease initiation and propagation. It is also possible that the susceptibility of autoantigens to cleavage by GrB reflects some unique protein conformation which has a determinative property in terms of immunogenicity. Definition of the structure of GrB cleavage sites and their interactions with components of the antigen-processing pathway is an important task.

14.5
Caspase-independent Cell Death: Role in Generating Unique Autoantigen Structure?

Several studies have demonstrated that the CTL granule pathway can efficiently activate downstream caspases and directly recruit the mitochondrial pathway through cleavage of Bid [74–78]. Many autoantigens are efficiently cleaved by caspases, which generate the default fragments elaborated during all forms of caspase-dependent death. For example, U1-70kDa is cleaved by both caspase-3 and GrB, at distinct sites. During CTL granule-induced death in normal target cells, almost all of the fragment generated is from caspase-3 activity [20, 22]. In contrast, when endogenous caspases in the target cells are specifically inhibited, generation of the caspase-mediated fragments is abolished and fragments directly generated by GrB are formed. Since several *in vivo* circumstances have been defined in which caspases are under profound endogenous or exogenous inhibition (e.g. through expression of viral or endogenous caspase inhibitors or Bcl-2 homologues), it is likely that the initial generation of unique autoantigen structure occurs under such circumstances. It is possible that other unique structural autoantigen modifications occur during such caspase-independent death pathways and that GrB-induced fragments are only one relevant example of a more general phenomenon. For example, the lack of glutathiolation of PDC-E2 observed in cholangiocytes (in which high-level expression of Bcl-2 abrogates this modification normally observed in other cells) may generate a unique form of this primary biliary cirrhosis autoantigen that is not generated at sites of tolerance development [25]. In this regard, it will be of interest to define whether unique forms of autoantigens are generated through other modifications during caspase-independent death.

14.6
Defects in Clearance of the Apoptotic Corpse in Tissues may be an Important Defect Underlying Systemic Autoimmune Diseases

There has recently been a rapid increase in understanding the mechanisms and immune consequences of apoptotic cell clearance (reviewed in [41]). Recent studies have emphasized that apoptotic cells are not immunologically inert, but rather are potently anti-inflammatory both *in vitro* and *in vivo* [41]. This process has a high degree of redundancy, and requires communication at multiple levels between the engulfed and engulfing cells (reviewed in [82–86]). Recent studies also indicate an important role for components of the innate immune system in defining the immune consequences of such phagocytic events [46, 62–65]. New evidence that abnormalities in those pathways involved with rapid clearance of apoptotic material, or expression of its anti-inflammatory activities, may initiate autoimmunity have underscored the importance of such innate immune pathways in determining the non-inflammatory, non-immune response to self. For example, the binding of C1q and pentraxins to the surface of apoptotic cells appears to enhance clearance of apoptotic cells [47, 87] and recruit additional components (e.g. CRP) to activate anti-inflammatory pathways of clearance [46]. The recent observation of increased numbers of uncleared apoptotic cells in the kidneys of an SLE-susceptible C1q-null mouse reinforces that abnormal clearance of apoptotic cells may play a role in the pathogenesis of SLE in these animals [47]. Exciting recent studies have further supported the notion that impairment in the phagocytosis and anti-inflammatory cytokine secretion induction by apoptotic cells play a central role in the initiation of autoimmunity. In studies by Matsushima *et al.*, functional deficiency in the Mer tyrosine kinase led to an impairment of phagocytosis of apoptotic cells and was associated with an increased prevalence of anti-DNA antibodies [88]. In animals deficient in the function of all three members of this Tyro-3 family (Tyro3, Axl and Mer), Lemke's group showed that there was marked activation of antigen-presenting cells, which secreted large amounts of pro-inflammatory cytokines [89]. It is therefore likely that the propensity to develop autoimmunity and lymphoproliferation in these animals represents both delayed clearance of apoptotic cells, and a markedly impaired secretion of anti-inflammatory cytokines. It is likely that delayed clearance of apoptotic cells changes (1) the compartmentation of autoantigens (allowing leakage of autoantigen during secondary necrosis, and access of soluble molecules to efficient macropinocytotic and endocytic antigen capture by dendritic cells) and/or (2) provides apoptotic cells and membrane-bound fragments access to different (pro-immune) populations of antigen presenting cells, from which they are normally excluded (see, e.g. [90]). The presence of a highly activated group of antigen-presenting cells secreting proinflammatory cytokines in the Tyro-3 family mutants appears to play a critical role [89].

A recent study in humans has demonstrated that the clearance of apoptotic lymphocytes and fragments by macrophages is impaired in some patients with SLE. In this work, there was significant heterogeneity in the level of impairment among patients, which did not appear to correlate with disease activity or therapy [91]. It remains unclear whether (1) this phenomenon results from abnormalities in recognition, binding or phagocytosis of apoptotic cells by SLE macrophages; (2) the phagocytosis of all types of apoptotic cell are similarly affected; and (3) the degree of impairment in a particular patient varies with disease activity. The marked heterogeneity of the degree of impairment observed in different patients predicts significant complexity in resolving this question, resulting from the contribution of multiple different clearance pathways, and the fact that the form of apoptotic cell, the phenotype and physiologic state of the antigen-presenting cells, and the relevant tissue microenvironment may play important roles [84, 92]. It is also of relevance that some of the antibodies elaborated in patients with systemic autoimmunity recognize the surface of apoptotic cells and might account for the apparently normal clearance of apoptotic cells seen in some patients. The critical difference between clearance of apoptotic cells through non-inflammatory pathways and clearance of apoptotic cells opsonized with autoantibodies is that the latter pathway effectively induces pro-inflammatory cytokine secretion by macrophages, potentially contributing to the propagation of the autoimmune response to other components in apoptotic cells, as well as to tissue damage [93, 94].

14.7 Model of Systemic Autoimmunity

The studies presented above have focused attention on unique forms of apoptosis as a candidate process that initiates and propagates systemic autoimmune diseases (summarized in Tab. 14.2). In the genetically susceptible individual (e.g. someone who has a defect in the ability to efficiently phagocytose and degrade apoptotic cells and debris or to mount an adequate anti-inflammatory response upon ingestion of apoptotic material), the confluence of several forces allows the generation of supra-threshold concentrations of non-tolerized structure in the presence of co-stimulatory signals and the access of this material to the MHC class II pathway of a population of antigen-presenting cells that efficiently initiate a primary immune response. The low frequency of this form of autoimmunity in the population likely reflects this need to simultaneously satisfy several very stringent criteria to initiate the primary immune response. The molecules targeted are unified by their susceptibility to modification during the perturbing process, likely revealing previously cryptic structure. Available data demonstrates that many (but not all) autoantigens are specifically

Tab. 14.2 Critical events in systemic autoimmunity.

Genetic susceptibility	genes may affect many aspects of immunoregulation, cell cycle, apoptotic signaling and effector pathways or clearance of apoptotic cells, amongst others
Primary immunization – a pro-immune apoptotic event	(1) unique environmental force (2) generation of non-tolerized structure (3) pro-immune context (4) specific immunizing microenvironment
Amplification	immune effector pathways generate antigen, which further drives antigen release and an autoamplifying cycle of immune system activation

cleaved by caspases during apoptosis. Furthermore, a similar but not identical subset of autoantigens is also directly cleaved by GrB, generating unique fragments not observed during any other forms of apoptosis studied to date. Once primary immunization has occurred, the repeated generation of apoptotic material (e.g. during sun exposure, viral infection or drug exposure) might efficiently rechallenge the primed immune system (the stringency of this secondary response being significantly lower than that of the primary response). Furthermore, the effector pathways activated by the primed immune system include several which themselves generate loads of apoptotic material (e.g. cellular cytotoxicity, myelomonocytic cell recruitment and apoptosis). The opsonization of apoptotic material by antiphospholipid and anti-β_2-glycoprotein I antibodies, may both increase the efficiency of apoptotic antigen capture, as well as induce the production of pro- rather than anti-inflammatory cytokines, potentially further driving the immune response. This capacity for immune-driven autoamplification may be one of the critical principles underlying severe systemic autoimmune disease.

14.8 Acknowledgments

This work was supported by National Institutes of Health grants AR44684 (L. C. R.) and DE12354 (A. R.), the SLE Foundation, the Lynchpin Foundation, and an Arthritis Foundation Maryland Chapter MARRC Grant. A. R. is supported by a Burroughs Wellcome Fund Translational Research Award.

14.9 References

1 TAN, E.M. Autoantibodies in pathology and cell biology. *Cell* **1991**, 67, 841–2.

2 VON MUHLEN, C.A. and TAN, E.M. Autoantibodies in the diagnosis of systemic rheumatic diseases. *Semin Arthitis Rheum* **1995**, 24, 328–58.

3 MOHAN, C., ADAMS, S., STANIK, V. and DATTA, S.K. Nucleosome: a major immunogen for pathogenic autoantibody-inducing T cells of lupus. *J Exp Med* **1993**, 177, 1367–81.

4 SHERO, J.H., BORDWELL, B., ROTHFIELD, N.F. and EARNSHAW, W.C. High titers of autoantibodies to topoisomerase I (Scl-70) in sera from scleroderma patients. *Science* **1986**, 231, 737–40.

5 MOROI, Y., PEEBLES, C., FRITZLER, M.J., STEIGERWALD, J. and TAN, E.M. Autoantibody to centromere (kinetochore) in scleroderma sera. *Proc Natl Acad Sci USA* **1980**, 77, 1627–31.

6 WIGLEY, F.M., WISE, R.A., MILLER, R., NEEDLEMAN, B.W. and SPENCE, R.J. Anticentromere antibody as a predictor of digital ischemic loss in patients with systemic sclerosis. *Arthritis Rheum* **1992**, 35, 688–93.

7 RADIC, M.Z. and WEIGERT, M. Genetic and structural evidence for antigen selection of anti-DNA antibodies. *Annu Rev Immunol* **1994**, 12, 487–520.

8 BURLINGAME, R.W., RUBIN, R.L., BALDERAS, R.S. and THEOFILOPOULOS, A.N. Genesis and evolution of antichromatin autoantibodies in murine lupus implicates T-dependent immunization with self-antigen. *J Clin Invest* **1993**, 91, 1687–96.

9 Diamond, B., Katz, J. B., Paul, E., Aranow, C., Lustgarten, D. and Scharff, M. D. The role of somatic mutation in the pathogenic anti-DNA response. *Annu Rev Immunol* **1992**, 10, 731–57.

10 Rosen, A. and Casciola-Rosen, L. Autoantigens as substrates for apoptotic proteases: implications for the pathogenesis of systemic autoimmune disease. *Cell Death Different* **1999**, 6, 6–12.

11 Sercarz, E. E., Lehmann, P. V., Ametani, A., Benichou, G., Miller, A. and Moudgil, K. Dominance and crypticity of T cell antigenic determinants. *Annu Rev Immunol* **1993**, 11, 729–66.

12 Gammon, G., Sercarz, E. E. and Benichou, G. The dominant self and the cryptic self: shaping the autoreactive T-cell repertoire. *Immunol Today* **1991**, 12, 193–5.

13 Lanzavecchia, A. How can cryptic epitopes trigger autoimmunity? *J Exp Med* **1995**, 181, 1945–8.

14 Lehmann, P. V., Forsthuber, T., Miller, A. and Sercarz, E. E. Spreading of T-cell autoimmunity to cryptic determinants of an autoantigen. *Nature* **1992**, 358, 155–7.

15 Mamula, M. J. Lupus autoimmunity: from peptides to particles. *Immunol Rev* **1995**, 144, 301–14.

16 Simitsek, P. D., Campbell, D. G., Lanzavecchia, A., Fairweather, N. and Watts, C. Modulation of antigen processing by bound antibodies can boost or suppress class II major histocompatibility complex presentation of different T cell determinants. *J Exp Med* **1995**, 181, 1957–63.

17 Watts, C. and Lanzavecchia, A. Suppressive effect of an antibody on processing of T cell epitopes. *J Exp Med* **1993**, 178, 1459–63.

18 Salemi, S., Caporossi, A. P., Boffa, L., Longobardi, M. G. and Barnaba, V. HIVgp120 activates autoreactive CD4-specific T cell responses by unveiling of hidden CD4 peptides during processing. *J Exp Med* **1995**, 181, 2253–7.

19 Casciola-Rosen, L. A., Anhalt, G. and Rosen, A. Autoantigens targeted in systemic lupus erythematosus are clustered in two populations of surface structures on apoptotic keratinocytes. *J Exp Med* **1994**, 179, 1317–30.

20 Casciola-Rosen, L., Andrade, F., Ulanet, D., Wong, W. B. and Rosen, A. Cleavage by granzyme B is strongly predictive of autoantigen status: implications for initiation of autoimmunity. *J Exp Med* **1999**, 190, 815–25.

21 Casciola-Rosen, L., Nicholson, D. W., Chong, T., Rowan, K R., Thornberry, N. A., Miller, D. K. and Rosen, A. Apopain/CPP32 cleaves proteins that are essential for cellular repair: a fundamental principle of apoptotic death. *J Exp Med* **1996**, 183, 1957–64.

22 Andrade, F., Roy, S., Nicholson, D., Thornberry, N., Rosen, A. and Casciola-Rosen, L. Granzyme B directly and efficiently cleaves several downstream caspase substrates: implications for CTL-induced apoptosis. *Immunity* **1998**, 8, 451–60.

23 Casciola-Rosen, L. A., Miller, D. K., Anhalt, G. J. and Rosen, A. Specific cleavage of the 70-kDa protein component of the U1 small nuclear ribonucleoprotein is a characteristic biochemical feature of apoptotic cell death. *J Biol Chem* **1994**, 269, 30757–60.

24 Casciola-Rosen, L. A., Anhalt, G. J. and Rosen, A. DNA-dependent protein kinase is one of a subset of autoantigens specifically cleaved early during apoptosis. *J Exp Med* **1995**, 182, 1625–34.

25 Odin, J. A., Huebert, R. C., Casciola-Rosen, L., LaRusso, N. F. and Rosen, A. Bcl-2-dependent oxidation of pyruvate dehydrogenase-E2, a primary biliary cirrhosis autoantigen, during apoptosis. *J Clin Invest* **2001**, 108, 223–32.

26 Casiano, C. A., Martin, S. J., Green, D. R. and Tan, E. M. Selective cleavage of nuclear autoantigens during CD95 (Fas/APO-1)-mediated T cell apoptosis. *J Exp Med* **1996**, 184, 765–70.

27 Casiano, C. A., Ochs, R. L. and Tan, E. M. Distinct cleavage products of nuclear proteins in apoptosis and necrosis revealed by autoantibody probes. *Cell Death Different* **1998**, 5, 183–90.

28 Utz, P. J., Hottelet, M., Le, T. M., Kim, S. J., Geiger, M. E., van Venrooij, W. J. and Anderson, P. The 72-kDa component of signal recognition particle is cleaved during apoptosis. *J Biol Chem* **1998**, 273, 35362–70.

29 Utz, P. J., Hottelet, M., Schur, P. H. and Anderson, P. Proteins phosphorylated during stress-induced apoptosis are common targets for autoantibody production in patients with systemic lupus erythematosus. *J Exp Med* **1997**, 185, 843–54.

30 Laman, S. D. and Provost, T. T. Cutaneous manifestations of lupus erythematosus. *Rheum Dis Clin N Am* **1994**, 20, 195–212.

31 Orteu, C. H., Sontheimer, R. D. and Dutz, J. P. The pathophysiology of photosensitivity in lupus erythematosus. *Photodermatol Photoimmunol Photomed* **2001**, 17, 95–113.

32 LeFeber, W. P., Norris, D. A., Ryan, S. R., Huff, J. C., Lee, L. A., Kubo, M., Boyce, S. T., Kotzin, B. L. and Weston, W. L. Ultraviolet light induces binding of antibodies to selected nuclear antigens on cultured human keratinocytes. *J Clin Invest* **1984**, 74, 1545–51.

33 Golan, T. D., Elkon, K. B., Gharavi, A. E. and Krueger, J. G. Enhanced membrane binding of autoantibodies to cultured keratinocytes of systemic lupus erythematosus patients after ultraviolet B/ ultraviolet A irradiation. *J Clin Invest* **1992**, 90, 1067–76.

34 Casciola-Rosen, L. and Rosen, A. Ultraviolet light-induced keratinocyte apoptosis: a potential mechanism for the induction of skin lesions and autoantibody production in LE. *Lupus* **1997**, 6, 175–80.

35 Fadok, V. A., Voelker, D. R., Campbell, P. A., Cohen, J. J., Bratton, D. L. and Henson, P. M. Exposure of phosphatidylserine on the surface of apoptotic lymphocytes triggers specific recognition and removal by macrophages. *J Immunol* **1992**, 148, 2207–16.

36 Koopman, G., Reutelingsperger, C. P., Kuitjen, G. A., Keehnen, R. M., Pals, S. T. and van Oers, M. H. Annexin V for flow cytometric detection of phosphatidylserine expression on B cells undergoing apoptosis. *Blood* **1994**, 84, 1415–20.

37 Casciola-Rosen, L., Rosen, A., Petri, M. and Schlissel, M. Surface blebs on apoptotic cells are sites of enhanced procoagulant activity: implications for coagulation events and antigenic spread in systemic lupus erythematosus. *Proc Natl Acad Sci USA* **1996**, 93, 1624–9.

38 Martin, S. J., Finucane, D. M., Amarante-Mendes, G. P., O'Brien, G A. and Green, D. R. Phosphatidylserine externalization during CD95-induced apoptosis of cells and cytoplasts requires ICE/CED-3 protease activity. *J Biol Chem* **1996**, 271, 28753–6.

39 Price, B. E., Rauch, J., Shia, M. A., Walsh, M. T., Lieberthal, W., Gilligan, H. M., O'Laughlin, T., Koh, J. S. and Levine, J. S. Anti-phospholipid autoantibodies bind to apoptotic, but not viable, thymocytes in a β_2-glycoprotein I-dependent manner. *J Immunol* **1996**, 157, 2201–8.

40 Casciola-Rosen, L., Pluta, A. F., Plotz, P. H., Cox, A. E., Morris, S., Wigley, F. M., Petri, M., Gelber, A. C. and Rosen, A. The DNA mismatch repair enzyme PMS1 is a myositis-specific autoantigen. *Arthritis Rheum* **2001**, 44, 389–96.

41 Cocca, B. A., Seal, S. N., D'Agnillo, P., Mueller, Y. M., Katsikis, P. D., Rauch, J., Weigert, M. and Radic, M. Z. Structural basis for autoantibody recognition of phosphatidylserine – $beta_2$ glycoprotein I and apoptotic cells. *Proc Natl Acad Sci USA* **2001**, 98, 13826–31.

42 Korb, L. C. and Ahearn, J. M. C1q binds directly and specifically to surface blebs of apoptotic human keratinocytes – complement deficiency and systemic lupus erythematosus revisited. *J Immunol* **1997**, 158, 4525–8.

43 Navratil, J. S., Watkins, S. C., Wisnieski, J. J. and Ahearn, J. M. The globular heads of C1q specifically recognize surface blebs of apoptotic vascular endothelial cells. *J Immunol* **2001**, 166, 3231–9.

44 Wener, M. H., Uwatoko, S. and Mannik, M. Antibodies to the collagen-like region of C1q in sera of patients with autoimmune rheumatic diseases. *Arthritis Rheum* **1989**, 32, 544–51.

45 Gershov, D., Kim, S., Brot, N. and Elkon, K. B. C-reactive protein binds to apoptotic cells, protects the cells from assembly of the terminal complement components, and sustains an antiinflammatory innate immune response: implications for systemic autoimmunity. *J Exp Med* **2000**, 192, 1353–64.

46 Botto, M., Dell'Agnola, C., Bygrave, A. E., Thompson, E. M., Cook, H.T., Petry, F., Loos, M., Pandolfi, P. P. and Walport, M. J. Homozygous C1q deficiency causes glomerulonephritis associated with multiple apoptotic bodies. *Nat Genet* **1998**, 19, 56–9.

47 Bowness, P., Davies, K. A., Norsworthy, P. J., Athanassiou, P., Taylor-Wiedeman, J., Borysiewicz, L. K., Meyer, P. A. R. and Walport, M. J. Hereditary C1q deficiency and systemic lupus erythematosus. *Q J Med* **1994**, 87, 455–64.

48 Hengartner, M. O. The biochemistry of apoptosis. *Nature* **2000**, 407, 770–776.

49 Thornberry, N. A. and Lazebnik, Y. Caspases: enemies within. *Science* **1998**, 281, 1312–6.

50 Kaufmann, S. H., Desnoyers, S., Ottaviano, Y., Davidson, N. E. and Poirier, G. G. Specific proteolytic cleavage of poly(ADP-ribose) polymerase: an early marker of chemotherapy-induced apoptosis. *Cancer Res* **1993**, 53, 3976–85.

51 Kaufmann, S. H. Induction of endonucleolytic DNA damage in human acute myelogenous leukemia cells by etoposide, camptothecin, and other cytotoxic anticancer drugs: a cautionary note. *Cancer Res* **1989**, 49, 5870–8.

52 Yamanaka, H., Willis, E. H., Penning, C. A., Peebles, C. L., Tan, E. M. and Carson, D. A. Human autoantibodies to poly(adenosine disphosphate-ribose) polymerase. *J Clin Invest* **1987**, 80, 900–4.

53 Reeves, W. H., Chaudhary, N., Salerno, A. and Blobel, G. Lamin B autoantibodies in sera of certain patients with systemic lupus erythematosus. *J Exp Med* **1987**, 165, 750–62.

54 Waterhouse, N., Kumar, S., Song, Q. H., Strike, P., Sparrow, L., Dreyfuss, G., Alnemri, E. S., Litwack, G., Lavin, M. and Watters, D. Heteronuclear ribonucleoproteins C1 and C2, components of the spliceosome, are specific targets of interleukin 1b-converting enzyme-like proteases in apoptosis. *J Biol Chem* **1996**, 271, 29335–41.

55 Haneji, N., Nakamura, T., Takio, K., Yanagi, K., Higashiyama, H., Saito, I., Noji, S., Sugino, H. and Hayashi, Y. Identification of α-fodrin as a candidate autoantigen in primary Sjögren's syndrome. *Science* **1997**, 276, 604–7.

56 Martin, S. J., O'Brien, G. A., Nishioka, W. K., McGahon, A. J., Mahboubi, A., Saido, T. C. and Green, D. R. Proteolysis of fodrin (non-erythroid spectrin) during apoptosis. *J Biol Chem* **1995**, 270, 6425–8.

57 Ayukawa, K., Taniguchi, S., Masumoto, J., Hashimoto, S., Sarvotham, H., Hara, A., Aoyama, T. and Sagara, J. La autoantigen is cleaved in the COOH terminus and loses the nuclear localization signal during apoptosis. *J Biol Chem* **2000**, 275, 34465–70.

58 Casciola-Rosen, L. A., Pluta, A. F., Plotz, P. H., Cox, A. E., Morris, S., Wigley, F. M., Petri, M., Gelber, A. C. and Rosen, A. The DNA mismatch repair enzyme PMS1 is a myositis-specific autoantigen. *Arthritis Rheum* **2001**, 44, 389–96.

59 Satoh, M., Ajmani, A. K., Ogasawara, T., Langdon, J. J., Hirakata, M., Wang, J. and Reeves, W. H. Autoantibodies to RNA polymerase II are common in systemic lupus erythematosus and overlap syndrome. Specific recognition of the phosphorylated (IIO) form by a subset of human sera. *J Clin Invest* **1994**, 94, 1981–1989.

60 Neugebauer, K. M., merrill, J. T., Wener, M. H., Lahita, R. G. and Roth, M. B. SR proteins are autoantigens in patients with systemic lupus erythematosus. *Arthritis Rheum* **2000**, 43, 1768–78.

61 Fadok, V. A., Bratton, D. L., Konowal, A., Freed, P. W., Westcott, J. Y. and Henson, P. M. Macrophages that have ingested apoptotic cells *in vitro* inhibit proinflammatory cytokine production through autocrine/paracrine mechanisms involving TGF-β, PGE_2, and PAF. *J Clin Invest* **1998**, 101, 890–8.

62 Voll, R. E., Herrmann, M., Roth, E. A., Stach, C., Kalden, J. R. and Girkontaite, I. Immunosuppressive effects of apoptotic cells. *Nature* **1997**, 390, 350–1.

63 Huynh, M. L., Fadok, V. A. and Henson, P. M. Phosphatidylserine-dependent ingestion of apoptotic cells promotes TGF-beta1 secretion and the resolution of inflammation. *J Clin Invest* **2002**, 109, 41–50.

64 Hoffmann, P. R., deCathelineau, A. M., Ogden, C. A., Leverrier, Y., Bratton, D. L., Daleke, D. L., Ridley, A. J., Fadok, V. A. and Henson, P. M. Phosphatidylserine (PS) induces PS receptor-mediated macropinocytosis and promotes clearance of apoptotic cells. *J Cell Biol* **2001**, 155, 649–59.

65 Henkart, P. A. Lymphocyte-mediated cytotoxicity: two pathways and multiple effector molecules. *Immunity* **1994**, 1, 343–346.

66 Poe, M., Blake, J. T., Boulton, D. A., Gammon, M., Sigal, N. H., Wu, J. K. and Zweerink, H. J. Human cytotoxic lymphocyte granzyme B: Its purification from granules and the characterization of substrate and inhibitor specificity. *J Biol Chem* **1991**, 266, 98–103.

67 Thornberry, N. A., Rano, T. A., Peterson, E. P., Rasper, D. M., Timkey, T., Garcia-Calvo, M., Houtzager, V. M., Nordstrom, P. A., Roy, S., Vaillancourt, J. P. *et al.* A combinatorial approach defines specificities of members of the caspase family and granzyme B. *J Biol Chem* **1997**, 272, 17907–11.

68 Harris, J. L., Peterson, E. P., Hudig, D., Thornberry, N. A. and Craik, C. S. Definition and redesign of the extended substrate specificity of granzyme B. *J Biol Chem* **1998**, 273, 27364–73.

69 Shi, L., Kam, C.-M., Powers, J. C., Aebersold, R. and Greenberg, A. H. Purification of three cytotoxic lymphocyte granule serine proteases that induce apoptosis through distinct substrate and target cell interactions. *J Exp Med* **1992**, 176, 1521–9.

70 Heusel, J. W., Wesselschmidt, R. L., Shresta, S., Russell, J. H. and Ley, T. J. Cytotoxic lymphocytes require granzyme B for the rapid induction of DNA fragmentation and apoptosis in allogeneic target cells. *Cell* **1994**, 76, 977–87.

71 Shresta, S., MacIvor, D. M., Heusel, J. W., Russell, J. H. and Ley, T. J. Natural killer and lymphokine-activated killer cells require granzyme B for the rapid induction of apoptosis in susceptible target cells. *Proc Natl Acad Sci USA* **1995**, 92, 5679–83.

72 Salvesen, G. and Dixit, V. M. Caspases: Intracellular signaling by proteolysis. *Cell* **1997**, 91, 443–6.

73 Li, H. L., Zhu, H., Xu, C. J. and Yuan, J. Y. Cleavage of BID by caspase 8 mediates the mitochondrial damage in the Fas pathway of apoptosis. *Cell* **1998**, 94, 491–501.

74 Pinkoski, M. J., Waterhouse, N. J., Heibein, J. A., Wolf, B. B., Kuwana, T., Goldstein, J. C., Newmeyer, D. D., Bleackley, R. C. and Green, D. R. Granzyme B-mediated apoptosis proceeds predominantly through a Bcl-2-inhibitable mitochondrial pathway. *J Biol Chem* **2001**, 276, 12060–7.

75 Alimonti, J. B., Shi, L., Baijal, P. K. and Greenberg, A. H. Granzyme B induces Bid-mediated cytochrome *c* release and mitochondrial permeability transition. *J Biol Chem* **2001**, 276, 6974–82.

76 Sutton, V. R., Davis, J. E., Cancilla, M., Johnstone, R. W., Ruefli, A. A., Sedelies, K., Browne, K. A. and Trapani, J. A. Initiation of apoptosis by granzyme B requires direct cleavage of bid, but not direct granzyme B-mediated caspase activation. *J Exp Med* **2000**, 192, 1403–14.

77 Heibein, J. A., Goping, I. S., Barry, M., Pinkoski, M. J., Shore, G. C., Green, D. R. and Bleackley, R. C. Granzyme B-mediated cytochrome *c* release is regulated by the Bcl-2 family members Bid and Bax. *J Exp Med* **2000**, 192, 1391–402.

78 Thomas, D. A., Du, C., Xu, M., Wang, X. and Ley, T. J. DFF45/ICAD can be directly processed by granzyme B during the induction of apoptosis. *Immunity* **2000**, 12, 621–32.

79 Nagaraju, K., Cox, A., Casciola-Rosen, L. and Rosen, A. Novel fragments of the Sjögren's syndrome autoantigens alpha-fodrin and type 3 muscarinic acetylcholine receptor are generated during cytotoxic lymphocyte granule-induced cell death. *Arthritis Rheum* **2001**, 44, 2376–86.

80 Gahring, L., Carlson, N. G., Meyer, E. L. and Rogers, S. W. Granzyme B proteolysis of a neuronal glutamate receptor generates an autoantigen and is modulated by glycosylation. *J Immunol* **2001**, 166, 1433–8.

81 Fadok, V.A., Bratton, D.L., Frasch, S.C., Warner, M.L. and Henson, P.M. The role of phosphatidylserine in recognition of apoptotic cells by phagocytes. *Cell Death Different* **1998**, 5, 551–62.

82 Fadok, V.A. and Henson, P.M. Apoptosis: getting rid of the bodies. *Curr Biol* **1998**, 8, R693–5.

83 Henson, P.M., Bratton, D.L. and Fadok, V.A. Apoptotic cell removal. *Curr Biol* **2001**, 11, R795–805.

84 Savill, J. Apoptosis – phagocytic docking without shocking. *Nature* **1998**, 392, 442–443.

85 Ren, Y. and Savill, J. Apoptosis: the importance of being eaten. *Cell Death Different* **1998**, 5, 563–8.

86 Taylor, P.R., Carugati, A., Fadok, V.A., Cook, H.T., Andrews, M., Carroll, M.C., Savill, J.S., Henson, P.M., Botto, M. and Walport, M.J. A hierarchical role for classical pathway complement proteins in the clearance of apoptotic cells *in vivo*. *J Exp Med* **2000**, 192, 359–66.

87 Scott, R.S., McMahon, E.J., Pop, S.M., Reap, E.A., Caricchio, R., Cohen, P.L., Earp, H.S. and Matsushima, G.K. Phagocytosis and clearance of apoptotic cells is mediated by MER. *Nature* **2001**, 411, 207–11.

88 Lu, Q. and Lemke, G. Homeostatic regulation of the immune system by receptor tyrosine kinases of the Tyro 3 family. *Science* **2001**, 293, 306–11.

89 Albert, M.L., Sauter, B. and Bhardwaj, N. Dendritic cells acquire antigen from apoptotic cells and induce class I restricted CTLs. *Nature* **1998**, 392, 86–9.

90 Herrmann, M., Voll, R.E., Zoller, O.M., Hagenhofer, M., Ponner, B.B. and Kalden, J.R. Impaired phagocytosis of apoptotic cell material by monocyte-derived macrophages from patients with systemic lupus erythematosus. *Arthritis Rheum* **1998**, 41, 1241–50.

91 Somersan, S. and Bhardwaj, N. Tethering and tickling: a new role for the phosphatidylserine receptor. *J Cell Biol* **2001**, 155, 501–4.

92 Miranda-Carus, M.E., Askanase, A.D., Clancy, R.M., Di Donato, F., Chou, T.M., Libera, M.R., Chan, E.K. and Buyon, J.P. Anti-SSA/Ro and anti-SSB/La autoantibodies bind the surface of apoptotic fetal cardiocytes and promote secretion of TNF-alpha by macrophages. *J Immunol* **2000**, 165, 5345–51.

93 Manfredi, A.A., Rovere, P., Galati, G., Heltai, S., Bozzolo, E., Soldini, L., Davoust, J., Balestrieri, G., Tincani, A. and Sabbadini, M.G. Apoptotic cell clearance in systemic lupus erythematosus – I. Opsonization by antiphospholipid antibodies. *Arthritis Rheum* **1998**, 41, 205–14.

15 Distinct Cleavage Products of Nuclear Autoantigens in Apoptosis and Necrosis: Implications for Autoimmunity

Carlos A. Casiano, Christine Molinaro, Sheldon Holder and Xiwei Wu

15.1 Introduction

Central to understanding the pathogenesis of systemic autoimmune diseases such as systemic lupus erythematosus (SLE), scleroderma, rheumatoid arthritis and Sjögren's syndrome is elucidating the mechanisms that contribute to the generation of antinuclear autoantibodies (ANA). There is strong evidence to support the hypothesis that ANA responses in systemic autoimmune diseases are driven by proteins and nucleic acids associated with nuclear and cytoplasmic particles [1, 2].

A fundamental question that remains unanswered is how normally sequestered self-intracellular antigens turn into evil immunogens capable of inciting and maintaining a vigorous and prolonged autoantibody response. A hypothesis that has emerged during the past decade is that dying cells serve as potential reservoirs of modified forms of autoantigens that could trigger autoantibody responses in susceptible individuals under appropriate conditions [3–5]. This hypothesis is supported by two main lines of evidence: (1) the link between impaired phagocytic clearance of apoptotic cells and systemic autoimmunity [6–8], and (2) the observation that intracellular autoantigens targeted by autoantibodies in systemic autoimmunity undergo post-translational modifications during cell death that might increase their immunogenicity [3, 4]. The sort of modifications sustained by autoantigens during cell death include, but are not limited to, proteolysis [9–11], changes in phosphorylation state [12, 13], oxidation [14, 15], transglutaminase crosslinking [16], citrunillation [17], and ubiquitin conjugation and deconjugation [4]. It has been hypothesized that this plethora of modifications may result in the presentation to the immune system of self-cryptic determinants for which tolerance has not been established and that upon repeated stimulation could sustain an autoantibody response [3, 4]. Thus far, the most characterized autoantigen modification associated with cell death, particularly apoptosis, is proteolytic cleavage. While the cleavage of intracellular autoantigens during apoptosis has been the focus of increasing attention during the past few years, information on the cleavage of autoantigens during non-apoptotic cell death is just beginning to emerge. Along with other groups, we have demonstrated recently that intracellular autoantigens also undergo cleavage during non-apoptotic cell death modalities, including primary and secondary necrosis and caspase-independent cell death. These non-apoptotic cleavages could also enhance the immunogenic properties of intracellular autoantigens and stimulate an immune response under the appropriate environment. In this chapter, we will first present a general overview of the major modes of cell death. This will be followed by a review of the current knowledge of the cleavage of intracellular autoantigens during various forms of cell death and a brief discussion of the implications of autoantigen cleavage for the induction of autoantibody responses.

15.2 The Multiple Faces of Cell Death

Traditionally, cell death has been considered as a type of Dr Jekyll and Mr Hyde, a process with two interrelated but morphologically distinct faces, apoptosis (the good fellow) and necrosis (the bad fellow) [18–21]. This bimodal classification relies mainly on the observation that apoptosis and necrosis are the major types of cell death associated with most physiological and pathological processes. Recent studies, however, have challenged this dual distinction based on the identification of novel forms of cell death whose morphological features do not fit into the classical apoptotic or necrotic morphologies [22, 23]. In this section, we will provide a general description of apoptosis, necrosis and caspase-independent cell death, which are the three cell death modalities that have been best studied within the context of autoimmunity.

15.2.1 Apoptosis

Apoptosis is a genetically regulated cell suicide process that is essential for the elimination of unwanted cells during organ development, immune system development and function, tumor regression and normal tissue turnover [18–21]. Defects in apoptosis have been associated with the pathogenesis of several human disease conditions, including autoimmunity, cancer, diabetes, liver disease and neurodegeneration [24–27]. Apoptotic cells display a distinctive morphology characterized by general cellular shrinkage, cytoskeleton disruption, cytoplasmic membrane blebbing, nuclear membrane solubilization and chromatin fragmentation. A hallmark of apoptosis is the fragmentation of the dying cell into numerous apoptotic bodies surrounded by a relatively intact cytoplasmic membrane. Retention of cytoplasmic membrane integrity not only facilitates swift phagocytic recognition but also prevents the release of potentially harmful intracellular contents, such as proteases and pro-inflammatory signals that could damage the surrounding tissue and provoke an inflammatory response [28]. Apoptotic cells or bodies that are not cleared eventually lose their cytoplasmic membrane integrity and develop secondary necrosis, resulting in the release of intracellular contents [28].

The activation of cysteine proteases of the caspase (cysteine aspartic acid-specific proteases) family has emerged as the central effector mechanism in apoptosis

[29, 30]. While approximately 14 mammalian caspases have been identified, only about half of them actively participate in the execution of apoptosis. Caspases are classified as initiators (caspase-2, -8, -9 and -10) or executioners (caspase-3, -6 and -7) of the apoptotic process. Caspases are typically found in the cell as inactive precursors and are activated at the onset of apoptosis by an autoaggregation process mediated by adaptor proteins which promote autocatalytic processing of the initiator caspases [29, 30]. Activation of initiator caspases can occur through two basic pathways: (1) engagement of death receptors, such as Fas and tumor necrosis factor (TNF) receptor 1 (TNF-RI), and (2) release of caspase-activating factors from the mitochondria [31, 32]. Once activated, initiator caspases process the executioner caspases, which in turn target a limited number of proteins, including autoantigens, involved in key cellular functions [33].

15.2.2
Necrosis

Necrosis has been traditionally considered as a non-suicidal process associated with a number of pathological conditions that develops in response to acute cell injury, including ischemia, hypoxia, oxidative stress, extreme heat, severe infections and exposure to high levels of chemicals or toxins [20, 21, 34–38]. In pathological and experimental conditions, necrosis often co-exists with apoptosis, arising either independently or as a secondary event following apoptotic cell death [20, 21, 37–43]. Under certain circumstances, extensive necrosis of cells in a particular region of a tissue may serve as a trigger of secondary tissue damage, via apoptosis, in surrounding areas [38, 44, 45]. It should be emphasized that necrosis and apoptosis can be induced *in vivo* and *in vitro* by the same insults, but the intensity of the insult determines which mode of cell death prevails [34]. While the role of necrosis in physiological processes is still not very clear, recent studies have implicated this mode of cell death in natural killer cell-mediated cytotoxicity [46, 47], interdigital cell death [48], development-associated regression of the human tail [49] and egg fertilization [50]. There is growing evidence to support the notion that there might be two types of necrosis, one that is physiological and is involved in programmed cell death, and another that is accidental and is associated with pathological conditions [22].

The morphology of necrotic cells is characterized by extensive cytoplasmic swelling and destruction, and nuclear shrinkage [20, 21]. A key feature that distinguishes necrosis from apoptosis is the rapid and early loss of cytoplasmic membrane integrity due to the swelling, which leads to extensive cytoplasmic damage with concomitant release of noxious intracellular contents, including pro-inflammatory signals. Interestingly, the cell nucleus stays relatively intact during necrosis [20, 21, 51], which could be associated with the preservation of lamin B integrity during this cell death process [11, 51, 52]. Lamin B is important for maintaining nuclear membrane integrity and its cleavage during apoptosis facilitates nuclear fragmentation [53].

While the role of proteases in the execution of apoptosis is well established, it is not clear whether the morphological changes observed during necrosis are driven by proteolysis. Activation of non-caspase proteases such as calpains and cathepsins (lysosomal proteases) has been implicated in necrotic cell death, both in experimental and pathological situations [38, 54]. There is evidence that elevation of intracellular free calcium during certain pathological processes leads to activation of calpains, phospholipases and endonucleases, alteration of membrane protein and lipid, generation of toxic reactive oxygen species, and mitochondrial disruption [38]. Excessive activation of calpains has been associated with lysosomal membrane disruption, leading to the release of cathepsins into the cytoplasm with resultant cell autolysis [38]. Understanding the mechanisms of self-digestion mediated by cathepsins would require identifying the individual cathepsins involved and their substrates.

15.2.3
Caspase-independent Cell Death

The term caspase-independent cell death has been used to describe a number of cell death pathways that occur in the absence of detectable caspase activity. This type of cell death could be considered as a back-up system that ensures the cell's demise in the event that the caspase activation program is rendered non-functional. Caspase-independent cell death is usually triggered in cell culture by inhibition of caspases with a pan-caspase inhibitor like benzyloxycarbonyl-Val-Ala-Asp-fluoromethyl ketone (zVAD-fmk), in the presence of a wide variety of apoptosis inducers such as cancer drugs [55, 56], death receptor ligands [57, 58], oncogenes [59], anti-CD2 antibodies [60], staurosporine (STS) [60], viral proteins [61] and expression of Bax-related proteins [59, 62]. It is unclear whether caspase-independent cell death occurs concurrently with apoptosis as a background pathway that is revealed or enhanced when the caspase activation program has been impaired or whether it is activated after the cell senses this impairment.

Some caspase-independent cell death pathways share morphological features with classical apoptosis whereas others are characterized by a necrotic morphology [22]. Whether a particular pathway is associated with apoptotic or necrotic morphology appears to depend on the cell type and the nature of the insult. For instance, apoptotic features such as chromatin condensation and DNA fragmentation, and phosphatidylserine exposure on the cell surface can be induced by release of apoptosis-inducing factor from the mitochondria in the presence of zVAD-fmk [63]. Other examples of cell death with morphological features of apoptosis (nuclear or cytoplasmic) in the presence of caspase inhibitors include colchicine-induced cerebellar granule cell death [64], CD47-induced death of B-chronic lymphocytic leukemia cells [65] and anti-CD2 or STS-induced cell death in activated human peripheral T lymphocytes [60]. It is unclear whether the apoptotic features associated with caspase-independent cell death are mediated by yet to be identified caspases that are insensitive to the known caspase inhibitors or by alternative mechanisms.

Most types of caspase-independent cell death described so far are associated with features of necrosis such as nuclear pyknosis, rapid loss of cytoplasmic membrane integrity and cytoplasmic fragmentation [22, 58, 66]. Examples include camptothecin-treatment of leukemia U-937 cells or STS treatment of human Jurkat T cells in the presence of zVAD-fmk [55, 67]. Recently, much attention has been devoted to a model system of caspase-independent cell death with necrotic morphology induced by the death receptors Fas and TNF-RI in susceptible cell lines, such as the murine L929 fibrosarcoma, exposed to either TNF or agonistic anti-Fas antibody in the presence of zVAD-fmk or other caspase inhibitors [68, 69]. In this system, L929 cells die by a swift and massive necrosis that appears to be mediated by the generation of reactive oxygen species since it can be partially inhibited by anti-oxidants such as butylated hydroxyanisole. Fas has also been shown to induce necrosis in a caspase 8-deficient subline (JB6) of Jurkat T cells [70]. Both TNF-RI and Fas appear to initiate the necrotic pathway through recruitment of the adaptor protein FADD, which is also required for initiating the caspase-8-mediated apoptotic pathway [70–73]. While there is no evidence that caspase-8 activation is required for death receptor-induced caspase-independent cell death with a necrotic phenotype, it appears that inactivation of caspase-8 favors this death process [70–73]. Whether inactivation of caspases other than caspase-8 also favors death receptor-induced necrosis remains to be investigated, although inhibition of caspase-3 has been implicated [69]. It is not clear how inactivation of caspases leads to TNF or Fas-induced necrosis, but it is possible that alternative proteolytic pathways, perhaps involving calpains and cathepsins, might drive this process when caspases are impaired. These observations suggest that inactivation of caspases removes a critical barrier yet to be identified that prevents the development of pro-inflammatory, necrotic cell death, thus implicating caspases as natural repressors of necrosis.

15.3 Autoantigen Cleavage in Apoptosis

The observation that the SLE-associated autoantigens poly(ADP-ribose) polymerase (PARP) and the 70-kDa protein of the U1 ribonucleoprotein particle (U1-70kDa) were proteolytically cleaved by caspases during apoptosis [74, 75] opened the door to the use of human autoantibodies in the systematic identification of other intracellular autoantigens cleaved during cell death. The value of autoantibodies in these studies is enhanced by their reactivity with multiple epitopes within a given autoantigen, which facilitates the identification of cleavage fragments that otherwise may escape detection when using sequence-specific, experimentally induced antibodies.

In the first systematic study on the cleavage of autoantigens during cell death, Casciola-Rosen *et al.* [9] screened over 200 human sera containing autoantibodies to intracellular proteins by immunoblotting against lysates of apoptotic cells. These investigators observed that a subset of intracellular autoantigens was specifically cleaved in various systems of apoptosis. This subset included PARP, U1-70kDa, the catalytic subunit of DNA-dependent protein kinase (DNA-PK$_{CS}$), the nuclear mitotic apparatus protein (NuMA), lamins A and B, and several other unidentified autoantigens. In a comparable study, investigators in our group used highly specific and well-characterized human autoantibodies to identify by immunoblotting autoantigens cleaved during Fas-mediated T cell apoptosis [10]. Only a subset (seven of 33) of the autoantigens examined underwent detectable proteolytic cleavage during apoptosis, as evidenced by the disappearance of bands corresponding to the intact protein concomitant with the appearance of lower molecular weight fragments in immunoblots. The cleaved autoantigens were PARP, lamin B, U1-70kDa, topoisomerases (Topo) I and II, NuMA, and the upstream binding factor of RNA polymerase I (UBF/NOR-90). More recent studies by various groups have reported the apoptotic cleavage of additional intracellular autoantigens such as actin [76, 77], the cancer-associated protein BARD1 [78], fodrin [79,

Tab. 15.1 Apoptotic cleavage of intracellular autoantigens associated with systemic autoimmunity.

Autoantigen	*Disease association*	*Intact protein size (kDa)*	*Major cleavage fragments (kDa)*
Actin	autoimmune hepatitis	45	15, 31
DFS70/LEDGF	AD, asthma, various inflammatory conditions	70	58, 65
DNA-PK	SLE, SSc, overlap syndromes	450	250
hnRNP C1 and C2	SSc, psoriasis	~40	disappearance
Fodrin	SS	240	120, 150
Golgin-160	SLE, SS	170	140, 163
Keratin 18	GVHD, DLE	45	19, 22, 26
Lamin B	SLE-like disease, APS, CFS	68–70	45
NuMA	SS	210–240	160, 180
PARP	SLE	116	85
RNA helicase A	SLE	140	120–130
SP1	UCTD	95, 105	45, 68
SRP72	PM/DM	72	6, 66
Topo I	SSc, PM	100	70
Topo II	SLE, fibrosing alveolitis	170/180	125–160
UBF/NOR-90	SS, SSc	90	24–55
U1-70kDa	SLE, SSc, MCTD	70	40
Vimentin	CFS	50	38, 41

AD = atopic dermatitis; APS = anti-phospholipid antibody syndrome; CFS = chronic fatigue syndrome; DFS70 = dense fine speckles protein of 70 kDa; DM = dermatomyositis; DNA-PK = DNA-dependent protein kinase; DLE = discoid lupus erythematosus; GVHD = graft-versus-host disease; hnRNP = heterogeneous nuclear ribonucleoprotein; LEDGF = lens epithelium derived growth factor; MCTD = mixed connective tissue disease; NuMA = nuclear mitotic apparatus protein; PARP = poly(ADP-ribose) polymerase; PM = polymyositis; SP1 = SP1 transcription factor; SS = Sjogren's syndrome; SRP72 = signal recognition particle protein of 72 kDa; SSc = systemic sclerosis/scleroderma; UBF/NOR–90 = upstream binding factor/nucleolar organizing region protein of 90 kDa; UCTD = undifferentiated connective tissue disease; U1-70kDa = U1 small nuclear RNA-associated protein of 70 kDa. The cleavage fragments listed are those detected by immunoblotting. The information provided on this table is based on [3, 9–11, 30, 33, 67, 74–94].

80], golgin-160 [81], heterogeneous nuclear ribonucleoproteins (hnRNP) C1 and C2 [82, 83], keratin [84, 85], the myositis associated antigens Mi-2 and PMS1 [86], RNA helicase A [87], the signal recognition particle protein of 72 kDa (SRP72) [88], SSB/La [89], transcription activator SP1 [90, 91], the transcription co-activator and survival protein DFS70/LEDGF [67, 92], and vimentin [93, 94]. The current information available on the apoptotic cleavage of autoantigens is summarized in Tab. 15.1.

It should be emphasized that many intracellular autoantigens frequently targeted by autoantibodies in systemic autoimmune diseases appear to remain intact during classical apoptosis, as assessed by the preservation of bands corresponding to the intact autoantigen and lack of putative cleavage products in immunoblots, as well as failure of caspases to cleave these autoantigens *in vitro* [9, 10, 95]. Among these 'uncleaved' autoantigens are alanyl tRNA synthetase, B23/nucleophosmin, C23/nucleolin, centromere protein B, fibrillarin, histidyl tRNA synthetase (Jo-1), histones, isoleucyl tRNA synthetase, DNA-binding protein Ku, the major mitochondrial autoantigens, PM-Scl, the proliferating cell nuclear antigen (PCNA), p80 coilin, ribosomal P proteins (rRNP), RNA polymerase II large subunit, Sm and SSA/Ro. These findings suggest that susceptibility to proteolytic cleavage is not a general property of intracellular autoantigens. However, it cannot be ruled out that some of the 'uncleaved' autoantigens may undergo limited proteolysis during apoptosis into small fragments that cannot be detected by immunoblotting. For instance, while the cleavage of the nucleolar autoantigen fibrillarin has not been detected in immunoblotting analyses performed by various groups [9, 10, 95], Pollard *et al.* [96] reported that apoptotic murine J774 macrophages, but not other murine cell lines, displayed limited cleavage of fibrillarin into small fragments that were detected only by immunoprecipitation. This observation raised the possibility that the apoptotic cleavage of a particular autoantigen might be cell type dependent. Another issue that needs to be borne in mind is that autoantigens that are not cleaved during apoptosis may sustain other types of post-translational modifications during cell death. For example, the ribosomal P proteins are not cleaved during apoptosis but undergo dephosphorylation during Fas-mediated apoptosis [13]. More detailed reviews of apoptotic autoantigen modifications are displayed elsewhere in this book.

15.4 Autoantigen Cleavage in Non-apoptotic Cell Death

During the past few years, investigators in our group and in other groups have used highly specific human autoantibodies and commercially available antibodies as tools in the systematic identification of intracellular autoantigens that undergo cleavage during non-apoptotic cell death. In these studies, cleaved forms of autoantigens are normally detected in immunoblots of whole-cell lysates from cells undergoing various types of non-apoptotic cell death, including primary necrosis, secondary necrosis and caspase-independent cell death.

15.4.1
Autoantigen Cleavage during Primary Necrosis

We observed that the cleavage of autoantigens during primary necrosis of Jurkat and HL-60 cells induced with high levels of mercury, ethanol, hydrogen peroxide or heat was associated with selective proteolysis of a subset of autoantigens [11, 67 and unpublished observations]. Cleaved autoantigens included DFS70/LEDGF, fodrin, NuMA, PARP, Topo I, UBF/NOR-90, U1-70kDa and vimentin (for examples, see Fig. 15.1). Interestingly, all these autoantigens are also cleaved during apoptosis, albeit into distinct fragments. On the other hand, autoantigens that had been reported previously as resistant to proteolysis during apoptosis [9, 10] appeared also to be resistant to cleavage during necrosis, as evidenced by the lack of proteolysis observed in the immunoblots for B23/nucleophosmin, fibrillarin, Jo-1, Ku, PM-Scl, PCNA, p80 coilin, rRNP, Sm and SSA/Ro. This suggested that specific caspase substrates may be highly susceptible to proteolysis during forms of cell death other than apoptosis, a notion that is supported by our studies on secondary necrosis and caspase-independent cell death ([67], and Sections 15.4.2 and 15.4.3). One exception is lamin B, which is cleaved during apoptosis but not during necrosis (Fig. 15.1). As in apop-

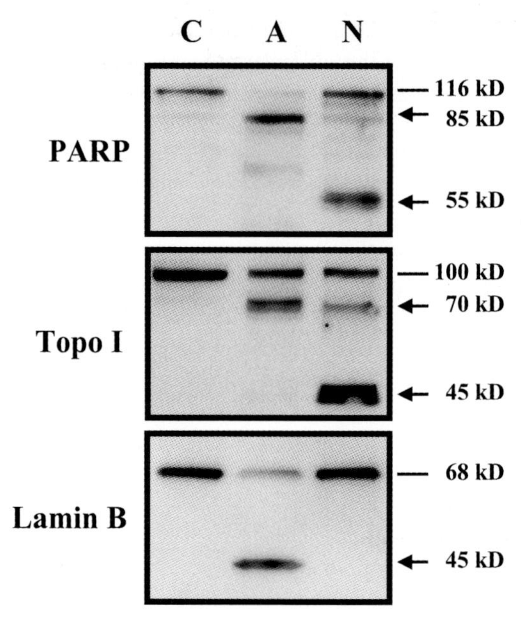

Fig. 15.1 Immunoblots of total protein from Jurkat T cells showing the cleavage of representative autoantigens during apoptosis and necrosis. Note that PARP and Topo I are cleaved into clearly distinct fragments in the two modes of cell death, whereas lamin B is cleaved only during apoptosis. C, control; A, apoptosis induced with 150 μM etoposide for 6 h; N, necrosis induced with 40 μM $HgCl_2$ for 6 h. Protein bands were detected with highly specific human autoantibodies. Intact proteins are indicated by lines, whereas proteolytic fragments are indicated by arrows.

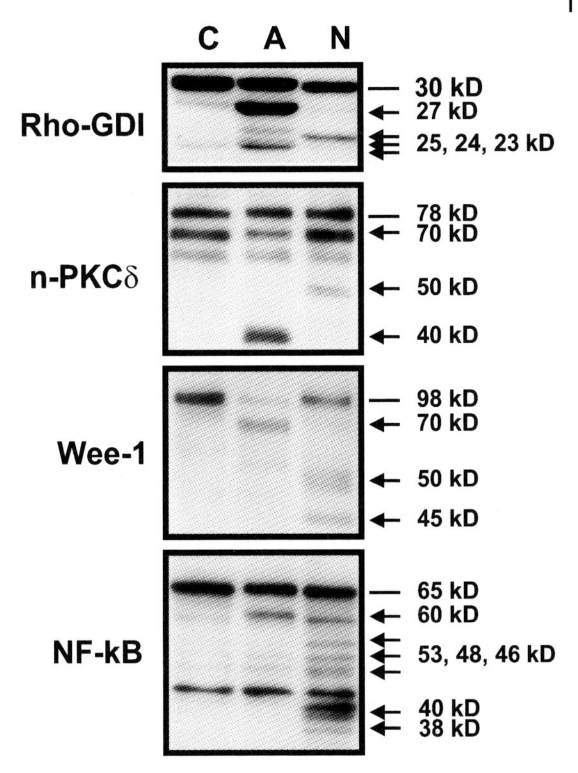

Fig. 15.2 Immunoblots of total protein from Jurkat T cells showing the cleavage of non-autoantigen caspase substrates during apoptosis and necrosis. The presence of additional protein bands in necrotic lysates is highly indicative of proteolytic degradation. Rho-GDI is a guanine nucleotide-dissociation inhibitor for the Rho family GTPase; n-PKCδ is protein kinase C $\delta\alpha$; Wee-1 is a tyrosine-specific protein kinase that phosphorylates the cell cycle-regulated protein Cdc2; NF-κB is a transcription factor consisting of a 65-kDa DNA-binding subunit (p65 Rel A) and an associated protein of 50 kDa. C, control; A, apoptosis; N, necrosis. Protein bands were detected with commercial antibodies obtained from Santa Cruz Biotechnology (Santa Cruz, CA). Intact proteins are indicated by lines, whereas proteolytic fragments are indicated by arrows.

tosis, one cannot rule out that some autoantigens that are resistant to proteolysis during necrosis may sustain in this cell death process other modifications or limited proteolysis into fragments that are not detectable by immunoblotting. This would be consistent with the observation of Pollard *et al.* [96] that fibrillarin undergoes limited cleavage during mercury-induced necrosis into a 19-kDa fragment that is detected only by immunoprecipitation.

We have observed that other caspase substrates that are not known autoantigens also undergo cleavage during necrosis (Fig. 15.2). In these studies, we used commercial antibodies that recognize apoptotic cleavage fragments of the substrates. These results indicated that cleavage during apoptosis or necrosis may not be an exclusive property of a subset of autoantigens and that specific intracellular proteins are highly sensitive to proteolysis in more than one cell death pathway.

Our studies on the selectivity of autoantigen cleavage during necrosis have been confirmed recently by Bortul *et al.* [97] who reported that specific intracellular autoantigens such as NuMA not only undergo selective cleavage during ethanol-induced necrosis in HL60 cells but also alterations in their nuclear distribution. These investigators also reported that SAF-A and SATB1, two nuclear matrix proteins which are caspase substrates and are not known autoantigens, also undergo cleavage during necrosis. They concluded that while the cleavage of NuMA, SAF-A and SATB1 could be linked to the nuclear matrix changes observed in necrotic cells, these cleavages did not appear to induce in necrosis the typical apoptotic nuclear fragmentation. This finding further pointed to the lack of proteolysis of lamin B as perhaps the critical event in the preservation of a non-fragmented nucleus in necrosis.

The proteolytic cleavages which take place during primary necrosis do not appear to be dependent on the activation of caspases, since they cannot be blocked by broad caspase inhibitors such as zVAD-fmk [11] or more specific caspase inhibitors such as AcDEVD-cmk (an inhibitor of caspase-3 and -7), VEID-CHO (an inhibitor of caspase-6), zIETD-fmk (an inhibitor of caspase-8) and zLEHD-fmk (an inhibitor of caspase-9) [97]. This would be in agreement with the notion that caspases may not be integral components of the proteolytic activities operating in necrosis. Consistent with this, we have not observed the proteolytic processing of caspases during necrosis (for examples, see Fig. 15.3).

Although very little is known about the nature of the proteases responsible for autoantigen cleavage during necrosis, a recent study by Gobeil *et al.* [98] impli-

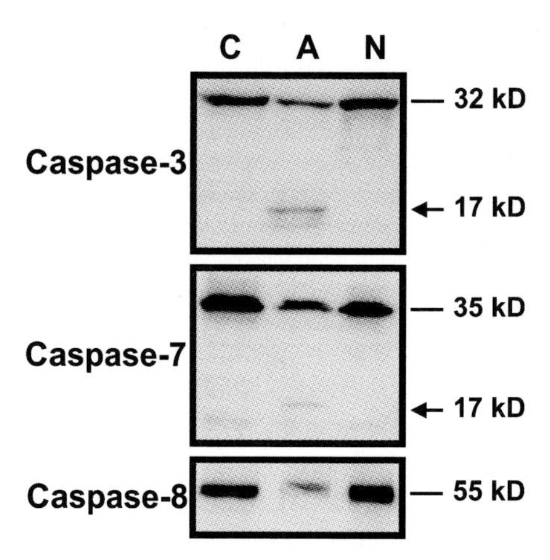

Fig. 15.3 Immunoblots of total protein from Jurkat T cells showing the lack of cleavage of caspases-3, -7 and -8 during necrosis. Note that the intensity of the bands corresponding to the pro-caspase decreases during apoptosis, consistent with the proteolytic processingof these proteases into their subunits during apoptosis. C, control; A, apoptosis; N, necrosis. Protein bands were detected with commercial antibodies obtained from PharMingen (San Diego, CA). Pro-caspases are indicated by lines, whereas subunits are indicated by arrows.

cated cathepsins in these cleavages. These investigators reported that lysosomal-rich fractions from Jurkat T cells promoted *in vitro* the cleavage of purified PARP into fragments identical to those found in lysates from Jurkat cells undergoing necrosis. Moreover, they observed that cathepsins B and G, but not D, were able to generate *in vitro* the necrotic PARP fragments. This was the first demonstration that lysosomal proteases are responsible for the cleavage of a specific autoantigen during necrotic cell death. The nature of the proteases involved in the cleavage of other intracellular autoantigens remains to be identified.

15.4.2
Autoantigen Cleavage during Secondary Necrosis

Apoptotic cells that are not removed by phagocytosis may ultimately lose their cytoplasmic membrane integrity and undergo secondary necrosis (also referred to as post-apoptotic necrosis or post-apoptotic cell lysis), with ensuing release of pro-inflammatory signals [28, 67]. Secondary necrosis occurs as a consequence of the disruption of mitochondrial function, ATP depletion, and the action of proteases and nucleases which take place during apoptosis [34–38].

An immunological consequence of defects in the phagocytic clearance of immune complexes and apoptotic cells in the rheumatic diseases would be the accumulation of cells in secondary necrosis [28, 99]. This would facilitate the exposure of intracellular autoantigens to the immune system under a pro-inflammatory environment. We reported recently that the transition from apoptosis to secondary necrosis is associated with post-translational modifications of specific autoantigens, which, if persistently exposed to the immune system under pro-inflammatory conditions, could potentially stimulate autoimmune responses [67]. In these studies, various cell lines were exposed for up to 60 h to apoptosis-inducing agents such as anti-Fas antibody, etoposide or STS. Under these conditions, cells underwent a rapid apoptosis that gradually progressed to secondary necrosis. This progression coincided with the loss of cytoplasmic membrane integrity, as assessed by Trypan blue exclusion, and irregular cellular fragmentation characteristic of late necrotic cell death.

Immunoblotting analysis indicated that the progression from apoptosis to secondary necrosis was associated with a second wave of proteolysis of specific intracellular autoantigens that are cleaved during apoptosis, including DFS70/LEDGF, PARP, SSB/La, Topo I and U1-70kDa (for examples, see Fig. 15.4). In these studies we observed that the autoantigens UBF and lamin B, which are cleaved during apoptosis, did not appear to sustain additional proteolysis during secondary necrosis. In contrast, the autoantigen Jo-1, which does not appear to be cleaved during apoptosis or primary necrosis [9–11], underwent a gradual degradation as apoptosis progressed to secondary necrosis. Many autoantigens that are not cleaved during apoptosis or primary necrosis (e.g. Ku, SSA/Ro, PCNA, Sm, PM-SCl, mitochondrial PDC-E2, and ribosomal proteins P0, P1 and P2) remained unaffected during the progression to secondary necrosis [67]. This further strengthened the notion that susceptibility to proteolysis in dying cells is not a general feature of autoantigens. These re-

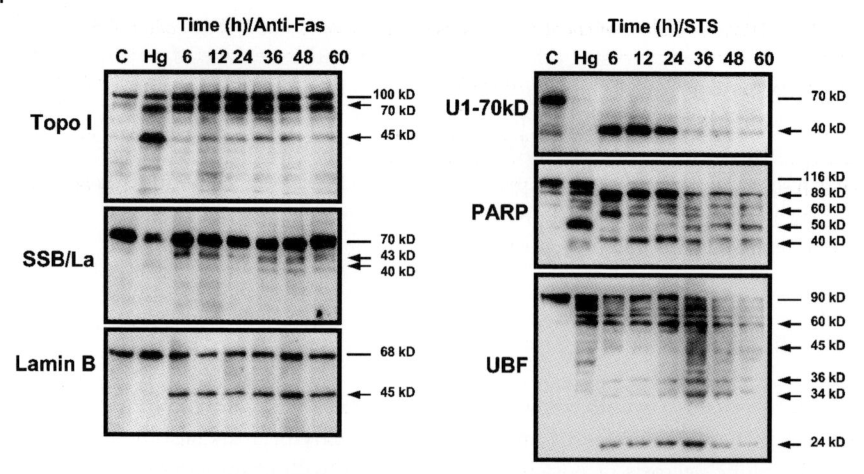

Fig. 15.4 Immunoblots showing the time-dependent progression of the cleavage of representative autoantigens in Jurkat T cells induced to undergo apoptosis with either anti-Fas antibody or STS. Control lysates from untreated cells (C) and cells induced to die by primary necrosis with 40 µM $HgCl_2$ for 6 h (Hg) were included. Progression of apoptosis into secondary necrosis was evident after 12 h of incubation in the presence of the apoptotic stimuli [67]. Note the progressive degradation of the autoantigens, except lamin B, during the cell death continuum. Intact proteins are indicated by lines, whereas proteolytic fragments are indicated by arrows.

sults also suggested that in the absence of phagocytosis, apoptotic cells are capable of undergoing secondary necrosis, a process that involves cellular fragmentation and additional degradation of specific autoantigens. Interestingly, although some of the cleavage fragments produced during secondary necrosis were also detected in primary necrosis, identical cleavage patterns were not observed in these pathways for all the autoantigens tested (Tab. 15.2). This could be attributed to differential compartmentalization of the proteases mediating these cleavages during upstream events leading to primary and secondary necrosis.

15.4.3
Autoantigen Cleavage during Caspase-independent Cell Death

As mentioned earlier, caspase-independent cell death can be induced in certain cell types by inactivation of the caspase cascade in the presence of a strong apoptotic stimulus. We investigated whether specific intracellular autoantigens undergo proteolysis during caspase-independent cell death of Jurkat T cells induced by STS in the presence of the broad caspase inhibitor zVAD-fmk [67]. In these studies, zVAD-fmk blocked the activation of effector caspases, but failed to block the eventual loss of cytoplasmic membrane permeability in the STS-treated Jurkat cells. While zVAD-fmk initially protected STS-treated cells against apoptosis, the inhibitor did not prevent a delayed cell death with features of necrosis. Immunoblotting analysis showed that zVAD-fmk partially blocked the cleavage of Topo I and PARP

Tab. 15.2 Cleavage profiles of specific autoantigens during apoptotic and necrotic cell death.

Autoantigen	*Intact protein size (kDa)*	*Cleavage fragments in apoptosis (kDa)*	*Cleavage fragments in primary necrosis (kDa)*	*Cleavage fragments in secondary necrosis (kDa)*
DFS70	75	58, 65	47	multiple (35, 40, 47)
Lamin B	68	45	none	45
Jo-1	50	none	none	progressively degraded; no cleavage fragments detected
PARP	116	89	multiple (45, 50, 60, 89)	multiple (45, 50, 60, 89)
SSB/La	48	43	none	40
Topo I	100	70	45, 70	70, 45
UBF/NOR-90	90	multiple (24, 34, 36, 60)	multiple (34, 36, 45, 60, 75)	multiple (24, 34, 36, 45, 60)
U1-70	70	40	partially degraded; no cleavage fragments detected	progressively degraded; no cleavage fragments detected

The cleavage fragments listed are those detected by immunoblotting only. See Table 15.1 for abbreviations. The information provided on this table is based on [11, 67, 97, 98].

during the initial hours of STS-treatment but failed to prevent the subsequent cleavage of these autoantigens into other fragments associated with secondary necrosis [67]. For instance, the 70- and 45-kDa fragments of Topo I and the 50-kDa fragment of PARP associated with both primary and secondary necrosis were observed in cells undergoing caspase-independent cell death. The kinetics of appearance of these fragments was essentially identical to the kinetics of appearance of similar fragments during STS-induced secondary necrosis. The presence of zVAD-fmk in STS-treated Jurkat cells blocked completely the cleavage of lamin B, consistent with the lack of cleavage of this autoantigen in primary necrosis [11, 97], but did not block the degradation of other autoantigens such as DFS70/LEDGF, UBF and U1-70kDa. In other experiments, we have observed that Topo I, but not lamin B, also undergoes cleavage into the necrotic signature 45-kDa fragment in murine fibrosarcoma L929 cells exposed to TNF in the presence of zVAD-fmk (Pacheco *et al.*, unpublished observations). Under these conditions, the L929 cells undergo a quick and massive cell death with necrotic morphology.

The observation that Topo I and PARP are cleaved into distinctive fragments (45 and 50 kDa, respectively), not observed in apoptosis, in various systems of primary necrosis, secondary necrosis, and caspase-independent cell death ([11, 67, 98] and Tab. 15.2) suggests that these cleavage products could potentially be used as markers to distinguish apoptotic from non-apoptotic cell death. Such distinction could be further strengthened by examining the integrity of lamin B, which, as mentioned previously, is cleaved in apoptosis but not in primary necrosis and caspase-independent cell death. The conservation of the cleavage patterns of these autoantigens in various systems of non-apoptotic cell death also suggests the presence of a relatively conserved proteolytic mechanism associated with non-apoptotic/necrotic cell death pathways. Since lysosomal proteases are released in massive amounts during necrosis and can induce *in vitro* the necrotic cleavage of PARP, it is very likely that these proteases are the primary mediators of autoantigen cleavage during non-apoptotic cell death.

One observation that still remains puzzling is the presence of autoantigen cleavage fragments of similar size in apoptosis, primary necrosis and caspase-independent cell death. These include the signature apoptotic fragments of Topo I (70 kDa) and PARP (89 kDa) which have also been observed in necrotic cells, even in the presence of caspase inhibitors [11, 67, 98]. One possibility in explaining this phenomenon is that some caspases may be insensitive to the caspase inhibitors used and undergo limited activation during necrosis. This could occur as a result of the cell's attempts to activate the apoptotic program during the initial exposure to the necrosis-inducing agent or by exposure of caspases to other cellular proteases during necrotic cellular fragmentation. Activation of caspases by calpains and cathepsins, proteases traditionally associated with necrosis, has been reported [100–106]. Alternatively, during non-apoptotic cell death, proteases other than caspases may target protease sensitive sites at the vicinity of the caspase cleavage sites, generating cleavage fragments similar to those produced by caspases. There is evidence that both calpains and cathepsins can directly cleave intracellular proteins during both apoptotic and non-apoptotic cell death [78, 107–111].

Caspase-independent cell death also occurs as an alternative killing mechanism during cytotoxic T lymphocyte (CTL)-mediated) cell death associated with inflammatory responses [112, 113]. CTL can induce apoptosis in their target cells either by activating the Fas-mediated pathway or by delivery of the granule protease granzyme B (GrB) into the target cells through perforin, a granule protein that forms pores in the cytoplasmic membrane [114]. GrB activates apoptosis through its ability to cleave upstream caspase-10 and -8 and downstream caspase-3 and -7 after specific aspartic acid residues [115, 116]. Investigators in our group showed previously that adenovirus-mediated GrB delivery into target cells is also associated with cleavage of specific autoantigens (PARP, U1-70kDa and lamin B) into their signature apoptotic fragments [116]. More recently, Casciola-Rosen *et al.* [117, 118] demonstrated that the majority of autoantigens targeted in human systemic autoimmune diseases are efficiently cleaved by GrB *in vitro* and during CTL-induced cell death, generating unique fragments not observed during apoptosis. Interestingly, GrB cleaved several autoantigens previously reported as not susceptible to cleavage during apoptosis or necrosis, such as Ku-70, Jo-1, CENP-B and PM-SCl, but failed to cleave other apoptotic/necrotic protease-resistant autoantigens, including SSA/Ro, Ku-80, ribosomal P proteins, histones and the Sm proteins [117, 118]. *In vivo* killing of target cells by CTLs generated low amounts of the unique autoantigen fragments produced by GrB *in vitro*, but favored the production of fragments corresponding to those generated by caspases during apoptosis, indicating that caspase-mediated proteolysis is the predominant pathway used during GrB-mediated apoptosis [117, 118]. However, the production of GrB-specific fragments was enhanced in the presence of the caspase-specific inhibitor Ac-DEVD-CHO, suggesting that GrB may facilitate cell death independent of caspase-activation by directly cleaving intracellular substrates [117, 118]. This implies that under conditions where caspase activation is blocked by either viral proteins or endogenous inhibitors, GrB may generate modified forms of autoantigens that might be immunostimulatory under a pro-inflammatory context. It was not clear from these studies whether the activity of GrB in the presence of caspase inhibitors induces cell death with necrotic morphology.

15.5 Implications of Autoantigen Cleavage for Autoimmunity

It is clear from the above discussion that cells undergoing apoptotic or non-apoptotic cell death have the potential to serve as sources of cleaved intracellular antigens. There is growing support for the hypothesis that these cleavages and other structural modifications associated with cell death could potentially target intracellular autoantigens for an autoantibody response in systemic autoimmune diseases, perhaps by revealing previously immunocryptic epitopes in individuals with the appropriate class II MHC molecules [3–5]. This hypothesis, however, has two limitations. First, it may not be sufficient to explain all ANA responses, since it is not clear whether all intracellular autoantigens are modified during cell death. Second, systemic autoimmunity is absent in most individuals, in spite of the constant exposure of their immune system to self-antigens (modified or not) derived from cells dying under a myriad of physiological situations (e.g. regulation of immune responses, fighting infections, aging, homeostasis, tissue turnover and remodeling, etc.). In this section, we will briefly discuss emerging evidence indicating that cell death-associated structural modifications, while potentially enhancing autoantigen immunogenicity, may not be sufficient to incite and sustain autoantibody responses unless they occur in conjunction with other conditions, such as defective cell death, a pro-inflammatory environment and breakdown of tolerance to self.

15.5.1 Immunogenicity of Apoptotic Cells

The ability of apoptotic cells to incite autoantibody responses to intracellular antigens has been examined recently by various groups. Mevorach *et al.* [119] reported that systematic immunization of normal mice with syngeneic thymocytes dying by ultraviolet (UV) irradiation-induced apoptosis led to the generation of antibodies that reacted predominantly against single-stranded DNA and cardiolipin. Mice immunized with control thymocyte lysates did not produce detectable levels of autoantibodies. Although this study suggested that material from apoptotic cells is immunogenic, the induced autoantibody levels were transient and generally lower than those arising naturally in autoimmune MRL/Fas-deficient (*lpr/lpr*) mice. The autoantibodies were also considered to be of low affinity and polyreactive. Very modest or no antibody responses were detected against double-stranded DNA, rheumatoid factor, nucleosomes or protein autoantigens targeted in systemic autoimmunity. While this report suggested that apoptotic cells may be immunogenic under certain circumstances, the authors emphasized that systemic exposure to syngeneic apoptotic cells is probably not sufficient for induction of high-titer, high-affinity, pathogenic autoantibodies and that additional immunoregulatory defects may be required for the induction of full autoimmune disorders [119]. In a more recent study, Gensler *et al.* [120] demonstrated that immunization of normal mice with apoptotic human Jurkat T cells yielded monoclonal antibodies targeting multiple autoantigens recognized in human systemic autoimmune diseases, including Ku, ribosomal P proteins and U small nuclear ribonucleoprotein (snRNP) proteins. While the material used for immunization was xenogeneic, the authors indicated that the generation of ANA in these immunization experiments was driven specifically by the apoptotic cells since previous immunization studies with lysed Jurkat cells had failed to produce antibodies to prominent autoantigens [120]. Ronchetti *et al.* [121] also reported that apoptotic, but not necrotic, tumor cells injected into normal mice elicited a moderate antitumor response that was 20-fold weaker than that elicited by non-replicating live tumor cells. These authors suggested that material from apoptotic cells is, though scarcely, immunogenic *in vivo*. It was also shown that immunization of non-autoimmune mice with β_2-glycoprotein I bound to apoptotic thymocytes induced antiphospholipid autoantibodies and lupus anticoagulant activity, but not ANA [122]. Taken together, these studies suggested that apoptotic cells have the potential to present immunostimulatory forms of intracellular antigens to the immune system. Detailed reviews of the immunogenicity of apoptotic cells can be found elsewhere in this book.

15.5.2
Immunogenicity of Necrotic Cells

There is increasing evidence indicating that signals derived from necrotic cells are capable of stimulating immune responses. For instance, Gallucci *et al.* [123] reported that dendritic cells (DCs) undergo maturation *in vitro* upon stimulation by signals from stressed, virally infected or necrotic cells, but not from healthy or apoptotic cells. These investigators also demonstrated that BALB/c mice immunized with a combination of ovalbumin and syngeneic necrotic cells stimulated a primary immune response to ovalbumin. However, syngeneic apoptotic cells did not elicit this response, suggesting that signals derived from necrotic but not apoptotic cells act as potent natural adjuvants [123]. These authors proposed that DCs engulf material derived from both necrotic and apoptotic cells, but signals released from the necrotic cells are primarily responsible for stimulating immune responses to antigens processed from dead cells or to local exogenous agents (bacteria, viruses or toxins) that may cause the cell damage. Along the same line, Sauter *et al.* [124] demonstrated that immature DCs efficiently ingest apoptotic and necrotic tumor cells, but only the latter provide maturation signals. Remarkably, in these studies only tumor cell lines, but not primary cells derived from normal tissues, induced maturation of DCs, suggesting that signals specifically associated with tumor cells may enhance the induction of DC maturation factors. These authors also observed that the cellular damage associated with necrosis has to be extensive enough to facilitate the release of those maturation factors. Other studies on the immunostimulatory properties of necrotic cells demonstrated that macrophages exposed to necrotic, but not apoptotic cells, expressed increased levels of co-stimulatory molecules and stimulated specific T cell responses, suggesting that these macrophages present antigens to T cells with greater efficiency than those that have ingested apoptotic cells [125, 126].

The nature of factors or so-called 'danger signals' released from dying cells that could induce DC maturation is not clear, although recent studies have implicated heat shock proteins released from necrotic cells. Melcher *et al.* [127] reported that *in situ* killing of tumor cells by necrotic, but not apoptotic, mechanisms was associated with a higher tumor immunogenicity that correlated with increased expression of heat shock protein 70 (Hsp70) in tumors that experienced high levels of necrotic cell death. This increased Hsp70 expression during necrotic tumor cell killing induced a T cell-mediated antitumor immune response characterized by infiltration of T cells, macrophages and DCs into the tumors as well as an intratumoral profile of pro-inflammatory cytokines [128]. It was also observed that Hsp70 released from dying cells not only targeted immature DC precursors to uptake released tumor antigens but was also taken up directly into DCs, suggesting that it could be involved in direct chaperoning of cellular antigens into DCs. It has been proposed that the combination of necrotic cell death/induction of Hsp70 may signal to the immune system the presence of an immunologically stressful situation against which an immune reaction should be raised [127–129].

Given the mounting evidence suggesting that necrotic cells are a source of 'danger' signals that could trigger a pro-inflammatory context and DC-mediated antigen specific immune responses, it would seem plausible that post-translational modifications associated with necrosis may also give rise, under the appropriate genetic background, to potentially immunostimulatory forms of intracellular autoantigens. While no systematic studies have been reported yet to explore this possibility, Pollard *et al.* [96] provided evidence that proteolytic fragments of autoantigens generated during necrosis could elicit an autoantibody response. These investigators demonstrated that a 19-kDa fragment of fibrillarin uniquely generated during mercury-induced non-apoptotic cell death was capable of inducing an antifibrillarin autoantibody response similar to that observed during mercury-induced autoimmunity in B10.S (H-2^s) mice. These results, combined with the previous observation by the same group that mercury modifies the structure of fibrillarin [15], suggested that an autoimmunity-inducing xenobiotic such as mercury might generate unique immunostimulatory fragments from a self-antigen most likely by a combination of chemical modification and necrosis-associated proteolysis.

15.5.3
Apoptosis, Necrosis and Immunity to Intracellular Antigens

Since apoptosis is a normal, non-inflammatory physiological cell death process, it would be difficult to envision that exposure of the immune system to modified forms of apoptotic self-antigens may be sufficient to elicit autoantibodies. Moreover, uptake of apoptotic cells by antigen-presenting cells under normal conditions not only stimulates the release of anti-inflammatory signals, but also serves as a mechanism to tolerize the immune system against intracellular antigens [130–134]. If apoptosis plays a major role in immune tolerance, then under what circumstances could apoptotic material provoke autoimmune responses? Sauter *et al.* [124] suggested that uptake of apoptotic material by DCs may lead to an immune response only if followed by a maturation signal provided under a pro-inflammatory context. This could be facilitated by excessive cell fragmentation (due to primary or secondary necrosis) induced by a tissue damaging insult (e.g. bacterial or viral infection, release of cytotoxic cytokines, trauma or toxicity) or other inflammatory products. This would be consistent with the report that opsonization of apoptotic cells with anti-β_2-glycoprotein I antibodies, considered as a pro-inflammatory context, influenced apoptotic cell uptake by DCs and enabled DCs to present apoptosis-derived antigens with higher efficiency and to secrete pro-inflammatory and maturation factors [135, 136].

Various models propose that the balance between apoptosis and necrosis, and the environment surrounding cell death, may determine the tolerogenic or immunogenic response of DCs [130–134, 137, 138]. Under these models, apoptosis associated with normal cellular turnover induces DCs or other phagocytic cells to produce anti-inflammatory molecules and cross-present apoptotic self-antigens to the immune system, leading to induction of peripheral tolerance. However, under pro-inflammatory conditions such as an infection, mature DCs may trigger efficiently the activation of antigen-specific T and B cells. For instance, the induction of massive apoptosis by a viral infection *in vitro* was found to be associated with DC-mediated activation of virus-specific CTLs [139]. Consistent with this observation, Salio *et al.* [140] reported that the capacity of necrotic or apoptotic cells to induce DC maturation *in vitro* was dependent on the presence of a mycoplasma infection, suggesting that cell death in the presence of an infectious agent provides the necessary pro-inflammatory signals for stimulating a DC-mediated immune response. Green and Beere [137] argued that extensive necrosis associated with an infection would induce an inflammatory response, leading the nearby DCs, which engulf both necrotic cells and infected apoptotic cells, to display co-stimulatory molecules and self-peptides derived from the dying cells, as well as foreign peptides from the infectious agent. Normally, an immune response will ensue against foreign antigens but not against self-antigens, because autoreactive lymphocytes are continuously silenced by the daily phagocytic clearance and processing of dead cells. It can be inferred from this model that self-antigens from apoptotic and necrotic cells could potentially be immunogenic only if displayed by DCs under pro-inflammatory conditions and presented to autoreactive lymphocytes that have escaped elimination. It could be also envisioned that abnormal interaction or aggregation of self-intracellular antigens with foreign antigens (e.g. infectious agents or xenobiotics) during cell death may generate cryptic or novel epitopes for which tolerance has not been established and that under pro-inflammatory conditions could initiate an autoimmune response.

15.5.4
Modified Autoantigens and Defective Phagocytic Function

Under normal circumstances, cells dying by apoptotic and post-apoptotic/secondary necrotic mechanisms are rapidly removed by phagocytic cells and presented to the immune system for induction of tolerance [28, 29, 130–138]. This rapid clearance prevents the accumulation of dying cells and the release of danger signals leading to presentation of cellular antigens to the immune system under a pro-inflammatory context [28, 29]. It is becoming evident that in systemic autoimmune diseases, increased presentation of intracellular autoantigens by mature DCs under a pro-inflammatory context could be triggered by excessive accumulation of dying cells due to exposure to tissue-damaging environmental agents (e.g. UV light, xenobiotics, anti-inflammatory drugs, and viral and bacterial infections, etc) combined with impaired clearance of apoptotic cells [5, 141–145]. As discussed elsewhere in this book, there is increasing evidence that defects in proteins involved in the phagocytic clearance of dying cells, including C1q, C-reactive protein (CRP), serum amyloid P component (SAP) and the Mer tyrosine kinase, may lead to impairment of phagocytic function, with resulting development of autoimmunity to intracellular antigens [145–151]. Defective phagocytic function is likely to be associated with excessive accumulation of cells in different stages of the cell death continuum, from early apoptosis to secondary necrosis, which may facilitate the release of pro-inflammatory signals that induce DC maturation and presentation of modified self-antigens from the dying cells. This would be consistent with studies by Manfredi *et al.* [152, 153] demonstrating *in vitro* that excessive apoptosis or delayed apoptosis leading to secondary necrosis, mimicking a failure of their *in vivo* clearance, were sufficient to trigger DC maturation and presentation of intracellular antigens. This could skew the outcome of cross-presentation of intracellular antigens to autoimmunity if intracellular antigens, particularly those that are modified or altered, are presented to autoreactive lymphocytes. Fig. 15.5 illustrates how defective clearance of dying cells could contribute to the induction of DC-mediated autoantibody responses to modified intracellular autoantigens.

15.6
Conclusions

The immunogens driving autoantibodies in systemic autoimmunity appear to be intracellular antigens associated with subcellular organelles or particles. Emerging evidence suggests a role for apoptotic and non-apoptotic/necrotic cell death path-

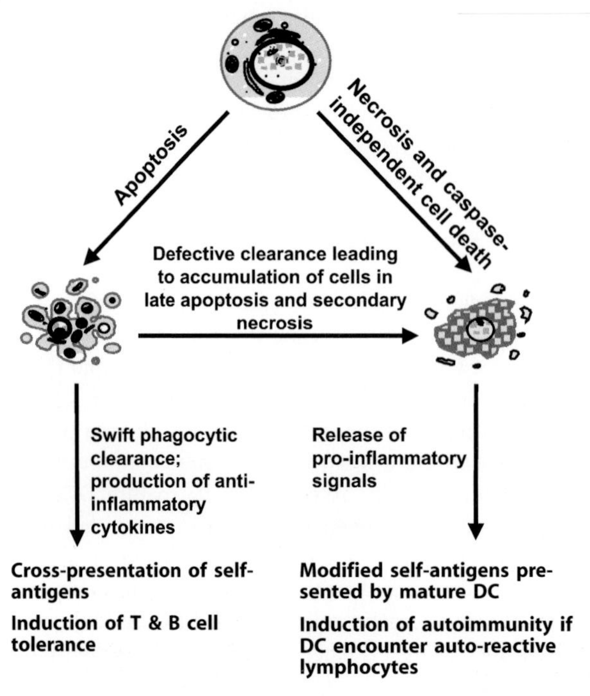

Fig. 15.5 The possible contribution of various cell death pathways to the induction of autoimmune responses to intracellular self-antigens. Induction of apoptosis during normal tissue turnover leads to swift phagocytic clearance of apoptotic cells under non-inflammatory conditions. Under these conditions, self-intracellular antigens are cross-presented to the immune system to induce tolerance. Induction of apoptosis, necrosis or caspase-independent cell death under conditions where phagocytic clearance is defective leads to the accumulation of dying cells, with subsequent development of a pro-inflammatory environment. This leads to the exposure of self-intracellular antigens (modified and non-modified) to the immune system under pro-inflammatory conditions. DC-mediated presentation of these antigens to autoreactive lymphocytes may trigger autoantibody responses.

ways in facilitating the presentation of cleaved forms of self-intracellular antigens to the immune system. Under a non-inflammatory context, cell death, especially by apoptosis, may serve as the leading mechanism to maintain immune tolerance to these antigens, including their modified forms. However, in the inflammatory environment associated with non-apoptotic/necrotic cell death (triggered by infection, exposure to xenobiotics or defective phagocytic clearance) self-intracellular antigens could be presented to autoreactive lymphocytes in a manner that stimulates a specific autoimmune response. Proteolytic cleavage and other modifications/alterations associated with cell death could potentially enhance the immunogenicity of autoantigens. Cell death-associated abnormal interactions of autoantigens with foreign antigens (e.g. viral or bacterial proteins), xenobiotics (e.g. mercury) and, potentially, self-proteins (e.g. Hsp70) may also expose cryptic or novel epitopes for which the immune system has not developed tolerance, resulting in autoimmunity.

15.7 Acknowledgments

This work was supported by NIH-NIAID grant AI44088 to C. A. C. and by basic science research grants from Loma Linda University School of Medicine.

15.8 References

1 van Venrooij WJ, Prujin GJM. Ribonucleoprotein complexes as autoantigens. *Curr Opin Immunol* **1995**, 7, 819–24.
2 Casiano CA, Tan EM. Recent developments in the understanding of antinuclear autoantibodies. *Int Arch Allergy Immunol* **1996**, 111, 308–13.
3 Rosen A, Casciola-Rosen L. Autoantigens as substrates for apoptotic proteases: implications for the pathogenesis of systemic autoimmune disease. *Cell Death Different* **1999**, 6, 6–12.
4 Utz PJ, Anderson P. Posttranslational protein modifications, apoptosis, and the bypass of tolerance to autoantigens. *Arthritis Rheum* **1998**, 41, 1152–60.
5 Rodenburg RJ, Raats JM, Pruijn GJ, van Venrooij WJ. Cell death: a trigger of autoimmunity? *BioEssays* **2000**, 22, 627–36.
6 Botto M. Links between complement deficiency and apoptosis. *Arthritis Res* **2001**, 3, 207–10.
7 Walport MJ. Lupus, DNase and defective disposal of cellular debris. *Nat Genet* **2000**, 25, 135–6.
8 Pickering MC, Walport MJ. Links between complement abnormalities and systemic lupus erythematosus. *Rheumatology* **2000**, 39, 133–41.
9 Casciola-Rosen LA, Anhalt G, Rosen A. DNA-dependent protein kinase is one of a subset of autoantigens specifically cleaved early during apoptosis. *J Exp Med* **1995**, 182, 1625–34.
10 Casiano CA, Martin SJ, Green DR, Tan EM. Selective cleavage of nuclear autoantigens during CD95 (Fas/APO-1)-mediated T cell apoptosis. *J Exp Med* **1996**, 184, 765–70.
11 Casiano CA, Ochs RL, Tan EM. Distinct cleavage products of nuclear proteins in apoptosis and necrosis revealed by autoantibody probes. *Cell Death Different* **1998**, 5, 183–90.
12 Utz PJ, Hottelet M, Schur PH, Anderson P. Proteins phosphorylated during stress-induced apoptosis are common targets for autoantibody production in patients with systemic lupus erythematosus. *J Exp Med* **1997**, 185, 843–54.
13 Zampieri S, Degen W, Ghiradello A, Doria A, van Venrooij WJ. Dephosphorylation of autoantigenic ribosomal P proteins during Fas-L induced apoptosis: a possible trigger for the development of the autoimmune response in patients with systemic lupus erythematosus. *Ann Rheum Dis* **2001**, 60, 72–6
14 Odin JA, Huebert RC, Casciola-Rosen L, LaRusso NF, Rosen A. Bcl-2-dependent oxidation of pyruvate dehydrogenase-E2, a primary biliary cirrhosis autoantigen, during apoptosis. *J Clin Invest* **2001**, 108, 223–32.
15 Pollard KM, Lee DK, Casiano CA, Blüthner M, Johnston MM, Tan EM. The autoimmunity-inducing xenobiotic mercury interacts with the autoantigen fibrillarin and modifies its molecular and antigenic properties. *J Immunol* **1997**, 158, 3521–8.
16 Piacentini M, Colizzi V. Tissue transglutaminase: apoptosis versus autoimmunity. *Immunol Today* **1999**, 20, 130–4.
17 van Venrooij WJ, Pruijn GJ. Citrullination: a small change for a protein with great consequences for rheumatoid arthritis. *Arthritis Res* **2000**, 2, 249–51.
18 Wyllie AH, Kerr JFR, Currie AC. Cell death: the significance of apoptosis. *Int Rev Cytol* **1980**, 68, 251–305.
19 Ellis RE, Yuan J, Horvitz HR. Mechanisms and function of cell death. *Annu Rev Cell Biol* **1991**, 7, 663–98.

20 Buja LM, Eigenbrodt ML, Eigenbrodt EH. Apoptosis and necrosis: basic types and mechanisms of cell death. *Arch Pathol Lab Med* **1993**, 117, 1208–14.

21 Majno G, Joris I. Apoptosis, oncosis, and necrosis: an overview of cell death. *Am J Pathol* **1995**, 146, 3–15.

22 Leist M, Jaattela M. Four deaths and a funeral: from caspases to alternative mechanisms. *Nat Rev Mol Cell Biol* **2001**, 2, 589–98.

23 Sperandio S, de Belle I, Bredesen DE. An alternative, nonapoptotic form of programmed cell death. *Proc Natl Acad Sci USA* **2000**, 97, 14376–81.

24 Lowe SW, Lin AW. Apoptosis in cancer. *Carcinogenesis* **2000**, 21, 485–95.

25 Mauricio D, Mandrup-Poulsen T. Apoptosis and the pathogenesis of IDDM: a question of life and death. *Diabetes* **1998**, 47, 1537–43.

26 Rust C, Gores GJ. Apoptosis and liver disease. *Am J Med* **2000**, 108, 567–74.

27 Mattson MP, Duan W. 'Apoptotic' biochemical cascades in synaptic compartments: roles in adaptive plasticity and neurodegenerative disorders. *Neurosci Res* **1999**, 58, 152–66.

28 Savill J, Fadok V. Corpse clearance defines the meaning of cell death. *Nature* **2000**, 407, 784–8.

29 Chang HY, Yang X. Proteases for cell suicide: functions and regulation of caspases. *Microbiol Mol Biol Rev* **2000**, 64, 821–46.

30 Earnshaw WC, Martins LM, Kaufmann SH. Mammalian caspases: structure, activation, substrates, and functions during apoptosis. *Annu Rev Biochem* **1999**, 68, 383–24.

31 Krammer PH. CD95(APO-1/Fas)-mediated apoptosis: live and let die. *Adv Immunol* **1999**, 71, 163–210.

32 Adrain C, Martin SJ. The mitochondrial apoptosome: a killer unleashed by the cytochrome seas. *Trends Biochem Sci* **2001**, 26, 390–7.

33 Cohen GM. Caspases: the executioners of apoptosis. *Biochem J* **1997**, 326, 1–16.

34 Leist M, Nicotera P. The shape of cell death. *Biochem Biophys Res Commun* **1997**, 236, 1–9.

35 Nicotera P, Lipton SA. Excitotoxins in neuronal apoptosis and necrosis. *J Cerebral Blood Flow Metab* **1999**, 19, 583–91.

36 Fiers W, Beyaert R, Declercq W, Vandenabeele P. More than one way to die: apoptosis, necrosis and reactive oxygen damage. *Oncogene* **1999**, 18, 7719–30.

37 Kaplowitz N. Mechanisms of liver cell injury. *Hepatology* **2000**, 32, 39–47.

38 Yamashima T. Implication of cysteine proteases calpain, cathepsin and caspase in ischemic neuronal death of primates. *Prog Neurobiol* **2000**, 62, 273–95.

39 Columbano A. Cell death: current difficulties in discriminating apoptosis from necrosis in the context of pathological processes *in vivo*. *J Cell Biochem* **1995**, 58, 181–90.

40 Nicotera P, Leist M, Ferrando-May E. Apoptosis and necrosis: different execution of the same death. *Biochem Soc Symp* **1999**, 66, 69–73.

41 Watson AJM. Necrosis and apoptosis in the gastrointestinal tract. *Gut* **1995**, 37, 165–7.

42 Snider BJ, Gottron FJ, Choi DW. Apoptosis and necrosis in cerebrovascular disease. *Ann NY Acad Sci* **1999**, 893, 243–53.

43 Kelly L, Reid L, Walker NI. Massive acinar cell apoptosis with secondary necrosis, origin of ducts in atrophic lobules and failure to regenerate in cyanohydroxybutene pancreatopathy in rats. *Int J Exp Pathol* **1999**, 80, 217–26.

44 Roy M, Sapolsky R. Neuronal apoptosis in acute necrotic insults: why is this subject such a mess? *Trends Neurosci* **1999**, 22, 419–22.

45 Zeng YS, Xu ZC. Co-existence of necrosis and apoptosis in rat hippocampus following transient forebrain ischemia. *Neurosci Res* **2000**, 37, 113–25.

46 Blom WM, De Bont HJ, Meijerman I, Kuppen PJ, Mulder GJ, Nagelkerke JF. Interleukin-2-activated natural killer cells can induce both apoptosis and necrosis in rat hepatocytes. *Hepatology* **1999**, 29, 785–92.

47 Gardiner CM, Reen DJ. Differential cytokine regulation of natural killer cell-mediated necrotic and apoptotic cytotoxicity. *Immunology* **1998**, 93, 511–7.

48 Chautan M, Chazal G, Cecconi F, Gruss P, Golstein P. Interdigital cell death can occur through a necrotic and caspase-independent pathway. *Curr Biol* **1999**, 9, 967–70.

49 Sapunar D, Vilovic K, England M, Saraga-Babic M. Morphological diversity of dying cells during regression of the human tail. *Ann Anat* **2001**, 183, 217–22.

50 Kaneko T, Iida H, Bedford JM, Mori T. Spermatozoa of the shrew, *Suncus murinus*, undergo the acrosome reaction and then selectively kill cells in penetrating the cumulus oophorus. *Biol Reprod* **2001**, 65, 544–53.

51 Bortul R, Zweyer M, Billi AM, Tabellini G, Ochs RL, Bareggi R, Cocco L, Martelli AM. Nuclear changes in necrotic HL-60 cells. *J Cell Biochem* **2001**, 36, 19–31.

52 Leist M, Single B, Castoldi AF, Kuhnle S, Nicotera P. Intracellular adenosine triphosphate (ATP) concentration: a switch in the decision between apoptosis and necrosis. *J Exp Med* **1997**, 185, 1481–6.

53 Rao L, Perez D, White E. Lamin proteolysis facilitates nuclear events during apoptosis. *J Cell Biol* **1996**, 135, 1441–55.

54 Nixon RA, Cataldo AM. The lysosomal system in neuronal cell death: a review. *Ann NY Acad Sci* **1993**, 679, 87–109.

55 Sane AT, Bertrand R. Caspase inhibition in camptothecin-treated U-937 cells is coupled with a shift from apoptosis to transient G_1 arrest followed by necrotic cell death. *Cancer Res* **1999**, 59, 3565–9.

56 Amarante-Mendes GP, Finucane DM, Martin SJ, Cotter TG, Salvesen GS, Green DR. Anti-apoptotic oncogenes prevent caspase-dependent and independent commitment for cell death. *Cell Death Different* **1998**, 5, 298–306.

57 Kawahara A, Ohsawa Y, Matsumura H, Uchiyama Y, Nagata S. Caspase-independent cell killing by Fas-associated protein with death domain. *J Cell Biol* **1998**, 143, 1353–60.

58 Denecker G, Vercammen D, Steemans M, Vanden Berghe T, Brouckaert G, Van Loo G, Zhivotovsky B, Fiers W, Grooten J, Declercq W, Vandenabeele P. Death receptor-induced apoptotic and necrotic cell death: differential role of caspases and mitochondria. *Cell Death Different* **2001**, 8, 829–40.

59 McCarthy NJ, Whyte MK, Gilbert CS, Evan GI. Inhibition of Ced-3/ICE-related proteases does not prevent cell death induced by oncogenes, DNA damage, or the Bcl-2 homologue Bak. *J Cell Biol* **1997**, 136, 215–27.

60 Deas O, Dumont C, MacFarlane M, Rouleau M, Hebib C, Harper F, Hirsch F, Charpentier B, Cohen GM, Senik A. Caspase-independent cell death induced by anti-CD2 or staurosporine in activated human peripheral T lymphocytes. *J Immunol* **1998**, 161, 3375–83.

61 Liu Y, Tergaonkar V, Krishna S, Androphy EJ. Human papillomavirus type 16 E6-enhanced susceptibility of L929 cells to tumor necrosis factor alpha correlates with increased accumulation of reactive oxygen species. *J Biol Chem* **1999**, 274, 24819–27.

62 Xiang J, Chao DT, Korsmeyer SJ. BAX-induced cell death may not require interleukin 1 beta-converting enzyme-like proteases. *Proc Natl Acad Sci USA* **1996**, 93, 14559–63.

63 Joza N, Susin SA, Daugas E, Stanford WL, Cho SK, Li CY, Sasaki T, Elia AJ, Cheng HY, Ravagnan L, Ferri KF, Zamzami N, Wakeham A, Hakem R, Yoshida H, Kong YY, Mak TW, Zuniga-Pflucker JC, Kroemer G, Penninger JM. Essential role of the mitochondrial apoptosis-inducing factor in programmed cell death. *Nature* **2001**, 410, 549–54.

64 Volbracht C, Leist M, Kolb SA, Nicotera P. Apoptosis in caspase-inhibited neurons. *Mol Med* **2001**, 7, 36–48.

65 Mateo V, Lagneaux L, Bron D, Biron G, Armant M, Delespesse G, Sarfati M. CD47 ligation induces caspase-independent cell death in chronic lymphocytic leukemia. *Nat Med* **1999**, 5, 1277–84.

66 Kitanaka C, Kuchino Y. Caspase-independent programmed cell death with necrotic morphology. *Cell Death Different* **1999**, 6, 508–15.

67 Wu X, Molinaro C, Johnson N, Casiano CA. Secondary necrosis is a source of proteolytically modified forms of specific intracellular autoantigens: implications

for systemic autoimmunity. *Arthritis Rheum* **2001**, 44, 2642–52.

68 Vercammen D, Brouckaert G, Denecker G, Van de Craen M, Declercq W, Fiers W, Vandenabeele P. Dual signaling of the Fas receptor: initiation of both apoptotic and necrotic cell death pathways. *J Exp Med* **1998**, 188, 919–30.

69 Vercammen D, Beyaert R, Denecker G, Goossens V, Van Loo G, Declercq W, Grooten J, Fiers W, Vandenabeele P. Inhibition of caspases increases the sensitivity of L929 cells to necrosis mediated by tumor necrosis factor. *J Exp Med* **1998**, 187, 1477–85.

70 Matsumura H, Shimizu Y, Ohsawa Y, Kawahara A, Uchiyama Y, Nagata S. Necrotic death pathway in Fas receptor signaling. *J Cell Biol* **2000**, 151, 1247–56.

71 Khwaja A, Tatton L. Resistance to the cytotoxic effects of tumor necrosis factor alpha can be overcome by inhibition of a FADD/caspase-dependent signaling pathway. *J Biol Chem* **1999**, 274, 36817–23.

72 Boone E, Vanden Berghe T, Van Loo G, De Wilde G, De Wael N, Vercammen D, Fiers W, Haegeman G, Vandenabeele P. Structure/function analysis of p55 tumor necrosis factor receptor and fas-associated death domain. Effect on necrosis in L929sA cells. *J Biol Chem* **2000**, 275, 37596–603.

73 Jones BE, Lo CR, Liu H, Srinivasan A, Streetz K, Valentino KL, Czaja MJ. Hepatocytes sensitized to tumor necrosis factor-alpha cytotoxicity undergo apoptosis through caspase-dependent and caspase-independent pathways. *J Biol Chem* **2000**, 275, 705–12.

74 Lazebnik YA, Kaufmann SH, Desnoyers S, Poirier GG, Earnshaw WC. Cleavage of poly (ADP-ribose) polymerase by a proteinase with properties like ICE. *Nature* **1994**, 371, 346–47.

75 Casciola-Rosen LA, Miller DK, Anhalt G, Rosen A. Specific cleavage of the 70-kDa protein component of the U1 small nuclear ribonucleoprotein is a characteristic biochemical feature of apoptotic cell death. *J Biol Chem* **1994**, 269, 30757–60.

76 Mashima T, Naito M, Tsuruo T. Caspase-mediated cleavage of cytoskeletal actin plays a positive role in the process of morphological apoptosis. *Oncogene* **1999**, 18, 2423–30.

77 Orth T, Gerken G, Kellner R, Meyer zum Buschenfelde KH, Mayet WJ. Actin is a target antigen of anti-neutrophil cytoplasmic antibodies (ANCA) in autoimmune hepatitis type-1. *J Hepatol* **1997**, 26, 37–47.

78 Gautier F, Irminger-Finger I, Gregoire M, Meflah K, Harb J. Identification of an apoptotic cleavage product of BARD1 as an autoantigen: a potential factor in the antitumoral response mediated by apoptotic bodies. *Cancer Res* **2000**, 60, 6895–900.

79 Nagaraju K, Cox A, Casciola-Rosen L, Rosen A. Novel fragments of the Sjögren's syndrome autoantigens alpha-fodrin and type 3 muscarinic acetylcholine receptor generated during cytotoxic lymphocyte granule-induced cell death. *Arthritis Rheum* **2001**, 44, 2376–86.

80 Janicke RU, Ng P, Sprengart ML, Porter AG. Caspase-3 is required for alpha-fodrin cleavage but dispensable for cleavage of other death substrates in apoptosis. *J Biol Chem* **1998**, 273, 15540–5.

81 Mancini M, Machamer CE, Roy S, Nicholson DW, Thornberry NA, Casciola-Rosen LA, Rosen A. Caspase-2 is localized at the Golgi complex and cleaves golgin-160 during apoptosis. *J Cell Biol* **2000**, 149, 603–12.

82 Waterhouse N, Kumar S, Song Q, Strike P, Sparrow L, Dreyfuss G, Alnemri ES, Litwack G, Lavin M, Watters D. Heteronuclear ribonucleoproteins C1 and C2, components of the spliceosome, are specific targets of interleukin 1β-converting enzyme-like proteases in apoptosis. *J Biol Chem* **1996**, 271, 29335–41.

83 Stanek D, Vencovsky J, Kafkova J, Raska I. Heterogenous nuclear RNP C1 and C2 core proteins are targets for an autoantibody found in the serum of a patient with systemic sclerosis and psoriatic arthritis. *Arthritis Rheum* **1997**, 40, 2172–7.

84 Caulin C, Salvesen GS, Oshima RG. Caspase cleavage of keratin 18 and reorganization of intermediate filaments during epithelial cell apoptosis. *J Cell Biol* **1997**, 138, 1379–94.

85 Borg AA. Antibodies to cytokeratins in inflammatory arthropathies. *Semin Arthritis Rheum* **1997**, 27, 186–95.

86 Casciola-Rosen LA, Pluta AF, Plotz PH, Cox AE, Morris S, Wigley FM, Petri M, Gelber AC, Rosen A. The DNA mismatch repair enzyme PMS1 is a myositis-specific autoantigen. *Arthritis Rheum* **2001**, 44, 389–96.

87 Takeda Y, Caudell P, Grady G, Wang G, Suwa A, Sharp GC, Dynan WS, Hardin JA. Human RNA helicase A is a lupus autoantigen that is cleaved during apoptosis. *J Immunol* **1999**, 163, 6269–74.

88 Utz PJ, Hottelet M, Le TM, Kim SJ, Geiger ME, van Venrooij WJ, Anderson P. The 72- kDa component of signal recognition particle is cleaved during apoptosis. *J Biol Chem* **1998**, 273, 35362–70.

89 Rutjes SA, Utz PJ, van der Heijden A, Broekhuis C, van Venrooij WJ, Pruijn GJ. The La (SS-B) autoantigen, a key protein in RNA biogenesis, is dephosphorylated and cleaved early during apoptosis. *Cell Death Different* **1999**, 6, 976–86.

90 Rickers A, Peters N, Badock V, Beyaert R, Vandenabeele P, Dorken B, Bommert K. Cleavage of transcription factor SP1 by caspases during anti-IgM-induced B-cell apoptosis. *Eur J Biochem* **1999**, 261, 269–74.

91 Spain TA, Sun R, Gradzka M, Lin SF, Craft J, Miller G. The transcriptional activator Sp1, a novel autoantigen. *Arthritis Rheum* **1997**, 40, 1085–95.

92 Ochs RL, Muro Y, Si Y, Ge H, Chan EK, Tan EM. Autoantibodies to DFS 70 kd/transcription coactivator p75 in atopic dermatitis and other conditions. *J Allergy Clin Immunol* **2000**, 105, 1211–20.

93 von Mikecz A, Konstantinov K, Buchwald DS, Gerace L, Tan EM. High frequency of autoantibodies to insoluble cellular antigens in patients with chronic fatigue syndrome. *Arthritis Rheum* **1997**, 40, 295–305.

94 Morishima N. Changes in nuclear morphology during apoptosis correlate with vimentin cleavage by different caspases located either upstream or downstream of Bcl-2 action. *Genes Cells* **1999**, 4, 401–14.

95 Martelli AM, Robuffo I, Bortul R, Ochs RL, Luchetti F, Cocco L, Zweyer M, Bareggi R, Falcieri E. Behavior of nucleolar proteins during the course of apoptosis in camptothecin-treated HL60 cells. *J Cell Biochem* **2000**, 78, 264–77.

96 Pollard KM, Pearson DL, Blüthner M, Tan EM. Proteolytic cleavage of a self-antigen following xenobiotic-induced cell death produces a fragment with novel immunogenic properties. *J Immunol* **2000**, 165, 2263–70.

97 Bortul R, Zweyer M, Billi AM, Tabellini G, Ochs RL, Bareggi R, Cocco L, Martelli AM. Nuclear changes in necrotic HL-60 cells. *J Cell Biochem* **2001**, suppl 36, 19–31.

98 Gobeil S, Boucher CC, Nadeau D, Poirier GG. Characterization of the necrotic cleavage of poly (ADP-ribose) polymerase (PARP-1): implication of lysosomal proteases. *Cell Death Different* **2001**, 8, 588–94.

99 Ren Y, Savill J. Apoptosis: the importance of being eaten. *Cell Death Different* **1998**, 5, 563–8.

100 Varghese J, Radhika G, Sarin A. The role of calpain in caspase activation during etoposide induced apoptosis in T cells. *Eur J Immunol* **2001**, 31, 2035–41.

101 Blomgren K, Zhu C, Wang X, Karlsson JO, Leverin AL, Bahr BA, Mallard C, Hagberg H. Synergistic activation of caspase-3 by *m*-calpain after neonatal hypoxia-ischemia: a mechanism of 'pathological apoptosis'? *J Biol Chem* **2001**, 276, 10191–8.

102 Nakagawa T, Yuan J. Cross-talk between two cysteine protease families. Activation of caspase-12 by calpain in apoptosis. *J Cell Biol* **2000**, 150, 887–94.

103 Hishita T, Tada-Oikawa S, Tohyama K, Miura Y, Nishihara T, Tohyama Y, Yoshida Y, Uchiyama T, Kawanishi S. Caspase-3 activation by lysosomal enzymes in cytochrome *c*-independent apoptosis in myelodysplastic syndrome-derived cell line P39. *Cancer Res* **2001**, 61, 2878–84.

104 Zhou Q, Salvesen GS. Activation of pro-caspase-7 by serine proteases includes a non-canonical specificity. *Biochem J* **1997**, 324, 361–4.

105 Vancompernolle K, Van Herreweghe F, Pynaert G, Van de Craen M, De Vos K, Totty N, Sterling A, Fiers W, Vandenabeele P, Grooten J. Atractyloside-induced release of cathepsin B, a protease with caspase-processing activity. *FEBS Lett* **1998**, 438, 150–8.

106 Ishisaka R, Utsumi T, Kanno T, Arita K, Katunuma N, Akiyama J, Utsumi K. Participation of a cathepsin L-type protease in the activation of caspase-3. *Cell Struct Funct* **1999**, 24, 465–70.

107 Yamakawa H, Banno Y, Nakashima S, Yoshimura S, Sawada M, Nishimura Y, Nozawa Y, Sakai N. Crucial role of calpain in hypoxic PC12 cell death: calpain, but not caspases, mediates degradation of cytoskeletal proteins and protein kinase C-alpha and -delta. *Neurol Res* **2001**, 23, 522–30.

108 Shim SR, Kook S, Kim JI, Song WK. Degradation of focal adhesion proteins paxillin and p130cas by caspases or calpains in apoptotic rat-1 and L929 cells. *Biochem Biophys Res Commun* **2001**, 286, 601–8.

109 Nath R, Davis M, Probert AW, Kupina NC, Ren X, Schielke GP, Wang KK. Processing of cdk5 activator p35 to its truncated form (p25) by calpain in acutely injured neuronal cells. *Biochem Biophys Res Commun* **2000**, 274, 16–21.

110 Biggs JR, Yang J, Gullberg U, Muchardt C, Yaniv M, Kraft AS. The human brm protein is cleaved during apoptosis: the role of cathepsin G. *Proc Natl Acad Sci USA* **2001**, 98, 3814–9.

111 Stoka V, Turk B, Schendel SL, Kim TH, Cirman T, Snipas SJ, Ellerby LM, Bredesen D, Freeze H, Abrahamson M, Bromme D, Krajewski S, Reed JC, Yin XM, Turk V, Salvesen GS. Lysosomal protease pathways to apoptosis. Cleavage of bid, not pro-caspases, is the most likely route. *J Biol Chem* **2001**, 276, 3149–57.

112 Sarin A, Williams MS, Alexander-Miller MA, Berzofsky JA, Zacharchuk CM, Henkart PA. Target cell lysis by CTL granule exocytosis is independent of ICE/Ced-3 family proteases. *Immunity* **1997**, 6, 209–15.

113 Trapani JA, Jans DA, Jans PJ, Smyth MJ, Browne KA, Sutton VR. Efficient nuclear targeting of granzyme B and the nuclear consequences of apoptosis induced by granzyme B and perforin are caspase-dependent, but cell death is caspase-independent. *J Biol Chem* **1998**, 273, 27934–8.

114 Trapani JA, Davis J, Sutton VR, Smyth MJ. Proapoptotic functions of cytotoxic lymphocyte granule constituents *in vitro* and *in vivo*. *Curr Opin Immunol* **2000**, 12, 323–9.

115 Martin SJ, Amarante-Mendes G, Tewari M, Shi L, Chuang TH, Casiano CA, Fitzgerald P, Tan EM, Bokoch GM, Dixit VM, Greenberg AH, Green DR. The cytotoxic cell protease granzyme B initiates apoptosis in a cell-free system by proteolytic processing and activation of the ICE/CED-3 family protease, CPP32, via a novel two-step mechanism. *EMBO J* **1996**, 15, 2407–16.

116 Talanian RV, Yang X, Turbov J, Seth P, Ghayur T, Casiano CA, Orth K, and Froelich CJ. Granule-mediated cell killing: pathways for granzyme B initiated apoptosis. *J Exp Med* **1997**, 186, 1323–31.

117 Andrade F, Roy S, Nicholson D, Thornberry N, Rosen A, Casciola-Rosen L. Granzyme B directly and efficiently cleaves several downstream caspase substrates: implications for CTL-induced apoptosis. *Immunity* **1998**, 8, 451–60.

118 Casciola-Rosen L, Andrade F, Ulanet D, Wong WB, Rosen A. Cleavage by granzyme B is strongly predictive of autoantigen status: implications for initiation of autoimmunity. *J Exp Med* **1999**, 190, 815–26.

119 Mevorach D, Zhou JL, Song X, Elkon KB. Systemic exposure to irradiated apoptotic cells induces autoantibody production. *J Exp Med* **1998**, 188, 387–92.

120 Gensler TJ, Hottelet M, Zhang C, Schlossman S, Anderson P, Utz PJ. Monoclonal antibodies derived from BALB/c mice immunized with apoptotic Jurkat T cells recognize known autoantigens. *J Autoimmun* **2001**, 16, 59–69.

121 Ronchetti A, Rovere P, Iezzi G, Galati G, Heltai S, Protti MP, Garancini

MP, Manfredi AA, Rugarli C, Bellone M. Immunogenicity of apoptotic cells *in vivo*: role of antigen load, antigen-presenting cells, and cytokines. *J Immunol* **1999**, 163, 130–6.

122 Levine JS, Subang R, Koh JS, Rauch J. Induction of anti-phospholipid autoantibodies by beta$_2$-glycoprotein I bound to apoptotic thymocytes. *J Autoimmun* **1998**, 11, 413–24.

123 Gallucci S, Lolkema M, Matzinger P. Natural adjuvants: endogenous activators of dendritic cells. *Nat Med* **1999**, 5, 1249–55.

124 Sauter B, Albert ML, Francisco L, Larsson M, Somersan S, Bhardwaj N. Consequences of cell death: exposure to necrotic tumor cells, but not primary tissue cells or apoptotic cells, induces the maturation of immunostimulatory dendritic cells. *J Exp Med* **2000**, 191, 423–34.

125 Reiter I, Krammer B, Schwamberger G. Cutting edge: differential effect of apoptotic versus necrotic tumor cells on macrophage antitumor activities. *J Immunol* **1999**, 163, 1730–2.

126 Barker RN, Erwig L, Pearce WP, Devine A, Rees AJ. Differential effects of necrotic or apoptotic cell uptake on antigen presentation by macrophages. *Pathobiology* **1999**, 67, 302–5.

127 Melcher A, Todryk S, Hardwick N, Ford M, Jacobson M, Vile RG. Tumor immunogenicity is determined by the mechanism of cell death via induction of heat shock protein expression. *Nat Med* **1998**, 4, 581–7.

128 Todryk S, Melcher AA, Hardwick N, Linardakis E, Bateman A, Colombo MP, Stoppacciaro A, Vile RG. Heat shock protein 70 induced during tumor cell killing induces T_h1 cytokines and targets immature dendritic cell precursors to enhance antigen uptake. *J Immunol* **1999**, 163, 1398–408.

129 Todryk SM, Melcher AA, Dalgleish AG, Vile RG. Heat shock proteins refine the danger theory. *Immunology* **2000**, 99, 334–7.

130 Rovere P, Fazzini F, Sabbadini MG, Manfredi AA. Apoptosis and systemic autoimmunity: the dendritic cell connection. *Eur J Histochem* **2000**, 44, 229–36.

131 Steinman RM, Turley S, Mellman I, Inaba K. The induction of tolerance by dendritic cells that have captured apoptotic cells. *J Exp Med* **2000**, 191, 411–6.

132 Heath WR, Kurts C, Miller JF, Carbone FR. Cross-tolerance: a pathway for inducing tolerance to peripheral tissue antigens. *J Exp Med* **1998**, 187, 1549–53.

133 Hirt UA, Gantner F, Leist M. Phagocytosis of nonapoptotic cells dying by caspase-independent mechanisms. *J Immunol* **2000**, 164, 6520–9.

134 Huang FP, Platt N, Wykes M, Major JR, Powell TJ, Jenkins CD, MacPherson GG. A discrete subpopulation of dendritic cells transports apoptotic intestinal epithelial cells to T cell areas of mesenteric lymph nodes. *J Exp Med* **2000**, 191, 435–44.

135 Rovere P, Manfredi AA, Vallinoto C, Zimmermann VS, Fascio U, Balestrieri G, Ricciardi-Castagnoli P, Rugarli C, Tincani A, Sabbadini MG. Dendritic cells preferentially internalize apoptotic cells opsonized by anti-β_2-glycoprotein I antibodies. *J Autoimmun* **1998**, 11, 403–11.

136 Rovere P, Sabbadini MG, Vallinoto C, Fascio U, Recigno M, Crosti M, Ricciardi-Castagnoli P, Balestrieri G, Tincani A, Manfredi AA. Dendritic cell presentation of antigens from apoptotic cells in a pro-inflammatory context: role of opsonizing anti-beta$_2$-glycoprotein I antibodies. *Arthritis Rheum* **1999**, 42, 1412–20.

137 Green DR, Beere HM. Apoptosis. Gone but not forgotten. *Nature* **2000**, 405, 28–29.

138 Rovere P, Sabbadini MG, Fazzini F, Bondanza A, Zimmermann VS, Rugarli C, Manfredi AA. Remnants of suicidal cells fostering systemic autoaggression. Apoptosis in the origin and maintenance of autoimmunity. *Arthritis Rheum* **2000**, 43, 1663–72.

139 Albert ML, Sauter B, Bhardwaj N. Dendritic cells acquire antigen from apoptotic cells and induce class I-restricted CTLs. *Nature* **1998**, 392, 86–9.

140 Salio M, Cerundolo V, Lanzavecchia A. Dendritic cell maturation is induced by mycoplasma infection but not by necrotic cells. *Eur J Immunol* **2000**, 30, 705–8.

141 Casciola-Rosen L, Rosen A. Ultraviolet light-induced keratinocyte apoptosis: a potential mechanism for the induction of skin lesions and autoantibody production in LE. *Lupus* **1997**, 6, 175–80.

142 Lorenz HM, Herrmann M, Winkler T, Gaipl U, Kalden JR. Role of apoptosis in autoimmunity. *Apoptosis* **2000**, 5, 443–9.

143 Herrmann M, Voll RE, Zoller OM, Hagenhofer M, Ponner BB, Kalden JR. Impaired phagocytosis of apoptotic cell material by monocyte-derived macrophages from patients with systemic lupus erythematosus. *Arthritis Rheum* **1998**, 41, 1241–50.

144 Kalden JR. Defective phagocytosis of apoptotic cells: possible explanation for the induction of autoantibodies in SLE. *Lupus* **1997**, 6, 326–7.

145 Rosen A, Casciola-Rosen L. Clearing the way to mechanisms of autoimmunity. *Nat Med* **2001**, 7, 664–5.

146 Botto M, Dell'Agnola C, Bygrave AE, Thompson EM, Cook HT, Petry F, Loos M, Pandolfi PP, Walport MJ. Homozygous C1q deficiency causes glomerulonephritis associated with multiple apoptotic bodies. *Nat Genet* **1998**, 19, 56–9.

147 Mitchell DA, Taylor PR, Cook HT, Moss J, Bygrave AE, Walport MJ, Botto M. Cutting edge: C1q protects against the development of glomerulonephritis independently of C3 activation. *J Immunol* **1999**, 162, 5676–9.

148 Taylor PR, Carugati A, Fadok VA, Cook HT, Andrews M, Carroll MC, Savill JS, Henson PM, Botto M, Walport MJ. A hierarchical role for classical pathway complement proteins in the clearance of apoptotic cells *in vivo*. *J Exp Med* **2000**, 192, 359–66.

149 Gershov D, Kim S, Brot N, Elkon KB. C-Reactive protein binds to apoptotic cells, protects the cells from assembly of the terminal complement components, sustains an antiinflammatory innate immune response. Implications for systemic autoimmunity. *J Exp Med* **2000**, 192, 1353–64.

150 Bickerstaff MC, Botto M, Hutchinson WL, Herbert J, Tennent GA, Bybee A, Mitchell DA, Cook HT, Butler PJ, Walport MJ, Pepys MB. Serum amyloid P component controls chromatin degradation and prevents antinuclear autoimmunity. *Nat Med* **1999**, 5, 694–7.

151 Scott RS, McMahon EJ, Pop SM, Reap EA, Caricchio R, Cohen PL, Earp HS, Matsushima GK. Phagocytosis and clearance of apoptotic cells is mediated by MER. *Nature* **2001**, 411, 207–11.

152 Rovere P, Vallinoto C, Bondanza A, Crosti MC, Rescigno M, Ricciardi-Castagnoli P, Rugarli C, Manfredi AA. Bystander apoptosis triggers dendritic cell maturation and antigen-presenting function. *J Immunol* **1998**, 161, 4467–71.

153 Rovere P, Sabbadini MG, Vallinoto C, Fascio U, Zimmermann VS, Bondanza A, Ricciardi-Castagnoli P, Manfredi AA. Delayed clearance of apoptotic lymphoma cells allows cross-presentation of intracellular antigens by mature dendritic cells. *J Leuk Biol* **1999**, 66, 345–9.

16
'Tissue' Transglutaminase and Autoimmunity

Alessandra Amendola and Mauro Piacentini

16.1
Introduction

The understanding of the mechanisms of cell death has increased dramatically in the last 10 years due to the identification of key components of the apoptotic program. Programmed cell death is a physiological process that plays a fundamental role in embryogenesis, tissue homeostasis and elimination of damaged cells [1]. It is well documented that faults in the regulation of apoptosis may be implicated in the cause or as a contributing factors in the pathogenesis of major human diseases, including autoimmunity [Hashimoto's thyroiditis, systemic lupus erythematosus (SLE) and rheumatoid arthritis] [2], neurodegenerative diseases (Alzheimer's and Huntington's diseases), cancer [3], infectious diseases (HIV, hepatitis viruses, human herpes viruses and bacterial infections) [4] and myocardial malfunctions [5], although the pathological mechanisms involved in such diseases are clearly distinct.

Extensive structural changes occur in cells undergoing apoptosis determined by post-translational modification of cellular proteins. This drastic remodeling of cellular components may lead to profound modification of the antigenic profile of dying cells. Evidence is accumulating that post-translational modifications of proto-autoantigens during apoptosis may promote the development of autoantibodies by passing the normal mechanisms of tolerance [6]. The potentially harmful consequences of this event are mostly prevented by the rapid clearance of cells undergoing apoptosis. Phagocytosis of apoptotic cells by macrophages and dendritic cells, in fact, is necessary to eliminate pro-inflammatory debris and neo-self-antigens. Therefore, phagocytosis deregulation might play an important pathogenetic role in the immune system, its efficiency being a key determinant in the suppression of tissue inflammation and prevention of autoimmunity. A common feature of autoimmune diseases is the break down of tolerance consequent to the production of autoantibodies reactive with multiple self-proteins. In this chapter we discuss how the impairment of 'tissue' transglutaminase (tTG), a cell death-associated gene product, might represent an important factor in autoimmunity, both by preventing the spreading of self-antigens and/or contributing to the post-translational modification of proteins in dying cells.

16.2 Transglutaminases

The transglutaminase family includes seven intracellular (the ubiquitous tTG and six different isoenzymes differentially expressed in the tissues) and two extracellular enzymes (Factor XIIIa and prostate transglutaminase) that catalyze Ca^{2+}-dependent reactions resulting in the post-translational modification of proteins at the level of glutamine and lysine residues [7–10]. This post-translational modification leads to the formation of the $\varepsilon(\gamma$-glutamyl)lysine crosslinks and/or to the covalent incorporation of polyamines into proteins [7–10]. Diamines and polyamines may also participate in crosslinking reactions through the formation of *N,N* -bis(γ-glutamyl)polyamine bonds [8–10]. The formation of these covalent crosslinks leads to oligomerization of substrate proteins which become resistant to physicochemical stress, such that the polypeptides released from the polymer(s) only by the proteolytic degradation of the protein chains [8–10]. Although transglutaminases act as crosslinking enzymes, in the absence of an appropriate acceptor peptide they may cause the deamidation of protein-bound glutamine residues.

16.3 tTG and Apoptosis

The onset of apoptosis is associated with the marked induction of the tTG gene, followed by its Ca^{2+}-dependent enzymatic activation [11–13]: tTG-dependent protein crosslinking has been observed in the most extensively characterized *in vivo* and *in vitro* models of apoptosis [11–15]. Although a definitive role for tTG in apoptosis has not yet been firmly established, the specific expression of the tTG gene observed in dying cells does not seem to be an epiphenomenon [1]. Mammalian cells transfected with a full-length tTG complementary DNA (cDNA) show a marked increase in the rate of spontaneous cell death [14, 16–17]. Conversely, stable transfectants containing segments of the human tTG cDNA, in *antisense* orientation, show a pronounced decrease both in spontaneous and induced apoptosis [16]. This suggests that the tTG-catalyzed irreversible crosslinking of intracellular proteins is an important biochemical event in apoptosis. Mice lacking transglutaminase, although not showing major defects during embryogenesis, present several dysfunctions related to anomalies occurring in the apoptotic program (unpublished observations).

It has been demonstrated that over-expression of tTG reduces the release of macromolecules [double-stranded (ds)DNA and lactate dehydrogenase (LDH)] that characterize tumor necrosis factor α-induced death of L929 cells [17]. These results suggest that the tTG-catalyzed intracellular protein crosslinking plays an important role in stabilizing apoptotic bodies and in preventing the leakage of their content [17]. This phenomenon may represent a key feature of apoptosis *in vivo*, since a controlled disposal of crosslinked apoptotic bodies is required to prevent the inflammatory response as well as the exposure of self-antigens, which may lead to the development of autoimmunity [13, 18].

Tab. 16.1 Proteins post-translationally modified by both tTG and thiol proteases.

tTG substrate	*Reference*	*Corresponding thiol protease*	*Reference*
Histone H2B	19	caspase	20
Retinoblastoma protein pRb	15	caspase	21
Actin	22	caspase	20
Huntingtin	23	caspase	17
Vimentin	24	caspase	25
Spectrin	26	calpain	18
Tau	27	calpain	28
Tubulin	10	calpain	29
Troponin	30	calpain	30

The identification of *ced3*, one of the pro-apoptotic genes of the *Caernohabditis elegans* genetic death pathway [19], as the cysteine protease interleukin-1β-converting enzyme (ICE) has led to the discovery of 14 cysteine proteases, classified as caspases [20, 21]; as well as to define the involvement of calpains in apoptosis [22]. Although a number of protein substrates have been shown to be cleaved by the cysteine proteases during apoptosis, how many of these proteins must be processed to establish the death phenotype is still unclear [20, 21]. It has been shown that several proteins act as substrates for both caspases and tTG during apoptosis, and, consistent with this observation, other well-characterized tTG substrate proteins are also cleaved by calpains (Tab. 16.1). These findings indicate that, during apoptosis, tTG and cysteine proteases act on a common set of target proteins (Tab. 16.1).

Although the physiological significance of 'cleaving and polymerizing' the same substrate proteins in the establishment of the apoptotic phenotype has yet to be defined, several hypothesis can be proposed. For example, the presence of a cysteine active site is essential for the catalytic activity of tTG, calpains and caspases [10, 21, 22]. Recently, tTG has been shown not only to catalyze the formation of protein crosslinks, but also to deaminate them, acting as a hydrolytic enzyme [23]. Hence, tTG may be considered to have evolved from a papain-like ancestral cysteine protease [24], since papain, like caspases, cleaves proteins at aspartate residues [21, 23, 25].

16.4 tTG and Autoimmunity

Alteration of the delicate balance between cell survival and death has been implicated in the pathogenesis of many autoimmune diseases [18, 26–28]. Defects in the CD95 receptor, involved in apoptosis induction after specific interaction with CD95 ligand (CD95L), have been shown to play an important role in autoimmune diabetes, thyroiditis and in the lymphoproliferation disorder described in MRL/Mp-*lpr/lpr* mice [26–28]. However, the complex pathogenesis of these autoimmune diseases cannot be fully explained by a malfunctional CD95/CD95L system alone [18]; in fact, malfunction of other gene products has also been hypothesized [18].

Tab. 16.2 Autoantibodies against tTG protein substrates in autoimmune diseases.

tTG substrate	*Reference*	*Autoimmune disease*	*Reference*
Actin	22	hepatitis	41
		necrobiosis lipoidica	40
		human lupus	43, 44
Keratins	5	hepatitis	41
		necrobiosis lipoidica	40
Histone H2B	19	murine and human lupus	42, 44
Lipocortin 1	45	murine lupus	46
Myosin	38	human lupus	44
		necrobiosis lipoidica	40
Troponin	30	necrobiosis lipoidica	40
tTG	23	human lupus	44
		celiac disease	37
Tubulin	10	hemolytic anemia	39
		human lupus	44

A deregulated tTG is present in autoimmunity-prone MRL-*lpr/lpr* mice: crosslinking activity of tTG is reduced; autoantibodies against tTG are produced followed by the abnormal accumulation of LDH and dsDNA in the blood [17]. The display of cryptic molecular determinants has been proposed to play an important role in the induction of the pathogenic autoimmune response [18]. The question then arises as to how the apoptotic process results in the formation of harmful neo-determinants.

In addition to the autoantibodies against tTG [17, 29], immunoglobulins against substrate proteins of tTG have been demonstrated in various autoimmune diseases [30–38] (Tab. 16.2). This phenomenon is particularly interesting considering that many tTG substrate proteins are also cleaved by caspases and calpains during apoptosis (Tab. 16.1). On this basis, it is tempting to hypothesize that the tTG-dependent polymerization of polypeptides that are generated by proteolytic action during the apoptosis execution phase represents a safety mechanism for the organism. In the absence of an active tTG, the accumulation of cleavage products, not normally present in viable cells, may represent a harmful event and may contribute to the development of autoimmunity. In fact, these novel polypeptides can be released into the extracellular space and/or be presented as neo-antigens with the consequent generation of autoimmune responses. Surface blebs, typically formed in apoptotic cells, represent important immunogenic elements in SLE [39]. In keeping with this hypothesis, recent findings show that tTG knockout mice have a defect in the clearance of apoptotic cells and develop nuclear autoantibodies with aging (unpublished observations). The pathogenic effects of impaired tTG-catalyzed crosslinking as a cofactor in the development of autoimmunity could be dramatic in the presence of the impairment of phagocytosis resulting in the accumulation of dead or dying cells [40]. tTG may influence the clearance of apoptotic bodies by its function of polymerization of protein substrates, thus contributing to the production of signals in dying cells, which may favor their clearance. Consequently, the lack of tTG-dependent polymerization could play a role in the accumulation of damaged cells in tissues, leading to presentation of neo-determinants [38] and activation of autoimmune reactions. In keeping with this assumption, recent data show that apoptotic cells injected into guinea pig skin are phagocytosed and cleared by infiltrating monocyte/macrophage 48 h after their injection. This event is paralleled by the release, from apoptotic cells, of a monocyte chemotactic factor [41] carrying out an important role in the phagocytic clearance of apoptotic cells. This factor was identified as the crosslinked homodimer of S19 ribosomal protein [41]. Interestingly the dimerization of the S19 protein is catalyzed by transglutaminase, thus confirming an important role for tTG in the apoptotic cells clearance [41]. Taken together these findings confirm that defective clearance of apoptotic cells may promote an autoimmune response in diseases characterized by increased rates of apoptosis such as AIDS and, eventually, SLE [42].

16.5 Celiac Disease (CD)

CD is a complex disease characterized by a wide spectrum of lesions in the intestinal mucosa that can ultimately lead to the atrophy of the villi [43]. Permanent intolerance to gluten, and in particular to gliadin, triggers the production of elevated levels of circulating anti-gliadin antibodies (AGA), anti-reticulin antibodies and anti-endomysium antibodies (EMA), and measurement of these IgA autoantibodies is generally used as a serological marker for clinical diagnosis of CD [44].

Although it has been hypothesized that autoimmune mechanisms are important in the pathogenesis of celiac disease [29, 45], this disorder is not a classical model of an autoimmune disease. In fact, IgA antibodies disappear and most patients' mucosae regenerate villi, when gluten is removed from the diet. However, the mucosal damage, that characterizes CD is considered to be immune-mediated [44]. It remains unknown if specific immune mechanisms are responsible for the tissue damage or whether they act in concert with gliadin, in a direct effect on susceptible mucosa. The sequence of immunological events is still not known, but binding of gliadin by HLA-DQ molecules, for antigen presentation to lymphoid cells, appears to be a central event in genetically predisposed individuals [45]. Dieterich *et al.* identified tTG as the predominant autoantigen recognized by EMA in patients with CD [29]. tTG seems to play an important role in CD by modifying the gliadin molecule and increasing its affinity for HLA-DQ molecule peptide-binding groove [29]. Molberg *et al.* demonstrated that *in vitro* deamination of gliadin by tTG creates new epitopes that bind to HLA-DQ and are recognized by gut-derived T cells [46]. This event results in T cell activation followed by lamina propria and mucosal transformation [47]. The modified gliadin binds strongly to tTG, probably favoring the gluten-dependent production of anti-tTG antibodies [29]. At the same time, HLA-DQ binding of tTG-modified gliadin induces T cell activation and, consequently, cytokine production that might also play a role in the intestinal changes [48]. In fact, production of antibodies directed to tTG has been shown to inhibit transforming growth factor-β and this, in turn, inhibits intestinal epithelial cell differentiation, necessary for maintaining villus integrity [49].

It has been shown that CD antigens are localized in the extracellular matrix (ECM) and are produced by fibroblasts [45, 50]. Although tTG is an intracellular protein, elevated tTG activity has been demonstrated in the ECM of CD patients, co-localizing with the structures that specifically bind IgA [51]. It has been proposed that the tTG-dependent crosslinking activity contributes to stabilization of the ECM and promotes cell–substrate interaction [52]. The tTG release from cells occurring during inflammation seems to be implicated in a number of pathological conditions, including fibrosis, atherosclerosis, neurodegenerative diseases, cancer metastasis and also CD [52].

Previous studies have demonstrated the involvement of tTG in diseases characterized by induction of apoptosis and a drastic remodeling of the affected tissue, particularly in ECM [53, 54]. In a recent study aimed to elucidate the role of cell death in the pathogenesis of various degenerative diseases affecting the intestinal mucosa, such as CD, Crohn's disease and ulcerative rectocolitis (UR), we demonstrated that the presence of anti-tTG antibodies is a general phenomenon related to mucosal lesions, and not correlated to the autoimmune nature hypothesized for these disorders [55]. In fact, we detected both circulating anti-tTG antibodies and induction of apoptosis in CD, Crohn's disease and UR, but not in other systemic autoimmune diseases (such as diabetes, thyroiditis, multiple sclerosis and SLE) [55]. In addition, considering that the enhanced apoptosis is paralleled by the induction of tTG both in physiological and pathological settings [1, 7, 11, 56], we have analyzed whether the lesions occurring in CD, as well as in Crohn's diseases, are associated with an abnormal onset of apoptosis and a tTG expression. Our results suggest that the accumulation of tTG in the enterocytes, as well as its release in the ECM, are a consequence of the induction of apoptosis in those regions undergoing the destruction, typical of severe CD-associated lesions (Fig. 16.1). In fact, the enterocytes become tTG-positive and show the morphological and biochemical features typical of apoptotic cells, including the expression of the CD95L in early stages of the diseases. Although defects in the CD95 pathway have been shown to play an important role in several autoimmune diseases [6, 57], very little is known about the CD95/CD95L pathway in CD. We have presented data suggesting the involvement of this receptor also in CD pathogenesis [55].

It is interesting to note that, similarly to the wound healing process and tissue repair occurring in CD lesions, in autoimmunity-prone MRL *lpr/lpr* mice a deregulated tTG activity is present, and the enzyme is also detected in the ECM [17]. Several studies indicated that tTG, by crosslinking a number of proteins such as collagens, fibronectin, laminin, nidogen and transforming growth factor-β might play an important role in the modification of the ECM occurring in degenerative diseases [54, 58]. The presence of the tTG in the ECM in CD lesions might represent *per se* an important pathogenic event. It is very likely that the tTG-mediated polymerization of ECM proteins may generate new self-antigens and thus contribute to eliciting the autoimmune response (Fig. 16.1). In keeping with this as-

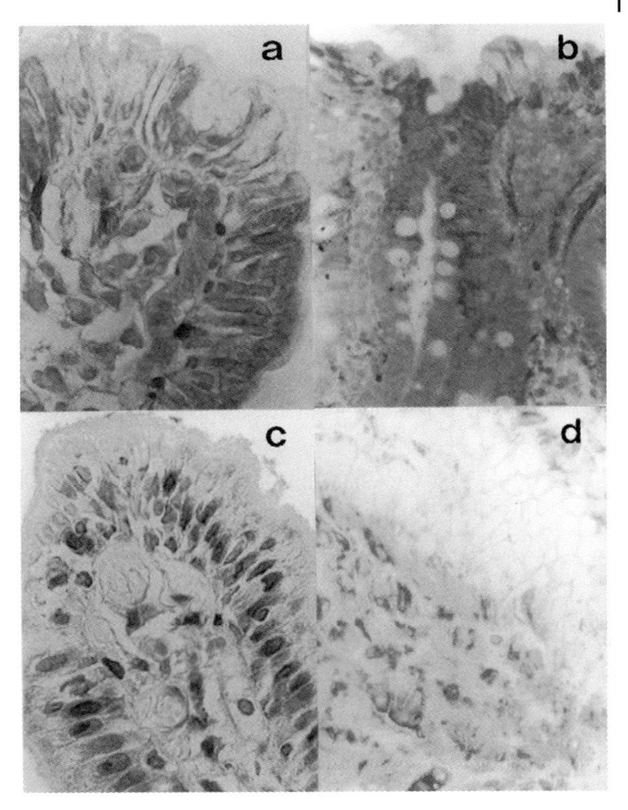

Fig. 16.1 Localization of tTG and $\varepsilon(\gamma$-glutamyl)lysine crosslinks by immunohistochemistry, and DNA fragmentation by the TUNEL technique in the intestinal mucosa of CD patients. Biopsies of intestinal mucosae from CD patients were taken, fixed and stained with antibodies against tTG (a and b), and $\varepsilon(\gamma$-glutamyl)lysine crosslinks (d). DNA fragmentation was detected by the TUNEL technique (c). Note that in stages of the CD disease in which the mucosa was still present (a and c), tTG staining was detected in the enterocytes localized in the upper part of the villi and in the fibroblasts lining the intestinal epithelium (a), while DNA fragmentation was confined to the enterocytes (c). In the acute stage of the disease (b and d), characterized by villus flattening, a large proportion of the tTG staining was detected in the ECM of the lesion and only a limited number of fibroblasts were stained (b). In the same lesions, very few cells were found positive to both TUNEL and CD95L staining (data not shown). Note the intense staining detected by both the tTG (b) and anti-$\varepsilon(\gamma$-glutamyl)lysine crosslink antibodies (d) in ECM of the intestinal areas showing a complete flattening of the mucosa. The intense staining observed with the antibody against $\varepsilon(\gamma$-glutamyl)lysine crosslinks in the ECM (d) indicate that the released tTG is active.

sumption is the demonstration that the binding of gliadin to reticular matrix components is Ca^{2+}-dependent, and it is inhibited by putrescine and by preincubation with antibodies against tTG: this means that binding of gliadin might be a tTG-mediated event [45, 59]. Taken together these findings highlight an important complex role carried out by tTG and apoptosis in the pathogenesis of CD.

16.6 Autoimmunity and Infectious Diseases

Both viral and bacterial infections can cause apoptosis in immune competent cells and are often accompanied by the appearance of autoimmunity [60, 61]. Although there is a consensus view that the onset of autoimmunity during microbial infection does not play a major role in the pathogenesis of infectious diseases, it is possible that deregulated apoptosis induced by pathogens is responsible to a substantial degree for this autoimmune response. It has been suggested that this may represent an immunopathogenic cofactor in AIDS and tuberculosis [60, 61].

Antigen-presenting cells (APC), such as dendritic cells and macrophages in the lymph nodes of HIV-infected individuals, express abnormally high levels of tTG [15]. Similarly, alveolar macrophages obtained from the lavage of the lung from HIV-infected patients with pulmonary tuberculosis display high levels of tTG in their cytoplasm [62]. The induction of tTG in APC infected with HIV and/or *Mycobacterium tuberculosis* undergoing apoptosis may not only prevent the generation of self antigens, but also the spreading of viable bacteria and viral particles [63, 64]. Consistent with this view, autoimmunity in *lpr* mice is associated with the accumulation of an inactive tTG in the APC of lymphoid tissues [17]. Interestingly, the polyclonal B cell activation and antibody production against self-epitopes observed in autoimmune *lpr* mice is accompanied by the appearance of high titers of antibodies binding retroviral proteins, including HIV [65]. The anti-HIV antibody repertoire detected in MRL *lpr/lpr* mice is very similar to that present in HIV-infected humans and in patients suffering with Sjögren's syndrome and SLE [66], thus suggesting that different pathogenetic events might determine a similar autoimmune response when apoptosis and APC function are impaired. In keeping with this assumption, a recent report demonstrates the presence of anti-tissue transglutaminase antibodies in HIV-infected patients [67].

16.7 Conclusions

In conclusion, we have discussed how the post-translational modification of proteins catalyzed by tTG may play an important role in preventing the neo-antigens presentation to the immune system during apoptosis. However, in the absence of an active clearance of apoptotic bodies, tTG-catalyzed modification of proteins may effectively represent a source of potentially pathogenetic neo-antigens. The tight control of the potentially harmful post-translational modifications of 'death protein substrates' during apoptosis is an essential step to achieve the 'immunologically silent' demise of the cell [39, 43, 68]. We believe that the advancement in the knowledge of the possible role of apoptosis and its regulation in the pathogenesis of autoimmune diseases may provide new insight into the therapeutics for the prevention and/or treatment of these diseases.

16.8 Acknowledgments

This work was partially supported by grants from EU 'V framework' grant 'Apoptosis Mechanisms'. Associazione Italiana Ricerca sul Cancro (AIRC) and Ministero Sanitá 'Progetto AIDS' e Ricerca corrente e finalizzata.

16.9 References

1 Piacentini, M. *Curr Topics Microbiol Immunol* **1995**, 200, 163–75.
2 Eguchi, K. *Intern Med* **2001**, 40, 275–84.
3 Gibson, R.M. *Br Med J* **2001**, 322, 1539–40.
4 Dockrell, D.H. *J Infect.* **2001**, 42, 227–34.
5 Schwartz, S.M., Duffy, J.Y., Pearl, J.M., Nelson, D. P. *Crit Care Med* **2001**, 29 (10 Suppl), S214–9.
6 Piacentini, M., Colizzi, V. *Immunol Today* **1999**, 20, 130–4.
7 Fesus, L., Davies, P.J.A., Piacentini, M. *Eur J Cell Biol*, **1991**, 56, 170–77.
8 Greenberg, C.S., Birckbichler, P., Rice, R.H. *FASEB J* **1991**, 5, 3071–7.
9 Piacentini, M., Martinet, N., Beninat, I.S., Folk, J. E. *J Biol Chem* **1988**, 26, 3790–4.
10 Folk, J.E. *Annu Rev Biochem* **1980**, 49, 517–31.
11 Fesus, L., Thomazy, V., Autuori, F., Cerú, M.P., Tarcsa, E., Piacentini, M. *FEBS Lett* **1989**, 245, 150–4.
12 Piacentini, M., Fesus, L., Farrace, M. G., Ghibelli, L., Piredda, L., Melino, G. *Eur J Cell Biol* **1991**, 5, 246–54.
13 Knight, C., Hand, D., Piacentini, M., Griffin, M. *Eur J Cell Biol* **1993**, 60, 210–7.
14 Gentile, V., Thomazy, V., Piacentini, M., Fesus, L., Davies, P. J. A. *J Cell Biol* **1992**, 119, 463–74.
15 Amendola, A., Gougeon, M.L., Poccia, F., Bondurand, A., Fesus, L., Piacentini, M. *Proc Natl Acad Sci USA* **1996**, 93, 1157–62.
16 Melino, G., Annichiarico-Petruzzelli, M., Piredda, L., Candi, E., Gentile, V., Davies, P. J. A., Piacentini, M. *Mol Cell Biol* **1994**, 14, 6584–92.
17 Piredda, L., Amendola, A., Colizzi, V., *et al. Cell Death Different* **1997**, 4, 463–72.
18 Theofilopoulos, A. N. *Immunol Today* **1995**, 16, 90–8 and 150–9.
19 Yuan, J., Shaham, S., Ledoux, S., Ellis, H.M., Horvitz, H. R. *Cell* **1993**, 75, 641–52.
20 Hale, A.J., Smith, C.A., Sutherland, L.C., *et al. Eur J Biochem* **1996**, 236, 1–26.
21 Cohen, G.M. *Biochem J* **1997**, 326, 1–16.
22 Squier, M.K.T., Cohen, J.J. *Cell Death Different* **1996**, 3, 275–83.
23 Parameswaran, K.N., Cheng, X.F., Chen, E.C., Velasco, P.T., Wilson, J.H., Lorand, L. *J Biol Chem* **1997**, 272, 10311–7.
24 Dosemeci, A., Reese, T.S. *Synapse* **1995**, 20, 91–7.
25 Kumar, S., Lavin, M.F. *Cell Death Different* **1996**, 3, 275–83.
26 Watanabe-Fukunaga, R., Brannan, C.I., Copeland, N.G., Jenkins, N.A., Nagata, S. *Nature* **1992**, 356, 314–7.
27 Benoist. C., Mathis, D. *Cell* **1997**, 89, 1–3.
28 Giordano, C., Stassi, G., De Maria, R., Todaro, M., Richiusa, P., Papoff, G., Ruberti, G., Bagnasco, M., Testi, R., Galluzzo, A. *Science* **1997**, 275, 960–3.
29 Dieterich, W., Ehnis, T., Bauer, M, *et al. Nature Med* **1997**, 3, 797–801.
30 Kang, S.J., Shin, K.S., Song, W.K., Ha, D. B., Chung, C.H., Kang, M.S. *J Cell Biol* **1995**, 130, 1127–36.
31 Hentati, B., Payelle-Brogard, B., Jouanne, C., Avrameas, S., Ternynck, T. *J Autoimmun* **1994**, 7, 425–39.
32 Haralambous, S., Blackwell, C., Mappouras, D.G., Weir, D.M., Kemmett, D., Lymberi, P. *Autoimmunity* **1995**, 20, 267–75.
33 Orth, T., Gerken, G., Kellner, R., Meyer zum Buschenfelde, K.H., Mayet, W.J. *J Hepatol* **1997**, 26, 37–47.

34 Elouaai, F., Lule, J., Benoist, H., Appolinaire-Pilipenko, S., Atanassov, C., Muller, S., Fournie, G.J. *Nephrol Dial Transplant* **1994**, 9, 362–6.
35 Burlingame, R.W., Boey, M.L., Starkebaum, G., Rubin, R.L. *J Clin Invest* **1994**, 94, 184–92.
36 Adyel, F.Z., Hentati, B., Boulila, A., Hachicha, J., Ternynck, T., Avrameas, S., Ayadi, H. *J Clin Lab Anal* **1996**, 10, 451–7.
37 McKanna, J.A. *Anat Rec* **1995**, 242, 1–10.
38 Ikai, K, Shimizu, K., Kanauchi, H., Ando, Y., Furukawa, F., Imamura, S. *Autoimmunity* **1992**, 12, 239.
39 Casciola–Rosen, L., Rosen, A., Petri, M., Schlissel, M. *Proc Natl Acad Sci USA* **1996**, 93, 1624–9.
40 Van Houten, N., Budd, R.C. *J Immunol* **1992**, 149, 2513–7.
41 Nishimura, T., Horino, K., Nishiura, H., Shibuya, Y., Hiraoka, T., Tanase, S., Yamamoto, T. *J Biochem* **2001**, 129, 445–54.
42 Mevorach, D., Zhou, J.L., Song, X., Elkon, K.B. *J Exp Med* **1998**, 188, 387–92.
43 Trier, J.S. *N Engl J Med,* **1991**, 325, 1709–19.
44 Ascher, A., Hahn-Zoric, M., Hanson, L. A., *et al. Scand J Gastroenterol* **1996**, 31, 61–7.
45 Maki, M., Collin, P. *Lancet* **1997**, 349, 1755–9.
46 Molberg, O., Mcadam, S.N., Corner, R., *et al. Nat Med* **1998**, 4, 713–7.
47 Schuppan, D. *Gastroenterology* **2001**, 120, 586–7.
48 Scott, H., Kett, K., Halstensen, T., Hvatum, M., *et al.* In: Marsh M.N., eds. *Coeliac Disease.* Oxford: Blackwell 1992, 239–82.
49 Sollid, L.M. *Annu Rev Immunol* **2000**, 18, 53–81.
50 Marttinen, A., Sulkanen, S., Maki, M. *Eur J Clin Invest* **1997**, 27, 135–40.
51 Sulkanen, S., Halttunen, T., Laurila, K., *et al. Gastroenterology* **1998**, 115, 1322–8.
52 Aeschlimann, D., Thomazy, V. *Connect Tissue Res* **2000**, 41, 1–27.
53 Piacentini, M., Farrace, M.G., Hassan, C., Serafini, B., Autuori, F. *J Pathol* **1999**, 189, 92–8.
54 Johnson, T.S., Griffin, M., Thomas, G.L., *et al. J Clin Invest* **1997**, 90, 2950–60.
55 Farrace, M.G., Picarelli, A., Di Tola, M., Sabbatella, L., Marchione, O.P., Ippolito, G., Piacentini, M. *Cell Death Different* **2001**, 8, 767–70.
56 Melino, G., Piacentini, M. *FEBS Lett* **1998**, 430, 59–63.
57 Thompson, C.B. *Science* **1995**, 267, 1456–62.
58 Mirza, A., Liu, S.L., Frizell, E., *et al. Am J Physiol* **1997**, 272, G281–288.
59 Uhlig, H., Osman, A.A., Tanevm, I.D., Viehweg, J., Mothes, T. *Autoimmunity* **1998**, 28, 185–195.
60 Gougeon, M.L., Colizzi, V., Dalgleish, A., Montagnier, L. *AIDS Res Hum Retroviruses* **1993**, 9, 287–9.
61 Monack, D.M, Raupach, B., Hromockyj, S., Falkow, S. *Proc Natl Acad Sci USA* **1996**, 93, 9833–8.
62 Placido, R., Mancino, G., Amendola, A., Mariani, F., Vendetti, S., Piacentini, M., Sanduzzi, A., Bocchino, M.L., Zembala, M., Colizzi, V. *J Pathol* **1997**, 181, 31–8.
63 Bergamini, A., Capozzi, M., Piacentini, M. *Immunol Lett* **1994**, 42, 35–40.
64 Fratazzi, C., Arbeit R.D., Carini, C., Remold, H.G. *J Immunol* **1997**, 158, 4320–4327.
65 Lombardi, V., Placido, R., Scarlatti, G., Romiti, M.., Mattei, M., Mariani, F., Poccia, F., Rossi, P., Colizzi, V. *J Infect Dis* **1993**, 167, 1267–73.
66 Fraziano, M., Montesanom, C., Lombardi, V.R., Sammarco, I., De Pisa, F., Mattei, M., Valesini, G., Pittoni, V., Colizzi, V. *AIDS Res Hum Retroviruses* **1996**, 12, 491–6.
67 Pereda, I., Bartolome-Pacheco, M.J., Martin, M., Lopez–Escribano, H., Echevarria, S., Lope-Hoyos, M. *J AIDS* **2001**, 27, 507–8.
68 Utz, P.J., Hottelet, M., Schur, P.H., Anderson, P. *J Exp Med* **1997**, 185, 843–54.
69 Itoh, G., Tamura, J., Suzuki, M., Suzuki, Y., Ikeda, H., Koike, M., Nomura, M., Jie, T., Ito, K. *Am J Pathol* **1995**, 146, 1325–31.
70 Gorza, L., Menabo, R., Vitadello, M., Bergamini, C.M., Di Lisa, F. Circulation **1996**, 93, 1896–904.

17
Modification of RNA Autoantigens in Apoptosis

Wolfgang Hueber and Paul J. Utz

17.1
Introduction

RNA specific antibodies, RNA antigens and RNA cleavage during apoptosis are topics of interest in the field of systemic autoimmunity. First, the hypothesis that protein *and* RNA neoepitopes may be the deleterious first signals in alerting the immune system to attack against self is gaining momentum [1]. Second, neoepitopes are generated during apoptosis, and the list of molecules modified during apoptosis is continuously growing [2]. Third, the role of viral nucleic acids (and proteins) interacting with host molecules is increasingly recognized as a potential mechanism generating neoepitopes within or outside the cell death program pathways. Last, the use of RNA molecules in the form of ribozymes, antisense oligonucleotides or aptamers as vehicles or therapeutics in various medical applications raises safety questions concerning the immunogenic potential of these compounds.

Nucleic acids have long been known to be prominent targets of autoantibodies in animal models of autoimmunity and in human autoimmune disease. However, the general paradigm that anti-DNA antibodies play a role in the pathogenesis of lupus nephritis and the diagnostic usefulness of these antibodies for the clinician have reduced awareness that anti-RNA antibodies in human autoimmune diseases are also prominent. Due to a lack of experimental evidence, these antibodies have never found a place in routine clinical testing, screening or follow-up for autoimmune diseases. Consequently, the topic has nearly disappeared – at least from a practitioner's point of view – into the 'obscurity' of more hardcore basic science publications. However, recent exciting findings suggest a more important role for RNA as a target for autoimmune responses in the context of programmed cell death.

Particularly intriguing for researchers in the field is the selectivity by which a small subset of RNA molecules is modified during apoptosis, whereas the vast majority of RNA molecules remain untouched. This selectivity became even more remarkable when it was discovered that antibodies to certain individual RNA epitopes are associated with more severe disease, implying a pathogenic role for these autoantibodies. Moreover, the rediscovery of certain viral infections in the pathogenesis of autoimmune disease will certainly spur new interest in the immunogenic potential of RNA. Similarly, the use of nucleic acids in gene therapy has already caused concerns about the potential induction of autoantibodies and autoimmune syndromes.

It is probably best to view the emergence of autoantibodies to RNA not as an isolated phenomenon, but as part of an autoimmune response directed towards large multi-molecular RNA-protein complexes. Interestingly, the autoimmune response selectively targets specific components of these macromolecular complexes, usually involving the biologically most active epitopes that play crucial roles for the function of the cell. It is increasingly recognized that irreversible events occurring during apoptosis that modify certain molecules vital to the function of the cell play a key role in the breakage of tolerance to self. During apoptosis, post-translational modifications of various proteins occur, and some of these altered proteins have been identified as autoantigens [2]. Only recently it was discovered that RNA modifications occur during apoptosis. Moreover, some of these modifications are substantiated by highly specific cleavage of selected nucleic acids. Cleavage of nucleic acids is dependent on caspases, the main enzymatic executioners of cell death [3]. Taken together, modifications of proteins as well as RNA molecules during apoptosis may contribute both to the formation of neoepitopes, to breakage of tolerance, and to development of disease-initiating or -perpetuating immune responses in autoimmune disorders. (The second prerequisite for induction of autoimmunity within the frame of this hypothesis – insufficient or defective clearance of apoptotic remnants – is discussed in detail in part III in this book.) The exact chronology of the evolution of neoepitopes and the sequence of events overriding T cell tolerance to an increasing array of self-epitopes on macromolecular complexes (a process referred to as epitope spreading) has been the focus of intense research in recent years [4, 5].

Although the role of naked RNA, its modifications during apoptosis and the time point of RNA (neo) epitope recruitment to this array of autoantigenic determinants that are eventually recognized by lymphocytes are incompletely understood, recent work clearly suggests a more prominent involvement of RNA antigens in the process of epitope spreading. Fig. 17.1 shows incorporation of such an immunogenic RNA epitope in one possible scenario of intermolecular epitope spreading.

This chapter is henceforth organized as follows. First, we will provide a historical framework, describing seminal work performed on anti-RNA antibodies in the late 1960s and early 1970s. We will then provide a brief overview of this research, which is essential for the understanding of modifications of RNA molecules during apoptosis. Second, we will review recent data on three subsets of RNAs that have been shown to be modified during apoptosis. Third, the role of viral infections as a triggering catalyst for the generation of altered RNA (and altered protein) molecules will be discussed. Finally, a synopsis of current hypotheses focusing on modified RNA antigens as the target of autoimmune responses will conclude this chapter.

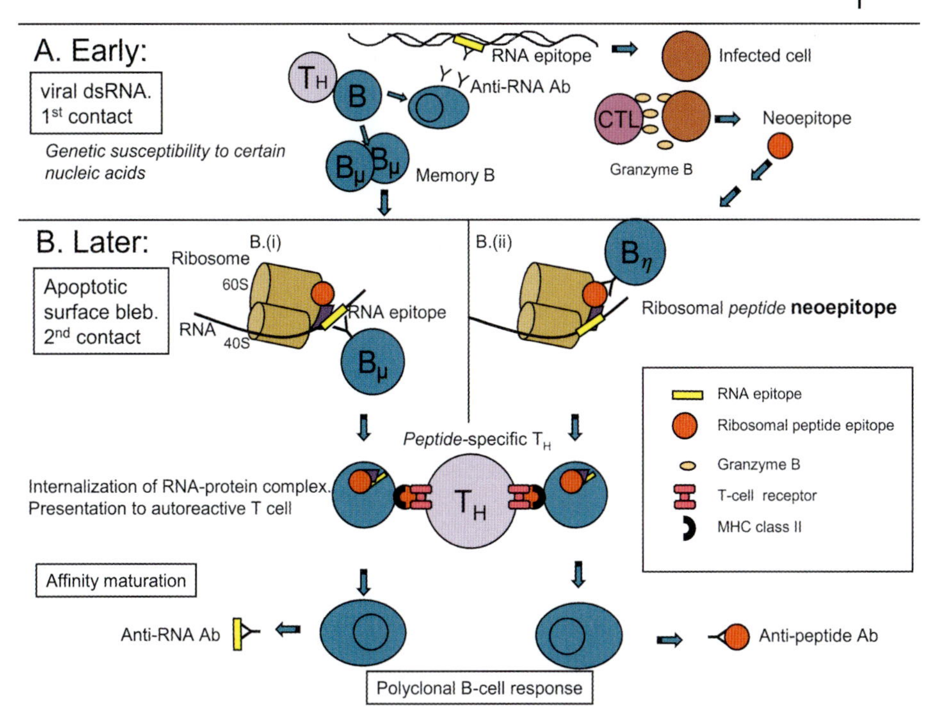

Fig. 17.1 Molecular mimicry, neoepitopes and epitope spreading. (A) Viral infection introduces foreign (viral) RNA into the bloodstream. RNA antibody production is induced with the help of T cells. This likely occurs in secondary lymphoid organs such as lymph nodes and spleen. Some B cells remain as memory B cells (Bμ). Virally infected cells become targets of CTL and NK cells. Granzyme B introduced into target cells generates novel cleavage fragments of prominent autoantigens. (B.i) RNA antibodies may cross-react with similar rRNA epitopes when these epitopes are exposed during apoptosis of virally infected cells: a memory B cell with anti-RNA specificity (Bμ) recognizes an RNA epitope (yellow) on the ribosome (molecular mimicry). (B.ii) A naïve B cell (Bη) recognizes a peptide neoepitope (red). Ribosomal particles relocate in apoptotic cell surface blebs, whereby genetically susceptible individuals develop polyclonal expansion of B cells, a subset of which generates antibodies that have undergone affinity maturation: a single T cell clone with specificity for one ribosomal antigen can activate B cells with multiple antibody specificities. B cells internalize the whole RNA-peptide complex and present one among several peptide antigens (red) to the reactive T cell, which in turn induces affinity maturation of multiple B cell clones via MHC II-'red peptide'-TCR ligation (epitope spreading).

17.2
RNA, Anti-RNA Antibodies and Associated Protein Complexes

17.2.1
RNAs

Native RNA comprises many different molecules. Their abundance, nuclear and cytoplasmic localization, and their association with nucleic acid binding proteins are shown in Tab. 17.1. It is important to remember that these molecules differ considerably in sequence, secondary and tertiary structure, and certainly function. Therefore, it was recognized long ago that detailed characterization of the antibody-RNA interaction might yield important clues to the nature and origin of immunizing antigens [6]. Moreover, the discovery by Cech and Altman that certain RNAs are catalytically active, dramatically expanded our knowledge of the role of RNA [7]. Briefly, catalytic RNAs or ribozymes are generated to specifically cleave or splice target mRNAs, resulting in altered expression of target molecules. For example, anti-tumor necrosis factor (TNF)-α ribozymes can inhibit TNF-α secretion [8] and human B cell clones producing pathogenic anti-DNA antibodies have been successfully targeted by antisense ribozyme to the nephritogenic V3-7 gene [9]. Gene therapy using this approach is a fast moving and exceptionally exciting field. However, whether or not these therapeutic RNA molecules have immunogenic properties may turn out to be one of several critical questions that must be answered prior to its potential use in human disease. This is not so unlikely a scenario, bearing in mind the widespread cross-reactivity among RNA molecules.

17.2.2
Anti-RNA Antibodies and Disease

In the late 1960s, two investigators independently described antibodies to RNAs in both human systemic lupus erythematosus and in NZB/NZW lupus-prone mice. Schur and Monroe, and Steinberg *et al.*, were the first to take advantage of the then recent availability of synthetic polynucleotides representing both single-stranded (ss) and double-stranded (ds) RNA [10, 11]. In Schur's investigation of RNA antibodies in the serum of patients with systemic lupus erythematosus (SLE), his group identified antibodies directed against synthetic ssRNA and dsRNA as well as viral dsRNA. Because of the low abundance of these antibodies in the serum, they may have escaped detection in earlier investigations. In the same study it was shown that these antibodies were frequently associated with anti-DNA and anti-ribosomal antibodies, and that anti-RNA antibodies were most frequently directed to dsRNA, i.e. poly-inosinic·poly-cytidylic acid (poly I–C), polyadenylic · poly-uridylic acid (poly A–U) and statalon viral RNA (derived from Polio virus). Interestingly, no antibodies against ssRNA and ribosomal RNA, which represent more than 95% of the mammalian RNA, were found in patients with SLE. The authors concluded that their findings suggest that these antibodies are primarily directed to foreign antigen in the form of a dsRNA virus.

Tab. 17.1 RNA and apoptosis.

RNA	*Abundance*	*Localization*	*Immunogenic?*	*Modified in apoptosis?*	*Associated proteins*	*Modified in apoptosis?*
mRNA	variable	nucleus and cytoplasm	*	no	multiple RNPs	no
tRNA	moderate	cytoplasm	yes	no	tRNA Synthetases	yes
5S rRNA	high	nucleus and cytoplasm	*	no	several	*
28S rRNA	high	nucleus and cytoplasm	yes	yes	Ribosomal P, L, S proteins	no
5.8S rRNA	high	nucleus and cytoplasm	*	*	several	*
18S RNA	high	nucleus and cytoplasm	*	yes	several	*
16S RNA	high	mitochondria	*	yes	several	*
7SL RNA	high	cytoplasm	*	no	several	SRP72
hnRNA	low	nucleus and cytoplasm	*	no	hnRNP/spliceosome	A1, C1, C2
snRNA	low	nucleus	U1 snRNA	yes	snRNP/spliceosome	U1-70 kDa, SRPs, SmF
snoRNA	low	nucleolus	*	no	snoRNP	fibrillarin, nucleolin, SRP
Y-RNA	low	cytoplasm	hY5 RNA	yes	Ro, La/Ro RNP proteins	La

* No data.
Abbreviations: mRNA, messenger RNA; rRNA, ribosomal RNA; hnRNA, heterogeneous nuclear RNA; snRNA, small nuclear RNA; hY5 RNA, human Y5 RNA; tRNA, transfer RNA; hnRNP, heterogeneous nuclear ribonucleoprotein; snRNP, small nuclear RNP; snoRNP, small nucleolar RNP; SRP72, signal recognition particle 72; SRP, serine/arginine phosphoprotein; SmF; Smith protein F.

In the study by Steinberg *et al.*, it was shown that dsRNA functions as a potent antigen in New Zealand mice. Synthetic dsRNA (and presumably also viral dsRNA) induce antibodies to dsRNA, and it was speculated that dsRNA induces anti-RNA antibodies and interferon that protect against viral infections. In their study, antibodies to DNA and RNA were induced in NZB/NZW F_1 female mice by injections of the interferon inducer poly I–C. Anti-RNA antibodies were not only found in induced animals, but also in 50% of female NZB/NZW mice, as well as in eight of 24 sera from patients with SLE. The authors found that RNA-treated animals produced antibodies to RNA and DNA earlier and in larger amounts, and had accelerated renal disease as compared to control mice.

It is likely that the anti-RNA reactivity described in these investigations represents a cross-reaction with antibody directed against some other RNA, since anti-nucleic acid antibodies are known to cross-react widely. The authors speculate that naturally occurring nucleic acids (e.g. viruses) act as stimuli to a genetically hyper-responsive immune system to produce anti-RNA antibodies (see Fig. 17.1).

17.2.3
Ribosomal RNAs (rRNAs) and Transfer RNAs (tRNAs)

17.2.3.1 Ribosome and rRNA

Ribosomes are the protein factories of the cell. The subunits of the ribosome on which synthesis takes place must be preloaded with auxiliary protein factors. Additionally, a single messenger RNA (mRNA) molecule and all the different tRNA molecules loaded with their respective amino acids come together to form a functioning ribosome. The precise stepwise movements necessary to decode mRNA are all catalyzed in this RNA-protein complex. Interestingly, the key catalytic activities are attributed to rRNA, whereas the heterogeneous auxiliary proteins appear to have only modulatory functions. The prokaryotic 70S ribosome is composed of two subunits, 50S and 30S, and the eukaryotic 80S ribosome is composed of a 60S and 40S subunit. The rRNA components make up more than 50% of the ribosome, and are classified as follows: (1) prokaryotic 50S subunit: 5S rRNA, 23S rRNA; (2) prokaryotic 30S subunit: 16S rRNA; (3) eukaryotic 60S subunit: 5S rRNA, 28S rRNA, 5.8S rRNA; and (4) eukaryotic 40S subunit: 18s rRNA.

The ribosomal phosphoproteins P0, P1 and P2, which form a pentameric complex that associates with 28S rRNA and indirectly with the 60S ribosomal subunit, are targeted by the immune response in lupus patients [12]. Some anti-ribosomal P autoimmune sera also contain antibodies to a fragment of 28S rRNA comprising a highly catalytic domain, the ribosomal GTPase center [13]. Antibodies to this highly conserved region of 28S rRNA were found to inhibit the interaction of elongation factors 1α and 2 with ribosomes, thus interfering with protein synthesis [14].

17.2.3.2 **tRNA**

Another important set of RNA molecules that are targets of autoantibodies is tRNA. Although not known to be modified during apoptosis, tRNAs are included in this review because antibodies to tRNA have been important tools in the elucidation of antibody specificity to nucleic acids in general [15]. Interest in antibodies to tRNA has primarily been fueled by their usefulness in the study of nucleic acid structure, cellular localization and biologic function. In a historical context it is interesting to know that the structure of tRNA was the first to be reported among all RNA types in 1975 [16].

Eilat *et al.* communicated the finding of an immunoglobulin fraction in NZB/NZW mice that specifically binds native tRNA in 1976 [17]. This immunoglobulin is spontaneously produced by NZB/NZW F_1 mice. These authors later reported that *Escherichia coli* tRNA was recognized by antibodies in the serum derived from SLE patients, but not from other connective tissue diseases [18]. As opposed to murine antibodies, antibodies occurring in human SLE sera appeared to have little conformation specificity for native tRNA, but contain predominantly sequence specificity. This is in line with later findings regarding the major role of light chains in determining the sequence specificity of disease-related anti-U1 RNA antibodies (see below and [19]).

Apart from antibodies to tRNA in NZB/NZW mice [17], direct reactivity with free native ssRNA has been observed only in the context of antibodies to ribosomal RNA in SLE [20] and uracil-specific anti-RNA antibodies in scleroderma [21]. Investigators could not find evidence for the involvement of secondary or tertiary structure in the recognition of the polynucleotide chain by the human ssRNA antibodies, which is in opposition to results obtained in NZB/NZW mice. In conclusion, in humans there is no clearly distinct antibody subpopulation that specifically recognizes tRNA within the pool of antibodies to native RNA. The specificity seems to be determined more by the different nucleotides rather than RNA conformation, and considerable multivalent binding may occur.

17.2.3.3 **Studies on the Specificities of Anti-RNA Antibodies**

Eilat *et al.* have studied anti-tRNA and anti-rRNA autoantibodies in the sera of SLE patients [18], and NZB/NZW F_1 mice [22]. They found that antibody populations that bound polyribonucleotides were distinct from populations that bound polydeoxyribonucleotides. Competition experiments showed that the anti-RNA antibodies preferentially bound naive ssRNA as compared with synthetic ss and ds homopolyribonucleotides, suggesting sequence and/or conformation specificity for these antibodies. Subsequent studies have demonstrated that anti-RNA antibodies from individual patients' sera recognize unique determinants in methionine and alanine-specific tRNAs [23]. This holds equally true for 28S ribosomal RNA [14] and U1 snRNA [24].

Of several monoclonal antibodies (mAb) that were later generated from different individual NZB/NZW mice, one antibody (BWR 5) differed from the previously studied immunoglobulins in that it showed a clear preference for ds polyribonucleotides in a sequence-independent manner [25]. The authors speculated that this mAb probably represented the SLE anti-dsRNA autoimmune response that was described by Schur and Monroe 20 years earlier. An explanation for this difference in antibody specificity to RNA in mice and human is still missing.

The analysis of epitopes recognized by individual anti-nucleic acid mAbs is facilitated nowadays by the recently introduced method by Tsai and Keene [26], whereby an epitope library comprising a pool of degenerate RNA transcripts is used in a direct immunoprecipitation assay to select RNA species capable of binding to autoantibodies present in a patient's serum. The sequence analysis of anti-RNA autoantibodies with fine specificity for GC-rich polyribonucleotides showed that the heavy chains of these mAbs are very similar to each other. The RNA-related V_H chains were combined with a variety of D, J_H, V_κ and J_κ elements. This suggests that RNA specificity is mostly dictated by the germline V_H gene. It is noteworthy that some heavy chains of mAbs with Sm and anti-ssDNA specificity were encoded by the presumed anti-RNA V_H gene [27–29]. However, many of these mAbs were of the IgM class and showed relative low affinity for ssDNA. Moreover, the authors speculated that some of the analyzed mAbs to Sm [30] and to other ribonucleoprotein particles [31] did, in fact, recognize determinants of naked RNA. Evidence to confirm this hypothesis was provided by several others and is particularly strong for U1 snRNA. In 1999, Dutch researchers described the isolation of the first human anti-U1 snRNA autoantibodies from a combinatorial IgG library made from the bone marrow of a systemic lupus erythematosus patient [19]. Their findings suggest an important role for the light chain for the determination of sequence specificity of these anti-U1 snRNA antibodies, with possible implications to the understanding of epitope spreading as a result of secondary light chain rearrangements.

For detailed information on genetic and structural studies dealing with the specificities of autoantibodies to nucleic acids, we refer the reader to the in-depth and exhaustive review by Eilat and Anderson [15]. In a nutshell, protein-nucleic acid interactions are fundamental to the understanding of key processes in autoimmune diseases.

17.2.4
Small Nuclear RNA (snRNA) and the U1 Small Nuclear Ribonucleoprotein (snRNP) Complex

Small U RNAs are RNAs of 250 nucleotides or less, which have arbitrarily been named U1, U2...U12 RNAs. The complexes of protein associated with these RNAs are designated snRNPs. The principle role of these complexes is RNA splicing, a process that removes the intron sequences from the primary mRNA transcript. RNA splicing requires the formation of a multi-protein complex, called the spliceosome. U1, U2, U5 and U4/6 snRNPs are components of the spliceosome. The U1 snRNP binds to the 5′ splice site guided by a nucleotide sequence in the U1 snRNA that forms base pairs complementary to the 9-nucleotide splice site consensus sequence. It is possible that the RNA component and the protein component of the snRNP complex could be responsible for catalyzing the reaction since RNA has enzymatic properties.

U1 snRNP is an autoantigen in SLE and mixed connective tissue disease (MCTD), whereby the 70-kDa component of U1 snRNP is almost always targeted in MCTD. Thus, autoantibodies to U1 snRNP have become a prerequisite for the diagnosis of, and the major serological classification criterion for, MCTD [32]. U1A, U1C and eight core proteins, named the Smith (Sm) complex, are also associated with this spliceosomal subunit and are prominent targets of autoantibodies in SLE. Anti-Sm antibodies were discovered in 1966 and were shown to have 99% specificity for SLE, now constituting one immunoserological criterion for the classification of SLE [33]. There is a notable relation of the Sm reactivity to the HLA-DQw6 or HLA-DR7 histocompatibility antigens, and racial differences are present in juvenile SLE.

Anti-U1 snRNA antibodies are seen in 49% of patients with SLE [34]. Correlations with disease features include pulmonary manifestations and Raynaud's syndrome, and abnormal findings on nail fold capillary microscopy, suggesting an SLE/scleroderma overlap syndrome in these patients [35]. Moreover, it was shown that almost 40% of patients with anti-U1 snRNP antibodies also had the anti-U1 snRNA reactivity, but surprisingly none of these double-positive sera demonstrated Sm reactivity. These findings already suggested an intimate link in the development of anti-U1 snRNA and anti-U1 snRNP autoantibodies [36]. Remarkably, anti-U1 RNA antibodies are associated with more severe disease in MCTD patients, and they correlate with disease activity [37]. Whether involvement of anti-U1 snRNA antibodies is important in the pathogenesis of subsets of connective tissue diseases remains to be seen.

Specific cleavage of U1-70kDa [38], but not of U1A, U1C and Sm-B/B′ [39] have been described during apoptosis. U1-70kDa is cleaved by both caspase-3 and granzyme B, and the U1-70kDa cleavage fragments are different. Granzyme B, a novel protease released by cytotoxic T lymphocytes (CTL) and natural killer (NK) cells, cleaves a broad spectrum of autoantigens and thus may be a crucial molecule in neoepitope formation (see Fig 17.1 A). Granzyme B substrates comprise a variety of functionally very different molecules involved in DNA repair (DNA-PK, PMS1, PMS2, Ku70), DNA binding (PARP), RNA synthesis (RNA polymerase I and II), RNA splicing (U1-70kDa), translation (tRNA synthetases), protein translocation (SRP72), and mitosis (NuMA), among several others [40]. The cleavage sites are unique for each molecule and their identification has become a scientific field in its own right (see Chapter 15). Similarly, the list of autoantigens screened for potential caspase-3 cleavages is continuously growing [2]. However, cleavage by caspases or granzyme B is only one mechanism by which posttranslational modifications are generated, and other death-associated autoantigen modifications may be equally important. One example is the recently described citrullination of filaggrin peptides by peptidylarginine deiminase, an intriguingly simple biochemical process with the potential to create immunogenic neoepitopes significant for the development of autoimmune disease [41]. The U1-snRNP autoantigen complex has also been shown by our lab to associate with a family of phosphoproteins called serine/arginine (SR) RNA splicing factors during apoptosis [42, 43]. SR proteins are recently described autoantigens that are recognized by autoantibodies in a phosphorylation-dependent context [44].

17.2.5 Y RNAs and the Ro RNP complex

Yet another subset of immunogenic RNA is found within the Y RNAs. Four types of human (h) Y RNA exist: hY1, hY3, hY4 and hY5 RNA. hY2 is a degraded form of hY1 RNA. Human Ro ribonucleoproteins are composed of one of the four Y RNAs and at least two proteins, Ro60 and La. Ro60 and the 48-kDa phosphoprotein La belong to the family of RNA-binding proteins characterized by a conserved RNA-binding domain, the so called RNA recognition motif (RRM). Another protein associated with this complex, Ro52, is completely unrelated to Ro60 and La, and does not bind RNA directly [45].

However, the autoimmune response to Ro60 and Ro52 is closely linked [46]. Ro52 is assumed to play a role in cell activation and transformation [47]. Y RNAs bind to the La protein through their oligo (U) stretch at their 3′ end, which is a common feature of all Y RNAs. The many important cytoplasmic and nuclear functions of the La protein, which is 50 fold more abundant than Ro60, will be discussed elsewhere in this book. For a detailed review, see [48]. Moreover, the discovery of a possible involvement of La in the translation of certain viral RNAs has to be added to the list of important functions [49, 50]. Interestingly, the La autoantigen is also a host substrate for poliovirus 3C protease, whereby proteolytic cleavage of La removes a nuclear localization motif, thus preventing shuttling of La back to the nucleus [51]. This cleavage certainly holds the potential to create neoepitopes. La is an *in vitro* substrate of granzyme B [40], and, in the context of apoptosis, La has been shown to be partly cleaved and rapidly dephosphorylated [52]. As opposed to La, Ro60 is not cleaved after induction of apoptosis [39].

The functions of Y RNA are largely unknown. Differences in secondary structure of Y3 and Y4 RNA led to the speculation that these two Y RNAs might have different cellular functions [53]. Y RNAs might act as co-factors in La-associated translational events or even regulate La functions. Autoantibodies are only found to Y5 RNA, but not to the other three subsets of Y RNA [54]. Correlations with clinical manifestations are unknown, although it is not clear if such correlations were ever investigated.

17.3
Modifications of Selected RNA Molecules during Apoptosis

17.3.1
Cleavage of hY RNAs during Apoptosis

Although all hY RNAs are polymerase III transcripts, only hY5 is the specific target of autoantibodies in human anti-Ro sera [54]. Following a variety of apoptotic stimuli, hY RNAs are rapidly cleaved and/or degraded, irrespective of cell type and type of apoptosis induction [55]. This is not the case for other polymerase III products, like 7SL RNA, 5S ribosomal RNA or transfer RNA. The hY RNA cleavage products have been characterized in detail, and have been shown to differ according to the association of the remaining fragments with the Ro and La protein, respectively, pointing to multiple endonucleolytic cleavage reactions. The same authors demonstrated that cleavage of hY RNA is almost completely inhibited by caspase-1, -3 and -8 inhibitors, suggesting the recruitment of caspase-activated RNase (CAR) to the site of RNA cleavage. In contrast, caspase-9 inhibitor only partially inhibits the cleavage of hY RNA. This is roughly analogous to the results obtained from U1 snRNA cleavage experiments (see below). In summary, experimental evidence suggests that modifications of hY RNA during apoptosis occur in a caspase-dependent manner, with the restriction that the caspase-9 pathway seems to play a less important role in this process.

17.3.2
Cleavage of U1 snRNA during Apoptosis

The same group of researchers in the Netherlands one year later described the specific cleavage of the U1 snRNA molecule during Fas-mediated apoptosis [56]. In Northern blotting experiments the authors examined the behavior of U1 snRNA and demonstrated the appearance of cleavage products as early as 2 h after induction of apoptosis. The first appearance of these shortened U1 snRNA products coincided with the appearance of the 40-kDa cleavage product of the U1-70kDa protein. It is known from earlier studies that U1-70kDa is specifically cleaved by caspase-3 at position ^{338}DGPD^{341}, resulting in an N-terminal 40-kDa (amino acids 1–341) and a C-terminal 20-kDa (amino acids 342–437) protein fragment [38]. The cleavage of U1 snRNA is also highly specific, as none of the other U RNAs are detectably cleaved over a time course of 8 h after apoptosis induction. The irreversible caspase inhibitors-1, -3 and -8, but not -9, markedly inhibit the cleavage of U1 snRNA. All caspase inhibitors inhibit cleavage of the 70-kDa protein. Furthermore, cleavage is not restricted to Fas-induced apoptosis and occurs in mammalian and mouse cell lines. It was also shown in this report that the 2,2,7-trimethylguanosine (TMG) cap is removed during apoptotic cleavage of U1 snRNA, and S1 mapping studies determined the exact cleavage site to be between nucleotides C5 and pseudouracil (ψ) 6, or between nucleotides ψ6 and ψ7.

From data presented in this first report it is not possible to determine which nucleolytic activity is responsible for the cleavage of U1 RNA. Despite caspase-3 being a candidate protease that may be responsible for activating a site-specific RNase, the fact that incomplete inhibition of U1snRNA is observed under conditions where the model substrate for caspase-3 (U1-70kDa) is almost completely inhibited, favors alternative mechanisms. One possibility is that an apoptosis-specific nuclease is only indirectly activated by caspases-1, -3, -8 or -9, as suggested by Degen *et al.* [1].

17.3.3
Modification of rRNAs during Apoptosis

The vast expansion of subsets of the variable or so-called divergent (D) domains of the 28S rRNA molecule in higher eukaryotes during phylogeny (about 50% of sequences are divergent in eukaryotes) has been taken as an indication of an important functional role of these domains. Possible roles are: (1) regulation and tuning of the translational machinery, (2) protein anchoring, and (3) RNA-RNA interactions. However, the precise function of D domains is still unclear. Importantly, in the 1990s it could be demonstrated that these key domains are the cleavage sites of enzymes activated during apoptosis. Houge *et al.* made the first observation regarding the cleavage of the 28S rRNA in 1993 [57]. In their studies the authors showed several distinct cleavage products generated by induction of apoptosis in a rat myeloid leukemia cell line (IPC-81). In other investigations, the same researchers performed fine mapping studies and reported two specific cleavage sites in the hypervariable regions of both the largest divergent domains D2 and D8 of the rat 28S rRNA [58]. Interestingly, these cleavage sites, D2c and D8b, are apoptosis-specific and are not found in necrotic cells [39]. Similarly, incubation with RNase T1 and RNase U2 yields different cleavage patterns. In a recent review of their findings, Houge *et al.* speculate that two alternative 28S rRNA cleavage pathways might be operative during apoptosis [59]. In pathway A, cleavage would be initiated in D8b and subsequently followed by cleavage in D2c. In pathway B, the first cleavage takes place 5′ upstream of the D2c subdomain, the second cleavage 3′ downstream of the D2c subdomain, resulting in the excision of D2c. In human cells, specific cleavage products have not yet been reported. However, the 28S rRNA was cleaved in some human leukemia cell lines (K562, Molt-3, Molt-4 and U937), whereas in the human HL-60 leukemia cell line, rRNA remains intact after induction of apoptosis [60]. This discrepancy has yet to be reconciled. Moreover, the 18S rRNA subunit appears to remain intact in rat leukemia cells, whereas in developing rat cerebellar cells clear evidence of cleavage of both 28S and 18S during methylazoxymethanol-induced apoptosis was found [61]. In conclusion, cleavage of 28S rRNA during apoptosis seems to be a cell-type dependent biochemical mechanism, a notion that raises important questions about the selectivity of these particular cleavage events. Intriguingly, the autoantigenic proteins associated with the ribosome, the so-called P proteins, are not susceptible to apoptotic cleavage by caspases and granzyme B [40]. This finding provides supportive evidence to the hypothesis that alterations of RNA rather than of proteins during apoptosis may in some circumstances be the initial immunogenic upstream event in the evolution of a broader autoimmune response (see Fig. 17.1 B.))

17.4 Viral Infection, Apoptosis and Autoimmunity

Several mechanisms have been proposed concerning virus-host interactions on a molecular level:

(1) Host response to a virally infected cell [cytotoxic T lymphocyte (CTL)-mediated killing]
(2) Disruption of normal cellular function by virally encoded factors
(3) Viral nucleic acid interactions with host proteins
(4) Interactions of viral and host proteins
(5) Direct modification of host proteins by viral enzymes.

Within the limits of this chapter, we would only like to briefly point out those mechanisms with the potential to involve nucleic acids. When do epitopes appear foreign to the organism? Truly novel epitopes are created when the powerful proteolytic cascade of caspases or granzyme B is activated by direct attack of virally infected host cells by NK cells or CTLs [62] (Fig. 17.1 A). This mechanism may be extremely important in the initial insult leading to novel peptide fragments, illustrating a host response to a virally infected cell. However, with regard to nucleic acids, an alternative mechanism is the specific interaction of viral nucleic acid with host proteins. Important examples include the La protein, which binds to hepatitis C virus, HIV and Epstein–Barr virus (EBV) RNA [49, 50, 63]. The interaction of La with EBV RNAs EBER 1 and EBER 2 is particularly intriguing [6], the more so with a recent albeit controversial report on the increased prevalence of EBV infection in young lupus patients [64]. Despite these insightful reports, the precise role for viral RNA as molecular mimics and for the generation of neoepitopes remains incompletely understood as of today. Fig. 17.1 attempts a synopsis of current models.

17.5 Summary

For many reasons, RNA antigens warrant closer observation by the community of (auto) immunologists and clinicians alike. First, the elegant work done by van Venrooij *et al.* convincingly suggests an important involvement of certain RNA molecules during apoptosis. These investigators show selective and highly specific cleavage of RNA subsets. Their new data support the notion that anti-RNA antibodies are associated with disease subsets and more severe disease, and thus may be pathogenic. Second, the rebirth of the viral hypothesis in autoimmune disease points to the involvement of RNA mimics in the development of autoimmune responses. Increasing evidence for interactions of viral protein/RNA with host protein/RNA corroborates the hypothesis of the emergence of neoepitopes in susceptible individuals, within or outside the context of apoptosis. Third, RNA antigens may be early targets of autoimmune responses. One might even speculate that RNA is involved in early events of intermolecular epitope spreading, events that eventually lead to pathogenic antibody formation and ultimately to autoimmune disease manifestations. Last, the advent of RNA molecules as therapeutics in the age of gene therapy waves a red flag to those aware of the immunogenic potential of (naked) nucleic acids.

It remains to be seen if cleavage of RNA during apoptosis might be less, equally, or even more important than protein modifications as a trigger in initiating an autoimmune response in susceptible individuals. Hence, the scientific endeavors ahead of us are a promising playground for the next generation of dedicated programmed cell death researchers.

17.6 Acknowledgments

This work was funded by a grant from the Arthritis National Research Foundation to W.H.; P.J.U. was supported by NIH grants U19DK61934-01 and K08AI01521, the Donald and Delia Baxter Foundation, and an Arthritis Investigator Award. We are grateful to members of the Utz lab and Dr W.J. van Venrooij for many insightful discussions.

17.7 References

1 Degen, W.G.J., *et al.*, Caspase-dependent cleavage of nucleic acids. *Cell Death Different* **2000**, 7, 616–27.

2 Utz, P.J., *et al.*, Death, autoantigen modifications, and tolerance. *Arthritis Res* **2000**, 2, 101–14.

3 Cohen, G.M., Caspases: the executioners of apoptosis. *Biochem J* **1997**, 326, 1–16.

4 McCluskey, J., *et al.*, Determinant spreading: lessons from animal models and human disease. *Immunol Rev* **1998**, 164, 209–29.

5 Craft, J. and S. Fatenejad, Self antigens and epitope spreading in systemic autoimmunity. *Arthritis Rheum* **1997**, 40, 1374–82.

6 Lerner, M., *et al.*, Two small RNAs encoded by Epstein-Barr virus and complexed with protein are precipitated by antibodies from patients with systemic lupus erythematosus. *Proc Natl Acad Sci USA* **1981**, 78, 805–9.

7 Kruger, K., *et al.*, Self-splicing RNA: autoexcision and autocyclization of the ribosomal RNA intervening sequence of *Tetrahymena*. *Cell* **1982**, 31, 147–57.

8 Kisich, K.O., *et al.*, Specific inhibition of macrophage TNF-α expression by *in vivo* ribozyme treatment. *J Immunol* **1999**, 163, 2008–16.

9 Suzuki, Y., *et al.*, Chemically modified ribozyme to V gene inhibits anti-DNA production and the formation of immune deposits caused by lupus lymphocytes. *J Immunol* **2000**, 165, 5900–5.

10 Schur, P.H. and M. Monroe, Antibodies to ribonucleic acid in systemic lupus

erythematosus. *Proc Natl Acad Sci USA* **1969**, 63, 1108–12.

11 Steinberg, A. D., S. Baron, and N. Talal, The pathogenesis of autoimmunity in New Zealand mice, I. induction of antinucleic acid antibodies by polyinosinic-polycytidylic acid. *Proc Natl Acad Sci USA* **1969**, 63, 1102–7.

12 Elkon, K. B., *et al.*, Lupus autoantibodies target ribosomal P proteins. *J Exp Med* **1985**, 162, 459–71.

13 Chu, J. L., *et al.*, Lupus antiribosomal P antisera contain antibodies to a small fragment of 28S rRNA located in the proposed ribosomal GTPase center. *J Exp Med* **1991**, 174, 507–14.

14 Uchiumi, T., *et al.*, A human autoantibody specific for a unique conserved region of 28 S ribosomal RNA inhibits the interaction of elongation factors 1 alpha and 2 with ribosomes. *J Biol Chem* **1991**, 266, 2054–62.

15 Eilat, D. and W. F. Anderson, Structure-function correlates of autoantibodies to nucleic acids. Lessons from immunochemical, genetic and structural studies. *Mol Immunol* **1994**, 31, 1377–90.

16 Sigler, P. B., An analysis of the structure of tRNA. *Annu Rev Biophys Bioeng* **1975**, 4, 477.

17 Eilat, D. and A. N. Schechter, Antibodies to native tRNA in NZB/NZW mice. *Nature*, **1976**, 259, 141–3.

18 Eilat, D., *et al.*, The reaction of SLE antibodies with native, single stranded RNA: radioassay and binding specificity. *J Immunol* **1978**, 120, 550–7.

19 Hoet, R. M. A., *et al.*, The importance of the light chain for the epitope specificity of human anti-U1 small nuclear autoantibodies present in systemic lupus erythematosus patients. *J Immunol* **1999**, 163, 3304–12.

20 Lamon, E. W. and J. C. Bennett, Antibodies to ribosomal ribonucleic acid (rRNA) in patients with systemic lupus erythematosus (SLE). *Immunology* **1970**, 19, 439–42.

21 Alarcon-Segovia, D., *et al.*, Uracil-specific anti-R. N. A. antibodies in scleroderma. *Lancet* **1975**, i(7903), 363–6.

22 Eilat, D., R. Asofsky, and R. Laskov, A hybridoma from an autoimmune NZB/NZW mouse producing monoclonal antibody to ribosomal-RNA. *J Immunol* **1980**, 124, 766–8.

23 Wilusz, J. and J. D. Keene, Autoantibodies specific for U1 RNA and initiator methionine tRNA. *J Biol Chem* **1986**, 261, 5467–72.

24 Deutscher, S. L. and J. D. Keene, A sequence-specific conformational epitope on U1 RNA is recognized by a unique autoantibody. *Proc Natl Acad Sci USA* **1988**, 85, 3299–303.

25 Eilat, D. and R. Fischel, Recurrent utilization of genetic elements in V regions of antinucleic acid antibodies from autoimmune mice. *J Immunol* **1991**, 147, 361–8.

26 Tsai, D. E. and J. D. Keene, *In vitro* selection of RNA epitopes using autoimmune patient serum. *J Immunol* **1993**, 150, 1137–45.

27 Kofler, R., *et al.*, Immunoglobulin κ light chain variable region gene complex organization and immunoglobulin genes encoding anti-DNA antibodies in lupus mice. *J Clin Invest* **1988**, 82, 852–9.

28 Tillman, D. M., *et al.*, Both IgM and IgG anti-DNA antibodies are the products of clonally selective B cell stimulation in (NZB $\times$ NZW) F_1 mice. *J Exp Med* **1992**, 176, 761–79.

29 Radic, M. Z. and M. Weigert, Genetic and structural evidence for antigen selection of anti-DNA antibodies. *Annu Rev Immunol* **1994**, 12, 487–520.

30 Bloom, D. D., *et al.*, V region gene analysis of anti-Sm hybridomas from MRL/Mp-*lpr/lpr* mice. *J Immunol* **1993**, 150, 1591–610.

31 Takeda, Y., *et al.*, Nucleotide sequence of immunoglobulin heavy and light chain V-regions from a monoclonal autoantibody specific for a unique set of small nuclear ribonucleoprotein complexes. *Nucleic Acids Res* **1992**, 20, 4099.

32 Amigues, J. M., *et al.*, Comparative study of 4 diagnosis criteria sets for mixed connective tissue disease in patients with anti-RNP antibodies. Autoimmune Group of the Hospitals of Toulouse. *J Rheumatol* **1996**, 23, 2055–62.

33 Tan, E. M., Antinuclear antibodies: diagnostic markers for autoimmune diseases

and probes for cell biology. *Adv Immunol* **1989**, 44, 93–151.

34 Kallenberg, C.G. M., *et al.*, Anti-U1 snRNA antibodies in SLE: clinical and serologic associations. *Lupus* **1992**, 1 (Suppl. 1), 103 (abstr.).

35 Groen, H., *et al.*, Pulmonary function in systemic lupus erythematosus is related to distinct clinical, serological and nail-fold capillary pattern. *Am J Med* **1992**, 93, 619–27.

36 Van Venrooij, W.J., *et al.*, Anti (U1) small nuclear ribonucleoprotein sera from patients with connective tissue disease. *J Clin Invest* **1990**, 86, 2154–60.

37 Hoet, R.M.A., *et al.*, Changes in anti-U1 RNA antibody levels correlate with disease activity in patients with systemic lupus erythematosus overlap syndrome. *Arthritis Rheum* **1992**, 35, 1202–10.

38 Casciola-Rosen, L.A., *et al.*, Specific cleavage of the 70-kDa protein component of the U1 small nuclear ribonucleoprotein is a characteristic biochemical feature of apoptotic cell death. *J Biol Chem* **1994**, 269, 30757–60.

39 Casiano, C.A., *et al.*, Selective cleavage of nuclear autoantigens during CD95 (Fas/APO-1)-mediated T cell apoptosis. *J Exp Med* **1996**, 184, 765–70.

40 Casciola-Rosen, L.A., *et al.*, Cleavage by granzyme B is strongly predictive of autoantigen status: Implications for initiation of autoimmunity. *J Exp Med* **1999**, 190, 815–25.

41 Schellekens, G., *et al.*, Citrulline is an essential constituent of antigenic determinants recognized by rheumatoid arthritis-specific autoantibodies. *J Clin Invest*, **1998**, 101, 273–81.

42 Utz, P.J., *et al.*, Proteins phosphorylated during stress-induced apoptosis are common targets for autoantibody production in patients with systemic lupus erythematosus. *J Exp Med* **1997**, 185, 843–54.

43 Utz, P.J., *et al.*, Association of phosphorylated serine/arginine (SR) splicing factors with the U1-small nuclear ribonucleoprotein (snRNP) autoantigen complex accompanies apoptotic cell death. *J Exp Med* **1998**, 187, 547–60.

44 Neugebauer, K.M., *et al.*, SR proteins are autoantigens in patients with systemic lupus erythematosus. Importance of phosphoepitopes. *Arthritis Rheum* **2000**, 43, 1768–8.

45 Boire, G., *et al.*, Purification of antigenically intact Ro ribonucleoproteins; biochemical and immunological evidence that the 52-kD protein is not a Ro protein. *Clin Exp Immunol* **1995**, 109, 32–40.

46 Keech, C.L., T.P. Gordon, and J. McCluskey, The immune response to 52-kDa Ro and 60-kDa Ro is linked in experimental autoimmunity. *J Immunol* **1996**, 157, 3694–9.

47 Saurin, A.J., *et al.*, Does this family have a RING? *Trends Biochem Sci* **1996**, 21, 208–14.

48 Maraia, R. J., La protein and the trafficking of nascent RNA polymerase III transcripts. *J Cell Biol* **2001**, 153, F13–8.

49 Ali, N. and A. Siddiqui, The La antigen binds 5′ noncoding region of the hepatitis C virus RNA in the context of the initiator AUG codon and stimulates internal ribosome entry site-mediated translation. *Proc Natl Acad Sci USA* **1997**, 94, 2249–54.

50 Chang, Y., *et al.*, Direct interactions between autoantigen La and human immunodeficiency virus leader RNA. *J Virol* **1995**, 68, 7008–20.

51 Shiroki, K., *et al.*, Intracellular redistribution of truncated La protein produced by poliovirus 3C pro-mediated cleavage. *J Virol* **1999**, 73, 2193–200.

52 Rutjes, S.A., *et al.*, The La (SS-B) autoantigen, a key protein in RNA biogenesis, is dephoshorylated and cleaved early during apoptosis. *Cell Death Different* **1999**, 6, 976–86.

53 Farris, A. D., *et al.*, Y3 is the most conserved small RNA component of Ro ribonucleoprotein complexes in vertebrate species. *Gene* **1995**, 154, 193–8.

54 Boulanger, C., *et al.*, Autoantibodies in human anti-Ro sera specifically recognize deproteinated hY5 Ro RNA. *Clin Exp Immunol* **1995**, 99, 29–36.

55 Rutjes, S.A., *et al.*, Rapid nucleolytic degradation of the small cytoplasmic Y RNAs during apoptosis. *J Biol Chem* **1999**, 274, 24799–807.

56 Degen, W.G.J., *et al.*, The fate of U1 snRNP during anti-Fas induced apoptosis: specific cleavage of the U1 snRNA

molecule. *Cell Death Different* **2000**, 7, 70–80.

57 Houge, G., *et al.*, Selective cleavage of 28S rRNA variable regions V3 and V13 in myeloid leukemia cell apoptosis. *FEBS Lett* **1993**, 315, 16–20.

58 Houge, G., *et al.*, Fine mapping of 28S rRNA sites specifically cleaved in cells undergoing apoptosis. *Mol Cell Biol* **1995**, 15, 2051–62.

59 Houge, G. and S.O. Doskeland, Divergence towards a dead end? Cleavage of the divergent domains of ribosomal RNA in apoptosis. *Experientia* **1996**, 52, 963–7.

60 Samali, A., *et al.*, The ability to cleave 28S ribosomal RNA during apoptosis is a cell-dependent trait unrelated to DNA fragmentation. *Cell Death Different* **1997**, 4, 289–93.

61 Lafarga, M., *et al.*, Apoptosis induced by methylazoxymethanol in developing rat cerebellum: organization of the cell nucleus and its relationship to DNA and rRNA degradation. *Cell Tissue Res* **1997**, 289, 25–38.

62 Andrade, F., *et al.*, Granzyme B directly and efficiently cleaves several downstream caspase substrates: implications for CTL-induced apoptosis. *Immunity* **1998**, 8, 451–60.

63 Meerovitch, K., J. Pelletier, and N. Sonenberg, A cellular protein that binds to the 5′ noncoding region of poliovirus RNA: implications for internal translation initiation. *Genes Dev* **1989**, 3, 1026–34.

64 James, J., *et al.*, An increased prevalence of Epstein-Barr virus infection in young patients suggests a possible etiology for systemic lupus erythematosus. *J Clin Invest* **1997**, 100, 3019–26.

Part 6
The Role of DNA Binding Proteins for Systemic Autoimmunity

18
Nucleosomes and Anti-Nucleosome Autoantibodies as Mediators of Glomerular Pathology in Systemic Lupus Erythematosus

Jürgen W.C. Dieker, Johan van der Vlag and Jo H.M. Berden

18.1
Introduction

Systemic lupus erythematosus (SLE) is an autoimmune disease characterized by the formation of autoantibodies directed against various autoantigens. These autoantibodies are often specific for antigens located in the nucleus, such as nucleosomes, histones and double stranded (ds) DNA, but also RNA-associated proteins, like SS-A, SS-B and Sm. Historically, antibodies against dsDNA are considered as a hallmark of the disease. Antibodies specific for dsDNA can be detected in about 70% of SLE patients and in about 96% of SLE patients with active disease [1]. Increasing anti-DNA antibody levels also precede exacerbations of the disease [2]. First, polyclonal B cell activation was regarded as the mechanism behind the formation of anti-dsDNA antibodies. Later, it became clear, however, that SLE is a T cell-dependent and autoantigen-driven autoimmune disease. Since naked DNA is a poor immunogen, the nucleosome, the basic structure of chromatin that consists of DNA and proteins, has been proposed as the major autoantigen [3]. The significance of the nucleosome as a major autoantigen in SLE has been further substantiated by the identification of nucleosome-specific T helper cells [4] and the high prevalence of anti-nucleosome autoantibodies [5]. In addition, nucleosomes have been detected in the circulation of SLE patients [6] and lupus mice [7]. During apoptosis nucleosomes are clustered in apoptotic blebs at the surface of apoptotic cells [8]. When the removal of apoptotic cells is impaired, nucleosomes may be released in the circulation due to the instability of these cells. Indeed, there is increasing evidence that apoptosis is disturbed in SLE [9]. Apoptosis-induced modifications of autoantigens targeted in SLE may make them more immunogenic [10]. This becomes particularly relevant if the removal of apoptotic cells is insufficient.

Often no distinction is made between anti-DNA and anti-nucleosome antibodies, which is confusing. Anti-DNA and anti-histone antibodies are specific for respectively DNA and histones, but also recognize nucleosomes. Anti-nucleosome antibodies are defined by their much higher specificity/reactivity towards the complete nucleosome compared to isolated histones or naked DNA. The total group of antibodies reacting with nucleosomes, histones and DNA are referred to as anti-chromatin antibodies. The prevalence of anti-nucleosome antibodies is even higher than that of anti-DNA antibodies. These anti-nucleosome antibodies are not exclusively found in SLE, but also in scleroderma and mixed connective tissue disease [11], although this latter finding was not confirmed in another study [12]. Anti-nucleosome antibodies are detected before anti-DNA antibodies are present in the circulation, which indicates that they most likely are responsible for the subsequent induction of both the anti-DNA and the anti-histone antibody responses via antigen spreading [13].

Apart from their role as autoantigen, nucleosomes also play a role in mediating tissue lesions, especially glomerulonephritis [14]. Nephritis is one of the most serious manifestations of SLE and develops in 40–50% of the SLE patients. Granular deposits of immunoglobulins and complement factors can be found in glomeruli in renal biopsies of SLE patients, which suggests an immune complex-mediated pathogenesis. Renal manifestations of the disease are often preceded by a rise in the level of anti-chromatin antibodies in the circulation, while these antibodies are also detected in immune deposits in the glomerulus [15]. Apparently, anti-chromatin antibodies play an important role in the development of renal disease during SLE. However, the mechanism that explains the induction of nephritis by anti-chromatin is still under debate. Until now two different models have been formulated. The first model explains the deposition of anti-chromatin antibodies in the kidney by direct interaction of these antibodies with molecules present in the glomerular basement membrane (GBM). The second, in our opinion more favorable model, attributes the glomerular pathology to nucleosome-mediated deposition of immune complexes either formed *in situ* or originating from circulating immune complexes. Both models will be discussed in more detail in the next sections.

18.2
Reactivity of Anti-Chromatin Antibodies with Glomerular Components

The causal relationship between autoantibody formation and nephritis in SLE has intrigued many investigators. A variety of autoantibody specificities has been assigned a pathogenic role in the development of lupus nephritis. Antibodies specific for known components of the GBM are not commonly found in SLE, but most sera are reactive towards the GBM, which led to the idea of cross-reactive recognition of glomerular components [16]. In particular, the observation that murine and human monoclonal antibodies specific for dsDNA showed a cross-reactivity to several proteins, has been the basis to explain the binding of autoantibodies to glomerular components, especially in the GBM [17]. This cross-reactivity has even been used to distinguish pathogenic anti-DNA antibodies in SLE sera from non-pathogenic anti-DNA antibodies present in normal sera [18]. The cross-reactivity of anti-DNA autoantibodies has been described for various biomolecules, such as membrane structures like cardiolipin [19–22] and phosphocholin [23], ribosomal protein P1 [24], A and D snRNP [25], nuclear proteins like nuclear envelope proteins [26] and the transcription factor EF-2 [27].

The various *in vitro* and *in vivo* studies showing reactivity of anti-DNA antibodies with glomerular components or the GBM are listed in Tabs 18.1 and 18.2. Immunoglobulins in serum from lupus mice and patients show reactivity with isolated glomeruli [28], which could be inhibited by the addition of DNA, but not by DNase I treatment of the glomeruli. This finding may implicate that DNA shares epitopes with GBM constituents, but does not mediate the glomerular binding. Monoclonal anti-DNA antibodies derived from several mouse models for human SLE, such as the MRL-*lpr*/*lpr*, (NZW×NZB)F_1 and graft-versus-host (GVH) mouse, have been shown to react *in vivo* with isolated glomeruli [29–34]. As with the human sera, DNA could in most cases inhibit the antibody binding to isolated glomeruli, while DNase I treatment did not affect this binding. These *in vitro* results are summarized in Tab. 18.1. To approach a more *in vivo* situation of antibody binding to the glomerulus, perfusion studies have been performed in which monoclonal antibodies were directly perfused via the renal artery after which glomerular binding could be determined by immunofluorescence (Tab. 18.2). Several mouse monoclonal anti-DNA antibodies, indeed, could bind to the glomerulus [32]. When anti-DNA antibodies that were complexed with DNA were perfused, no binding could be observed. Also these more *in vivo* experiments revealed that human anti-DNA antibodies could bind to the GBM. After perfusion of IgG, isolated from human SLE sera by Protein A-Sepharose, glomerular binding also could be observed [32]. Interestingly, IgG from SLE patients with nephritis resulted in a higher intensity of glomerular binding, than IgG from SLE patients without nephritis. The binding to the GBM could be inhibited by pre-incubating the isolated IgG fraction with DNA, but not by pre-treatment with DNase I. A different *in vivo* approach is intraperitoneal injection in immune-deficient mice of hybridoma cells producing a monoclonal anti-DNA antibody. This procedure leads to circulating monoclonal antibodies, which can bind to the GBM. Indeed the injection of anti-DNA hybridomas derived from MRL-*lpr*/*lpr* [29, 33–36], (NZW×NZB)F_1 [29, 32, 36] and SNF1 mice [34] resulted in glomerular deposits in the capillary wall and in the mesangium. In some cases the staining of the mesangium was more intense than the staining of the capillary walls. In another study, i.v. injection of a mouse anti-DNA monoclonal antibody also resulted in a glomerular staining pattern [33].

In conclusion, monoclonal anti-DNA antibodies, either from lupus mice or SLE patients, seem to bind to the GBM both *in vitro* and *in vivo*. However, not all antibodies are able to react with the GBM and the exact features of the antibodies, which are responsible for the GBM-associated 'cross-reactivity' still remain unknown. The molecules, to which the anti-DNA antibodies bind, have been identified in some cases only.

A major component of the GBM is heparan sulfate (HS), which is the negatively charged glycosaminoglycan side chain of HS proteoglycan (HSPG). HS is responsible for the majority of the anionic sites in the GBM, which are the most important determinants for the charge-selective permeability of the GBM. Loss of anionic sites leads to proteinuria, while injection of antibodies specific for HS in rats instantly induces an acute selective albuminuria [37]. In both human and murine SLE sera anti-HS reactivity has been found [38]. In an early report anti-

Tab. 18.1 *In vitro* evidence for cross-reactivity of anti-DNA antibodies with glomerular constituents.

Species[a]	*Antibody source*[b]	*Antibody type*	*Detection method*	*Ligand*	*Binding inhibited by*[c]	*DNase I treatment*[c]	*Antibody purification*	*Reference*
M	GVH	DNA mAb	immuno-fluorescence	isolated glomeruli	ND	ND	Protein A-Sepharose	29
M	MRL-*lpr/lpr*	DNA mAb	radioactive assay	isolated glomeruli	ND	no effect	anti-Ig-Sepharose	33
M	MRL-*lpr/lpr*	DNA mAb	ELISA	laminin	DNA	ND	none	39
M	MRL-*lpr/lpr*	DNA mAb	ELISA	endothelial cell	DNA	ND	anti-Ig-Sepharose	35
M	MRL-*lpr/lpr*	DNA mAb	ELISA	HS	DNA	no effect	Protein A-Sepharose	29, 38
M	MRL-*lpr/lpr*	DNA mAb	immuno-fluorescence	isolated glomeruli	ferritin	ND	Protein A-Sepharose	29
M	MRL-*lpr/lpr*	serum/kidney eluate	ELISA	HS	DNA	no effect	none	38
M	(NZW × NZB)F_1	DNA mAb	immunoprecipi-tation	α-actinin	ND	no effect	Protein A-Sepharose	36
M	(NZW × NZB)F_1	DNA mAb	immunofluores-cence	isolated glomeruli	ND	ND	Protein A-Sepharose	29
H	SLE	serum	ELISA	HS	DNA	no effect	none	38
H	SLE	serum	immunofluores-cence	isolated glomeruli	DNA	no effect	none	28

a) Antibodies derived from mouse models (M) or SLE patients (H).
b) Mouse models are GVH, (NZW × NZB)F_1 or MRL-*lpr/lpr*.
c) ND, no data.

Tab. 18.2 *In vivo* evidence for cross-reactivity of anti-DNA antibodies with glomerular constituents.

Species[a]	*Antibody source*[b]	*Antibody type*	*Method of antibody transfer*	*Site of deposition*[c]	*Proteinuria*[d]	*Ligand*[d]	*Antibody purification*	*Reference*
M	GVH	DNA mAb	hybridoma inoculation	glomerular	ND	HS	Protein A-Sepharose	29
M	MRL-*lpr/lpr*	DNA mAb	hybridoma inoculation	capillary/mesangial	ND	ND	anti-Ig-Sepharose	33
M	MRL-*lpr/lpr*	DNA mAb	antibody injection	capillary/mesangial	ND	ND	anti-Ig-Sepharose	33
M	MRL-*lpr/lpr*	DNA mAb	hybridoma inoculation	capillary/mesangial	yes	ND	anti-Ig-Sepharose	34
M	MRL-*lpr/lpr*	DNA mAb	hybridoma inoculation	capillary/mesangial	yes	5 and 108 kDa protein	anti-Ig-Sepharose	35
M	MRL-*lpr/lpr*	DNA mAb	hybridoma inoculation	glomerular	ND	HS	Protein A-Sepharose	29
M	MRL-*lpr/lpr*	DNA mAb	hybridoma inoculation	capillary/mesangial	yes	α-actinin	Protein A-Sepharose	36
M	NZB	DNA mAb	hybridoma inoculation	none	no	ND	anti-Ig-Sepharose	34
M	(NZW × NZB)F_1	DNA mAb	perfusion	capillary/mesangial	yes	ND	Protein A-Sepharose	32
M	(NZW × NZB)F_1	DNA mAb	hybridoma inoculation	glomerular	ND	HS	Protein A-Sepharose	29
M	(NZW × NZB)F_1	DNA mAb	hybridoma inoculation	capillary/mesangial	yes	α-actinin	Protein A-Sepharose	36
M	SNF1	DNA mAb	hybridoma inoculation	intranuclear (glomerulus)	no	ND	anti-Ig-Sepharose	34
H	SLE	DNA mAb	hybridoma inoculation	capillary/mesangial	yes	ND	Protein G + high salt	38
H	SLE	serum IgG	perfusion	glomerular	yes	ND	Protein A-Sepharose	35

a) Antibodies derived from mouse models (M) or SLE patients (H).
b) Mouse models are graft versus host (GVH), (NZW×NZB)F1, NZB or MRL-*lpr/lpr*.
c) The site of deposition is listed as glomerular if it is not specified further. If deposition occurred along the glomerular capillary wall it is listed as capillary. Mesangial represents exclusive mesangial deposition.
d) ND, no data.

HS reactivity has been demonstrated in 30 out of 33 SLE patients that were positive for anti-DNA antibodies in the Farr assay and the anti-DNA ELISA. The observed HS binding in ELISA could not be attributed to a much higher IgG concentration in the SLE patients because the IgG concentration was only 2-fold higher compared to the control sera, while the reactivity towards HS was about 100-fold higher. To exclude any possible DNA contamination of the HS, which could easily explain the apparent binding of anti-DNA antibodies, the reactivity of human SLE sera has also been tested with HS that was pre-treated with DNase I. However, no difference in reactivity towards HS could be observed with or without DNase I pre-treatment [38]. The titers of anti-HS-reactive antibodies showed a significant correlation with the titers of anti-DNA antibodies both in sera from SLE patients and MRL-*lpr/lpr* mice [38]. To obtain additional evidence that the binding of anti-DNA antibodies to the GBM was mediated by HS, ELISA blocking experiments with DNA and HS have been performed [29]. The binding of sera from SLE patients and MRL-*lpr/lpr* mice to DNA in ELISA could be inhibited by the addition of either HS or DNA. *Vice versa*, the HS reactivity in ELISA could also be inhibited in a dose-dependent manner by the addition of either HS or DNA. The degree of inhibition varied from 60 to 100% [29]. In addition, antibodies eluted from MRL-*lpr/lpr* kidneys and from a kidney of a SLE patient showed reactivity towards HS, which again could be inhibited by the addition of DNA. The observed HS binding is very unlikely caused by entrapment of serum IgG in the eluate, because the reactivity was much higher in the eluates than in the corresponding serum [29]. Murine monoclonal anti-DNA antibodies, that are able to bind to the GBM, also showed reactivity to HS or HSPG in ELISA. Again this could be inhibited dose dependently with DNA or HS [29]. Binding of anti-DNA antibodies to isolated GBM loops also could be completely prevented by pre-incubation of the GBM loops with cationic ferritin, while pre-treatment with heparitinase (i.e. enzymatic removal of HS chains) reduced the binding considerably, but not completely. The remaining HS might be responsible for the residual binding of anti-DNA antibodies, but the result might also indicate that other anionic molecules than HS within the GBM, such as laminin, can serve as binding sites for anti-DNA antibodies [39]. The glomerular binding was also not due to non-specific charge interaction, since a non-related monoclonal antibody with a high isoelectric point (above 9.0) did not bind to the GBM. The anti-DNA antibodies that resulted in glomerular deposits and showed reactivity in ELISA towards HS, were used for *in vivo* studies. Hybridomas that produced non-HS-binding anti-DNA antibodies only resulted in deposits in the mesangium. Injection of HS-binding anti-DNA producing hybridomas also led in most cases to proteinuria in normal mice. This was not the case when non-relevant antibody-producing hybridomas were used [29]. Similar experiments with hybridomas that produce human monoclonal anti-DNA antibodies did also result in a capillary wall and mesangial staining and proteinuria [40]. All *in vivo* results that relate anti-DNA antibodies to glomerular deposition are summarized in Tab. 18.2. All previously discussed results seem to support the theory that the most prominent autoantibody in SLE, anti-DNA, can cross-react with the negatively charged component of the GBM, HS [29]. As will be outlined further this 'cross-reactivity' of anti-DNA with HS was not due to a direct binding, but mediated via nucleosomes bound to the antibody.

Screening of a random peptide phage display library is a more recent approach to identify structural motifs [41] that can explain the reactivity of anti-DNA antibodies with the GBM. This technique can be used to find mimotopes for anti-DNA (or anti-nucleosome) antibodies, which are peptide motifs that mimic the epitopes which are recognized by the antibody. The random peptide phage display technology has proven to be a powerful technique, especially for antibodies that by definition do not recognize a protein-based epitope, such as anti-DNA autoantibodies [42, 43]. Briefly, the autoantibody is coated on a carrier (i.e. a plate or beads) and subsequently incubated with a phage library, which contains numerous different phages that display a short random peptide fused with a phage coat protein. Finally, the phages that bind to the anti-DNA autoantibody with the highest affinity are selected. *In silico*, the selected motifs of course can be compared to all known proteins sequences, including GBM proteins. A positive match between the sequence of the selected motif and the sequence of a GBM protein then readily explains the observed reactivity of the anti-DNA antibody with the GBM.

Several DNA mimotopes with a motif that mimics the epitope of anti-DNA autoantibodies have been selected so far with this random peptide phage display [43–45]. In one study anti-DNA antibodies from sera of SLE patients, which were positive in the Farr assay, have been purified by a γ-bind Sepharose column followed by a DNA-Sepharose column [44]. The purified antibodies have been used for the screening of a phage library with a random sequence of 15 amino acids fused to the pVIII phage coat protein. After several rounds of screening a common motif (amino acid sequence: RLTSSLRYNP) could be deduced from the selected sequences. Despite the negative charge of DNA, this motif surprisingly contained some positive charges. The motif showed only homology with a chromosome-associated polypeptide. The peptide that was synthesized on the basis of the motif did react with 88% of the anti-DNA positive SLE patient sera. The reaction of the IgG used for the selection with the peptide could be inhibited by the addition of dsDNA, native RNA, denatured DNA, and most efficiently with ssDNA. Apparently, the peptide mimics a common epitope present in dsDNA, ssDNA and RNA [44].

Random peptide phage display has also been applied to several monoclonal anti-DNA antibodies. Screening of a 16-mer random peptide phage display library with three anti-DNA antibodies derived from $(NZB \times NZW)F_1$ mice did result in mimotopes for two antibodies [45]. The motif obtained by one of the antibodies (F14.6) contained two cysteines, which could form a loop, negatively charged amino acids and some aromatic residues. The motif selected by the other antibody was different, but also contained two cysteines, negatively charged and aromatic amino acids. No significant homology could be found with any other proteins [45]. Screening of a 10mer pIII phage library with a mouse monoclonal anti-DNA antibody (R4A) that causes glomerular deposits in non-autoimmune mice, did also result in a mimotope [43]. This mimotope contains a net negative charge, like DNA, and has two aromatic amino acids, which may mimic the sugar backbone structure or one of the base pairs. It was evaluated whether the mimotope-based synthetic peptide could inhibit the deposition of the R4A antibody to the GBM *in vivo*. Administration of only the anti-DNA antibody caused deposits in the glomerulus in SCID mice. Administration of both the mimotope-based peptide and the R4A antibody reduced the deposits in the glomerulus. Binding of the antibody to the GBM seems likely, because of the inhibition of glomerular binding by the mimotope-based peptide. Therefore, one could expect or argue that the mimotope should mimic a GBM component or structure. However, on the protein level only homology with collagen XIII could be found (C. Putterman, personal communication), a molecule that is normally not present in the GBM. The DWEYS peptide on a MAP backbone has been used to immunize 4- to 6-week-old BALB/c mice [46]. The immune response resulted in antibodies against DNA, histones and cardiolipin. Most importantly, the mice developed a SLE-like disease with glomerular deposits of IgG and complement C3 [46]. In a recent study the reactivity of R4A with the NR2 glutamate receptor was found [47].

Comparison of the amino acid sequences of the different DNA mimics that have been identified thus far [43–45] reveals that there is no homology between the mimotopes at all. At first, this finding may seem rather unexpected because the basic structure of naked DNA, i.e. without associated proteins, is a double helix. The differences between the several reported DNA mimics may be explained by different characteristics of the original DNA autoantigen such as different higher-order structures of DNA (supercoils) or just differences in the nucleotide sequence. Most importantly, on the amino acid level the DNA mimotopes do not show homology with any known protein in the GBM. However, it cannot be excluded that the DNA mimotopes mimic a structural feature of the GBM that cannot be easily deduced from the amino acid sequences of known proteins.

In summary, several studies suggest that certain anti-nuclear antibodies can bind to glomerular components and induce glomerular deposits. However, from these studies it is difficult to conclude which characteristics of the antibody determine glomerular binding. After a decade of research by several groups, the ligands in the glomerulus that could serve as cross-reactive epitopes for anti-nuclear antibodies still have not been identified. So the direct binding of anti-DNA antibodies to the GBM cannot yet be explained by a look-a-like of DNA present in the GBM.

After our initial observation that anti-DNA antibodies could cross-react with HS and GBM loops [38], we showed that this binding was mediated by nucleosomes bound to the anti-DNA antibodies [48]. As will be outlined in the subsequent section, removal of these nucleosomal components completely abrogated the binding to HS *in vitro* and to the GBM *in vivo* [14]. A similar observation has been made with respect to other cross-reactive molecules [31, 49]. Given these results, in our opinion, it is likely that the majority of the observed cross-reactions, as discussed above, are mediated by nucleosomal components bound to the antibodies. Let us assume that this is the case, is it possible to explain the results described in Tabs 18.1 and 18.2. A common feature of all these experiments, is that DNA can block the binding to the GBM and DNase I treatment has no effect. This led to the conclusion that binding of DNA to the antigen binding sites of the antibody prevents cross-reactive recognition of the GBM ligand. The lack of effect of DNase I treatment was regarded as proof for the absence of DNA in the antigen binding site. However, these findings can be explained differently. When an anti-DNA antibody is complexed to nucleosomes, addition of DNA to these complexed antibodies will lead to binding of DNA to the positively charged histone tails, thereby neutralizing these positive charges. This masking will lead to the abrogation of binding to HS in the GBM. Addition of DNase I to the antibodies complexed to nucleosomes will lead to the removal of DNA from the nucleosome, but will not remove the bound histone part, as we have documented [50]. These remaining histone molecules are still able to bind HS in the GBM.

18.3 The Nucleosomes as Mediator of Autoantibody Binding to the GBM

The nucleosome is the fundamental unit of chromatin, which is defined as the total of compacted DNA and DNA-associated biomolecules in the eukaryotic cell nucleus. In the nucleosome two superhelical turns of 146 base pairs of DNA are complexed to the pairs of four different core-histones that form an octamer (Fig. 18.1). The core histones are histones H2A, H2B, H3 and H4, while histone H1 is bound outside of the nucleosome. The core histones have a molecular weight that ranges from about 11 to 15 kDa, a positive charge and an isoelectric point of about 11. The basic residues are clustered in the flexible N-terminal parts of the core histones, which are located outside of the histone octamer. Nucleosomes are exclusively formed during apoptosis by enzymatic cleavage of the linker dsDNA regions of chromatin. The observation that apoptosis is disturbed in SLE makes the nucleosome an important candidate autoantigen. The antigenicity of nucleosomes can probably be enhanced by modifications of the histones or DNA during apoptosis. Several types of modifications associated with apoptosis and/or chromatin condensation have been described so far. These include (1) phosphorylation or dephosphorylation of specific sites on the N-terminus of the core histones [51], (2) methylation of certain amino acids on the N-terminus of the core histones, (3) hyperacetylation or deacetylation of histones [52, 53], (4) ubiquitination so far described for topoisomerase II and histone H2A [54], (5) citrullination, the selective deamination of arginine to citrullin [55], and (6) transglutaminase-facilitated crosslinking of proteins like histone H2B [56]. In addition to protein modifications, (nucleosomal) DNA can be altered by methylation.

As outlined in the introduction there is now convincing evidence that the nucleosome is the driving autoantigen in SLE. Nucleosomes are not only important in the induction phase of SLE, but they also play an important role in the development of tissue lesions. Several studies provided the evidence that nucleosomes can act as a mediator for the binding of autoantibodies to the GBM. As outlined before we found that certain anti-DNA monoclonal antibody derived from MRL-*lpr/lpr*, (NZB×NZW)F_1 and GVH mice bound to HS isolated from bovine kidney

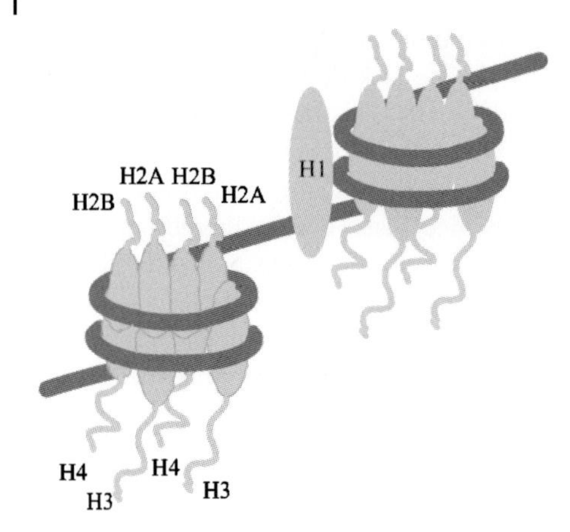

Fig. 18.1 Composition of the nucleosome. The core particle consists of an octamer of four pairs of core histones H2A, H2B, H3 and H4, and around 146 bp of dsDNA, histone H1 is attached to the nucleosomes.

in ELISA [38]. The majority of these HS positive anti-DNA antibodies also showed reactivity with HSPG isolated from human kidneys. The antibodies showed no 'cross-reactivity' with other GBM components such as fibronectin, laminin or collagen IV. However, when the culture supernatants of the anti-DNA antibody-producing hybridomas were treated with DNase I, the binding to HS or HSPG was strongly reduced in both ELISA and Western blot. This could mean that DNA mediates the binding of anti-DNA antibodies to HS or HSPG. The residual binding then might be explained by the insufficient removal of DNA form the antibody preparations by DNase I treatment, therefore further purification of the antibodies on a Protein A-Sepharose column under dissociating high salt conditions was carried out. Strikingly, after this extended purification procedure all previously HS positive anti-DNA antibodies lost their ability to bind to HS as could be determined in the anti-HS ELISA, whereas they showed an unaltered anti-nuclear antibody reactivity in immunofluorescence on rat liver sections. The reactivity of these purified antibodies with HS and HSPG could be restored by the addition of the effluent of the Protein A–Sepharose column. Figure 18.2(A) depicts a representative experiment with the monoclonal antibody clone #32, which later appeared to be a genuine anti-nucleosome antibody. The purified anti-DNA antibody was used for immunoprecipitation of the component that could restore the binding to HS, from the Protein A effluent. After SDS-PAGE, the precipitated material clearly showed multiple bands with a molecular weight between 10 and 15 kDa, which were identical to calf thymus histones (Fig. 18.2B). In addition, the presence of DNA in the unpurified antibody preparation could be detected by agarose gel electrophoresis after DNA isolation (Fig. 18.2C). DNase I treatment of the Protein A-Sepharose column effluent abolished the reconstitutive effect of the effluent with respect to HS binding of purified anti-DNA antibodies, which provided additional evidence for the presence of bound DNA in unpurified anti-DNA antibodies. The presence of both DNA and histones was required for the binding of anti-DNA antibodies to HS because the separate addition of histones or DNA

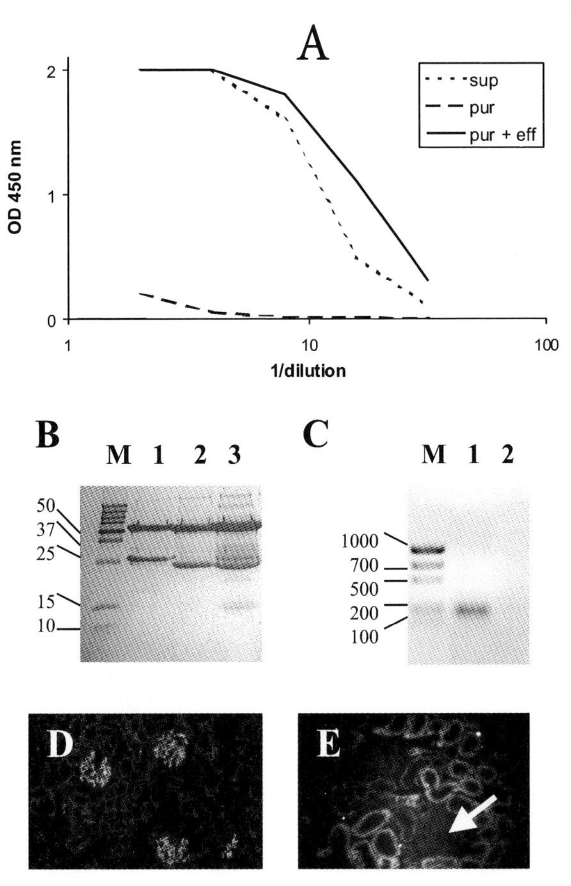

Fig. 18.2 Reactivity with HS and the GBM of anti-nucleosome antibody #32. The purification procedure included DNase I treatment and Protein A-Sepharose chromatography under dissociative high salt conditions. (A) Clone #32 binds in the HSPG ELISA before purification (sup); after purification under dissociative condition on a Protein A column this reactivity is lost (pur), while addition of the Protein A column effluent to the purified monoclonal antibody restores the binding activity (pur + eff). (B) SDS–PAGE followed by silver staining of antibody fractions before purification (lane 3) and after purification (lane 1 and 2). Before purification histone bands are visible. Marker (M) is a protein marker (in kDa). (C) Agarose gel electrophoresis after DNA extraction of antibody fractions before (lane 1) and after purification (lane 2). Marker (M) is a DNA marker (in bp). The gel is stained with ethidium bromide. (D) Immunofluorescence (stained with FITC-labeled anti-IgG) of rat kidney sections. After successive renal perfusion of histones, DNA and purified monoclonal antibody #32 into the left kidney. A strong staining of glomeruli can be detected. (E) Similar to (D), but now histones, DNA and a non-relevant antibody (WT32) have been successively perfused. No glomerular staining is seen (glomeruli indicated with arrow). Note that due to prolonged exposure auto-fluorescence of the tubuli is observed.

to the purified antibody could not restore the HS binding. Apparently, a DNA-histone complex was already bound to the unpurified anti-DNA antibody. Indeed, during hybridoma culture nucleosomes are released [57]. We therefore concluded that the binding of anti-DNA antibodies to HS and HSPG is mediated by DNA-histone complexes bound to the anti-DNA antibody [48].

In addition to the ELISA studies, *in vitro* binding studies have been performed with isolated GBM loops [58]. Anti-DNA antibodies extensively purified as described above did not bind *in vitro* to the GBM loops isolated from human kidney. Pre-incubation of the anti-DNA antibodies with both DNA and histones led to a strong granular binding. Addition of histones or DNA alone to the anti-DNA antibodies did not lead to binding to GBM loops. Pre-incubation of GBM loops with cationic ferritin prevented subsequent binding of histone-DNA-antibody complexes, which indicated that anionic sites like HS in the GBM were important for this binding.

To analyze whether this *in vitro* binding to HS could also take place *in vivo*, renal perfusion studies in normal rats have been performed. In this approach histones, DNA and purified antibody were subsequently perfused via the renal artery (Fig. 18.2D and E). Perfusion of non-complexed purified antibodies did not lead to glomerular binding, however, after successive perfusion of histones, DNA and several anti-DNA antibodies, immunofluorescence analysis revealed an intense staining of the glomeruli, arteries and peritubular capillaries. Immunoelectron microscopy analysis showed large deposits at the cell membranes of glomerular endothelial cells and smaller complexes in the GBM. In the case of a high-avidity anti-DNA monoclonal antibody, perfusion of only DNA and subsequently an anti-DNA antibody did only result in mesangial deposits, but not in deposits in the capillary loop.

A completely different glomerular deposition pattern could be observed when the high avidity anti-DNA antibody was administered into the tail vein 1 h after the perfusion of histones and DNA into the left kidney via the renal artery and subsequent restoration of the renal circulation [58]. Deposits now could be detected along the capillary wall in a membranous pattern, while variable amounts were present in the mesangium. Immunoelectron microscopy revealed the presence of immune deposits in the GBM located in the slit pores and under the foot processes of the podocytes. In contrast to the direct perfusion method, no binding to the endothelial cells could be observed. Most likely the 1 h time interval between intrarenal perfusion of histones-DNA and the intravenous administration of the anti-DNA antibody allows for the penetration of DNA-histones into the glomerular capillary wall, where they then serve as planted antigens for the monoclonal anti-DNA antibody. The direct perfusion method also resulted in an increased urinary excretion of albumin, which suggests a damaged glomerular filter. Control experiments with non-related monoclonal antibodies of the same isotype and at the same concentration did not result in any glomerular binding. In conclusion, anti-DNA antibodies can bind via complexes of histones and DNA to the GBM, which results in subendothelial or subepithelial deposits and proteinuria. Only in the case of high-avidity anti-DNA antibodies was the presence of DNA alone sufficient for glomerular binding, which, however, was exclusively localized in the mesangium. These studies revealed that the DNA-histone complex served as a planted antigen in the GBM for anti-DNA antibodies [58]. Histones alone can also act as planted antigens for artificially prepared DNA-anti-DNA complexes as has been shown in a study in rats, in which prior to the administration of DNA-anti-DNA complexes, histones were perfused through the renal artery [59].

In vivo histone-DNA complexes exist as nucleosomes, therefore, the perfusion experiments described above were repeated with antibodies complexed with nucleosomes [50]. The applied monoclonal antibodies were partially purified from the hybridoma culture supernatant by Protein A-Sepharose under physiological salt conditions, which yields antibodies still complexed to nucleosomes. Control perfusions were performed with purified antibodies free of nucleosomes (see also Fig. 18.2B and C). This latter purification method included DNase I treatment of the hybridoma culture supernatant, a Protein A-Sepharose column under high salt conditions and in some cases a DNA-cellulose column, which removes antibodies still bound to histones. The purity of the antibody fractions was analyzed by SDS-PAGE and agarose gel electrophoresis for the presence of histones and DNA, respectively. In contrast to unpurified antibodies and antibodies purified under physiological conditions, all extensively purified antibodies did not contain DNA and/or histones. The purified antibodies have been tested in ELISA for their reactivity towards DNA, histones and nucleosomes. It then turned out that only some anti-DNA antibodies reacted with DNA. Most likely, the reactivity with dsDNA of the formerly classified anti-DNA antibodies was due to histones within the nucleosome bound to the antibody. The 'genuine' anti-DNA antibodies reacted after purification with both DNA and nucleosomes. The reactivity of 'anti-DNA antibodies', which were in fact anti-nucleosome antibodies, with HS in ELISA could only be demonstrated when these antibodies were not purified. In agreement with this latter finding, the purified anti-nucleosome antibodies did not show glomerular binding, while unpurified anti-nucleosome antibodies did show glomerular binding, as could be seen by immunofluorescence after renal perfusion experiments. As expected, reconstitution of purified anti-nucleosome antibodies with purified calf thymus nucleosomes restored glomerular binding. In conclusion, both immune complexes consisting of anti-DNA-nucleosome and anti-nucleosome-nucleosome are able to bind to the glomerulus *in vitro* and *in vivo* [50]. To investigate the significance of HS in the GBM for glomerular binding of nucleosome-complexed autoantibodies, heparinase was perfused intrarenally prior to administration of the immune complexes. After heparinase perfusion the staining of HS side chains of HSPG within the GBM almost disappeared completely, while the staining of the HSPG core protein was not affected. Staining of other GBM components, like laminin and collagen IV also was not affected by heparinase perfusion. Perfusion of nucleosome-autoantibody complexes after the heparinase treatment did result in a decreased binding to the GBM, but this binding did not disappear totally [50]. This observation suggested that HS is not the only ligand that is responsible for the binding of anti-nucleosomes-nucleosome immune complexes to the GBM. The binding of nucleosome-containing immune complexes to other GBM components has indeed been reported [60]. In this latter study, the binding of sera from MRL-*lpr*/*lpr* mice to the GBM could be competitively inhibited by DNA or histones and collagenase treatment of the GBM. The binding of the antibodies could only be restored after the addition of collagen IV. With purified GBM components coated in ELISA antibody binding could only be observed when DNA was coated on collagen I; however, no or little binding was observed when DNA was coated on collagen IV, laminin or fibronectin. Coating of histones prior to DNA coating on collagen IV resulted in a positive binding of the MRL/*lpr* sera. These latter findings suggested that nucleosomes can bind to collagen IV via their histone parts, analogous to the situation with HS [60].

Other evidence that HS is the major target in the GBM for nucleosome-mediated autoantibody binding came from our studies in which we used heparin, which is structural very similar to HS [61]. We hypothesized that heparin could bind to the positively charged N-terminal parts of the core histones within the nucleosomes, thereby preventing histone binding to HS. Indeed, in ELISA heparin and non-coagulant heparinoids could inhibit dose-dependently the binding of nucleosome-complexed autoantibodies to HS or DNA (Fig. 18.3A and B). Similarly, in renal perfusion experiments heparin could prevent glomerular binding of nucleosome-complexed autoantibodies, while dextran showed no effect (Fig. 18.3C). *In vivo* in MRL-*lpr*/*lpr* mice daily subcutaneous administration of heparin or non-coagulant heparin derivatives prevented the development of proteinuria and glomerular lesions (Fig. 18.3D and F). Immunohistology of the kidneys revealed that in heparin(oid)-treated mice only mesangial deposits were present and in 80% of the heparin(oid)-treated animals no glomerular lesions could be observed (Fig. 18.3E) [61]. These results support that immune complex deposition in lupus mice is mediated for the greatest part by the interaction of cationic histones with anionic HS. To further document that the N-termini of the core histones are important for the binding to HS, we performed *in vivo* inoculation studies with hybridoma cells in BALB/c nude mice [62]. We hypothesized that anti-histone producing hybridomas would generate less nephritogenic nucleosome-autoantibody complexes than anti-nucleosome or anti-DNA antibodies. Our hypothesis was based on the fact that the epitopes for anti-histone antibodies reside within these N-termini. Binding of anti-histone antibodies to the cationic tails would reduce the ability of histones to bind to the anionic HS in the GBM, i.e. because of masking of the positive charges on these N-termini. Binding of anti-nucleosome or anti-DNA antibodies would hardly influence these charges. Inoculation of the different hybridoma cells led to detectable levels of antibodies in ascites only against the part of the nucleosome, to which the antibody produced by the hybridoma was directed (Fig. 18.4A). Antibody reactivities in plasma samples were comparable to those obtained in ascites indicating that the antibodies were transferred to the systemic circulation. Indeed, inoculation of mice with three different anti-histone hybridomas induced only in 15% of the animals glomerular deposits, while with three different anti-nucleosome and three anti-DNA antibodies 60% of the animals had GBM deposits (Fig. 18.4C and D) [62]. Apparently, the differences in glomerular

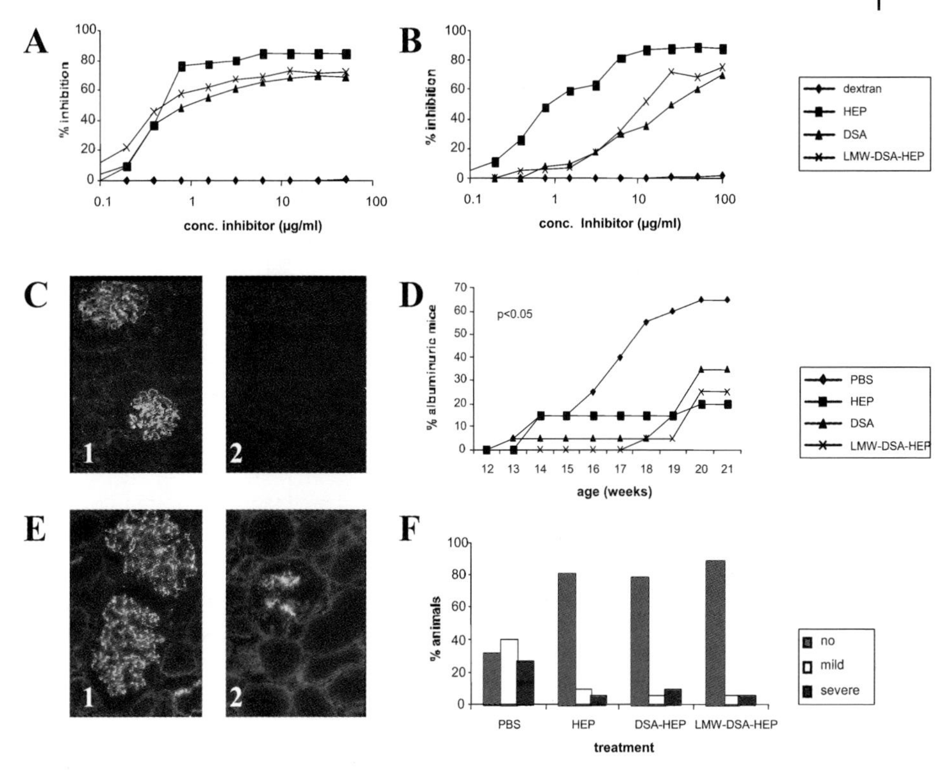

Fig. 18.3 Effect of heparin and non-coagulant heparinoids on binding of nucleosome/anti-nucleosome complexes to HS or the GBM. *In vitro* experiments: inhibition of the binding of the complexed anti-nucleosome antibody #34 to coated HS (A) or DNA (B) in the inhibition ELISA. Binding of complexed #34 to both HS (A) and DNA (B) was inhibited dose dependently by heparin (HEP), *N*-desulfated/acetylated heparin (DSA-HEP) and low-molecular-weight desulfated/acetylated heparin (LMW-DSA-HEP). Dextran could not inhibit the binding. *In vivo* experiments: (C) Direct immunofluorescence (with FITC-labeled anti-IgG) of glomeruli of BALB/c mice after perfusion of complexed anti-nucleosome #34 mixed with dextran (1) or heparin (2). After perfusion of complexed antibody #34 mixed with dextran clear binding of the complex to the glomerulus was observed (1). However, the addition of heparin to the complexed antibody #34 could completely prevent this binding (2). (D) Cumulative incidence of albuminuria in the various groups of MRL-*lpr*/*lpr* mice. Starting at the age of eight weeks MRL-*lpr*/*lpr* mice were treated once daily with either 50 µg HEP, DSA-HEP, LMW-DSA-HEP or with PBS as control. Each group consisted of 15 animals. Albuminuria was considered to be present when the urinary albumin excretion exceeded 300 µg/18 h (mean +2 SD of albuminuria in non-SLE normal control mice). This mean value is 100 µg/18 h, $P<0.05$. (E) Representative examples of the immunofluorescence findings of glomeruli of MRL-*lpr*/*lpr* mice either treated with PBS (1) or HEP (2). PBS-treated mice show extensive deposition of IgG, mainly along the capillary wall. In HEP-treated animals the deposits were confined to the mesangium. Thus, HEP treatment prevented deposition in the capillary loop. (F) Severity of glomerular lesions in MRL-*lpr*/*lpr* mice treated with phosphate-buffered saline (PBS), HEP, DS-HEP or LMW-HEP. The severity of the glomerulus was scored as normal (no), mild glomerulonephritis or severe glomerulonephritis.

A

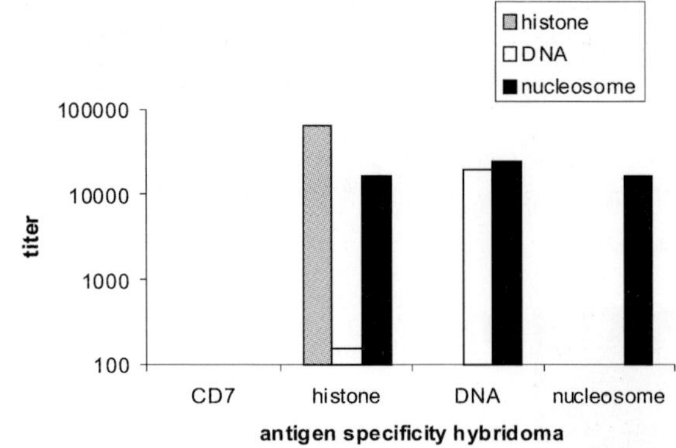
histone
DNA
nucleosome
100000
10000
1000
100
titer
CD7
histone
DNA
nucleosome
antigen specificity hybridoma

B

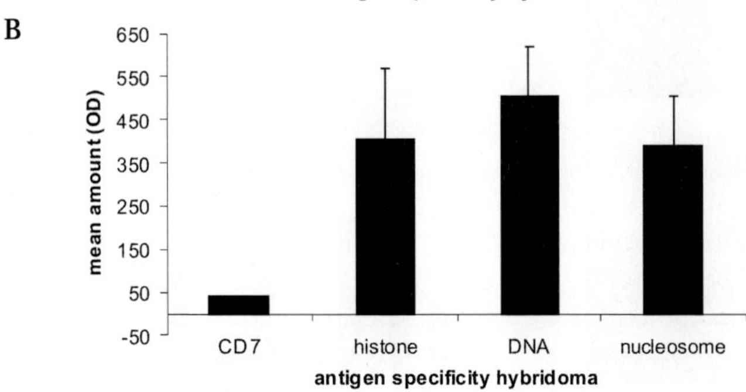
650
550
450
350
250
150
50
-50
mean amount (OD)
CD7
histone
DNA
nucleosome
antigen specificity hybridoma

C

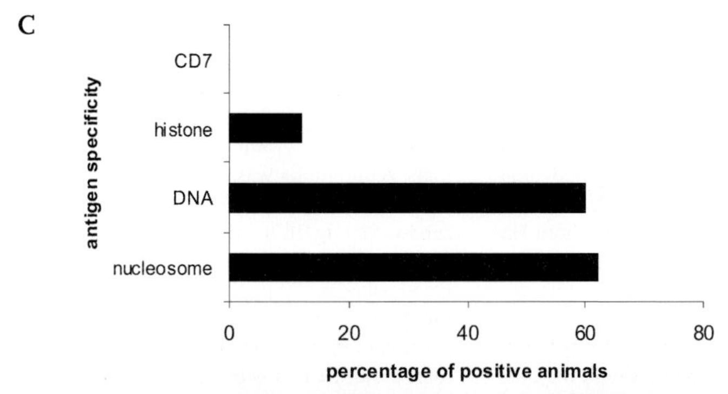
CD7
histone
DNA
nucleosome
antigen specificity
0
20
40
60
80
percentage of positive animals

D

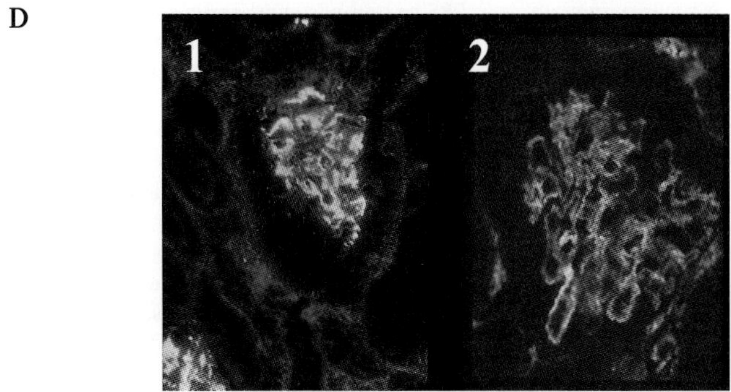
1
2

binding between the antibodies were not due to differences in the levels of nucleosome-autoantibody complexes, since these were similar (Fig. 18.4B).

These latter observations may seem in contrast with previous results that show direct binding of anti-DNA antibodies to the GBM, which were explained by cross-reactivity. However, there is a good explanation for this apparent contradiction. A major point of concern is the purity of the antibody fractions that have been used in the experiments as outlined earlier. During the culture of hybridoma cells apoptosis occurs spontaneously and nucleosomal material is released [57]. This material will subsequently bind to the monoclonal antibodies that are present in the supernatant and which are directed against nucleosomal material. This undoubtedly leads to the formation of nucleosome-antibody complexes [50, 63]. In studies describing cross-reactivity mostly unpurified antibodies from culture supernatants or only partial purified antibodies by Protein A/G-Sepharose columns are used. However, this latter purification step under physiological conditions does not dissociate the bound DNA and especially the bound histones from the antibody, as we have shown [50]. Sometimes a DNase I treatment of the culture supernatant is used, but in most cases not in combination with high salt-Protein A/G-Sepharose. In our experience anti-nucleosome-anti-DNA antibodies have to be purified by DNase I treatment, high salt-Protein A/G-Sepharose and a DNA cellulose column (which removes residual histone-containing complexes). Many reports do not show or discuss the purity of the applied anti-DNA, anti-histone or anti-nucleosome antibody preparations, which is of prime importance for studies that aim to unravel the mechanism of glomerular deposition in SLE.

18.4 Evidence for Nucleosome-Mediated Binding in Lupus Nephritis

So far, we have presented experimental data for the role of nucleosomes as mediators for autoantibody binding to the GBM. However, if this concept is true, nucleosomes and anti-nucleosome antibodies should be present in the glomeruli of pa-

Fig. 18.4 Glomerular binding after intraperitoneal inoculation of hybridomas producing anti-nuclear antibodies. (A) Reactivity of ascites of mice inoculated with anti-DNA, anti-histone, anti-nucleosome or control hybridoma (producing anti-CD7) in ELISA towards the different nuclear antigens: histones, DNA and nucleosomes. The titer in ELISA was defined as the reciprocal of the dilution giving an OD of 1.0 at 450 nm and expressed per mg of Ig. The results show that the appropriate specificity was detected in the ascites. (B) Nucleosome-IgG complex assay performed on ascites from mice inoculated with anti-DNA ($n=3$), anti-histone ($n=3$), anti-nucleosome ($n=3$) or anti-CD7 control hybridomas ($n=3$). Values are given as mean ± SEM. No statistically significant difference was found between the amounts of complexes formed for the different anti-nuclear antibodies. (C) Immunofluorescence analysis of kidney sections of mice inoculated with anti-DNA ($n=15$), anti-histone ($n=13$), anti-nucleosome ($n=13$) or control hybridomas ($n=6$). The results are expressed as percentage of mice with GBM deposits. (D) Representative examples of kidney sections of mice inoculated with anti-DNA #42 (1) or anti-nucleosome #32 (2).

A

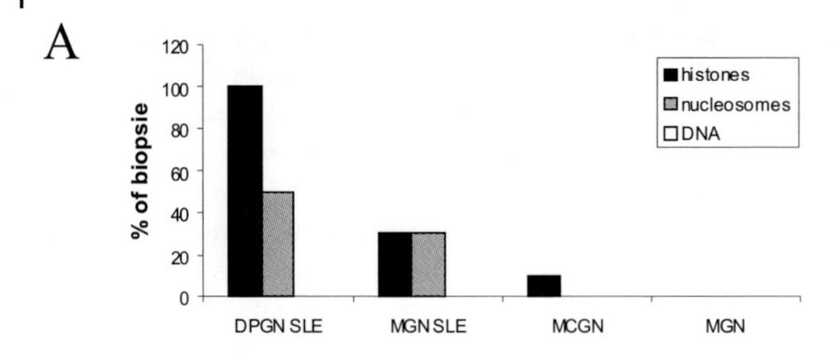

B

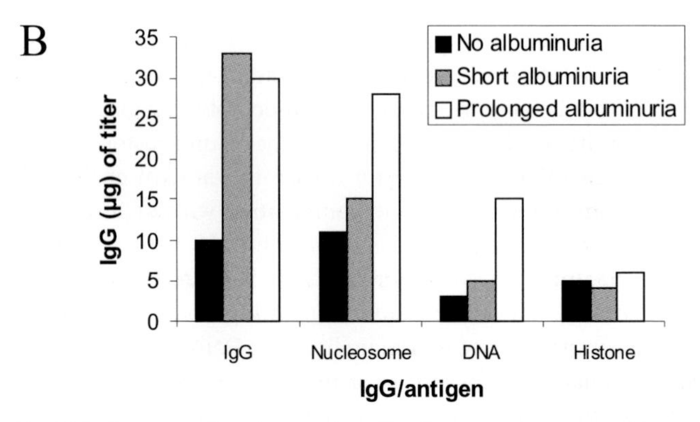

Fig. 18.5 Presence of nucleosomes and antinuclear antibodies in glomerular deposits in lupus nephritis. (A) Presence of nucleosomal antigens in GBM deposits in kidney biopsies of SLE patients with either DPGN or MGN and non-SLE controls with either mesangiocapillary glomerulonephritis (MCGN) or idiopathic MGN. Staining of histones with polyclonal rabbit anti-H3 1-21 serum, staining of nucleosomes with monoclonal antibody #34, LG8-1 and LG10-1, staining for DNA with antibody #42 was negative in all biopsies. (B) Immunoglobulin concentration and titers towards nucleosomal antigens of glomerular eluates from MRL-*lpr*/*lpr* mice. The 18- to 24-week-old mice either had no albuminuria, short albuminuria of less than 7 days or prolonged albuminuria for 14–21 days.

tients and mice with spontaneous lupus nephritis. We therefore set up a number of studies to investigate whether these proteins and antibodies were present in the glomerulus in lupus nephritis. First, we searched for the presence of nucleosomes/nucleosomal antigens in glomerular deposits. A polyclonal anti-histone H3 serum which reacts with the N-terminal amino acids 1–21 was used to probe histones in kidney biopsies of 11 SLE patients with diffuse proliferative glomerulonephritis (DPGN) and six patients with lupus membranous glomerulonephritis (MGN) [64]. Histones could be detected in all of the eleven SLE patients with DPGN and two out of the six SLE patients with MGN (Fig. 18.5A). A histone monoclonal antibody (KM-2), which reacts with the N-terminal parts of both histone H2A and H4, only demonstrated the presence of histones in three patients with DPGN and none of the patients with MGN. In addition a nucleosomal staining could be detected by using three different nucleosome-specific antibodies (LG10-1, LG8-1 and #34) in five patients with DPGN and two patients with MGN (the same two patients that stained positively for histones). Anti-DNA antibodies (#36, #42 and #56) did not reveal DNA in glomerular deposits. This latter finding seems remarkable since nucleosomes (that contain DNA per definition) could be detected by anti-nucleosome antibodies. We assume that all epitopes within DNA recognized by the probing anti-DNA antibodies were masked by autoantibodies present within the glomerular deposits. This also explains why the histone epitopes were much more easily to detect, because when these epitopes are covered by anti-histone antibodies the complex will not bind to the GBM as outlined above. In non-SLE glomerulonephritis none of the applied monoclonal antibodies could stain glomerular deposits. However, the presence of DNA in immune deposits associated with human lupus nephritis has been demonstrated at the electron microscopy level by an anti-DNA antibody labeled with Protein A-gold and DNase I-gold complexes [65]. Histones and nucleosomes have also been detected in immune deposits in two models of murine SLE, (NZB×NZW)F_1 and GVH mice [66].

In order to evaluate the sequence of deposition of the different autoantibody specificities in MRL-*lpr*/*lpr* mice, we eluted glomeruli from 18- to 24-week-old mice with either no proteinuria, short duration proteinuria (within 1 week of onset) or heavy proteinuria (for more than 3 weeks) [15]. The onset of proteinuria was accompanied by a 3-fold higher IgG content of the eluate, which did not increase further in mice with heavy proteinuria. The analysis of the antigen specificities in the eluates showed that anti-nucleosome antibodies deposited first, while the highest reactivity for anti-DNA antibodies was found in mice with heavy proteinuria. Anti-histone reactivity was low and did not increase further when proteinuria developed or progressed (Fig. 18.5B).

Evidence for the presence of nucleosome-immune complexes in lupus nephritis was obtained rather indirectly. Using monoclonal antibodies against HS and HSPG core protein [37, 67] we found an almost complete absence of HS staining in the GBM in renal biopsies from patients with lupus nephritis, while the staining for the core protein remained unaltered [68]. Since MRL/*lpr* mice also showed this decrease in HS staining at advanced states of nephritis [69], we could study the responsible mechanisms in more detail. In these lupus mice the decrease in HS staining correlated inversely with the amount of immunoglobulin deposits in the GBM. The pathophysiological significance of the reduction of HS staining was underlined by the inverse correlation with albuminuria. Also in human lupus nephritis an inverse correlation was found between HS staining and the amount of histone deposits in the GBM [64]. The decrease in HS staining was not due to decrease in HS content since the amount of HS within the GBM was normal [69]. Together, these data suggest that the decrease in HS staining is due to masking of HS by nucleosome-immune complexes. Indeed, nucleosome autoantibody complexes were able to inhibit dose dependently the binding of the anti-HS monoclonal antibody JM403 to HS in ELISA, whereas non-complexed autoantibodies had no effect [69]. As outlined above, heparin treatment in MRL/*lpr* mice prevented formation of deposits in the GBM. In these heparin-treated animals a decrease in HS staining was not observed [61]. Therefore, in SLE nephritis HS is

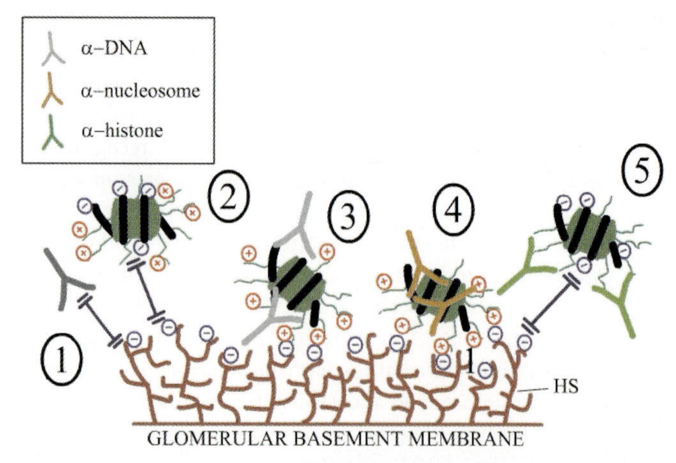

Fig. 18.6 Schematic representation of the binding of nucleosome-anti-nucleosome complexes to the GBM. (Left to right) No binding will occur of non-complexed anti-nuclear antibodies (1) or free nucleosomes (2). Binding of anti-nucleosome (3) or anti-DNA (4) will decrease the density of negative charges of the nucleosome – this will enhance binding of the complex to the negatively charged GBM. In contrast to this, binding of anti-histone antibodies (5) to the nucleosome will decrease the amount of positive charges, which reduces the capacity to bind to HS in the GBM.

masked by deposited nucleosome-autoantibody complexes. This masking may have the same functional consequences as reduction of HS, since it may lead to an enhanced permeability of the GBM as shown by the induction of albuminuria by the anti-HS monoclonal [37].

18.5 Conclusion

Nucleosome-mediated autoantibody binding to HS in the GBM is responsible for the formation of immune deposits in the GBM in lupus nephritis. As outlined in this review, a large part of the evidence for direct binding of autoantibodies to the GBM due to 'cross-reactivity' is questionable and is explained by the use of unpurified autoantibodies, more specifically the use of autoantibodies already complexed with nucleosomes. In our model, detailed in Fig. 18.6, autoantibody binding or deposition to the anionic GBM is mediated by the binding of the cationic and protruding N-termini of the histones to the GBM. Furthermore, the binding of anti-nucleosome or anti-DNA antibodies to the nucleosome will mask its negative charge and will facilitate the binding of the positively charged N-terminal histone tails to the GBM, which ultimately leads to the deposition of immune complexes. On the other hand the binding of anti-histone antibodies to the N-terminal histone parts will mask their positive charge, which will lead to reduction of deposition of immune complexes (see Fig. 18.6 for details).

18.6 References

1 Swaak AJ, Aarden LA, Statius van Eps LW, Feltkamp TE. Anti-dsDNA and complement profiles as prognostic guides in systemic lupus erythematosus. *Arthritis Rheum* **1979**, 22, 226–35.

2 ter Borg EJ, Horst G, Hummel EJ, Limburg PC, Kallenberg CG. Measurement of increases in anti-double-stranded DNA antibody levels as a predictor of disease exacerbation in systemic lupus erythematosus. A long-term, prospective study. *Arthritis Rheum* **1990**, 33, 634–43.

3 Tax WJ, Kramers C, van Bruggen MC, Berden JH. Apoptosis, nucleosomes, and nephritis in systemic lupus erythematosus. *Kidney Int* **1995**, 48, 666–73.

4 Mohan C, Adams S, Stanik V, Datta SK. Nucleosome: a major immunogen for pathogenic autoantibody-inducing T cells of lupus. *J Exp Med* **1993**, 177, 1367–81.

5 Amoura Z, Piette JC, Bach JF, Koutouzov S. The key role of nucleosomes in lupus. *Arthritis Rheum* **1999**, 42, 833–43.

6 Rumore PM, Steinman CR. Endogenous circulating DNA in systemic lupus erythematosus. Occurrence as multimeric complexes bound to histone. *J Clin Invest* **1990**, 86, 69–74.

7 Licht R, van Bruggen MC, Oppers-Walgreen B, Rijke TP, Berden JH. Plasma levels of nucleosomes and nucleosome-autoantibody complexes in murine lupus: effects of disease progression and lipopolysacharide administration. *Arthritis Rheum* **2001**, 44, 1320–30.

8 Casciola-Rosen LA, Anhalt G, Rosen A. Autoantigens targeted in systemic lupus erythematosus are clustered in two populations of surface structures on apoptotic keratinocytes. *J Exp Med* **1994**, 179, 1317–30.

9 Berden JH, Licht R, van Bruggen MC, Tax WJ. Role of nucleosomes for induction and glomerular binding of autoantibodies in lupus nephritis. *Curr Opin Nephrol Hypertens* **1999**, 8, 299–306.

10 Utz PJ, Anderson P. Posttranslational protein modifications, apoptosis, and the bypass of tolerance to autoantigens. *Arthritis Rheum* **1998**, 41, 1152–60.

11 Amoura Z, Koutouzov S, Chabre H, Cacoub P, Amoura I, Musset L, Bach JF, Piette JC. Presence of antinucleosome autoantibodies in a restricted set of connective tissue diseases: antinucleosome antibodies of the IgG3 subclass are markers of renal pathogenicity in systemic lupus erythematosus. *Arthritis Rheum* **2000**, 43, 76–84.

12 Bruns A, Blass S, Hausdorf G, Burmester GR, Hiepe F. Nucleosomes are major T and B cell autoantigens in systemic lupus erythematosus. *Arthritis Rheum* **2000**, 43, 2307–15.

13 Burlingame RW, Boey ML, Starkebaum G, Rubin RL. The central role of chromatin in autoimmune responses to histones and DNA in systemic lupus erythematosus. *J Clin Invest* **1994**, 94, 184–92.

14 Berden JH. Lupus nephritis. *Kidney Int* **1997**, 52, 538–58.

15 van Bruggen MC, Kramers C, Hylkema MN, Smeenk RJ, Berden JH. Significance of anti-nuclear and anti-extracellular matrix autoantibodies for albuminuria in murine lupus nephritis, a longitudinal study on plasma and glomerular eluates in MRL/l mice. *Clin Exp Immunol* **1996**, 105, 132–9.

16 Eilat D. Cross-reactions of anti-DNA antibodies and the central dogma of lupus nephritis. *Immunol Today* **1985**, 6, 123–7.

17 Foster MH, Cizman B, Madaio MP. Nephritogenic autoantibodies in systemic lupus erythematosus: immunochemical properties, mechanisms of immune deposition, and genetic origins. Lab Invest **1993**, 69, 494–507.

18 Sabbaga J, Pankewycz OG, Lufft V, Schwartz RS, Madaio MP. Cross-reactivity distinguishes serum and nephritogenic anti-DNA antibodies in human lupus from their natural counterparts in normal serum. *J Autoimmun* **1990**, 3, 215–35.

19 Gilbert D, Lopez B, Parain J, Koutouzov S, Tron F. Overlap of the anti-cardiolipin and anti-nucleosome responses of the (NZW × BXSB)F_1 mouse strain: a new pattern of cross-reactivity for lupus-related autoantibodies. *Eur J Immunol* **2000**, 30, 3271–80.

20 Koike T, Maruyama N, Funaki H, Tomioka H, Yoshida S. Specificity of mouse hybridoma antibodies to DNA. II. Phospholipid reactivity and biological false positive serological test for syphilis. *Clin Exp Immunol* **1984**, 57, 345–50.

21 Lafer EM, Rauch J, Andrzejewski C, Jr, Mudd D, Furie B, Schwartz RS, Stollar BD. Polyspecific monoclonal lupus autoantibodies reactive with both polynucleotides and phospholipids. *J Exp Med* **1981**, 153, 897–909.

22 Shoenfeld Y, Rauch J, Massicotte H, Datta SK, Andre-Schwartz J, Stollar BD, Schwartz RS. Polyspecificity of monoclonal lupus autoantibodies produced by human-human hybridomas. *N Engl J Med* **1983**, 308, 414–20.

23 Sharma A, Isenberg DA, Diamond B. Crossreactivity of human anti-dsDNA antibodies to phosphorylcholine: clues to their origin. *J Autoimmun* **2001**, 16, 479–84.

24 Sun KH, Liu WT, Tsai CY, Tang SJ, Han SH, Yu CL. Anti-dsDNA antibodies cross-react with ribosomal P proteins expressed on the surface of glomerular mesangial cells to exert a cytostatic effect. *Immunology* **1995**, 85, 262–9.

25 Koren E, Koscec M, Wolfson-Reichlin M, Ebling FM, Tsao B, Hahn BH, Reichlin M. Murine and human antibodies to native DNA that cross-react with the A and D SnRNP polypeptides cause direct injury of cultured kidney cells. *J Immunol* **1995**, 154, 4857–64.

26 Herrera-Diosdado R, Avalos-Diaz E, Herrera-Esparza R. Cross-reactivity of anti-nDNA antibodies with nuclear envelope proteins. Isolation of a cDNA encoding the 70 kDa annular protein recognized by autoantibodies from patients with systemic lupus erythematosus. *Rev Rheum Engl Ed* **1997**, 64, 82–8.

27 Alberdi F, Dadone J, Ryazanov A, Isenberg DA, Reichlin M. Cross-reaction of lupus anti-dsDNA antibodies with protein translation factor EF-2. *Clin Immunol* **2001**, 98, 293–300.

28 Budhai L, Oh K, Davidson A. An *in vitro* assay for detection of glomerular binding IgG autoantibodies in patients with systemic lupus erythematosus. *J Clin Invest* **1996**, 98, 1585–93.

29 Berden JH, Termaat RM, Brinkman K, Smeenk RJ, Swaak AJ, Faaber P, Assmann K, van de Heuvel LP, Veerkamp JH. Binding of anti-DNA antibodies to glomerular heparan sulfate: a new clue for the pathogenesis of SLE nephritis? *Nephrologie* **1989**, 10, 127–32.

30 Madaio MP, Yanase K. Cellular penetration and nuclear localization of anti-DNA antibodies: mechanisms, consequences, implications and applications. *J Autoimmun* **1998**, 11, 535–8.

31 Jacob L, Viard JP, Allenet B, Anin MF, Slama FB, Vandekerckhove J, Primo J, Markovits J, Jacob F, Bach JF, *et al.* A monoclonal anti-double-stranded DNA autoantibody binds to a 94-kDa cell-surface protein on various cell types via nucleosomes or a DNA-histone complex. *Proc Natl Acad Sci USA* **1989**, 86, 4669–73.

32 Raz E, Brezis M, Rosenmann E, Eilat D. Anti-DNA antibodies bind directly to renal antigens and induce kidney dysfunction in the isolated perfused rat kidney. *J Immunol* **1989**, 142, 3076–82.

33 Madaio MP, Carlson J, Cataldo J, Ucci A, Migliorini P, Pankewycz O. Murine monoclonal anti-DNA antibodies bind directly to glomerular antigens and form immune deposits. *J Immunol* **1987**, 138, 2883–9.

34 Vlahakos DV, Foster MH, Adams S, Katz M, Ucci AA, Barrett KJ, Datta SK, Madaio MP. Anti-DNA antibodies form immune deposits at distinct glomerular and vascular sites. *Kidney Int* **1992**, 41, 1690–700.

35 D'Andrea DM, Coupaye-Gerard B, Kleyman TR, Foster MH, Madaio MP. Lupus autoantibodies interact directly with distinct glomerular and vascular cell surface antigens. *Kidney Int* **1996**, 49, 1214–21.

36 Mostoslavsky G, Fischel R, Yachimovich N, Yarkoni Y, Rosenmann E,

Monestier M, Baniyash M, Eilat D. Lupus anti-DNA autoantibodies cross-react with a glomerular structural protein: a case for tissue injury by molecular mimicry. *Eur J Immunol* **2001**, 31, 1221–7.

37 van den Born J, van den Heuvel LP, Bakker MA, Veerkamp JH, Assmann KJ, Berden JH. A monoclonal antibody against GBM heparan sulfate induces an acute selective proteinuria in rats. *Kidney Int* **1992**, 41, 115–23.

38 Faaber P, Rijke TP, van de Putte LB, Capel PJ, Berden JH. Cross-reactivity of human and murine anti-DNA antibodies with heparan sulfate. The major glycosaminoglycan in glomerular basement membranes. *J Clin Invest* **1986**, 77, 1824–30.

39 Sabbaga J, Line SR, Potocnjak P, Madaio MP. A murine nephritogenic monoclonal anti-DNA autoantibody binds directly to mouse laminin, the major non-collagenous protein component of the glomerular basement membrane. *Eur J Immunol* **1989**, 19, 137–43.

40 Ravirajan CT, Rahman MA, Papadaki L, Griffiths MH, Kalsi J, Martin AC, Ehrenstein MR, Latchman DS, Isenberg DA. Genetic, structural and functional properties of an IgG DNA-binding monoclonal antibody from a lupus patient with nephritis. *Eur J Immunol* **1998**, 28, 339–50.

41 Scott JK, Smith GP. Searching for peptide ligands with an epitope library. *Science* **1990**, 249, 386–90.

42 Valadon P, Nussbaum G, Boyd LF, Margulies DH, Scharff MD. Peptide libraries define the fine specificity of anti-polysaccharide antibodies to *Cryptococcus neoformans*. *J Mol Biol* **1996**, 261, 11–22.

43 Gaynor B, Putterman C, Valadon P, Spatz L, Scharff MD, Diamond B. Peptide inhibition of glomerular deposition of an anti-DNA antibody. *Proc Natl Acad Sci USA* **1997**, 94, 1955–60.

44 Sun Y, Fong KY, Chung MC, Yao ZJ. Peptide mimicking antigenic and immunogenic epitope of double-stranded DNA in systemic lupus erythematosus. *Int Immunol* **2001**, 13, 223–32.

45 Sibille P, Ternynck T, Nato F, Buttin G, Strosberg D, Avrameas A. Mimotopes of polyreactive anti-DNA antibodies identified using phage-display peptide libraries. *Eur J Immunol* **1997**, 27, 1221–8.

46 Putterman C, Diamond B. Immunization with a peptide surrogate for double-stranded DNA (dsDNA) induces autoantibody production and renal immunoglobulin deposition. *J Exp Med* **1998**, 188, 29–38.

47 DeGiorgio LA, Konstantinov KN, Lee SC, Hardin JA, Volpe BT, Diamond B. A subset of lupus anti-DNA antibodies cross-reacts with the NR2 glutamate receptor in systemic lupus erythematosus. *Nat Med* **2001**, 7, 1189–93.

48 Termaat RM, Brinkman K, van Gompel F, van den Heuvel LP, Veerkamp JH, Smeenk RJ, Berden JH. Cross-reactivity of monoclonal anti-DNA antibodies with heparan sulfate is mediated via bound DNA/histone complexes. *J Autoimmun* **1990**, 3, 531–45.

49 Lefkowith JB, Gilkeson GS. Nephritogenic autoantibodies in lupus: current concepts and continuing controversies. *Arthritis Rheum* **1996**, 39, 894–903.

50 Kramers C, Hylkema MN, van Bruggen MC, van de Lagemaat R, Dijkman HB, Assmann KJ, Smeenk RJ, Berden JH. Anti-nucleosome antibodies complexed to nucleosomal antigens show anti-DNA reactivity and bind to rat glomerular basement membrane *in vivo*. *J Clin Invest* **1994**, 94, 568–77.

51 Utz PJ, Hottelet M, Schur PH, Anderson P. Proteins phosphorylated during stress-induced apoptosis are common targets for autoantibody production in patients with systemic lupus erythematosus. *J Exp Med* **1997**, 185, 843–54.

52 Lee E, Furukubo T, Miyabe T, Yamauchi A, Kariya K. Involvement of histone hyperacetylation in triggering DNA fragmentation of rat thymocytes undergoing apoptosis. *FEBS Lett* **1996**, 395, 183–7.

53 Allera C, Lazzarini G, Patrone E, Alberti I, Barboro P, Sanna P, Melchiori A, Parodi S, Balbi C. The condensation of chromatin in apoptotic thymocytes shows a specific structural change. *J Biol Chem* **1997**, 272, 10817–22.

54 Elouaai F, Lule J, Benoist H, Appolinaire-Pilipenko S, Atanassov C, Mul-

LER S, FOURNIE GJ. Autoimmunity to histones, ubiquitin, and ubiquitinated histone H2A in NZB×NZW and MRL-*lpr/lpr* mice. Anti-histone antibodies are concentrated in glomerular eluates of lupus mice. *Nephrol Dial Transplant* **1994**, 9, 362–6.

55 SCHELLEKENS GA, DE JONG BA, VAN DEN HOOGEN FH, VAN DE PUTTE LB, VAN VENROOIJ WJ. Citrulline is an essential constituent of antigenic determinants recognized by rheumatoid arthritis-specific autoantibodies. *J Clin Invest* **1998**, 101, 273–81.

56 PIACENTINI M, COLIZZI V. Tissue transglutaminase: apoptosis versus autoimmunity. *Immunol Today* **1999**, 20, 130–4.

57 FRANEK F, DOLNIKOVA J. Nucleosomes occurring in protein-free hybridoma cell culture. Evidence for programmed cell death. *FEBS Lett* **1991**, 284, 285–7.

58 TERMAAT RM, ASSMANN KJ, DIJKMAN HB, VAN GOMPEL F, SMEENK RJ, BERDEN JH. Anti-DNA antibodies can bind to the glomerulus via two distinct mechanisms. *Kidney Int* **1992**, 42, 1363–71.

59 MORIOKA T, WOITAS R, FUJIGAKI Y, BATSFORD SR, VOGT A. Histone mediates glomerular deposition of small size DNA anti-DNA complex. *Kidney Int* **1994**, 45, 991–7.

60 BERNSTEIN KA, VALERIO RD, LEFKOWITH JB. Glomerular binding activity in MRL *lpr* serum consists of antibodies that bind to a DNA/histone/type IV collagen complex. *J Immunol* **1995**, 154, 2424–33.

61 VAN BRUGGEN MC, WALGREEN B, RIJKE TP, CORSIUS MJ, ASSMANN KJ, SMEENK RJ, VAN DEDEM GW, KRAMERS K, BERDEN JH. Heparin and heparinoids prevent the binding of immune complexes containing nucleosomal antigens to the GBM and delay nephritis in MRL/*lpr* mice. *Kidney Int* **1996**, 50, 1555–64.

62 VAN BRUGGEN MC, WALGREEN B, RIJKE TP, TAMBOER W, KRAMERS K, SMEENK RJ, MONESTIER M, FOURNIE GJ, BERDEN JH. Antigen specificity of anti-nuclear antibodies complexed to nucleosomes determines glomerular basement membrane binding *in vivo*. *Eur J Immunol* **1997**, 27, 1564–9.

63 FOURNIE GJ. Detection of nucleosome-IgG immune complexes in ascites from mice transplanted with anti-DNA antibody-secreting hybridomas and in plasma from MRL-*lpr/lpr* mice. *Clin Exp Immunol* **1996**, 104, 236–40.

64 VAN BRUGGEN MC, KRAMERS C, WALGREEN B, ELEMA JD, KALLENBERG CG, VAN DEN BORN J, SMEENK RJ, ASSMANN KJ, MULLER S, MONESTIER M, BERDEN JH. Nucleosomes and histones are present in glomerular deposits in human lupus nephritis. *Nephrol Dial Transplant* **1997**, 12, 57–66.

65 MALIDE D, LONDONO I, RUSSO P, BENDAYAN M. Ultrastructural localization of DNA in immune deposits of human lupus nephritis. *Am J Pathol* **1993**, 143, 304–11.

66 SCHMIEDEKE T, STOECKL F, MULLER S, SUGISAKI Y, BATSFORD S, WOITAS R, VOGT A. Glomerular immune deposits in murine lupus models may contain histones. *Clin Exp Immunol* **1992**, 90, 453–8.

67 VAN DEN BORN J, VAN DEN HEUVEL LP, BAKKER MA, VEERKAMP JH, ASSMANN KJ, BERDEN JH. Monoclonal antibodies against the protein core and glycosaminoglycan side chain of glomerular basement membrane heparan sulfate proteoglycan: characterization and immunohistological application in human tissues. *J Histochem Cytochem* **1994**, 42, 89–102.

68 VAN DEN BORN J, VAN DEN HEUVEL LP, BAKKER MA, VEERKAMP JH, ASSMANN KJ, WEENING JJ, BERDEN JH. Distribution of GBM heparan sulfate proteoglycan core protein and side chains in human glomerular diseases. *Kidney Int* **1993**, 43, 454–63.

69 VAN BRUGGEN MC, KRAMERS K, HYLKEMA MN, VAN DEN BORN J, BAKKER MA, ASSMANN KJ, SMEENK RJ, BERDEN JH. Decrease of heparan sulfate staining in the glomerular basement membrane in murine lupus nephritis. *Am J Pathol* **1995**, 146, 753–63.

Glossary

(NZW×NZB)F1 Murine model for human SLE.

2,2,7-trimethylguanosine (TMG) cap Removed during apoptotic cleavage of U1 snRNA. The exact cleavage site is between nucleotides C5 and ψ6, or between nucleotides ψ6 and ψ7.

28S rRNA, rat Contains two specific cleavage sites in the hypervariable regions of the largest divergent domains D2 and D8. Both sites are apoptosis specific and are not found in necrotic cells.

ABC transporters One of the largest families of membrane proteins. They drive transport of a wide variety of substrates across cell membranes in an ATP-dependent fashion. Structurally they contain a pair of nucleotide-binding domains (NBD) and two sets of membrane-anchoring domains (TM), typically composed by six transmembrane α helices. To date 48 ABC genes have been fully characterized in the human genome (www.humanabc.org). Seven subclasses have been defined, named A–G, on the basis of sequence homologies and structural peculiarities. A diagnostic combination of consensus signatures has been defined as the hallmark of the family. This is located in the NBD and associates the Walker A and B motifs, shared by several ATP binding proteins, with a specific C motif located just upstream of the B site.

ACAMP Apoptotic cell-associated molecular pattern.

Ac-DEVD-cmk Inhibitor of caspase-3 and -7.

AIF Initiates apoptosis by activation of caspase-3.

ALPS Autoimmune lymphoproliferative syndrome [synonym for Canale–Smith Syndrome (CSS)].

ALPS 0 Refers to complete Fas expression defect.

ALPS Ia Defines functional Fas deficiency (with slightly diminished or normal Fas expression).

ALPS Ib Circumscribed to a FasL defect.

ALPS II Used to describe Fas-induced apoptosis defect in the absence of Fas mutations.

ALPS III Designates patients presenting with ALPS symptoms but with a normal *in vitro* Fas-induced apoptosis.

Altered 'self' During apoptosis protein epitopes get modified post-translationally in various ways (e.g. phosphorylation, glutathiolation, citrullination, transglutamination, or formation of novel protein–protein or protein–nucleic acid complexes).

Anti-chromatin antibodies The total group of antibodies reacting with nucleosomes, histones and DNA are referred to as anti-chromatin antibodies.

Anti-nucleosome antibodies Defined by their much higher specificity/reactivity towards the complete nucleosome compared to histones or naked DNA.

Anti-phospholipid antibodies (aPL) Autoantibodies recognizing various phospholipids like phosphatidylserine or cardiolipin. A hallmark for the anti-PL syndrome (Hughes syndrome).

Anti-Sm autoantibodies (aSm) Directed against eight spliceosomal core proteins, named the Smith (Sm) complex, were discovered in 1966 and were shown to have 99% specificity for SLE.

Apaf-1 knockout mice In contrast to the dramatic effect of deficiency on neuronal development, T cell development appears to be normal. The knockout caused a perinatal lethality starting at post-conception day 16.5.

APO3 ligand (APO-3L; TWEAK) Binds to DR3.

Apoptosis and necrosis A key distinction between apoptosis and necrosis is the lack of inflammation following apoptotic cell death.

Apoptosis Although there are no fixed criteria that define apoptosis, a number of changes in cellular morphology can help distinguish it from necrotic cell death. In the apoptotic cell, chromosomes condense, the nucleus fragments, cytoplasmic volume decreases, organelles compact, the cell membrane fuses with the endoplasmic reticulum and the cell finally fragments into numerous 'apoptotic bodies', which are engulfed by surrounding cells.

Apoptosis (PCD) Major form of programmed cell death defined by a series of unique changes which include the compaction of nuclear chromatin into dense masses that move to the edge of the intact nuclear envelope. Apoptosis is associated with extensive DNA degradation through the activation of endonucleases, fragmentation of the chromatin masses and condensation of the cytoplasm with shrinkage of the cell. Finally, the cell is fragmented into pieces called apoptotic bodies and blebs which are still enclosed by intact cell membrane, and contain the remaining antigen of the cell. *In vivo* apoptotic cells are rapidly cleared before they enter the late stages of cell death.

Apoptosis, complement activation
The complement activation of apoptotic cells may also be regulated by binding to apoptotic cells of proteins such as natural or autoimmune antibodies, β_2-glycoprotein 1 (β_2GP-1), C-reactive protein (CRP), serum amyloid P (SAP), mannose binding lectin (MBL) and others.

Apoptosis, definition
The term apoptosis that was coined in 1972 by Kerr *et al.* means in Greek 'falling leaves' and describes a de-adhesiveness of cells that undergo morphological changes like cell shrinkage, plasma membrane blebbing and chromatin condensation with intact membrane.

Apoptosis, DNA fragmentation
Induced by cleavage of ICAD.

Apoptosis, key steps
There are four key steps in the death program: (1) initiation of death either through a professional death receptor or through the mitochondria, (2) activation of effector caspases, (3) execution of death including activation of nucleases and cell membrane changes, and (4) phagocytosis and removal of the corpse.

Apoptosis, membrane blebbing
Induced by cleavage of fodrin and gelsolin.

Apoptosis, nuclear budding
Induced by digestion of nuclear lamin.

Apoptosis, phagocytosis
Under normal circumstances, cells dying by apoptotic and post-apoptotic/early necrotic mechanisms are rapidly removed by phagocytic cells and presented to the immune system for induction of tolerance.

Apoptosome
Pro-apoptotic structure made of caspase-9, cytochrome *c*, dATP and Apaf-1.

Apoptotic tumor cells
Usually display poor immunogenicity.

Autoantigen, La
In the context of apoptosis, La is partly cleaved and rapidly dephosphorylated. La is an *in vitro* substrate of granzyme B.

Autoantigen, Ro60
Not cleaved after induction of apoptosis.

Autoantigens, caspase resistant
The following autoantigens were recognized by a high titer autoantibody response and were not cleaved in apoptosis: CENP-B (targeted in scleroderma), fibrillarin and B23, PMS1 (targeted in myositis), and the type-3 muscarinic receptor (targeted in Sjögren's syndrome) U1A, U1C and Sm-B/B.

Autoantigens, caspase sensitive DFS70/LEDGF, fodrin, vimentin, NuMA, PARP, Topo I, UBF/NOR-90, U1-70kD, SSA/La, DNA-PK_{cs}, SRP72, NuMA, Sm, topoisomerase I, NOR-90, fodrin, hnRNP C1/C2, PMS2, La, CENP-C, Mi-2, Jo-1, Ku, PM-Scl, PCNA, p80 coilin, rRNP and SSA/Ro.

Autoantigens, granzyme B cleaved Granzyme B cleaves several autoantigens that are not susceptible to cleavage during apoptosis or necrosis, such as Ku-70, Jo-1, CENP-B and PMScl, but failed to cleave other apoptotic/necrotic protease-resistant autoantigens, including SSA/Ro, Ku-80, ribosomal P proteins, histones and the Sm proteins.

Autolysis Excessive activation of calpains is associated with lysosomal membrane disruption, leading to the release of cathepsins into the cytoplasm with the resultant cell lysis.

Bacterial DNA Contains statistically random amounts of unmethylated CpG motifs. Ligand for TLR-9.

Bak knockout mice Viable with only mild apoptotic defects.

Bcl-2 family, anti-apoptotic Bcl-2 and the large splice variant of Bcl-x (Bcl-x_L) have powerful anti-apoptotic effects.

Bcl-2 family, pro-apoptotic Bak and Bax have powerful pro-apoptotic effects.

Bcl-2 knockout mice Deficiency of the anti-apoptotic Bcl-2 protein results in early postnatal lethality, but initially normal lymphocyte development.

Bcl-x knockout mice Lack both the pro-apoptotic Bcl-x_S and the anti-apoptotic Bcl-x_L. They are embryonic lethal with severe excessive neuronal apoptosis and reduced survival of immature lymphocytes.

Bid knockout mice Develop normally but are resistant to Fas-mediated hepatocyte apoptosis.

Bid, truncated Induces mitochondria pore formation via Bak or Bax, resulting in the unleashing of pro-apoptotic molecules such as cytochrome c and Smac/DIABLO.

C1q The first component of the classical pathway of complement activation.

C5a Anaphylatoxin, which reduces neuronal apoptosis in hippocampal pyramidal layer *in vivo* and in *in vitro* cultured primary neurons.

C5–C9 Terminal complement components [membrane attack complex (MAC)].

Caenorhabditis elegans, apoptotic cells CED-1, -2, -5, -6, -7, -10 and -12 participate in the removal of apoptotic bodies. CED-2, -5, -10 and -12 are involved in organizing and controlling cytoskeletal rearrangement during cell migration and engulfment of apoptotic cells. The CED-2 mammalian homolog was identified as crkII, CED-5 encodes a homolog of human DOCK 180, a protein involved in receptor signaling and surface extension, while CED-10 and -12 analogs are rac and elmo, respectively. CED-1, CED-6/gulp and CED-7/abc1 may be related to recognition functions. CED-1 was cloned and identified to be a transmembrane protein sharing sequence homology with the human scavenger receptor SREC. However, SREC fails to show any homology to the cytoplasmic region of CED-1. CED-6 encodes an adaptor molecule that mediates engulfment of apoptotic cells. CED-7 is similar to the ABC1 cassette transporter.

***Caenorhabditis elegans*, necrotic cells** CED-1, -2, -5, -6, -7, -10 and -12 are also required for the clearance of necrotic cells.

Calpain The release of calcium from intracellular stores, which is associated with ER stress, activates this protease, which in turn activates caspase-12 and leads to apoptosis.

Calpains Non-caspase lysosomal proteases involved in necrotic cell death.

Canale-Smith Syndrome (CSS) More than 30 years ago, Canale and Smith reported a condition characterized by non-malignant lymphadenopathies associated with autoimmune features in children. It turned out that these patients and a number of newly described ones have a genetic disorder caused by mutations of the Fas-encoding gene. Fas deficiency is observed in cells from all carriers of heterozygous mutations. Thus, on a functional point of view, mutations are fully penetrant. In contrast, only 70% of the carriers of heterozygous Fas mutations develop clinical symptoms. A paradoxical decrease in lymph node has been observed during viral infection. Splenomegaly, lymphadenopathies and hypergammaglobulinemia are frequently observed in the patients. Synonym: ALPS (autoimmune lymphoproliferative syndrome).

Cardiolipin (CL) A phospholipid that shares some similarities in structure and charge with PS, was shown to activate complement via the classical pathway in an antibody-independent mechanism. CL is a target of anti-phospholipid antibodies (aPL).

Caspase-1 (ICE) The first member of the caspase family to be cloned. It is a cytoplasmic protease that is capable of converting the 34-kDa inactive precursor of IL-1 to its mature 17-kDa form and can also process the cytokine precursor of IL-18 (IFN-inducing factor).

Caspase-1 knockout mice Mice are resistant to the effects of LPS-induced shock. They do not have any developmental defects; instead, they are deficient in IL-1β and IL-18 production.

Caspase-2 knockout mice In facial motor neurons, where both caspase-2_L and -2_S are thought to be expressed, the knockout mice showed reduced cell numbers during late embryogenesis, implying an increase in apoptosis; an example of a caspase knockout with evidence of both pro- and anti-apoptotic effects. B lymphoblasts were resistant to apoptosis induced by granzyme B/perforin, but not to anti-Fas, doxorubicin, etoposide, γ-irradiation or staurosporine.

Caspase-2 An interesting property is its ability to either induce or antagonize apoptosis through alternatively spliced forms. Caspase-2_L, the prevailing form, is pro-apoptotic. On the other hand, caspase-2_S, a truncated form generated by insertion of an early stop codon, is anti-apoptotic. Caspase-2 possesses a long prodomain and participates in the formation of an apoptosome-like complex. Its prodomain contains a CARD, which allows it to efficiently interact with the adapter molecule CRADD/RAIDD, though the physiologic significance of this interaction remains unclear.

Caspase-3 A central executioner caspase in mammals, though not as indispensable as its counterpart CED-3 which is absolutely required for programmed cell death in *C. elegans*. An effector caspase, which is transiently activated through the mitochondrial pathway during erythroblast differentiation and cleaves proteins involved in nucleus integrity (lamin B) and chromatin condensation (acinus) without inducing cell death and cleavage of GATA-1.

Caspase-3 knockout hepatocytes Activation of caspase-6 and -7, not seen in the wild-type.

Caspase-3 knockout mice Can survive to birth, but they exhibit perinatal mortality as a result of defects in brain development that correlate with a decrease in levels of apoptosis. They display normal T cell and B cell development. The lack of defects in negative selection may be due to activation of other alternative caspases, such as caspase-6 and -7. Activated T cells show a dramatic deficiency in AICD.

Caspase-8 A critical initiator caspase for transducing apoptosis signals from death receptors.

Caspase-8 knockout mice In contrast to Fas-, TNF-RI- or TNF-RII-deficient mice, which all develop normally to adulthood, caspase-8 deficiency causes prenatal lethality with two particularly striking features: impaired heart muscle development and congested accumulation of erythrocytes (hyperemia). The phenotype is very similar to that of FADD knockout mice.

Caspase-9 A key requirement for its activation is the association with the protein cofactors, Apaf-1 and cytochrome c, and with d*ATP*.

Caspase-9 knockout hepatocytes Activation of caspase-2 and -6, not seen in the wild-type.

Caspase-9 knockout mice
In contrast to the dramatic effect of deficiency on neuronal development, T cell development appears to be normal. The knockout caused a perinatal lethality starting at post-conception day 16.5.

Caspase-10
Appears to be involved in the apoptotic cascade of all known receptors inducing lymphocyte apoptosis. Some caspase-10 mutations are associated with ALPS.

Caspase-11
Absolutely required for caspase-1 activity. Caspase-11 of mice is most homologous to human caspase-4. Overexpression of caspase-11 in Rat-1 and HeLa cells induces apoptosis, which can be inhibited by CrmA and Bcl-2. The expression of caspase-11 is highly inducible by LPS.

Caspase-11 knockout mice
Mice show a survival advantage in LPS-induced shock.

Caspase-12
Localizes to the ER, raising the possibility of involvement in the ER stress response.

Caspase-12 knockout mice
Neurons are resistant to β-amyloid-induced apoptosis, implying that β-amyloid exerts its effects, at least in part, by causing ER stress. The mice are viable with no apparent developmental abnormalities. Knockout thymocytes were normally susceptible to apoptosis induced by anti-Fas and dexamethasone and, similarly, knockout embryonic fibroblasts are normally susceptible to apoptosis from anti-Fas, TNF-α and staurosporine. When exposed to ER stress-inducing stimuli, such as brefeldin A, tunicamycin and thapsigargin, however, caspase-12 knockout embryonic fibroblasts showed resistance to apoptosis.

Caspase activation
May be an early physiological response in viable, stimulated lymphocytes and appears to be involved in early steps of lymphocyte activation.

Caspase-activated RNase (CAR)
Cleaves RNA substrates during apoptosis.

Caspase-independent cell death
Term used to describe a number of cell death pathways that occur in the absence of detectable caspase activity. This type of cell death could be considered as a back-up system that ensures the cell's demise in the case that the caspase activation program is rendered non-functional.

Caspase in erythropoiesis
Exposure of erythroid progenitors to mature erythroblasts or death-receptor ligands resulted in caspase-mediated degradation of the transcription factor GATA-1.

Caspase recruitment domain (CARD)
The long prodomains of caspases contain either DEDs or CARDs.

Caspase, specific cleavage site Upstream activating caspases (group III) prefer Ile/Val in P_4 and Glu in P_3.

Caspases Cysteine proteases that cleave after aspartic acid. Many caspases are activated during apoptosis. They exist in cells as inactive monomeric zymogens, consisting of an N-terminal prodomain, a large subunit and a small subunit. Processing occurs at sites between the domains that, consistent with the ability of pro-caspases to autoactivate or to be activated by other caspases, contain aspartate residues. Cleavage of two zymogen precursors at these sites releases the subunits, which then form the functional heterotetrameric enzyme, a complex of two large and two small subunits with two separate active sites. Caspases have among the most stringent substrate specificities of all proteases, always cleaving on the carboxyl side of aspartate residues.

Caspases, adaptors Found in the cell as inactive precursors and are activated at the onset of apoptosis by an autoaggregation process mediated by adaptor proteins which promotes autocatalytic processing of the initiator caspases.

Caspases, executioners Caspase-3, -6 and -7.

Caspases, initiators Caspase-2, -8, -9-, and -10.

Caspases, mammalian Only about half of the 14 mammalian caspases actively participate in the execution of apoptosis.

Caspases, pro-enzymes Synthesized as latent proenzymes made of an N-terminal CARD, and a large (p20) and a small (p10) protease subunit.

Caspases with long prodomains Frequently been termed 'initiators' and, once activated, are thought to cleave caspases with short domains, termed 'effectors'. The long prodomains have been found to contain either DEDs or CARDs that are capable of interacting with adapter molecules and result in clustering of pro-caspases. This close proximity can enhance a low, intrinsic autocatalytic activity of the zymogen, allowing it to cleave and activate itself.

Caspases with short prodomains Frequently been termed 'effectors'. They get activated by caspases with long prodomains that have been termed 'initiators'.

Cathepsins Non-caspase lysosomal proteases involved in necrotic cell death.

CD14 Human CD14 is a 356-amino-acid, 53- to 55-kDa protein that is glycosylated at *O*- and at *N*-linked sites. CD14 is expressed strongly by most monocytes and some tissue macrophages, and weakly by granulocytes. In both its soluble (sCD14) and membrane-anchored (mCD14) forms, CD14 has been shown to bind LPS and to generate LPS responsiveness. The 152 N-terminal amino acids are sufficient for the activation of inflammatory signaling pathways. CD14/LPS binding is enhanced by the LPS-binding protein, LBP, which catalyses transfer of LPS from micellar aggregates of the glycolipid to CD14 to form CD14-LPS complexes.

CD14, clearance of apoptotic cells Macrophages, which are immunosuppressive following apoptotic cell clearance do express CD14 and DCs, which are functional in antigen presentation and immune activation following apoptotic cell engulfment, do not.

CD14, ligands Its prototypic and still its best-known function is as a receptor for the endotoxin of Gram-negative bacteria, LPS. In addition, phospholipids have proven capacity to bind directly to mCD14. The pulmonary surfactant proteins, SP-A and SP-D, and mannose-binding protein A are known to interact directly with CD14. Apoptotic cells directly interact with murine CD14. However, CD14 may require additional co-operating molecules that are necessary for generating CD14-dependent responses to apoptotic cells. The monoclonal antibody MEM-18, known to block LPS binding to CD14, inhibits apoptotic cell clearance by macrophages to the same extent as monoclonal antibody 61D3. 61D3 appears to bind to the same region of CD14 as LPS. Furthermore, numerous examples of proteins, lipids and phospholipids of viral, fungal, yeast or mammalian origin are known to interact directly or indirectly with CD14, in some cases to generate pro-inflammatory responses and, in others, to prevent such responses. Lipoteichoic acid (LTA), peptidoglycan (PGN), lipoarabinomannan (LAM), rhamnose glucose polymers, uronic acid polymers, glycolipids and lipoproteins are molecularly defined ligands of CD14. Glycan moieties at apoptotic-cell surfaces may be functional as CD14 ligands. HSP-60 and -70 have been shown to interact with CD14, but as yet have no known role in apoptotic cell clearance.

CD14, soluble Present in plasma at around 4 μg/ml.

CD30 knockout mice Mice with a partial deficiency in eliminating self-reactive transgenic $CD8^+$ T cells in male mice.

CD4 T cells Can cause cell death of target cells by FasL (apoptosis) or by the release of TNF-*α* (apoptosis or necrosis).

CD8 T cells and NK cells Can cause cell death of target cells by the perforin/granzyme pathway (apoptosis or necrosis), FasL (apoptosis) or by the release of TNF-*α* (apoptosis or necrosis).

CED-1 Protein of *C. elegans* similar to the human scavenger receptor SREC. It is normally expressed on the membrane of all cells, but during the engulfment process it redistributes on the membrane

	and clusters at the phagocytic extensions formed around the dying cell.
CED-2	Protein of *C. elegans* similar to mammalian crkII.
	Pprotein of *C. elegans* similar to ICE, caspase-3.
CED-4	Protein of *C. elegans* similar to Apaf-1.
CED-5	Protein of *C. elegans* similar to human DOCK 180.
CED-6	Adaptor molecule of *C. elegans* involved in engulfment of apoptotic cells.
CED-7	Protein of *C. elegans* similar to the ABC1 cassette transporter.
CED-8	Protein of *C. elegans* similar to XK a putative membrane transport protein.
CED-9	protein of *C. elegans* similar to Bcl-2.
CED-10	Protein of *C. elegans* similar to human rac.
CED-12	Protein of *C. elegans* similar to human elmo.
Celiac disease (CD)	A complex disease characterized by a wide spectrum of lesions in the intestinal mucosa that can ultimately lead to the atrophy of the villi.
Cell death, complement-induced	Necrotic in essence with possible involvement of apoptotic features produced by external or secondary (e.g. oxygen radicals) factors.
Collectins	Lectins with partial collagenous structure like C1q, SP-A, and SP-D.
Complement activation	Complement can be activated in the absence of an antibody by nucleic acid, phospholipids, LPS, and apoptotic cells.
Complement opsonins	C1q, mannose binding lectin, and iC3b are referred to as opsonins.
Complement receptors	CD11b/CD18 and other complement receptors may actually be immunosuppressive by down-regulating IL-12 and IFN-γ production by human monocytes.
Complement receptors, phagocytes	C1qR/CD91, CR3 (CD11b/CD18) and/or CR4 (CD11c/CD18) are complement receptors on phagocytes.
C-reactive protein (CRP)	An acute-phase reactant synthesized following tissue injury and inflammation. It binds to nucleated cells and activates complement activation that is restricted to formation of the C3 convertase. C5 convertase formation and MAC generation are probably blocked by factor H that binds directly to CRP.
Cryptic epitopes	Antigenic determinants, which are generated at subthreshold concentrations during normal antigen processing of whole protein anti-

gens. T cells recognizing the 'cryptic' self never encounter their antigen during natural antigen presentation and are therefore not tolerized.

Cytochrome *c* Can form a complex with Apaf-1 and pro-caspases 9 in the cytoplasm to form the apoptosome. Once released into the cytoplasm, it interacts with the adaptor protein Apaf-1 and with pro-caspase-9 leading to caspase-9 activation and induction of an apoptosis signal.

Cytochrome *c* knockout mice Animals die prenatally at embryonic day 8.5, presumably due to the defect in aerobic metabolism, but analysis of embryo-derived cell lines reveal a resistance to apoptosis induced by UV irradiation, serum withdrawal and staurosporine. Interestingly, however, cytochrome c knockout cells appear more sensitive to TNF-α.

Cytotoxic T cells (CTLs) Induce apoptosis in their target cells either by activating the Fas-mediated pathway or by delivery of the granule protease granzyme B (GrB) into the target cells through perforin, a granule protein that forms pores in the cytoplasmic membrane.

Dendritic cells (DCs) Antigen-presenting cells (APCs) that are found in virtually all organs. They exist as interstitial DCs in peripheral tissues, interdigitating cells in lymph nodes and as circulating APCs in the blood stream. Langerhans cells and dermal DCs reside in epidermis and dermis, respectively.

DcR-1 (TRAIL-R3) A GPI-linked receptor, like that binds TRAIL.

DcR-3 Decoy receptor of FasL.

Death effector domains (DEDs) The long prodomains of caspases contain either DEDs or CARDs.

Death receptors A subset of TNF-R family members that includes Fas, TNF-RI, DR3 (TRAMP, wsl-1, APO-3, LARD), DR4 (TRAIL-R1, APO-2), DR5 (TRAIL-R2, TRICK2, KILLER) and DR6; the receptors are involved in transducing signals that result in cell death.

Decoy receptors Receptors that kept ligand binding properties, but lack much of the functions of the corresponding receptors.

Defense collagens Molecules with partial collagenous structure that are as adaptor molecules involved in the defense reactions of the innate immune system.

Deletion of autoreactive thymocytes Occurs normally in *lpr* (Fas mutation), Fas knockout, TNF-RI knockout, and TNF-RII knockout mice, and is only partially impaired in CD30 knockout mice.

DNA, human Contains reduced numbers of the CpG motif. Most CpG motifs are methylated in human DNA. No ligand for TLR-9.

DNase I From serum enters the cells through complement-induced pores.

DR6 DD-containing receptor on mononuclear cells.

FADD deficiency Causes decreased activation in peripheral T cells which is associated with a defective costimulatory response.

FADD knockout mice Phenotype is very similar to that of caspase-8 knockout mice.

FADD Fas-associated death domain protein does not lead exclusively to apoptosis, but under certain circumstances can promote cell survival and proliferation.

FADD, dominant negative mutant Transgenic mice expressing a dominant-negative mutant FADD in the thymus provided the surprising result of enhanced thymocyte negative selection.

Fas (Apo-1, CD95, TNFRSF6) Type I transmembrane glycoprotein that belongs to the TNF-R/NGF-R family. Fas is expressed on various types of cells, including activated lymphocytes and certain transformed cells. Fas molecules were previously trimerized through interactions of a N-terminal domain called PLAD (pre-ligand association domain). Fas activation can induce cell proliferation in some experimental systems.

Fas-deficient mice (*lpr*) Symptoms develop only in homozygous animals, despite a potential defective Fas-induced apoptosis in heterozygous animals. They accumulate with age $CD4^{-}CD8^{-}$ TCR $\alpha\beta$ T cells in the periphery. This phenotype is likely the consequence of the CD8 downregulation on mature peripheral lymphocytes.

Fas ligand (FasL/CD178) A 40-kDa type II transmembrane protein that is homologous to TNF and its binding to Fas transmits an apoptotic signal to susceptible target cells.

Fas, agonist Antibodies to Fas might provide co-stimulation for human T cells *in vitro*.

Fas, type I cells Induction of apoptosis is accompanied by activation of large amounts of caspase-8 by the DISC. Caspase-8 rapidly cleaves and activates caspase-3, leading to the effector stage of apoptosis.

Fas, type II cells DISC formation is strongly reduced and activation of caspase-8 and -3 occurs following the loss of mitochondrial transmembrane potential. In this variant of the caspase cascade, low levels of caspase-8 cut and activate the pro-apoptotic Bid.

GluR3 An autoantigen in Rasmussen's encephalitis. Whereas fully glycosylated GluR3 is relatively resistant to cleavage by GrB, the under-glycosylated form of GluR3 generated during inflammatory states allows the cleavage of GluR3 by granzyme B and the generation of previously cryptic peptide fragments.

Granzyme B (GrB)
A serine protease found in the cytoplasmic granules of CTLs and NK cells, and is involved in the induction of target cell death. It cleaves ICAD allowing activation of the caspase-activated DNase (CAD). GrB also cleaves Bid and promotes mitochondrial cell death. GrB cleaves caspase-2 *in vitro*. GrB facilitates cell death independent of caspase-activation by directly cleaving intracellular substrates. However, caspase-mediated proteolysis is the predominant pathway used during GrB-mediated apoptosis. GrB activates apoptosis through its ability to cleave upstream caspase-10 and -8 and downstream caspase-3 and -7 after specific aspartic acid residues.

Granzyme B, specific cleavage site
Some amino acids in the P_2 and P_3 positions that are preferred exclusively by GrB and not tolerated by the caspases (e.g. proline in P_2, and glycine or serine in P_3).

Heat shock protein 70 (Hsp70)
Released from dying cells. It is targeted to immature DC precursors. Increased Hsp70 expression during necrotic tumor cell killing induced a T cell-mediated anti-tumor immune response characterized by infiltration of T cells, macrophages and granulocytes.

Heat shock proteins (HSP)
Highly conserved chaperones designed for the cellular transport and folding of proteins.

Human Y RNA
Four types of human Y RNA exist: hY1, hY3, hY4 and hY5 RNA, which are cleaved during apoptosis. The cleavage of hY RNA is inhibited by caspase-1, -3 and -8 inhibitors, suggesting the recruitment of caspase-activated RNase (CAR) during apoptosis.

ICAM-3, apoptosis
During apoptosis the function of ICAM-3 is radically changed from its pre-apoptotic role as adhesion molecule. ICAM-3 becomes qualitatively altered such that it is unable to interact with its prototypic counter-receptor LFA-1. Instead, it appears that ICAM-3 gains the capacity to interact with macrophage molecules that are functional in apoptotic-cell clearance.

Interleukin (IL)-10, phagocytosis
Both increased and suppressed release of IL-10 following ingestion of apoptotic cells by human macrophages was found.

Lipopolysaccharide (LPS)-binding protein (LBP)
Catalyses the transfer of LPS from micellar aggregates to CD14 to form CD14-LPS complexes.

MAC high copy number
Lytic MAC doses are known to produce DNA fragmentation induced by entrance of DNase I. The morphology of complement-mediated cell death at the ultrastructural level is characterized by swelling of mitochondria, dilation of the rough endoplasmic reticulum, disruption of the Golgi complex and of the plasma and nuclear membranes and heterochromatin disappearance. The terminal complement complexes C5b-7, C5b-8 and MAC are known to bind

to cell membranes directly via hydrophobic regions in the absence of a known receptor.

MAC low copy number At low copy numbers per cell (non-lytic or sublytic doses), the MAC is known to have stimulatory activities on many cell types.

MDM2 Oncoprotein, which suppresses p53 degradation, accumulates to high levels and contributes to the release from cell cycle control in malignant tissue.

Membrane attack complex (MAC) Terminal complement components C5–C9.

MER A member of the Tyro 3 receptor tyrosine kinase family.

Mimotopes Artificial epitopes.

MRL-*lpr/lpr* Murine model for human SLE.

Natural adjuvants Signals derived from necrotic but not apoptotic cells act as potent adjuvants.

Necrosis Non-suicidal process associated with a number of pathological conditions that develops in response to acute cell injury, including ischemia, hypoxia, oxidative stress, extreme heat, severe infections and exposure to high levels of chemicals or toxins. Necrosis can be induced with high levels of mercury, ethanol, hydrogen peroxide or heat.

Necrosis, secondary The transition from apoptosis to secondary necrosis is associated with post-translational modifications of specific autoantigens.

Necrosis marker Preservation of lamin B integrity. Necrotic cells but not apoptotic ones release the pro-inflammatory nuclear protein HMGB1.

Nuclear factor (NF)-κB A ubiquitous transcription factor that plays an important role in inflammatory gene transcription. IκB family members, including IκBα, IkBβ and IkBγ, regulate the DNA binding and subcellular localization of NF-κB proteins by masking a nuclear localization signal. NF-κB activation occurs through signal-induced degradation of IκB in the cytoplasm, allowing the translocation of NF-κB to the nucleus. Immunohistochemistry and electromobility shift assays show that the transcription factor is highly activated in rheumatoid synovium as well as animal models of arthritis. Proteasome inhibitors inhibit NF-κB activation by preventing IκB degradation.

p53 When DNA is damaged, p53 induces cell cycle arrest at G_1/S and G_2/M interphase through the transcriptional activation of several genes, including $p21^{WAF}$, or apoptosis. The p53 protein is normally detected only at very low levels but its expression is induced in response to various stimuli, such as DNA damage (by ionizing radiation, UV radiation, and chemotherapeutic drugs, etc), hypoxia, heat shock, viral infection, growth factor deprivation and oncogene activation. Since wild-type p53 protein has a relatively short half-life (less than 20 min), it is usually not detectable in normal tissues by immunohistochemistry. In the presence of synovial inflammation, including murine collagen-induced arthritis or rat adjuvant arthritis, p53 protein expression increases dramatically. p53 may mediate negative selection through activation of Apaf1 and caspase-9. It may also mediate death by neglect at the pre-TCR stage.

PAMP
Pathogen-associated molecular pattern.

Pentraxins, short
C-reactive protein (CRP) or the serum amyloid P component (SAP) is produced in the liver in response to several cytokines, in particular IL-6.

Phagocytosis of apoptotic cells
Shown to suppress the release of GM-CSF, IL-1β, IL-8, TNF-α and thromboxane B_2, but not TGF-β and prostaglandin E_2.

Phosphatidylserine (PS)
Anionic phospholipid (PL) that can interact with both mCD14 and sCD14; it does so with lower affinity than PI, PC or PE. PS is exposed by apoptotic and necrotic cells. PS binds various adaptor molecules (e.g. MFG-E8, β2GP-1, Gas-6 and C3bi) and receptors (e.g. PS-R, CD36, SR-A and CD14).

Phospholipase D (PLD)
Cleaves GPI anchored proteins from membranes.

Protein translocation
60-kDa Ro, La, the snRNPs, Ku and poly(ADP-ribose)polymerase (PARP), which normally have a diffuse nuclear distribution, become concentrated as a rim around the condensing chromatin in early apoptosis. Lupus autoantigens, which are not restricted to any specific subcellular compartment in control cells, are strikingly redistributed in apoptotic cells, such that they become clustered and concentrated within small surface blebs and apoptotic bodies. Thus, small surface blebs (which contain fragmented rough ER) are highly enriched in 52-kDa Ro, ribosomal autoantigens, as well as those autoantigens found within the ER lumen (e.g. calreticulin).

PTEN
A tumor suppressor gene. Mutations in the PTEN gene have been detected in various types of malignancies.

PTX3
Prototypic long pentraxin structurally related to CRP and SAP. It is produced at the site of acute cell death *in vivo* and behaves as a 'do not eat me' signal for apoptotic cells.

Random peptide phage display
Powerful technique to produce antibodies especially against non-protein epitopes. Briefly, an autoantibody is coated on a carrier and incubated with a phage library containing numerous phages that display a short random peptide fused with a phage coat protein. Fi-

nally, the phages that bind to the autoantibody with the highest affinity are selected. *In silico* the selected motifs are compared to known proteins sequences.

RIP2/CARDIAK/RICK The serine/threonine kinase can bind and activate caspase-1 processing, and provides a possible mechanism for caspase-1 regulation.

RNS Reactive nitrogen species.

ROS Reactive oxygen species.

rRNA RNA components make up more than 50% of the ribosome, and are classified as follows: (I) prokaryotic 50S subunit: 5S rRNA, 23S rRNA; (II) prokaryotic 30S subunit: 16S rRNA; (III) eukaryotic 60S subunit: 5S rRNA, 28S rRNA, 5.8S rRNA; and (IV) eukaryotic 40S subunit: 18s rRNA.

S19 dimer Monocyte chemotactic factor released from apoptotic cells, carrying out an important role in their phagocytic clearance. The dimerization of the S19 protein is catalyzed by tTG, thus confirming an important role for this enzyme in the apoptotic cells clearance.

SAP (serum amyloid protein) Anti-opsonin; SAP-coated microorganisms evade recognition *in vivo*, suggesting an anti-opsonic role of SAP.

SATB1 A transcription factor that has been shown to be critical for T and B cell differentiation, but is also processed by caspase-6 in activated T cells.

Scavenger receptors Multiligand receptors involved in mammals in the recognition/uptake of dying corpses.

Sentrin A ubiquitin-like protein that binds to the death domain of Fas/APO-1 and TNF-RI, and protects cells from both anti-Fas- and TNF-induced cell death.

sFasL A 26-kDa soluble form of FasL, which is cleaved from 40-kDa membrane FasL by the action of a metalloproteinase. sFasL has been detected in the serum of patients with various diseases.

SNF1 mice Murine model for human SLE.

snRNPs Complexes of protein associated with these U RNAs are designated small nuclear ribonucleoproteins (snRNPs).

Spliceosome U1, U2, U5 and U4/6 snRNPs are components of the spliceosome, which are frequently targeted autoantigens in SLE.

Surfactant protein A (SP-A) Collectin lectin with partial collagenous structure.

Surfactant protein D (SP-D) Collectin lectin with partial collagenous structure.

Systemic autoimmune disease of unknown origin.	**Systemic lupus erythematosus (SLE)**
T helper cells which produce cytokines like IFN-γ favoring cellular immune responses.	**T_h1**
T helper cells which produce cytokines like IL-4 and IL-10 favoring humoral immune responses.	**T_h2**
An enzyme is associated with apoptotic cell death. It is crosslinking a number of proteins such as collagens, fibronectin, laminin, nidogen and TGF-β and might play an important role in the modification of the ECM occurring in degenerative diseases.	**Tissue transglutaminase (tTG)**
Binds to DR3.	**TL1-A**
DD-containing receptors have been found on peripheral blood mononuclear cells.	**TNF-RI**
Activated T cells have prolonged survival. They may potentially accumulate in aging animals, consistent with the autoimmune syndromes.	**TNF-RI/Fas double knockout mice**
Involved in CD14-dependent responses to non-LPS microbial components such as peptidoglycan and lipoteichoic acid.	**Toll-like receptor-2 (TLR-2)**
Critical for physiological LPS responses.	**Toll-like receptor-4 (TLR-4)**
Prototypic pattern recognition receptors, which are involved in proinflammatory responses.	**Toll-like receptors (TLRs)**
Binds to DR4 and DR5.	**TRAIL (APO-2 ligand, APO-2L)**
TNF-related apoptosis-inducing ligand-R1. DD-containing receptor on mononuclear cells.	**TRAIL (DR4/APO-2)**
DD-containing receptor on mononuclear cells.	**TRAIL-R2 (DR5/Trick/Killer)**
TNF-R-related apoptosis mediating protein. DD-containing receptor on mononuclear cells.	**TRAMP (DR3/wsl1/APO-3/LARD)**
Family of enzymes includes seven intracellular (the ubiquitous tTG and six different isoenzymes differentially expressed in the tissues) and two extracellular enzymes (Factor XIIIa and prostate transglutaminase).	**Transglutaminases**
Mice that have a defect in the clearance of apoptotic cells and develop nuclear autoantibodies with aging.	**tTG knockout mice**
Bind to TNF-RI.	**Tumor necrosis factor (TNF) and lymphotoxin**

Type I cells Thymocytes and peripheral T cells.

Type II cells Hepatocytes.

U RNAs, small RNAs of 250 nucleotides or less, which have arbitrarily been named U1, U2...U12 RNAs. The complexes of protein associated with these RNAs are designated snRNPs.

U1-70kDa Specifically cleaved by caspase-3 at position ^{338}DGPD^{341}, resulting in an N-terminal 40-kDa (amino acids 1–341) and a C-terminal 20kDa (amino acids 342–437).

VEID-CHO Inhibitor of caspase-6.

z-IETD-fmk Inhibitor of caspase-8.

z-LEHD-fmk Inhibitor of caspase-9.

z-VAD-fmk Pan-caspase inhibitor that initially protected staurosporine treated cells against apoptosis. However, the inhibitor did not prevent delayed cell death with features of necrosis.

Subject Index

d